Item Response Theory

Additional Volumes in Preparation

Item Response Theory
Parameter Estimation Techniques
Second Edition, Revised and Expanded

Frank B. Baker
University of Wisconsin
Madison, Wisconsin, U.S.A.

Seock-Ho Kim
The University of Georgia
Athens, Georgia, U.S.A.

CRC Press
Taylor & Francis Group
Boca Raton London New York

CRC Press is an imprint of the
Taylor & Francis Group, an **informa** business

CRC Press
Taylor & Francis Group
6000 Broken Sound Parkway NW, Suite 300
Boca Raton, FL 33487-2742

First issued in paperback 2022

© 2004 by Taylor & Francis Group, LLC
CRC Press is an imprint of Taylor & Francis Group, an Informa business

No claim to original U.S. Government works

ISBN 13: 978-1-03-247792-3 (pbk)
ISBN 13: 978-0-8247-5825-7 (hbk)

DOI: 10.1201/9781482276725

Library of Congress Cataloging-in-Publication Data

Catalog record is available from the Library of Congress

To Dr. Frederic M. Lord

who, over a thirty-year period, shepherded item response theory from
its infancy to its acceptance as the modern test theory

Preface to the Second Edition

In the years since the publication of the first edition of this book, "Item Response Theory" has continued to be a dynamic field of inquiry. The literature has continued to grow as the breadth of the field has expanded (e.g., Ackerman, 1996; Drasgow, 1995; Embretson & Reise, 2000; Fischer & Molenaar, 1995; Johnson & Albert, 1999; Junker & Sijtsma, 2001; Reckase, 1997; van der Linden & Hambleton, 1997) and as the applications of the theory to educational and psychological testing have become widespread (e.g., Drasgow & Olson-Buchanan, 1999; Holland & Wainer, 1993; Kolen & Brennan, 1995; Thissen & Wainer, 2001; van der Linden & Glas, 2000; Wainer, 2000). Because of this, the authors were faced with the difficult task of selecting the topics to be included in the second edition. Our selection criterion was that they possessed continuity with the underlying theme of the first edition, namely, procedures for the estimation of item and ability parameters. Three lines of inquiry were selected that fulfilled this requirement. The first two extend marginal maximum likelihood estimation to situations that are of great interest to practitioners. The third, Gibbs sampler, is a new and unique procedure for estimating item and ability parameters.

A common situation is one in which the data set contains multiple, identifiably different groups of examinees who have taken the same instrument. The parameter estimation procedures for this situation allow one to compare groups, differentiated on the basis of some criterion, on a common metric. They also provide a means of determining whether items in an instrument exhibit differential item functioning across the several groups of examinees. Finally, these procedures can be used to detect item parameter drift of an instrument. The marginal maximum likelihood estimation procedures for the multiple group situation are presented in Chapter 10.

A situation of great interest to instrument developers is that of estimating item parameters of instruments which contain items based upon more than one item response model; for example, where dichotomously-scored and polytomously-scored items are intermixed within a given instrument. Developers have an interest in this as it adds variety to an instrument and allows them to tailor the item to the characteristics of the topic. Marginal maximum likelihood estimation procedures for a given mixed item type case are presented in Chapter 11.

The Markov chain Monte Carlo (MCMC) method for estimating parameters has its origins far removed from item response theory and can be traced to the work of Metropolis and Ulam (1949). Within the framework of MCMC, the Gibbs sampler procedure was developed by Geman and Geman (1984) in the context of image processing. In a groundbreaking paper, Albert (1992) showed how Gibbs sampler could be used to simultaneously estimate the item parameters of an instrument and the ability parameters of the examinees. Since this paper, considerable research into the characteristics of the parameter estimates obtained via MCMC procedures has been conducted. In addition, different MCMC techniques have been explored and implemented. In that the fundamental approach was the Gibbs sampler, its estimation procedures are presented in Chapter 12.

In the first edition, each chapter in the book was accompanied by an appendix containing a BASIC program that implemented the particular procedure. The rationale behind these programs was to show how the mathematics of the estimation procedure actually worked. Historically, the parameter estimation procedures employed under item response theory have been computationally intensive. In particular, the MCMC procedures are very intensive. However, since the publication of the first edition, the computing power of personal computers has increased dramatically, so that BASIC programs for the three additional chapters are contained in their corresponding appendices.

The authors are indebted to Terry Ackerman, Allan S. Cohen, R. J. De Ayala, Jeffrey Douglas, Fritz Drasgow, Eiji Muraki, Mark D. Reckase, and Steven P. Reise for their thoughtful suggestions on the new topics to be included in the second edition. The authors are also indebted to Julien C. Sprott for the use of BASIC subroutines and Michael Harwell for his contribution to Chapters 6 and 7. The second co-author wishes to acknowledge James H. Albert, R. Darrell Bock, Robert J. Mislevy, Nambury S. Raju, Hariharan Swaminathan, and David Thissen for their advice on estimation techniques over the years. He also would like to thank Stephen F. Olejnik, Tae-Je Seong, Ronald C. Serlin, Michael J. Subkoviak, and Joung-kyu Whang for their encouragement.

Frank B. Baker

Seock-Ho Kim

References

Ackerman, T. (Guest Ed.). (1996). Special issue: Developments in multidimensional item response theory. *Applied Psychological Measurement, 20*(4).

Albert, J. H. (1992). Bayesian estimation of normal ogive item response curves using Gibbs sampling. *Journal of Educational Statistics, 17*, 251–269.

Drasgow, F. (Guest Ed.). (1995). Special issue: Polytomous item response theory. *Applied Psychological Measurement, 19*(1).

Drasgow, F., & Olson-Buchanan, J. B. (Eds.). (1999). *Innovations in computerized assessment.* Mahwah, NJ: Erlbaum.

Embretson, S. E., & Reise, S. P. (2000). *Item response theory for psychologists.* Mahwah, NJ: Erlbaum.

Fischer, G. H., & Molenaar, I. W. (Eds.). (1995). *Rasch models: Foundations, recent developments, and applications.* New York: Springer.

Geman, S., & Geman, D. (1984). Stochastic relaxation, Gibbs distributions, and the Bayesian restoration of images. *IEEE Transactions on Pattern Analysis and Machine Intelligence, 6,* 721–741.

Holland, P. W., & Wainer, H. (Eds.). (1993). *Differential item functioning.* Hillsdale, NJ: Erlbaum.

Johnson, V. E., & Albert, J. H. (1999). *Ordinal data modeling.* New York: Springer.

Junker, B. W., & Sijtsma, K. (Guest Eds.). (2001). Special issue: Nonparametric item response theory. *Applied Psychological Measurement, 25*(3).

Kolen, M. J., & Brennan, R. L. (1995). *Test equating: Methods and practices.* New York: Springer.

Metropolis, N., & Ulam, S. (1949). The Monte Carlo method. *Journal of the American Statistical Association, 44,* 335–341.

Reckase, M. D. (1997). The past and future of multidimensional item response theory. *Applied Psychological Measurement, 21,* 25–36.

Thissen, D., & Wainer, H. (Eds.). (2001). *Test scoring.* Mahwah, NJ: Erlbaum.

van der Linden, W. J., & Glas, C. A. W. (Eds.). (2000). *Computerized adaptive testing: Theory and practice.* Dordrecht, The Netherlands: Kluwer.

van der Linden, W. J., & Hambleton, R. K. (Eds.). (1997). *Handbook of modern item response theory.* New York: Springer.

Wainer, H. (Ed.). (2000). *Computerized adaptive testing: A primer* (2nd ed.). Mahwah, NJ: Erlbaum.

Preface to the First Edition

The process by which a field of inquiry acquires its name is an interesting one. Often a label is offered, such as "criterion referenced testing," which is universally accepted within a short time. In most cases, a field develops slowly and many different labels are employed before consensus is reached as to an acceptable name. Such is the case for the modern test theory that is the focus of this book. The basic concepts of the field have been developed across three-fourths of a century, and a diverse group of persons have made contributions. For many years there was little agreement as to what label should be employed to identify the field. Much of the literature in the 1950s through the 1970s used the term "latent trait theory" as it reflected the use of an underlying hypothetical variable. The latent trait terminology is due to Lazarsfeld (1954) and carries with it connotations of psychological scaling procedures. For a short period, Lord (1977) referred to the field as "item characteristic curve theory" in recognition of the major role of such curves. While item characteristic curve theory is an appropriate label, it tends to be associated with those concepts and procedures based upon dichotomously scored items. The use of this label becomes a bit strained when items are scored on a graded or nominal basis as well as when used within the context of other recent developments. In recent years, the label "item response theory," also due to Lord (1980), has gained acceptance as the name of the field since it reflects the dependence of the theory upon an examinee's responses to items. This label has sufficient generality to encompass the many facets of the field as well as to reflect the basic concepts involved. It is also true that the degree to which a set of concepts and procedures are accepted in practice is often determined by their label and this particular title has meaning to both the researcher and the practitioner.

One of the salient features of item response theory is that it is at once very old and very new even though it is often referred to as the "modern" test theory. The chapters by Allan Birnbaum in Lord and Novick's classic *Statistical Theories of Mental Test Scores* (1968) made item response theory visible. Although the basic theory has been available for a generation, only within the past ten years has a significant level of publication been achieved. A large number of articles dealing with the theory and application of item response theory have appeared in the measurement-oriented journals. A key publication was the Summer 1977 issue of the *Journal of Educational Measurement* under the editorship of Richard Jaeger. This issue spanned the range from theoretical concerns to applications. Many basic papers have appeared in *Psychometrika* and much of the research exploring and expanding the theory has appeared in the journal *Applied Psychological Measurement*. In addition to articles, a number of books have appeared: *Item Response Theory: Application to Psychological Measurement* (Hulin, Drasgow, & Parsons, 1983), *Advances in Psychological and Educational Measurement* (de Gruijter & van der Kamp, 1976), *Einführung in die Theorie Psychologisher Tests: Grundlagen und Anwendungen* (Fischer, 1974). Much of the interest in item response theory is due to the efforts of Benjamin Wright. The American Educational Research Association presessions conducted by Dr. Wright over a 20-year period and the book *Best Test Design* (Wright & Stone, 1979) have made the theory accessible to practitioners. A significant event in the history of item response theory was the publication of Frederic Lord's book *Applications of Item Response Theory to Practical Testing Problems* (1980). This book collects into a single source the contents of his numerous Educational Testing Service research memoranda and papers that have appeared in a variety of places. Because of Lord's pivotal role within the field, this book has had a considerable impact on the use of the theory in applied settings. The book *Applications of Item Response Theory* (Hambleton, 1983) also deals with the use of the theory in practice. Another important book is *Item Response Theory: Principles and Applications* (Hambleton & Swaminathan, 1985), which provides broad coverage of the whole of the theory. It relates classical test theory and item response theory under the assumption that the reader is familiar with both. The book by Baker (1985), *The Basics of Item Response Theory*, provided an introduction to the basic concepts of item response theory and is unique in that it has an accompanying computer program that allows one to explore the facets of the theory. Since there are a variety of sources for learning about item response theory as a test theory, the present book assumes the reader is familiar with the basic constructs of the theory.

One result of the available books and articles is that as a field of item response theory inquiry has become very large. This in turn makes it difficult for a single book to provide in-depth coverage of the whole field. As a result, there is a need for more specialized books dealing with specific aspects of the theory. Because of the mathematical sophistication and the data processing demands of item response theory test analysis procedures, computer program implementation is necessary. Thus, from a practical point of view, the history of item response theory is inextricably intertwined with the available computer programs such as LOGIST, BICAL, BILOG, and MULTILOG. Proper specification of the analyses to be performed and the interpretation of the outputs of these programs depends upon an intimate understanding of the parameter estimation procedures they implement. Thus I believed that a book dealing only with these procedures would be a valuable resource for both those who use item response theory in practice and those who need a basic understanding of these procedures.

The implementation of item response theory rests on the statistical techniques for estimating the parameters of test items and of examinee ability. However, in striving to present the full panorama of the theory, the existing books have necessarily provided only the outlines of these estimation procedures. Thus, presentation of the underlying logic and the mathematical details of these estimation procedures would complement the existing broader coverage. A unification of the existing literature would also be achieved by presenting the mathematics of these estimation procedures in a consistent manner. Finally, providing the full mathematical detail of these estimation procedures would make them accessible to a wider audience.

To meet these goals, this book has been organized in a systematic manner. In the first chapter, the underlying logic of the item characteristic curve concept and the several models for these curves are presented. Chapters 2 and 3 develop the two building blocks from which several other estimation paradigms are constructed. Chapter 2 shows the mathematics of the maximum likelihood procedures for estimating the parameters of the item characteristic curve under the various models. Chapter 3 presents the estimation of an examinee's ability score under these models. In Chapter 4, these two procedures are put together in the Birnbaum paradigm for the joint maximum likelihood estimation of item and ability parameters that is the basis for several widely employed computer programs. Owing to its importance, a separate chapter was devoted to the Rasch model. Georg Rasch's original work is presented in a notation consistent with the rest of the book. The estimation procedures developed by Wright and his co-workers are shown, as is the extension of the Rasch model in the form of the linear logistic test model. Chapter 6 provides the mathematical details of the Bock and Aitkin marginal maximum likelihood/EM approach to the estimation of item and examinee parameters. Chapter 7 presents the marginalized Bayesian approach to estimating item parameters and two Bayesian approaches to estimating an examinee's ability. Chapters 8 and 9 are concerned with the maximum likelihood estimation procedures for parameter estimation when graded and nominally-scored items are employed. The BASIC computer programs in the appendixes illustrate how the various estimation procedures are implemented. I hope that the approach taken provides a gradual increase in sophistication of the material such that the reader can progress easily from the basics to the complex estimation procedures. In addition, it should provide a context within which to place further developments.

The lack of a standardized notation within the item response theory literature posed a particularly difficult problem in writing this book. Only a few symbols have been used consistently and the literature has been very lax in differentiating between parameters and their sample estimators. In the early chapters, an attempt has been made to present the mathematics using a symbol system that corresponds to normal statistical usage in this regard. In the later chapters, the symbol system is more closely linked to that used in the item response theory literature. This was done to provide better access to the mathematics of the original articles. In addition, it was occasionally necessary to attach different meanings to the same symbols used in the early and later chapters. I trust that by later chapters the reader will have developed some notational flexibility. Nonetheless, a serious effort was made to make the notation as consistent as possible.

The writing of the first draft of this book was supported by a grant from the Research Committee of the Graduate School, University of Wisconsin. The typing from my atrocious handwriting was done by the secretaries of the Laboratory of Experimental Design: Connie Schlehammer, Beth Brown, Cathy Tobin, Chris Kringle, and Cheryl Houge. The difficult task of converting the manuscript from the NBI word processor to WordPerfect on the IBM PC computer was accomplished with great skill by Lauri Koch, who also typed the final version of the manuscript. Without the skill and diligence of these persons, the book would not have been possible. The contributions of several generations of graduate students are also acknowledged. Their probing questions revealed many places where the manuscript was unclear and in some cases incorrect. They contributed much to the final version. Chapters 6 and 7 were the result of a collaboration with Michael Harwell of the University of Pittsburgh in which the mysteries of marginal maximum likelihood and Bayesian estimation were unraveled.

Madison, WI   *Frank B. Baker*

References

Baker, F. B. (1985). *The basics of item response theory.* Portsmouth, NH: Heinemann.

de Gruijter, D. N. M., & van der Kamp, L. J. Th. (1976). *Advances in psychological and educational measurement.* London: Wiley.

Fischer, G. H. (1974). *Einführung in die Theorie Psychologisher Tests: Grundlagen und Anwendungen.* Bern: Huber.

Hambleton, R. K. (1983). *Applications of item response theory.* Vancouver, BC: Educational Research Institute of British Columbia.

Hambleton, R. K., & Swaminathan, H. (1985). *Item response theory: Principles and applications.* Boston, MA: Kluwer-Nijhoff Publishing.

Hulin, C. L., Drasgow, F., & Parsons, C. K. (1983). *Item response theory: Application to psychological measurement.* Homewood, IL: Dow Jones-Irwin.

Lazarsfeld, P. F. (1954). A conceptual introduction to latent structure analysis. In P. F. Lazarsfeld (Ed.), *Mathematical thinking in the social sciences* (pp. 349–387). Glencoe, IL: The Free Press.

Lord, F. M. (1977). Practical applications of item characteristic curve theory. *Journal of Educational Measurement, 14,* 117–138.

Lord, F. M. (1980). *Applications of item response theory to practical testing problems.* Hillsdale, NJ: Erlbaum.

Lord, F. M., & Novick, M. R. (1968). *Statistical theories of mental test scores.* Reading, MA: Addison-Wesley.

Wright, B. D., & Stone, M. H. (1979). *Best test design.* Chicago: MESA Press.

Contents

1. The Item Characteristic Curve: Dichotomous Response

1.1 Introduction

The intelligence testing movement of the first four decades of the 20th century and its attendant controversies lead to the development of classical test theory (CTT). Many of the familiar constructs, such as true score, reliability, and validity, arose from Spearman's work in providing a mathematical underpinning for his theory of intelligence. Since the dominant statistical theory of the time was Pearsonian statistics, CTT rests heavily on correlational concepts. Gulliksen's (1950) book contained a comprehensive presentation of the theory within a Pearsonian framework. Subsequently, Lord and Novick (1968) reformulated the basic constructs of the theory using a modern mathematical statistical approach. The basic element in this theory was the test score. Items and their characteristics played a minor role in the structure of the theory. Over the years, both the psychometric theoretician and the practitioner became dissatisfied with the discontinuity between roles of items and test scores in the theory. It seemed intuitively reasonable that a test theory should start with the characteristics of the items composing a test rather than with the resultant score. The origins of such an item-based test theory can be seen in the work of Binet and Simon (1916). They used a tabular presentation of the functional relation between the proportion of correct response to an item and chronological age to place items within their intelligence test. Terman (1916) and Terman and Merrill (1937) used this same kind of information to plot curves relating the two variables. In modern parlance, they were using item characteristic curves. For many years, the item characteristic curve approach was considered simply as an alternative item analysis technique. The work of Lawley (1943) marks the beginning of a test theory based upon the items of a test. In a remarkable paper, Lawley showed how to obtain maximum likelihood estimates of the parameters of the item characteristic curve, defined the true score in terms of the items of a test, and showed that the classical reliability coefficient can also be expressed as a function of these item parameters. Thus, what had been understood intuitively in the past had come to pass. A major extension of Lawley's work was due to Lord (1952), who showed that a wide range of additional classical test theory constructs could be expressed as functions of the parameters of the item characteristic curves of the test items. The work of these two men established the basic concepts of the psychometric theory based on items now known as item response theory (IRT).

The history of the item characteristic curve and the methods for estimating its parameters is threaded through psychophysics, quantal-response bioassay, educational measurement, and psychological scaling. Each of these fields has a different perspective, but they share a common set of concepts, models, and mathematics. The item characteristic curve will be presented in the context of test theory, with liberal borrowing from the other fields. Although IRT ultimately deals with sets of items constituting a test, the present chapter deals only with a single item. To a large degree, the historical development of the item characteristic curve concept was based upon the single item, and working with the single item facilitates the presentation as well as avoids certain theoretical considerations. For purposes of introducing the item characteristic curve, the situation will be further simplified by assuming that a criterion variable score, distinct from the test score, is known for each examinee. These two simplifications will be employed in this and the next chapter.

1.2 The Item Characteristic Curve

The item characteristic curve will be introduced by describing how Binet selected items for inclusion in the 1911 Binet-Simon Intelligence Scale. The procedure involved a number of steps: A criterion variable of chronological age, ranging from 5 to 12 years in increments of a year, was specified. An item of interest was administered to a number of subjects at each of the age levels. The free response of each subject to the item was scored as right or wrong, that is, dichotomously scored. (Such items are often referred to as binary items.) The proportion of correct response at each age level was obtained and presented in tabular form. In his 1916 revision of the Binet-Simon scale, Terman plotted the proportion of correct response as a function of age and fitted a smooth line to these points by graphical means. This fitted line, shown in Figure 1.1, is called the item characteristic curve (ICC).

Binet placed an item within his intelligence test at the age level at which the proportion of correct response was .75 for subjects at that age level, age 11 in the present example. The proportion of correct response used to allocate an item to an age level is completely arbitrary, and any convenient value could be used; for example, Terman (1916) used .50. Both Terman and Binet used the principle that with increasing age, there should be a corresponding increase in the proportion of correct response, what we now call item discrimination. Thus, the steeper the item characteristic curve, the greater the increase in the proportion of correct response as a function of age and the better the item is able to discriminate between adjacent age levels.

From this example, it can be seen that the item characteristic curve is the functional relationship between the proportion of correct response to an item

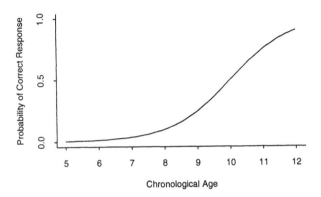

Fig. 1.1. An item characteristic curve.

and a criterion variable. This relationship is characterized by the location of the item on the criterion variable scale and by its discriminating power. It is important to note that the criterion variable, chronological age, had a known unit of measurement and the age of each subject was known before the item was administered. Also, the proportion of correct response to the item was calculated separately for each point (age level) along the criterion variable scale.

In the example above, the empirical data were used and a curve was fitted by graphical means. Let us next formalize the item characteristic curve in population terms using appropriate statistical notation. The criterion variable will be denoted by θ (theta). In IRT, the criterion variable θ is considered to be an unobserved hypothetical variable (a latent trait), such as intelligence, arithmetic ability, or scholastic ability. Generically, this latent trait is referred to as "ability" although it must be kept in mind that a wide variety of cognitive capabilities or physical skills can be labeled as ability. For present purposes, it will be assumed that the criterion variable (ability) score θ_j is known for each subject in the population, $j = 1, 2, \ldots, N$. The metric of the ability scores is assumed to have a midpoint of zero and a unit of measurement equal to one. A test theory can be constructed in which the response to an item is simultaneous function of several criterion variables. However, for both practical and theoretical reasons, only a single criterion variable will be used here. When dealing with test results, there will be a frequency distribution of the examinees' scores over the ability scale, which will have mean μ_θ and variance σ_θ^2. When the ability score for each subject is known, the form of the frequency distribution of the θ scores over the population is of limited concern. When such is not the case, it will be noted. When the criterion variable is used without designating any fixed person, it is denoted as θ (i.e., without a subscript).

Working in population terms, $P_i(\theta)$ will denote the probability of correct response at any point on the ability scale and i denotes an item, $i = 1, 2, \ldots, n$. The item characteristic curve can be defined as a member of a family of two-parameter (location and scale) monotonic functions of the ability variable. The exact form of this function will be given in, for example, equation (1.2). The two item parameters are defined as follows:

β_i is the location parameter, expressed in units of θ, that indicates the point on the ability scale at which the probability of correct response is .50, $-\infty \leq \beta_i \leq \infty$.

α_i is the scale parameter that indexes the discriminating power of an item, $-\infty \leq \alpha_i \leq \infty$.

Although this particular notation is a bit unfortunate, it has been chosen to correspond to that used by Lord and Novick (1968) and Lord (1980), which has become a de facto standard (see Baker, 1977, p. 159). For unstated reasons, Lord and Novick (1968, p. 366) used sample notation b_i, a_i to represent the item parameters. However, such a practice is not in concert with standard statistical notation. Therefore, in the initial chapters, α_i, β_i will be used to denote the parameters of the item characteristic curve, and $\hat{\alpha}_i$, $\hat{\beta}_i$ to denote the sample estimates of these parameters.

The location parameter β_i indicates where the median of the item characteristic curve falls on the ability scale. In psychophysics, β_i is known as the limen value and in bioassay, it is called the median lethal dose LD_{50} or X_{50}. In psychometrics, a number of authors (see Richardson, 1936; Lord & Novick, 1968; Hambleton & Cook, 1977) refer to β_i as item difficulty. While there is some connection between the location parameter β_i and classical test theory item difficulty, the concepts are not completely interchangeable. Classical item difficulty is the proportion of the correct response to the item in the total population. However, the use of the term "difficulty" in conjunction with the item parameter β_i is well entrenched in IRT. Thus, a difficult item will be defined as one whose value of β_i is greater than μ_θ. It will be up to the reader to maintain the distinction between the definitions of difficulty under IRT and under classical test theory. Fortunately, no such conceptual confusion exists in regard to α_i, and it can be considered the item discrimination parameter. Although the theoretical range of α_i is $-\infty \leq \alpha_i \leq \infty$, the value of α_i is usually positive for the correct response to an item, and the values seen in practice are typically less than 2.5.

Formally, the item characteristic curve can be defined by the mathematical function

$$P_i(\theta) = P(\beta_i, \alpha_i, \theta). \tag{1.1}$$

The function depends upon the values of the two item parameters and the ability. A number of different families of two-parameter functions could be used as the model for the item characteristic curves. For present purposes, it will suffice to indicate that all these functions are similar in appearance to a cumulative distribution function (an ogive).

Although it is standard practice to discuss only the item characteristic curve yielded by the correct response to an item, a dichotomously scored item actually yields two item characteristic curves, one for each response category. The probability of incorrect response $Q_i(\theta)$ also can be plotted as a function of the ability scores. This curve will be the mirror image of the item characteristic curve for the correct response. Its location parameter β_i will have the same value as that of the correct response. Its slope parameter α_i will have the same numerical value but be of opposite sign. Figure 1.2 shows the item characteristic curves for both the correct response ($\beta_i = 0.5$, $\alpha_i = +1.0$) and the incorrect response ($\beta_i = 0.5$, $\alpha_i = -1.0$) of an item. In the case of a dichotomously scored item, this mirror-image characteristic results from the fact that $P_i(\theta) + Q_i(\theta) = 1$ at each ability level. Samejima (1969) uses the terminology "item response category characteristic curves" to signify that there is an item characteristic curve associated with each response category. While this generalized label is to be preferred from a technical and conceptual point of view, a limited view of the item characteristic curve will be retained when items are dichotomously scored. However, the reader should not lose sight of the fact that each item response category yields an item characteristic curve.

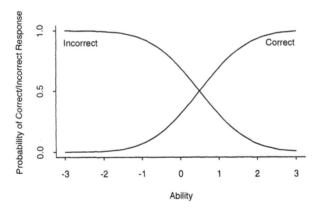

Fig. 1.2. Item characteristic curves for the correct ($\beta_i = 0.5$, $\alpha_i = +1.0$) and incorrect ($\beta_i = 0.5$, $\alpha_i = -1.0$) responses to an item.

1.3 Two Item Characteristic Curve Models

While fitting smooth functions to the observed proportion of correct response can be done by graphical means, the resulting curves lack mathematical rigor. If appropriate mathematical functions could be found that both fit the observed data and have reasonable mathematical properties, the theoretical aspects of the ICC could be advanced. It was shown above that the empirically obtained item characteristic curves have the appearance of a cumulative distribution function. In addition, only two properties of the item characteristic curves, its location and its "steepness," are needed to describe and item's technical characteristics. Although a large number of possible families of two-parameter functions exist, only two, the cumulative normal and the cumulative logistic distribution functions, have been widely used as mathematical models for the item characteristic curve. One of the reasons for restricting the present section to single items and a known ability is that it greatly simplifies the justification of these two mathematical models for the item characteristic curve (Birnbaum, 1968, p. 399).

1.3.1 The Normal Ogive Model

Since the normal distribution is a keystone of statistical theory, it is not surprising that the normal ogive has been used as a model. As early as 1860, Fechner fitted the normal ogive to the observed proportions of response obtained from psychophysical experiments. In addition, a considerable literature dating from the work of Müller (1904), Urban (1910), and Thomson (1919) deals with methods for estimating the parameters of the normal ogive that fits the observed data. The justification for the use of the normal ogive as a model for the item characteristic curve can be approached from two points of view—pragmatically and theoretically. Authors such as Richardson (1936), Ferguson (1942), and Finney (1944) justified the use of the normal ogive on pragmatic grounds. Basically, they assumed that the item characteristic curve could be modeled by the normal ogive. Then, they fitted the normal ogive to the observed proportions of correct response along the ability scale. Chi-square tests were used to assess the goodness of fit of the normal ogive and the observed data. In typical sets of item response data, the normal ogive has proved to be a workable model.

Before proceeding, the mathematics of the cumulative normal distribution function will be examined. Let

$$P_i(\theta) \equiv P(\mu_i, \sigma_i, \theta) = \Phi(Z_i) = \int_{-Z_i = -(\theta - \mu_i)/\sigma_i}^{\infty} \frac{1}{\sqrt{2\pi}} e^{-z^2/2} dz, \qquad (1.2)$$

where μ_i is the mean, σ_i is the standard deviation, θ is the ability, and Z_i is the normal deviate. Note that $\Phi(Z_i) = 1 - \Phi(-Z_i)$ due to the symmetric nature of the normal distribution. It should be noted that this function is being used as the equation of a curve and not as a cumulative distribution function that would be used in statistical theory.

In the paragraphs above, β_i was used as the location parameter and α_i as the discrimination parameter. When the normal ogive model is used, these item parameters can be expressed easily in terms of the parameters of the cumulative normal distribution function. In the case of the normal ogive, the median and the mean are identical. Since β_i was defined as the point on the ability scale at which the probability of correct response was .5, this corresponds to the mean of the normal ogive; hence, $\beta_i = \mu_i$. The parameter α_i was used as an index of steepness of the item characteristic curve. The standard deviation σ_i is a measure of the spread of normal distribution; thus, when σ_i is large, the normal ogive will be rather flat and not very steep near β_i. When σ_i is small, the normal ogive will be rather steep in the region of β_i. Thus, the steeper the middle section of a normal ogive, the smaller the value of σ_i. The inverse relationship exists between the steepness of the item characteristic curve and the standard deviation of the normal ogive model. If one lets $\alpha_i = 1/\sigma_i$, then α_i will increase as σ_i decreases and the index is logically correct. In many situations, the standard score transformation is used to convert the original measurements into a new scale of measurement having mean zero and standard deviation one; the unit of measurement is a standard deviation. As a result, statisticians often speak of σ_i as the scale parameter and, in the present case, α_i is a function of σ_i. It should be noted that the σ_i term here is the parameter of the normal ogive and has no relationship to the standard deviation σ_θ of the examinees' abilities over the ability scale. The net result is that the two-parameters β_i, α_i of the item characteristic curve are one-to-one functions of the *location* and *scale* parameters of the cumulative normal distribution function.

A second and somewhat more theoretical justification for the use of the normal ogive item characteristic curve model has been given by Lord and Novick (1968, Chapter 16). The justification rests upon the regression of an item variable, denoted by Γ_i, on the ability variable θ. The item variable Γ_i is a hypothetical continuous random variable over a population of subjects representing a subject's propensity to respond correctly to the item. It has the range $-\infty < \Gamma_i < \infty$. Large positive values of Γ_i indicate that a subject has a high propensity to answer the item correctly, and large negative values the converse. A critical value γ_i (i.e., threshold) can be selected such that, when $\Gamma_i \geq \gamma_i$, the item is scored as correct, $u_i = 1$; when $\Gamma_i < \gamma_i$, the item is scored as incorrect, $u_i = 0$. Although the item variable is unobservable, when a subject j's item response is scored as correct or incorrect, the value of U_i for subject j, that is, u_{ij}, can be obtained.

Following usual regression theory, a number of assumptions are involved.

1. The regression of Γ_i on θ is linear.
2. At each value of θ, there exists a conditional distribution of Γ_i that is normal with mean $\mu_j|\theta$ and variance $\sigma_j^2|\theta$.

3. The variance $\sigma_j^2|\theta$ is independent of θ and constant for all values of θ, the usual homogeneity of variance assumption.
4. The θ are treated as fixed quantities having no measurement error.

When one uses simple linear regression to depict the functional relationship between two variables, there is no need to specify the frequency distribution of either variable Γ_i or θ over the population of subjects. The regression is on the variable θ not upon a frequency distribution.

Figure 1.3 depicts the situation defined above. The means of the conditional distributions of Γ_i lie along the regression line $\mu_i'|\theta$. At a given value of θ, the conditional distribution is intersected by the line representing the critical value γ_i. Thus, the area of the conditional distribution above γ_i represents the probability of correct response $P_i(\theta)$ for subjects having ability score θ. This process can be repeated over the range of θ and the corresponding probabilities of correct response obtained. Now if the resulting $P_i(\theta)$ are plotted as a function of θ the resultant curve will be a normal ogive such as shown in Figure 1.4. It should be noted, that a very similar logic was used by Walker and Lev (1953) to derive the biserial correlation between an item variable and a criterion variable.

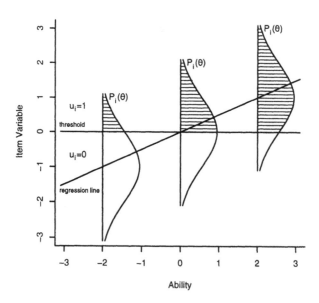

Fig. 1.3. Regression of item variable on ability, after Lord and Novick (1968).

There is one additional issue involved in this justification of the normal ogive as the model for the item characteristic curve. The metric of the ability θ, that is, its origin and unit of measurement, must be such that the process described above results in a normal ogive. To illustrate the issue, perform some nonlinear transformation on θ, say a logarithmic transformation. It would compress portions of the θ scale and elongate others. Thus, when the obtained $P_i(\theta)$ are plotted against the transformed θ, a normal ogive would not be obtained. In most practical situations, this is not a problem; however, situations can occur in which the original metric is inappropriate. In some of these cases, a transformation can be used to achieve an appropriate metric.

Figure 1.3 can also be used to illustrate the conceptual difference between the location parameter β_i and classical test theory item difficulty, P_i. The value of the location parameter β_i will be the value of θ at the intersection of the line representing the critical value γ_i and the regression line $\mu_i'|\theta$. Classical item difficulty can be obtained from the probability of correct response $P_i(\theta)$ at each value of θ as follows. Let n_j represent the number of examinees having ability score θ_j, $j = 1, \ldots, k$; then $r_j = n_j P_i(\theta_j)$ is the number of such persons answering the item correctly. The size of the population N is the sum of all the n_j. The total number of persons answering the item correctly, r, is the sum of the r_j; hence, item difficulty is

$$P_i = \frac{r}{N} = \frac{\sum_{j=1}^{k} r_j}{\sum_{j=1}^{k} n_j}. \tag{1.3}$$

To be more precise, the summation would be achieved via integration since θ is a continuous random variable. For a rigorous presentation of the appropriate mathematics, see Lord and Novick (1968, p. 376). The result is that the IRT concept of the location of the item characteristic curve is not interchangeable with item difficulty under classical test theory. This confusion does not exist in psychological scaling and bioassay in which the location parameter is called the limen value and the median lethal dose, respectively. Baker (1961) used X_{50} as the notation for the location parameter of an item characteristic curve to convey its proper meaning. For compatibility with the IRT literature, however, the term "difficulty" will be used in conjunction with β_i.

The two justifications for the use of the normal ogive model for the item characteristic curve are compatible. In the pragmatic approach, it is simply assumed that the functional relationship between the probability of correct response and the ability variable θ follows the normal ogive. Lord and Novick's justification approaches the issue from the opposite direction. They establish a set of conditions and assumptions that, if true, will yield the normal ogive model. This latter approach is advantageous in that it focuses attention up theoretical issues implicitly involved in the pragmatic approach but only attended to indirectly. With either justification, one employs a normal ogive model for the item characteristic curve. In practice, one uses curve-fitting techniques to match a member of the family of normal ogives to the observed proportions of correct response. One hopes that the fit will be adequate in a sufficiently large number of situations to justify faith in the model.

Before we proceed, tables of the cumulative normal distribution function will be used to plot an item characteristic curve having the parameters $\beta_i = 0.6$ and $\alpha_i = 1.2$. From equation (1.2),

$$Z_i = \frac{\theta - \mu_i}{\sigma_i}$$

but $\beta_i = \mu_i$ and $\alpha_i = 1/\sigma_i$. Then the normal deviate can be written as follows:

$$Z_i = \frac{\theta - \mu_i}{\sigma_i} = \alpha_i(\theta - \beta_i) = 1.2(\theta - 0.6). \tag{1.4}$$

To obtain the probability of correct response at a given value of θ, the tables of the normal ogive are entered with Z_i obtained from evaluating equation (1.4). To simplify the procedures, thirteen points were selected along the θ scale ranging from -3 to $+3$ in increments of 0.5. At a value of $\theta = -2.0$, the entering argument for the normal ogive tables would be $Z_i = 1.2(-2.0 - 0.6) = -3.12$, and the probability of correct response would be .0009. Table 1.1 presents the full set of calculations and resulting probabilities. The item characteristic curve corresponding to Table 1.1 is plotted in Figure 1.4.

Table 1.1. Item Characteristic Curve Based on Normal Ogive Model, $\beta_i = 0.6$, $\alpha_i = 1.2$

θ	$\theta - \beta_i$	Z_i	$P_i(\theta)$
-3.0	-3.6	-4.32	.0000
-2.5	-3.1	-3.72	.0001
-2.0	-2.6	-3.12	.0009
-1.5	-2.1	-2.52	.0059
-1.0	-1.6	-1.92	.0274
-0.5	-1.1	-1.32	.0934
0.0	-0.6	-0.72	.2358
0.5	-0.1	-0.12	.4522
1.0	0.4	0.48	.6844
1.5	0.9	1.08	.8599
2.0	1.4	1.68	.9535
2.5	1.9	2.28	.9887
3.0	2.4	2.88	.9980

One feature of the ICC is worth noting. In the region of β_i, the ICC is nearly linear with a slope of $\alpha_i/\sqrt{2\pi}$ (Lord & Novick, 1968, p. 368). Thus, a rough interpretation of the discrimination parameter is that it is proportional

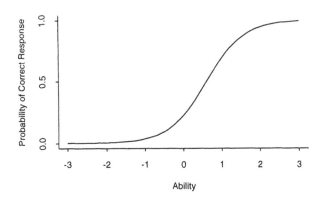

Fig. 1.4. Item characteristic curve based on normal ogive model, $\beta_i = 0.6$, $\alpha_i = 1.2$.

to the slope of the ICC at the point on the θ scale corresponding to β_i. Using one's general knowledge of the shape of the normal ogive and the numerical values of β_i, α_i for an item, an approximate ICC can be easily sketched as follows. Locate β_i along the θ scale, and plot the probability of correct response as .5. Then use the value of $\alpha_i/\sqrt{2\pi}$ or $0.3989\alpha_i$ to establish the slope of the linear segment of the ICC through the point $(\beta_i, .5)$. The remainder of the curve can be sketched in easily as it is asymptotically 1 or 0. The reader interested the relationship between the appearance of the item characteristic curve and the item parameters is referred to the book and computer program due to Baker (1985). The latter provides the capability to manipulate the numerical values of the item parameters and observe the resulting item characteristic curve on the computer monitor.

    Having adopted a normal ogive model for the item characteristic curve model, the principles underlying the techniques used to estimate the item parameters can be established. The function $P_i(\theta)$ defines a nonlinear relationship between the probability of correct response and θ. If a normal deviate transformation is applied to the $P_i(\theta)$, the result will be a linear relationship between the new variable Z_i and θ. The data in Table 1.1 will be used to illustrate the effect of this transformation. At a given θ value, say $\theta = -0.5$, the probability of correct response is .0934, and the corresponding normal deviate is $Z_i = -1.32$. Thus, by entering the tables of the normal ogive with $P_i(\theta)$, the corresponding normal deviate Z_i can be obtained, and a normal deviate transformation of $P_i(\theta)$ has been performed. Instead of plotting $P_i(\theta)$ as a function of θ, the Z_i of Table 1.1 can be plotted as a function of θ and the result of shown in Figure 1.5. The effect of this transformation is to convert the normal ogive into a linear regression line. The equation of this line is

$$Z_i' = \zeta_i + \lambda_i \theta, \tag{1.5}$$

where ζ_i is the intercept, λ_i is the slope, and θ is ability.

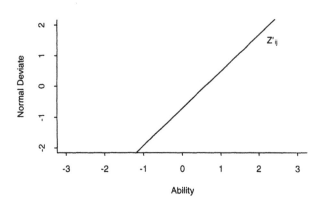

Fig. 1.5. Normal deviate transformation of the item characteristic curve, $\beta_i = 0.6$, $\alpha_i = 1.2$ ($\zeta_i = -0.72$, $\lambda_i = 1.2$).

In equation (1.4), Z_i was expressed as a function of the parameters of the item characteristic curve; hence,

$$Z_i' = \alpha_i(\theta - \beta_i) = \zeta_i + \lambda_i \theta. \tag{1.6}$$

Then,

$$\alpha_i \theta - \alpha_i \beta_i = \zeta_i + \lambda_i \theta$$

and equating by parts;

$$\alpha_i \theta = \lambda_i \theta$$

$$\alpha_i = \lambda_i. \tag{1.7}$$

Thus, the item discrimination parameter α_i is the slope of the linear regression line relating the normal deviate transformation of the probability of correct response and ability θ. Similarly,

$$-\alpha_i \beta_i = \zeta_i$$

and substituting for α_i;

$$-\lambda_i \beta_i = \zeta_i$$

$$\beta_i = \frac{-\zeta_i}{\lambda_i}. \tag{1.8}$$

The location parameter is the negative of the ratio of the intercept to the slope of the regression line.

It is here that Lord and Novick's (1968) unfortunate choice of a_i, b_i as the item parameters becomes apparent. Although notation is arbitrary, traditionally the equation of a regression line is $Y' = \alpha + \beta X$. Lord and Novick have reversed the roles of α and β; what they call a_i is β in a regression equation and $b_i = -\alpha/\beta$. Again, in the interests of conformity with the existing literature, their notation will be employed, except that α_i, β_i rather than a_i, b_i will be used to indicate population parameters.

Having shown that the normal ogive can be transformed into a linear regression line, the basic features of the estimation of the parameters of the item's characteristic curve from observed data can be developed. The task is to fit a linear regression line to the normal deviate transformations of the observed proportions of correct response at each θ value. Thus, a procedure such as least squares or maximum likelihood can be used to obtain the estimates of slope and intercept. Equations (1.7) and (1.8) can then be used to obtain $\hat{\alpha}_i$ and $\hat{\beta}_i$. Historically, the item parameters have been estimated via this regression approach (see Thomson, 1919; Ferguson, 1942; Finney, 1944).

One final observation on the normal ogive model. Under Lord and Novick's (1968) justification of the normal ogive model, no assumptions were made as to the bivariate frequency distribution of the population over the item and criterion variables. If one assumes that the distributions are bivariate normal, an interesting connection with classical test theory can be made. The critical value γ_i now dichotomizes the normally distributed item variable Γ_i. When $\Gamma_i \geq \gamma_i$, $u_i = 1$ and, when $\Gamma_i < \gamma_i$, $u_i = 0$. Richardson (1936) and Tucker (1946) have shown that under these conditions, the following holds:

$$\rho_{\theta U_i} = \frac{\alpha_i}{\sqrt{1 + \alpha_i^2}} \qquad -1 \leq \rho_{\theta U_i} \leq 1, \tag{1.9}$$

where $\rho_{\theta U_i}$ is the biserial correlation between ability θ and the item response variable U_i. In addition, under these same normality assumptions, Tucker (1946) has expressed classical item difficulty, P_i as a function of the item parameters α_i, β_i:

$$P_i = \frac{1}{\sqrt{2\pi}} \int_{\delta_i}^{\infty} e^{-y^2/2} dy, \tag{1.10}$$

where

$$\delta_i = \beta_i \rho_{\theta U_i} = \beta_i \frac{\alpha_i}{\sqrt{1 + \alpha_i^2}}.$$

Thus, under these normality assumptions, one can establish a ready transformation from the item parameters β_i, α_i to the classical theory item parameters P_i, $\rho_{\theta U_i}$. Baker (1959), Baker and Martin (1969), Jensema (1976), and Urry (1974) have employed Tucker's results to implement an item parameter estimation procedure. The item response data are analyzed using classical item analysis procedures, and then the $\hat{P}_i$, $\hat{\rho}_{\theta U_i}$ are transformed to $\hat{\alpha}_i$, $\hat{\beta}_i$ via equations (1.9) and (1.10). Much of the early IRT literature was expressed in terms of the classical theory parameters (see for example Lord, 1952). This conformity to past parameterization been abandoned, however, and only the IRT notation is currently used.

1.3.2 The Logistic Ogive Model

The second model for the item characteristic curve in common use is the logistic ogive, which is also a family of two-parameter (location and scale) cumulative distribution functions. The form of the logistic ogive is very similar in appearance to the normal ogive. In fact, the justification of the use of the logistic ogive as an item characteristic curve model is usually based upon this similarity. The use of the logistic ogive as an ICC model is attributed to Birnbaum (1957), and the first published account was due to Maxwell (1959). Although this use of the logistic function is relatively recent, its use in the field of biology has a long history. The logistic ogive has been used as a model for the growth of plants, people, populations, etc., for over a century. Pearl (1922) employed the logistic ogive to model the growth of human populations and attributed the mathematical formulation to Verhulst (1844). The logistic function has been advocated in quantal response bioassay by Berkson (1944, 1951, 1955a), who has derived several methods for estimating its parameters. The cumulative form of the logistic function, often called the logistic law, is

$$P_i(\theta) = P(\alpha_i^*, \beta_i, \theta) = \Psi(Z_i) = \frac{e^{Z_i}}{1 + e^{Z_i}} = \frac{1}{1 + e^{-Z_i}}, \tag{1.11}$$

where $Z_i = \alpha_i^*(\theta - \beta_i)$ and is called the logit; β_i is the location parameter and is the point on ability scale at which $P_i(\theta) = .5$; α_i^* is the discrimination (scale) parameter and is the reciprocal of the standard deviation of the logistic function. It was stated above, that the justification for the use of the logistic model was its close agreement with the normal model. Let us examine that agreement. In Figure 1.4 a normal ogive having parameters $\beta_i = 0.6$, $\alpha_i = 1.2$ was plotted; this ogive is replotted in Figure 1.6. Using these same parameter values ($\beta_i = 0.6$, $\alpha_i^* = 1.2$), the logistic ogive was calculated using equation (1.11). The obtained values of $P_i(\theta)$ are presented in Table 1.2 and the corresponding ICC plotted in Figure 1.6.

Inspection of Figure 1.6 reveals that the general forms of the two curves are similar; they share a common location parameter, but the slopes of the curves are different. Both the normal ogive and logistic ogives have a point of inflection at $\beta_i = \mu_i$ (Maxwell, 1959); hence, the numerical value of β_i will have the same meaning under both models. However, the same numerical value of α_i yields different slopes under the two models. This comes about due to the differences in variances of the normal and logistic functions. If one establishes a unit normal density function $N(0, 1)$, the corresponding logistic function will have mean zero and variance $\pi^2/3$. Thus, the logistic function

Table 1.2. Item Characteristic Curve Based on Logistic Ogive Model, $\beta_i = 0.6$, $\alpha_i^* = 1.2$

θ	$\theta - \beta_i$	Z_i	$P_i(\theta)$
−3.0	−3.6	−4.32	.0131
−2.5	−3.1	−3.72	.0237
−2.0	−2.6	−3.12	.0423
−1.5	−2.1	−2.52	.0745
−1.0	−1.6	−1.92	.1279
−0.5	−1.1	−1.32	.2108
0.0	−0.6	−0.72	.3274
0.5	−0.1	−0.12	.4700
1.0	0.4	0.48	.6177
1.5	0.9	1.08	.7465
2.0	1.4	1.68	.8429
2.5	1.9	2.28	.9072
3.0	2.4	2.88	.9468

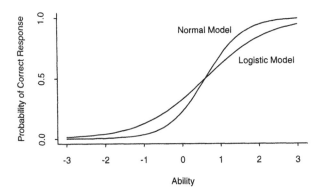

Fig. 1.6. Item characteristic curves based on normal ($\beta_i = 0.6$, $\alpha_i = 1.2$) and logistic ($\beta_i = 0.6$, $\alpha_i^* = 1.2$) models.

will have more spread. Since $\alpha_i = 1/\sigma_i$, the slope of the corresponding logistic ogive will be less than that of the normal ogive, a result observed in Figure 1.6. In order to obtain good agreement between the normal and logistic ogives, an adjustment to the entering argument Z_i is needed. Haley (1952) has shown that, if the logistic deviate $Z_i = \alpha_i^*(\theta - \beta_i)$ is expressed as having a scaling constant 1.702 and used as the entering argument in equation (1.11), the absolute difference between $P_i(\theta)$ of the normal ogive and the logistic ogive is less than .01 over the full range of θ. Thus,

$$|\Phi(Z_i) - \Psi(1.702Z_i)| < .01 \quad \text{for } -\infty < \theta < \infty. \tag{1.12}$$

The logistic deviate, $Z_i = \alpha_i^*(\theta - \beta_i)$, having a value of $1.702Z_i = 1.702\alpha_i(\theta_j - \beta_i)$, where α_i^* is the discrimination parameter under the logistic ogive model and α_i is a numerical value in normal ogive metric, will result in good agreement of the two ogives. The net result is that, in order to match the normal ogive, the numerical value of the logistic discrimination parameter $\alpha_i^* = 1.702\alpha_i$ is used as the entering argument in equation (1.11). For example, in Table 1.2, at $\theta = -0.5$, the logit Z_i based upon $\alpha_i^* = 1.2$ is -1.32, and the probability of correct response is .2108. Yet, in Table 1.1 at $\theta = -.5$, the normal deviate is also -1.32, but the probability of correct response is .0934. Now, if the logit Z_i is multiplied by 1.702, the value -2.246 yields $P_i(\theta) = .0956$, which is very close to the normal model value at $\theta = -.5$. Thus, it is clear that equal numerical values of α_i under the normal model and of α_i^* under the logistic model represent different slopes. However, when $\alpha_i^* = 1.702\alpha_i$ is used, under the logistic model, a very close approximation to the normal model is obtained. The use of the 1.702 multiplier is a carry over from an earlier time when the normal ogive was the standard item characteristic curve model. Using a numerical value of the logistic discrimination parameter matching that under a normal model aided interpretation. However, in recent years the logistic has become the model of choice, and the logistic discrimination parameter also will be denoted as α_i. The context will determine if α_i is to be interpreted in normal or logistic model terms. In equation (1.11) the logistic discrimination parameter was denoted by α_i^* just to alert the reader; henceforth the superscript * will not be used.

From both a practical and a mathematical point of view, the most important characteristic of the logistic function is that its cumulative distribution is a closed form. Recall that, in the case of the normal ogive,

$$P_i(\theta) = \frac{1}{\sqrt{2\pi}} \int_{-\infty}^{Z_i} e^{-z^2/2} dz$$

and, in order to obtain $P_i(\theta)$ for a given value of Z_i, the function must be integrated. Even on a digital computer, the integration of a continuous function or the use of an approximation to the integral is both time-consuming and expensive. In the case of the logistic function, $P_i(\theta)$ does not involve an integral, and it can be computed directly. Since many inexpensive modern pocket calculators contain an e^x function, $P_i(\theta)$ can be calculated readily for a given value of Z_i.

The logistic function is also related to the logarithm of the odds of getting the item correct. At any point on ability scale θ, the probability of correct response is given by $P_i(\theta)$, and the probability of incorrect response is $Q_i(\theta) = 1 - P_i(\theta)$. The odds of making a correct response are $P_i(\theta)/Q_i(\theta)$. If the natural logarithm (denoted by log) of this ratio is taken, the following holds:

$$\log\left[\frac{P_i(\theta)}{Q_i(\theta)}\right] = \log P_i(\theta) - \log Q_i(\theta) = Z_i. \tag{1.13}$$

This can be shown as follows. Under the logistic model,

$$P_i(\theta) = \frac{1}{1 + e^{-Z_i}},$$

$$e^{-Z_i} = \frac{1 - P_i(\theta)}{P_i(\theta)} = \frac{Q_i(\theta)}{P_i(\theta)}$$

and

$$Z_i = \log\left[\frac{P_i(\theta)}{Q_i(\theta)}\right] = \alpha_i(\theta - \beta_i) = \zeta_i + \lambda_i\theta. \tag{1.14}$$

The relationship between the logistic deviate and the proportion of correct response is also know as the logistic transformation. For purposes of item parameter estimation, it is convenient to use the transformation to obtain a linear representation of the ICC. With the transformation given in equation (1.14), the logistic ogive relating $P_i(\theta)$ and θ becomes a linear regression line relating Z_i and θ. In Figure 1.7, the ICC of Table 1.2 under a logistic model has been transformed to its linear counterpart. Although the resulting regression line appears to be identical to that for the normal ogive shown in Figure 1.5, it is important to recognize that the vertical axes have different units of measurement. In the present case, the vertical axis is in units of $\log[P_i(\theta)/Q_i(\theta)]$ and, in the previous case, it was in units of a normal deviate transformation of $P_i(\theta)$.

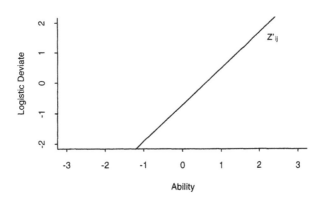

Fig. 1.7. Logistic deviate transformation of the item characteristic curve, $\beta_i = .6$, $\alpha_i^* = 1.2$.

Although both the normal ogive and logistic ogive models for the item characteristic curve are used, the logistic model predominates in the recent literature. This transition from the normal to the logistic ogive is based on a number of factors. When the 1.702 adjustment factor is used, the two ogives are in close agreement for all practical purposes. Because of its closed form, the evaluation of the logistic cumulative distribution function is quick and inexpensive relative to the normal. The logistic model can be extended to handle item scoring procedures beyond dichotomous scoring, such as ordered and nominal response, in a very straightforward fashion. Finally, mathematical operations involving the logistic density function and its cumulative distribution are much simpler than those based upon the normal distribution. Despite the superiority of the logistic model, it is often necessary to use the normal ogive model to illustrate some facet of the development of IRT and to provide linkages to classical test theory. Thus, for the next two chapters, both models shall be used.

1.4 Extension of the Item Characteristic Curve Models: Dichotomous Scoring

1.4.1 Birnbaum's Three-parameter Model

An interesting phenomenon has been noted when the empirical probability of correct response to a multiple choice item has been plotted as a function of ability. In the case of difficult items ($\beta_i > \mu_\theta$), it has been observed that the lower tail of the empirical item characteristic curve sometimes becomes asymptotic to a value greater than zero. This phenomenon has been interpreted in several different ways. Birnbaum (1968) conjectured that low-ability subjects select the correct response by chance. Lord (1974a) noted that the observed asymptote is often lower than the chance level $1/m$, where m is the number of response alternatives. He attributes this to clever item writing, where the distractors are particularly attractive to the examinee with partial or misinformation. This asymptotic behavior is probably also present for easy items ($\beta_i < \mu_\theta$) but not readily observed in the empirical data since the left tail of the ICC is not fully represented. In the item characteristic curves presented above, no allowance was made for this asymptotic behavior. Birnbaum (1968) proposed an item characteristic curve in which the probability of correct response to an item is given by

$$P_i(\theta) = c_i + (1 - c_i)\Phi[\alpha_i(\theta - \beta_i)]$$

for the normal ogive model or

$$P_i(\theta) = c_i + (1 - c_i)\Psi[\alpha_i(\theta - \beta_i)] \tag{1.15}$$

for the logistic ogive model, where c_i is the asymptotic probability of correct response for $\theta \to -\infty$. Since it appeared in the Lord and Novick (1968) book, the three-parameter model for the item characteristic curve has been credited to Birnbaum. However, this model was widely used in bioassay at an earlier date, and parameter estimation techniques for it appeared in Finney's (1952) book. In bioassay, the c_i parameter is known as Abbott's correction for natural mortality (see Abbott, 1925). A common practice in bioassay studies is to draw two independent groups of animals. One is administered the drug and is used to obtain the response data. The other is used simply to determine how many animals will die naturally during the course of the study. The obtained proportion is then the estimate of parameter c_i.

In the literature (see Hambleton & Cook, 1977; Waller, 1974), c_i is commonly referred to as the "guessing" parameter. However, Lord (1968, 1970, 1974b) indicates that c_i should not be interpreted as a guessing parameter. He prefers to consider it simply as the lower bound for the item characteristic curve. The data of Table 1.2 will be used to illustrate Birnbaum's three-parameter logistic model. Arbitrarily, c_i will be set at .15. Using equation (1.15), the values of $P_i(\theta)$ were recalculated. For example, at $\theta = -1.0$,

$$P_i(-1.0) = .15 + (1 - .15)(.1279) = .15 + .1087 = .2587.$$

Table 1.3 presents the complete set of values, and Figure 1.8 depicts the corresponding item characteristic curve. Several features of Figure 1.8 are of interest. First, the item characteristic curve appears to asymptote at $P_i(\theta) = .15$ for $\theta \to -\infty$. Second, the upper limit of the item characteristic curve will remain at 1.0. Finally, the values of α_i, β_i define an item characteristic curve, the $\Psi[\alpha_i(\theta_j - \beta_i)]$ of equation (1.15), which is constrained by the upper limit of 1 and the lower limit of c_i. As a result, the item difficulty parameter β_i is no longer defined as the point on the ability scale at which the probability of correct response is .5. Rather it is the point on the ability scale where the probability of correct response is $c_i + (1 - c_i)(.5)$. When $c_i = 0$, β_i returns to its original definition. However, when $c_i \neq 0$, the value of β_i will depend on the value of c_i and, hence, can correspond to levels of $P_i(\theta)$ other than .5. For example, in the item characteristic curve of Table 1.3 and Figure 1.8, the value of $P_i(\theta)$ at $\beta_i = \theta = .6$ is $.575 = .15 + .425$. Consequently, the interpretation of β_i depends on the value of c_i, and its relation to a fixed value of $P_i(\theta)$ will not be consistent from item to item when the values of c_i vary across items.

The justification for Birnbaum's three-parameter model is a pragmatic one. Birnbaum (1968) simply indicates that his model is more plausible than the random guessing model of classical test theory involving formula scoring. Very strong support for the three-parameter model was provided by Lord's (1970) study. In this study, the basic data were the item responses of 103,275 students to five items from the verbal section of the Scholastic Aptitude Test (SAT). Using a rather involved procedure, Lord was able to plot the item characteristic curves based upon the response data for this very large group of examinees. Maximum likelihood estimators of the item parameters under the three-parameter logistic model were obtained for a sample of 2962 examinees. The empirically obtained proportion of correct response curves and those based upon the model were plotted on the same graph. Visual inspection showed very close agreement between the two curves. Lord (1970, p. 49) concluded: "the agreement between the two methods of estimating the

Table 1.3. Item Characteristic Curve Based on Birnbaum's Three-Parameter (Logistic) Model, $\beta_i = 0.6$, $\alpha_i = 1.2$, $c_i = .15$

θ	$\theta - \beta_i$	Z_i	Two-Parameter $P_i(\theta)$	Three-Parameter $P_i(\theta)$
−3.0	−3.6	−4.32	.0131	.1612
−2.5	−3.1	−3.72	.0237	.1701
−2.0	−2.6	−3.12	.0423	.1859
−1.5	−2.1	−2.52	.0745	.2133
−1.0	−1.6	−1.92	.1279	.2587
−0.5	−1.1	−1.32	.2108	.3292
0.0	−0.6	−0.72	.3274	.4283
0.5	−0.1	−0.12	.4700	.5495
1.0	0.4	0.48	.6177	.6751
1.5	0.9	1.08	.7465	.7845
2.0	1.4	1.68	.8429	.8665
2.5	1.9	2.28	.9072	.9211
3.0	2.4	2.88	.9468	.9548

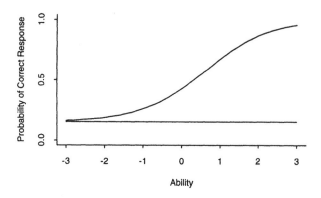

Fig. 1.8. Item characteristic curve under Birnbaum's three-parameter (logistic) model, $\beta_i = 0.6$, $\alpha_i = 1.2$, $c_i = .15$.

item characteristic curves is impressive, particularly so in view of the very different characters of the two methods used." Thus, as was the case with the previous two item characteristic curve models, the primary justification of the three-parameter model, either normal or logistic, rests upon pragmatic grounds. Although Lord (1970) was able to show that the three-parameter model was appropriate for his data set, this is not always the case. McKinley and Reckase (1980) examined 50 items taken from the Iowa Tests of Educational Development for a sample of 1999 examinees. Of the 50 items, only 8 showed empirical evidence of a lower tail, suggesting that a three-parameter model would be needed.

As was the case for the two-parameter item characteristic curve models, it is convenient for parameter estimation purposes to transform the ICC to its linear regression form. In the case of the three-parameter model, this transformation is given by

$$P_i(\theta) = c_i + (1 - c_i)\Psi[\alpha_i(\theta - \beta_i)],$$

where $Z_i = \alpha_i(\theta - \beta_i)$. Then,

$$P_i(\theta) = c_i + (1 - c_i)\frac{1}{1 + e^{-Z_i}}.$$

Because

$$e^{-Z_i} = \frac{1 - c_i}{P_i(\theta) - c_i} - 1 = \frac{Q_i(\theta)}{P_i(\theta) - c_i},$$

$$Z_i = \log\left[\frac{P_i(\theta) - c_i}{Q_i(\theta)}\right].$$

Thus, the transformation is not the same as the logistic transformation under a two-parameter model. Again, the plotted linear regression will be identical to that shown in Figures 1.5 and 1.7. However, the vertical axis is in units of $\log\{[P_i(\theta) - c_i]/Q_i(\theta)\}$.

Although this model is routinely called the three-parameter logistic model, Birnbaum (1968, Chapter 20) was explicit in indicating that it did not possess the "nice" mathematical properties of the usual logistic function. The introduction of the additional parameter c_i removed the function from the two-parameter family of logistic functions. Consequently, it is in actuality a separate model with its own set of mathematical and psychometric properties rather than a logistic model. Unfortunately, most of the literature has not recognized this distinction and refers to the model as a logistic model. A more precise label would be Birnbaum's three-parameter model.

1.4.2 The One-parameter Logistic Model—The Rasch Model

Probably no item characteristic curve model has generated as much interest among practitioners as the Rasch model. Since an entire chapter will be devoted to the Rasch model, the presentation here will be limited to the item characteristic curve employed under this model in IRT. The one-parameter logistic (Rasch) model is defined as

$$P_i(\theta) = \frac{e^{(\theta - \beta_i)}}{1 + e^{(\theta - \beta_i)}} = \frac{1}{1 + e^{-(\theta - \beta_i)}} \tag{1.16}$$

which is exactly the two-parameter logistic ICC model with location parameter β_i, a discrimination parameter α_i fixed at unity (Birnbaum, 1968), and an ability variable of θ. Much of the controversy surrounding the Rasch model centers on this restriction of the discrimination parameter since items typically exhibit varying degrees of discrimination.

1.5 Summary

In this chapter the item characteristic curve has been presented within the context of a single item and a known criterion variable. The item characteristic curve has been defined as the function relating the probability of correct response to the ability scale. It is important to emphasize that this definition does not involve the distribution of the examinees over the ability scale. Two basic mathematical functions, the normal ogive and logistic ogive, have been used to model this relationship. Under these models, a specific item characteristic curve is defined by the values of a difficulty (location) parameter and a discrimination (scale) parameter. The former, denoted by β_i, locates the median of the item characteristic curve relative to ability scale and, under IRT, is used to describe the difficulty of the item. The latter, α_i, indexes the steepness of the item characteristic curve. It was shown that the discrimination parameter is actually the slope of the linear regression of a transformation of the probability of correct response and the ability scale. Two additional item characteristic models were also presented. Birnbaum's three-parameter model used the parameter c_i to take into account the observed phenomena that, for some items, low-ability subjects achieve a higher probability of correct response than indicated by a two-parameter model. Although commonly referred to as the three-parameter logistic model, it is not a member of the family of logistic functions and is actually a separate mathematical model. The Rasch model is a special case of the two-parameter logistic model in which the discrimination parameter is fixed at a value of unity. Under all of these models, it is important to have a metric for ability scale that allows one to use the particular ICC model simultaneously for all items in an instrument.

Since item response theory is item-based, the parameters of the ICC under various models and scoring procedures are the keystone of the theory. Because of this, all of the remaining chapters in this book relate to either estimating the values of these parameters via various techniques or to estimating an examinee's ability that is a function of these parameters.

2. Estimating the Parameters of an Item Characteristic Curve

2.1 Introduction

One of the salient characteristics of item response theory is that the major constructs of the theory are based upon items and the parameters of a model for the item characteristic curve. In the previous chapter, the several models for dichotomously scored items and their parameterization were introduced. Given the responses of examinees to the items of a test, the task is then to characterize the items via the numerical values of the item parameters. To accomplish this, a selected item characteristic curve model must be fitted to the response data for each item in a test. While the parameters of the items in a test could be estimated simultaneously, this leads to computationally demanding procedures that are not economically feasible. Thus, in most of the IRT test analysis procedures the item parameters are estimated on a one item at a time basis. Because of this, a technique for estimating the parameters of a single item is one of the two basic building blocks of test analysis under IRT. In the present chapter, it will be assumed that the ability scores of the examinees responding to a test item are known. This assumption greatly simplifies the estimation process and it is employed in most of the approaches to be discussed in later chapters. In addition, it places the estimation procedures within the context of the well known statistical techniques for quantal-response bioassay.

Assume there are several groups of examinees, with known ability scores, positioned along the ability scale from low to high ability. At low ability levels, a small proportion of the examinees in a group will answer the dichotomously scored item correctly. At high ability levels, a large proportion will answer it correctly. These observed proportions of correct response can then be plotted as a function of the groups along the ability scale as is depicted in Figure 2.1. The basic task then is one of finding the item characteristic curve, under a given model, that fits the observed proportions of correct response. The solid line in Figure 2.1 represents that curve for this set of data.

From a statistical point of view, this fitting process becomes one of estimating the parameters of a cumulative distribution function, that is, the ICC, that "best" fits the data. The statistical procedures for estimating the parameters of a cumulative distribution function fitted to the observed proportions of response have received considerable attention in the literature.

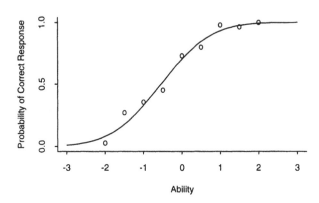

Fig. 2.1. Observed proportion of correct response and fitted normal ogive $\hat{\alpha}_i =$ .9110, $\hat{\beta}_i = -.5373$.

In psychophysics, the method of weighted least squares, called the "constant process" (see Thomson, 1919), was widely used around the turn of the 20th century to fit the normal ogive to the observed proportions of response in psychological experiments. Ferguson (1942) subsequently applied this procedure to the fitting of item characteristics curves to item response data. In quantal-response bioassay, groups of animals are administered ordered dosages of a drug to determine the number of animals that die at each dosage. A cumulative distribution function is then fitted to the observed proportions of death. The median of this function, X_{50}, is a measure of the toxicity of the drug. In this context, Berkson (1944, 1949, 1955b) has advocated minimum transform χ^2 estimation procedures which are essentially the same as those of the "constant process." Maximum likelihood estimation has been used in quantal-response bioassay to estimate the median lethal dose since the mid-1930s (see Bliss, 1935; Garwood, 1941). In Psychometrics, maximum likelihood estimation of item parameters, under the normal ogive ICC model, was independently introduced by Lawley (1943) and Finney (1944). Finney (1944) showed that from a technical point of view, estimating the median lethal dose in quantal response bioassay and fitting item characteristic curves involve the same estimation process. In the recent psychometric literature (see Mislevy & Bock, 1982a; Harwell, Baker, & Zwarts, 1988), the estimation of the parameters of a single item when the ability of the examinees is known is referred to as the "bioassay solution" in recognition of this equivalence.

In the paragraphs below, the presentation of the maximum likelihood estimation procedures will be in terms of the two-parameter normal and logistic item characteristic curve models for a single item with known ability scores for the examinees. Although, the normal ogive ICC model is not used in practice, it holds an important position in the history of the development of IRT (see Lord & Novick, 1968). This model also provides the framework for interpreting the numerical values of the item parameter estimates yielded by the standard IRT computer programs. In addition, many of the concepts important to estimation of item parameters can be developed via the normal ogive model. The estimation procedures for the two-parameter logistic ICC model are also presented as this is actually the model of choice. Once the two-parameter estimation procedures have been derived, the procedures for the three-parameter logistic ICC model will be presented. Berkson's minimum logit χ^2 estimation procedure will also be shown as it is computationally attractive due to its being a non-iterative procedure.

2.2 Maximum Likelihood Estimation: Normal Ogive Model

In the previous chapter, the two versions of the parameterization of an ICC were introduced. When describing items in an applied setting, the parameterization of item discrimination as α and of item difficulty as β is widely used. The equivalent parameterization in terms of the slope λ and intercept ζ is widely employed in the technical IRT literature and in the computer programs for IRT test analysis. Because the mathematics of the estimation procedures are somewhat clearer, the slope/intercept form ($Z_j = \zeta + \lambda\theta_j$) will be employed in the estimation procedures for the two-parameter ICC models.

Although the estimation procedures could be derived on the basis of individual examinees, the logic of the estimation process is simpler if examinees are grouped according to ability, hence, grouped examinees will be used below. Suppose that k groups of f_j subjects possessing known ability scores θ_j are drawn at random from a population of persons and $j = 1, \ldots, k$. Each subject has responded to a single dichotomously scored item. Because only a single item will be used here, the item subscript, i, will not be included in order to simplify the notation. Out of the f_j subjects having ability θ_j, r_j gave the correct response and $f_j - r_j$ the incorrect response. Let $R = (r_1, \ldots, r_k)$ be the vector of the observed number of correct responses. The observed proportion of correct response at ability θ_j is

$$p(\theta_j) = p_j = \frac{r_j}{f_j},$$

and the observed proportion of incorrect response is

$$q(\theta_j) = q_j = \frac{f_j - r_j}{f_j}.$$

It will be assumed that the observed r_j at each ability θ_j are binomially distributed with parameters f_j, P_j, where P_j is the true probability of correct response. Then $E(r_j) = f_j P_j$ and $\text{Var}(r_j) = f_j P_j Q_j$, where $Q_j = 1 - P_j$. Hence,

$$E(p_j) = E\left(\frac{r_j}{f_j}\right) = \frac{1}{f_j} E(r_j) = P_j.$$

The value of P_j depends upon the parameters ζ and λ of the item characteristic curve model as well as the value of θ_j. Under a normal ogive model P_j is given by

$$P_j = P(\zeta, \lambda, \theta_j) = \int_{-Z_j}^{\infty} \frac{1}{\sqrt{2\pi}} e^{-t^2/2} dt = \int_{-\infty}^{Z_j} \frac{1}{\sqrt{2\pi}} e^{-t^2/2} dt$$

with $Z_j = \zeta + \lambda \theta_j$.

It should be noted that the limits on the integral sign are needed to obtain the proper value of P_j. For example, when Z_j has a large positive value, the limits $\int_{Z_j}^{\infty}$ would yield a small value of P_j while $\int_{-Z_j}^{\infty}$ would yield a large value as would $\int_{-\infty}^{Z_j}$. The limits employed were chosen to correspond to those determining the shaded areas (P_j) of the conditional distributions shown in Figure 1.3.

At this point several observations are in order. There is no requirement that the number of subjects f_j be the same at all ability score levels. The random sampling also implies that the obtained values of θ_j would be a random sample of values. Finally, the sampling distribution of the observed r_j at each ability score level could be assumed to follow any of a number of distributions. However, the choice of the binomial distribution is basic to a number of important concepts in IRT.

Before proceeding further, it will be useful to present some derivatives associated with the normal ogive model. The ordinate of the normal density function at $Z_j = \zeta + \lambda \theta_j$ is

$$h_j = \frac{1}{\sqrt{2\pi}} e^{-Z_j^2/2}.$$

The derivatives of h_j with respect to ζ and λ are also needed and are obtained as follows: Let $h_j' = dh_j/dZ_j$. Then differentiation shows that

$$h_j' = h_j(-Z_j), \qquad \frac{\partial Z_j}{\partial \zeta} = 1, \qquad \text{and} \qquad \frac{\partial Z_j}{\partial \lambda} = \theta_j.$$

Hence, by the chain rule,

$$\frac{\partial P_j}{\partial \zeta} = h_j$$

and

$$\frac{\partial P_j}{\partial \lambda} = h_j \theta_j.$$

Now, taking all response categories and ability score groups into account, the probability of R, is given by the likelihood function

$$\text{Prob}(R) = \prod_{j=1}^{k} \frac{f_j!}{r_j!(f_j - r_j)!} P_j^{r_j} Q_j^{f_j - r_j}$$

and the natural logarithm of the likelihood is

$$L = \log \text{Prob}(R) = \text{constant} + \sum_{j=1}^{k} r_j \log P_j + \sum_{j=1}^{k} (f_j - r_j) \log Q_j. \quad (2.1)$$

The maximum likelihood estimates of the unknown parameters ζ and λ are based upon the information provided by the sample and are the values which satisfy the equations:

$$\frac{\partial L}{\partial \zeta} = 0 \quad \text{and} \quad \frac{\partial L}{\partial \lambda} = 0. \quad (2.2)$$

Taking the appropriate first partial derivatives,

$$\frac{\partial L}{\partial \zeta} = \sum_{j=1}^{k} r_j \frac{1}{P_j} \frac{\partial P_j}{\partial \zeta} + \sum_{j=1}^{k} (f_j - r_j) \frac{1}{Q_j} \frac{\partial Q_j}{\partial \zeta}$$

$$= \sum_{j=1}^{k} \frac{r_j h_j}{P_j} - \sum_{j=1}^{k} \left[\frac{(f_j - r_j) h_j}{Q_j} \right]$$

$$= \sum_{j=1}^{k} h_j \left[\frac{r_j}{P_j} - \frac{(f_j - r_j)}{Q_j} \right]$$

$$= \sum_{j=1}^{k} h_j \left[\frac{(1 - P_j) r_j - (f_j - r_j) P_j}{P_j Q_j} \right]$$

$$= \sum_{j=1}^{k} \frac{h_j}{P_j Q_j} (r_j - f_j P_j).$$

Substituting $f_j p_j$ for r_j,

$$\frac{\partial L}{\partial \zeta} = \sum_{j=1}^{k} \frac{h_j}{P_j Q_j} (f_j p_j - f_j P_j) = \sum_{j=1}^{k} \frac{f_j h_j}{P_j Q_j} (p_j - P_j).$$

Now, multiplying numerator and denominator of the right hand expression by h_j,

$$L_1 = \frac{\partial L}{\partial \zeta} = \sum_{j=1}^{k} f_j \frac{h_j^2}{P_j Q_j} \left(\frac{p_j - P_j}{h_j} \right).$$

Similarly,

$$\frac{\partial L}{\partial \lambda} = \sum_{j=1}^{k} r_j \frac{1}{P_j} \frac{\partial P_j}{\partial \lambda} + \sum_{j=1}^{k} (f_j - r_j) \frac{1}{Q_j} \frac{\partial Q_j}{\partial \lambda}$$

$$= \sum_{j=1}^{k} \frac{r_j h_j \theta_j}{P_j} - \sum_{j=1}^{k} \left[\frac{(f_j - r_j) h_j \theta_j}{Q_j} \right]$$

$$= \sum_{j=1}^{k} h_j \left[\frac{r_j}{P_j} - \frac{(f_j - r_j)}{Q_j} \right] \theta_j$$

$$= \sum_{j=1}^{k} \frac{h_j}{P_j Q_j} (r_j - f_j P_j) \theta_j.$$

$$= \sum_{j=1}^{k} \frac{f_j h_j}{P_j Q_j} (p_j - P_j) \theta_j.$$

Again after multiplication by h_j in the numerator and denominator of the right side,

$$L_2 = \frac{\partial L}{\partial \lambda} = \sum_{j=1}^{k} f_j \frac{h_j^2}{P_j Q_j} \left(\frac{p_j - P_j}{h_j} \right) \theta_j.$$

Now setting both derivatives equal to zero, the resulting simultaneous equations, in theory, can be solved for the estimates of ζ and λ that maximize the likelihood function. These are called the likelihood equations:

$$\sum_{j=1}^{k} f_j \frac{h_j^2}{P_j Q_j} \left(\frac{p_j - P_j}{h_j} \right) = 0$$

(2.3)

$$\sum_{j=1}^{k} f_j \frac{h_j^2}{P_j Q_j} \left(\frac{p_j - P_j}{h_j} \right) \theta_j = 0.$$

The $h_j^2 P_j Q_j$ term is known as the Urban-Müller weight (Thomson, 1919; Berkson, 1955a), and it plays an important role in IRT as well as in parameter estimation. Under a binomial distribution of the observed proportion of correct response at each ability score, p_j has a true variance of $P_j Q_j / f_j$ and for large f_j the variance of the normal deviate, z_j, corresponding to p_j is asymptotically $P_j Q_j / f_j h_j^2$. This variance will be different at each ability score level. In the previous chapter, and the presentation of regression in most elementary statistics textbooks, it is assumed that the variances of the conditional distributions of the dependent variable are constant over all ability score levels. In the present case, however, the variances are clearly unequal and this must be taken into account when fitting the regression line. Thus, the homogeneity of variance assumption in Lord's justification of the normal ogive model presented in Chapter 1 depends upon the asymptotic normality of the binomial distribution. At each ability level the reciprocal of the variance of Z_j is used to weight the contribution of the data point and we have $f_j h_j^2 / P_j Q_j = f_j W_j$, where $W_j = h_j^2 / P_j Q_j$ is the weighting factor. Thus, the maximum likelihood item parameter estimation procedure is actually a weighted procedure for fitting a linear regression line to the normal deviate transformations of the observed proportions of correct response. Because of this, it is known as the probit method in bioassay (see Bliss, 1935; Finney, 1944; Finney, 1952).

In equation (2.3), the term in the bracket is considered to be the difference between the "working" deviate and the "true" deviate at a given ability score level. Let z_j be the working deviate and Z_j be the true deviate then the following (Garwood, 1941) holds:

$$z_j - Z_j \approx \frac{p_j - P_j}{h_j}. \tag{2.4}$$

This equation can be motivated rather easily. Let

$$P_j = \int_{-\infty}^{Z_j} \frac{1}{\sqrt{2\pi}} e^{-t^2/2} dt$$

$$p_j = \int_{-\infty}^{z_j} \frac{1}{\sqrt{2\pi}} e^{-t^2/2} dt$$

and then

$$p_j - P_j = \int_{Z_j}^{z_j} \frac{1}{\sqrt{2\pi}} e^{-t^2/2} dt.$$

Now, from the mean value theorem,

$$p_j - P_j \approx (z_j - Z_j)h_j$$

and

$$\frac{p_j - P_j}{h_j}$$

is a reasonable expression for the difference, especially when p_j and P_j are close in value. Substituting $(z_j - Z_j)$ for $(p_j - P_j)/h_j$ and W_j for h_j^2/P_jQ_j in equation (2.3) yields

$$\sum_{j=1}^{k} f_j W_j (z_j - Z_j) = 0$$

$$\sum_{j=1}^{k} f_j W_j (z_j - Z_j)\theta_j = 0. \tag{2.5}$$

In this form, it is clear that the process is fitting a linear regression line via a weighted process.

The likelihood equations shown in equation (2.5) cannot be solved directly to obtain the item parameter estimates. The problem being that the true deviates Z_j depend upon the values of the unavailable item parameters ζ and λ.

However, this form of the likelihood equations is useful for comparative purposes since the likelihood equations based upon other item characteristic curve models and estimation procedures can be expressed in this form. Once this is done, the differences and similarities of the several approaches can be seen.

In order to solve the likelihood equations, an iterative procedure based upon a Taylor series is employed. Since, the maximum likelihood estimates of the parameters are functions of the sample, $\hat{\zeta} = \hat{\zeta}(\theta_1, \ldots, \theta_k)$, $\hat{\lambda} = \hat{\lambda}(\theta_1, \ldots, \theta_k)$; if approximations $\hat{\zeta}_1$, $\hat{\lambda}_1$ to $\hat{\lambda}$, $\hat{\lambda}$ can be found by some crude means, then $\hat{\zeta} = \hat{\zeta}_1 + \Delta\hat{\zeta}$ and $\hat{\lambda} = \hat{\lambda}_1 + \Delta\hat{\lambda}$ so that $\Delta\hat{\zeta}$, $\Delta\hat{\lambda}$ are errors in the approximations. Using a Taylor series expansion for the first equation in (2.2)

$$\frac{\partial L}{\partial \zeta} = \frac{\partial L}{\partial \zeta}(\hat{\zeta}_1, \hat{\lambda}_1) + \Delta\hat{\zeta}_1 \frac{\partial^2 L}{\partial \zeta^2}(\hat{\zeta}_1, \hat{\lambda}_1) + \Delta\hat{\lambda}_1 \frac{\partial^2 L}{\partial \zeta \partial \lambda}(\hat{\zeta}_1, \hat{\lambda}_1) +$$

terms involving higher-order powers of $\Delta\hat{\zeta}_1$ and $\Delta\hat{\lambda}_1$.

The notation $(\partial L/\partial \zeta)(\hat{\zeta}_1, \hat{\lambda}_1)$, $(\partial^2 L/\partial \zeta^2)(\hat{\zeta}_1, \hat{\lambda}_1)$, and so forth, indicates that the derivatives of the likelihood function are to be evaluated using the estimates $\hat{\zeta}_1$, $\hat{\lambda}_1$. Thus, the Taylor series allows the derivatives with respect to unknown parameters to be evaluated using estimates and the only unknowns are the $\Delta\hat{\zeta}_1$, $\Delta\hat{\lambda}_1$. Setting $(\partial L/\partial \zeta)(\zeta, \lambda) = 0$ and ignoring terms involving $\Delta\hat{\zeta}_1^2$, $\Delta\hat{\lambda}_1^2$ and higher order powers yields:

$$0 = \frac{\partial L}{\partial \zeta}(\hat{\zeta}_1, \hat{\lambda}_1) + \Delta\hat{\zeta}_1 \frac{\partial^2 L}{\partial \zeta^2}(\hat{\zeta}_1, \hat{\lambda}_1) + \Delta\hat{\lambda}_1 \frac{\partial^2 L}{\partial \zeta \partial \lambda}(\hat{\zeta}_1, \hat{\lambda}_1).$$

Let

$$L_1 = \frac{\partial L}{\partial \zeta}(\hat{\zeta}_1, \hat{\lambda}_1); \quad L_{11} = \frac{\partial^2 L}{\partial \zeta^2}(\hat{\zeta}_1, \hat{\lambda}_1); \quad L_{12} = \frac{\partial^2 L}{\partial \zeta \partial \lambda}(\hat{\zeta}_1, \hat{\lambda}_1).$$

Then the series can be written as

$$-L_1 = L_{11}\Delta\hat{\zeta}_1 + L_{12}\Delta\hat{\lambda}_1.$$

Similarly, for the second equation in (2.2)

$$0 = \frac{\partial L}{\partial \lambda}(\hat{\zeta}_1, \hat{\lambda}_1) + \Delta\hat{\lambda}_1 \frac{\partial^2 L}{\partial \lambda^2}(\hat{\zeta}_1, \hat{\lambda}_1) + \Delta\hat{\lambda}_1 \frac{\partial^2 L}{\partial \zeta \partial \lambda}(\hat{\zeta}_1, \hat{\lambda}_1) + \cdots.$$

Let

$$L_2 = \frac{\partial L}{\partial \lambda}(\hat{\zeta}_1, \hat{\lambda}_1); \quad L_{22} = \frac{\partial^2 L}{\partial \lambda^2}(\hat{\zeta}_1, \hat{\lambda}_1); \quad L_{21} = L_{12} = \frac{\partial^2 L}{\partial \lambda \partial \zeta}(\hat{\zeta}_1, \hat{\lambda}_1).$$

The equation can then be written as

$$-L_2 = L_{22}\Delta\hat{\lambda}_1 + L_{21}\Delta\hat{\zeta}_1.$$

Thus, the simultaneous equations to be solved for $\Delta\hat{\zeta}_1$ and $\Delta\hat{\lambda}_1$ are

$$-L_1 = L_{11}\Delta\hat{\zeta}_1 + L_{12}\Delta\hat{\lambda}_1$$
$$-L_2 = L_{21}\Delta\hat{\zeta}_1 + L_{22}\Delta\hat{\lambda}_1. \tag{2.6}$$

In matrix algebra terms,

$$\begin{bmatrix} L_1 \\ L_2 \end{bmatrix} = - \begin{bmatrix} L_{11} & L_{12} \\ L_{21} & L_{22} \end{bmatrix} \begin{bmatrix} \Delta\hat{\zeta}_1 \\ \Delta\hat{\lambda}_1 \end{bmatrix}.$$

Solving for $\Delta\hat{\zeta}_1$ and $\Delta\hat{\lambda}_1$ yields

$$\begin{bmatrix} \Delta\hat{\zeta}_1 \\ \Delta\hat{\lambda}_1 \end{bmatrix} = - \begin{bmatrix} L_{11} & L_{12} \\ L_{21} & L_{22} \end{bmatrix}^{-1} \begin{bmatrix} L_1 \\ L_2 \end{bmatrix}. \tag{2.7}$$

Since the $\hat{\zeta}_1$, $\hat{\lambda}_1$ were crude approximations and $\Delta\hat{\zeta}_1$, $\Delta\hat{\lambda}_1$ are estimates, the obtained values of the parameters $\hat{\zeta}_2 = \hat{\zeta}_1 + \Delta\hat{\zeta}_1$ and $\hat{\lambda}_2 = \hat{\lambda}_1 + \Delta\hat{\lambda}_1$ are only second-order approximations. Thus, the process must be repeated until $\Delta\hat{\zeta}_t$ and $\Delta\hat{\lambda}_t$ become sufficiently small, where t indexes the iteration. An iterative procedure to accomplish this can be established in the following fashion: Let, temporarily,

$$\hat{\zeta}_t = \hat{\zeta}_1 \quad \text{and} \quad \hat{\lambda}_t = \hat{\lambda}_1,$$

then

$$\hat{\zeta}_{t+1} = \hat{\zeta}_t + \Delta\hat{\zeta}_t$$

$$\hat{\lambda}_{t+1} = \hat{\lambda}_t + \Delta\hat{\lambda}_t.$$

The term $\hat{Z}_{jt} = \hat{\zeta}_t + \hat{\lambda}_t\theta_j$ is considered to be the "true" deviate at iteration t. In matrix algebra terms with the t as a subscript,

$$\begin{bmatrix} \hat{\zeta} \\ \hat{\lambda} \end{bmatrix}_{t+1} = \begin{bmatrix} \hat{\zeta} \\ \hat{\lambda} \end{bmatrix}_t + \begin{bmatrix} \Delta\hat{\zeta} \\ \Delta\hat{\lambda} \end{bmatrix}_t. \tag{2.8}$$

Substituting equation (2.7) for the last term yields

$$\begin{bmatrix} \hat{\zeta} \\ \hat{\lambda} \end{bmatrix}_{t+1} = \begin{bmatrix} \hat{\zeta} \\ \hat{\lambda} \end{bmatrix}_t - \begin{bmatrix} L_{11} & L_{12} \\ L_{21} & L_{22} \end{bmatrix}_t^{-1} \begin{bmatrix} L_1 \\ L_2 \end{bmatrix}_t, \tag{2.9}$$

which is known as the Newton-Raphson equation (see Kendall & Stuart, 1967, Vol. 2, p. 49). The matrix of second derivatives is known in mathematics as the "Hessian."

In order to solve equation (2.9) the second partial derivatives of L also are needed. These are obtained as follows:

$$L_{11} = \frac{\partial^2 L}{\partial\zeta^2} = \frac{\partial}{\partial\zeta} \left\{ \sum_{j=1}^{k} f_j h_j \left[\frac{(p_j - P_j)}{P_j Q_j} \right] \right\}$$

$$= \sum_{j=1}^{k} f_j \left\{ P_j Q_j \frac{\partial}{\partial\zeta} [h_j(p_j - P_j)] - \left[h_j(p_j - P_j)\frac{\partial}{\partial\zeta}(P_j Q_j) \right] \right\} / P_j^2 Q_j^2$$

$$= \sum_{j=1}^{k} f_j \left\{ P_j Q_j [h'_j p_j - (h_j^2 + P_j h'_j)] - [h_j (p_j - P_j)][-P_j h_j + Q_j h_j] \right\} / P_j^2 Q_j^2$$

$$= \sum_{j=1}^{k} f_j \left\{ P_j Q_j [-h_j Z_j p_j - h_j^2 + h_j Z_j P_j] - [h_j^2 (p_j - P_j)(Q_j - P_j)] \right\} / P_j^2 Q_j^2$$

$$= \sum_{j=1}^{k} f_j h_j \left\{ P_j Q_j [-Z_j (p_j - P_j) - h_j] - [h_j (p_j - P_j)(Q_j - P_j)] \right\} / P_j^2 Q_j^2$$

$$= \sum_{j=1}^{k} f_j h_j \left\{ -Z_j P_j Q_j (p_j - P_j) - h_j P_j Q_j - h_j (p_j - P_j)(Q_j - P_j)] \right\} / P_j^2 Q_j^2$$

$$= \sum_{j=1}^{k} \left[\frac{f_j h_j (p_j - P_j)[-Z_j P_j Q_j - h_j (Q_j - P_j)]}{P_j^2 Q_j^2} \right] - \sum_{j=1}^{k} \left[\frac{f_j h_j^2 P_j Q_j}{P_j^2 Q_j^2} \right].$$

$$L_{11} = \sum_{j=1}^{k} \frac{f_j h_j (p_j - P_j)}{P_j Q_j} \left[-Z_j - \frac{h_j}{P_j} + \frac{h_j}{Q_j} \right] - \sum_{j=1}^{k} \frac{f_j h_j^2}{P_j Q_j}.$$

Since $(\partial P_j / \partial \lambda) = (\partial P_j / \partial \zeta)\theta$, L_{22} can be written readily from L_{11}, yielding

$$L_{22} = \sum_{j=1}^{k} \frac{f_j h_j (p_j - P_j)}{P_j Q_j} \left[-Z_j - \frac{h_j}{P_j} + \frac{h_j}{Q_j} \right] \theta_j^2 - \sum_{j=1}^{k} \frac{f_j h_j^2 \theta_j^2}{P_j Q_j}.$$

Also,

$$L_{21} = L_{12} = \frac{\partial}{\partial \zeta} \left\{ \sum_{j=1}^{k} \frac{f_j h_j}{P_j Q_j} (p_j - P_j)\theta_j \right\}$$

$$= \sum_{j=1}^{k} f_j \theta_j \left\{ P_j Q_j \frac{\partial}{\partial \zeta} [h_j (p_j - P_j)] - [h_j (p_j - P_j)] \frac{\partial}{\partial \zeta} (P_j Q_j) \right\} / P_j^2 Q_j^2$$

$$= \sum_{j=1}^{k} f_j \theta_j \left\{ P_j Q_j [h'_j p_j - (h_j^2 + P_j h'_j)] - [h_j (p_j - P_j)][-P_j h_j + Q_j h_j] \right\} / P_j^2 Q_j^2.$$

Since the terms within the {} are identically those obtained for L_{11}, the algebraic simplification is the same and

$$L_{21} = L_{12} = \sum_{j=1}^{k} \frac{f_j h_j (p_j - P_j)}{P_j Q_j} \left[-Z_j - \frac{h_j}{P_j} + \frac{h_j}{Q_j} \right] \theta_j - \sum_{j=1}^{k} \frac{f_j h_j^2 \theta_j}{P_j Q_j}.$$

In order to evaluate the second derivatives at the same point as the first derivatives, the observed p_j in the second derivatives are replaced by their expectations, the P_j. This causes the terms involving $(p_j - P_j)$ to disappear. Under maximum likelihood, this substitution is based upon Fisher's concept of information where the variance of a maximum likelihood estimator is the reciprocal of the negative expectation of the second derivative of the log-likelihood function with respect to the parameter. A detailed justification for this substitution is given by Kendall and Stuart (1967, Vol. 2, Chapter 18) and by Garwood (1941) and it is a standard practice under maximum likelihood procedures. Let

$$E\left(\frac{\partial^2 L}{\partial \zeta^2}\right) = \Lambda_{11}, \quad E\left(\frac{\partial^2 L}{\partial \zeta \partial \lambda}\right) = \Lambda_{12} = \Lambda_{21}, \quad E\left(\frac{\partial^2 L}{\partial \lambda^2}\right) = \Lambda_{22}.$$

Then,

$$\Lambda_{11} = -\sum_{j=1}^{k} \frac{f_j h_j^2}{P_j Q_j} = -\sum_{j=1}^{k} f_j W_j,$$

where $W_j = h_j^2 / P_j Q_j$,

$$\Lambda_{12} = \Lambda_{21} = -\sum_{j=1}^{k} \frac{f_j h_j^2 \theta_j}{P_j Q_j} = -\sum_{j=1}^{k} f_j W_j \theta_j,$$

and

$$\Lambda_{22} = -\sum_{j=1}^{k} \frac{f_j h_j^2 \theta_j^2}{P_j Q_j} = -\sum_{j=1}^{k} f_j W_j \theta_j^2.$$

It should be noted that all of these expectations are negative. Substituting the expected values Λ's for the L's in equation (2.6) and let $v_j = (p_j - P_j)/h_j$ in L_1, L_2, the equations to be solved for $\Delta \hat{\zeta}_t$ and $\Delta \hat{\lambda}_t$ are:

$$\Delta \hat{\zeta}_t \sum_{j=1}^{k} f_j W_j + \Delta \hat{\lambda}_t \sum_{j=1}^{k} f_j W_j \theta_j = \sum_{j=1}^{k} f_j W_j v_j$$

$$\Delta \hat{\zeta}_t \sum_{j=1}^{k} f_j W_j \theta_j + \Delta \hat{\lambda}_t \sum_{j=1}^{k} f_j W_j \theta_j^2 = \sum_{j=1}^{k} f_j W_j v_j \theta_j. \tag{2.10}$$

Making the substitution in equation (2.7) results in

$$\begin{bmatrix} \Lambda_{11} & \Lambda_{12} \\ \Lambda_{21} & \Lambda_{22} \end{bmatrix} = \begin{bmatrix} -\sum_{j=1}^{k} f_j W_j & -\sum_{j=1}^{k} f_j W_j \theta_j \\ -\sum_{j=1}^{k} f_j W_j \theta_j & -\sum_{j=1}^{k} f_j W_j \theta_j^2 \end{bmatrix}$$

which after factoring out the negative sign is known as the information matrix. Then equation 2.9 becomes

$$\begin{bmatrix} \hat{\zeta} \\ \hat{\lambda} \end{bmatrix}_{t+1} = \begin{bmatrix} \hat{\zeta} \\ \hat{\lambda} \end{bmatrix}_t - \begin{bmatrix} -\sum_{j=1}^{k} f_j W_j & -\sum_{j=1}^{k} f_j W_j \theta_j \\ -\sum_{j=1}^{k} f_j W_j \theta_j & -\sum_{j=1}^{k} f_j W_j \theta_j^2 \end{bmatrix}_t^{-1} \begin{bmatrix} \sum_{j=1}^{k} f_j W_j v_j \\ \sum_{j=1}^{k} f_j W_j v_j \theta_j \end{bmatrix}_t .(2.11)$$

It should be noted here that the negative sign in equation (2.9) can become a positive sign due to the negative sign being extracted from the matrix of expectations. The inverse of the matrix of the negative expectations in equation (2.11) is the variance-covariance matrix of the item parameter estimates.

The iterative solution of equation (2.11) is known as Fisher's "method of scoring for parameters." Kale (1962) has shown that, for large N, this method usually converges to a solution faster than does the Newton-Raphson procedure of equation (2.9). An anomaly of the IRT literature is that, under maximum likelihood, Fisher's method is the standard approach to estimating parameters. However, in IRT, it is usually referred to as the Newton-Raphson procedure and the terms will be used somewhat interchangeably.

Working only with the second term on the right hand side of equation (2.11) yields

$$-\begin{bmatrix} -\sum_{j=1}^{k} f_j W_j & -\sum_{j=1}^{k} f_j W_j \theta_j \\ -\sum_{j=1}^{k} f_j W_j \theta_j & -\sum_{j=1}^{k} f_j W_j \theta_j^2 \end{bmatrix}_t^{-1} \begin{bmatrix} \sum_{j=1}^{k} f_j W_j v_j \\ \sum_{j=1}^{k} f_j W_j v_j \theta_j \end{bmatrix}_t .$$

Canceling the two negative signs, taking the inverse by the method of determinants, and multiplying by the vector of first derivatives yields the vector

$$\left(\frac{1}{\left(\sum_{j=1}^{k} f_j W_j \sum_{j=1}^{k} f_j W_j \theta_j^2 - \left(\sum_{j=1}^{k} f_j W_j \theta_j\right)^2 \right)_t} \right) \times$$

$$\begin{bmatrix} \sum_{j=1}^{k} f_j W_j \theta_j^2 & -\sum_{j=1}^{k} f_j W_j \theta_j \\ -\sum_{j=1}^{k} f_j W_j \theta_j & \sum_{j=1}^{k} f_j W_j \end{bmatrix}_t \begin{bmatrix} \sum_{j=1}^{k} f_j W_j v_j \\ \sum_{j=1}^{k} f_j W_j v_j \theta_j \end{bmatrix}_t$$

$$= \left(\frac{1}{\sum_{j=1}^{k} f_j W_j \sum_{j=1}^{k} f_j W_j \theta_j^2 - \left(\sum_{j=1}^{k} f_j W_j \theta_j \right)^2} \right)_t \times$$

$$\left[\begin{array}{l} \sum_{j=1}^{k} f_j W_j \theta_j^2 \sum_{j=1}^{k} f_j W_j v_j - \sum_{j=1}^{k} f_j W_j \theta_j \sum_{j=1}^{k} f_j W_j v_j \theta_j \\ - \sum_{j=1}^{k} f_j W_j \theta_j \sum_{j=1}^{k} f_j W_j v_j + \sum_{j=1}^{k} f_j W_j \sum_{j=1}^{k} f_j W_j v_j \theta_j \end{array} \right]_t .$$

Then,

$$\Delta \hat{\zeta}_t = \frac{\sum_{j=1}^{k} f_j W_j \theta_j^2 \sum_{j=1}^{k} f_j W_j v_j - \sum_{j=1}^{k} f_j W_j \theta_j \sum_{j=1}^{k} f_j W_j v_j \theta_j}{\sum_{j=1}^{k} f_j W_j \sum_{j=1}^{k} f_j W_j \theta_j^2 - \left(\sum_{j=1}^{k} f_j W_j \theta_j \right)^2} \qquad (2.12)$$

$$\Delta \hat{\lambda}_t = \frac{\sum_{j=1}^{k} f_j W_j \sum_{j=1}^{k} f_j W_j v_j \theta_j - \sum_{j=1}^{k} f_j W_j \theta_j \sum_{j=1}^{k} f_j W_j v_j}{\sum_{j=1}^{k} f_j W_j \sum_{j=1}^{k} f_j W_j \theta_j^2 - \left(\sum_{j=1}^{k} f_j W_j \theta_j \right)^2} . \qquad (2.13)$$

If the numerator and denominator of the expression (2.13) for the increment in the estimated slope parameter are divided by

$$\Delta \hat{\lambda}_t = \frac{\sum_{j=1}^{k} f_j W_j v_j \theta_j - \left[\sum_{j=1}^{k} f_j W_j \theta_j \sum_{j=1}^{k} f_j W_j v_j / \sum_{j=1}^{k} f_j W_j \right]}{\sum_{j=1}^{k} f_j W_j \theta_j^2 - \left[\left(\sum_{j=1}^{k} f_j W_j \theta_j \right)^2 / \sum_{j=1}^{k} f_j W_j . \right]} . \qquad (2.14)$$

Visual inspection of this equation reveals that it has exactly the same form as that for the usual least squares regression coefficient. However, because of the presence of the $f_j W_j$ terms, a formula parallel to that for a weighted least squares estimate is obtained.

In equation (2.12) the numerator and denominator can be divided by $\sum_{j=1}^{k} f_j W_j$. Then, letting

$$\bar{v} = \frac{\displaystyle\sum_{j=1}^{k} f_j W_j v_j}{\displaystyle\sum_{j=1}^{k} f_j W_j} \quad \text{and} \quad \bar{\theta} = \frac{\displaystyle\sum_{j=1}^{k} f_j W_j \theta_j}{\displaystyle\sum_{j=1}^{k} f_j W_j}$$

and substituting these in equation (2.12) yields

$$\Delta\hat{\zeta}_t = \frac{\bar{v}\displaystyle\sum_{j=1}^{k} f_j W_j \theta_j^2 - \bar{\theta}\displaystyle\sum_{j=1}^{k} f_j W_j v_j \theta_j}{\displaystyle\sum_{j=1}^{k} f_j W_j \theta_j^2 - \left[\left(\displaystyle\sum_{j=1}^{k} f_j W_j \theta_j \right)^2 \bigg/ \displaystyle\sum_{j=1}^{k} f_j W_j \right]} = \bar{v} - \Delta\hat{\lambda}_t \bar{\theta} \quad (2.15)$$

which has the same form as the usual intercept term in linear regression. With the expressions for $\Delta\hat{\zeta}_t$ and $\Delta\hat{\lambda}_t$ in hand, the iterative process defined by equation (2.11) can be described.

In order to start the iterative estimation process, initial estimates of the item parameters are needed. Kale (1962) has shown that these initial estimates must be within a "neighborhood" of the actual parameter values and that these initial values be based on consistent statistics in order for the iterative process to converge. As a result, a number of different schemes are used to obtain these initial estimates. A common one is to compute the classical test theory item difficulty and item-test score biserial correlation and use equations (1.9) and (1.10) to convert these into the item parameters $\hat{\alpha}$ and $\hat{\beta}$. Under a two-parameter ICC model, the initial crude estimates of the slope and intercept parameters, 1 and 0 will often suffice. At each ability score level $\hat{Z}_j = \hat{\zeta}_1 + \hat{\lambda}_1 \theta_j$ is computed. $\hat{Z}_j$ is used as the entering argument to the normal ogive and the values of $\hat{P}_j$, $\hat{Q}_j$ obtained. In addition, the value of $\hat{h}_j$ is obtained from the normal density. These calculations are performed at each level of the ability. Then the P_j, Q_j, and h_j terms in equations (2.14) and (2.15) are replaced by their "true values," $\hat{P}_j$, $\hat{Q}_j$, and $\hat{h}_j$, under the assumption that the parameters $\hat{\zeta}_1$ and $\hat{\lambda}_1$ were the "true" parameter values. $\Delta\hat{\zeta}_1$ and $\Delta\hat{\lambda}_1$ are obtained by evaluating equation (2.10). Then $\hat{\zeta}_2 = \hat{\zeta}_1 + \Delta\hat{\zeta}_1$ and $\hat{\lambda}_2 = \hat{\lambda}_1 + \Delta\hat{\lambda}_1$ are obtained. The whole process is repeated until $\Delta\hat{\zeta}_t$ and $\Delta\hat{\lambda}_t$ are both arbitrarily small, say less than .005 in absolute value. Once the regression parameters ζ and λ have been estimated, the item parameter estimates $\hat{\alpha} = \hat{\lambda}$ and $\hat{\beta} = -\hat{\zeta}/\hat{\lambda}$ are readily obtained.

This process works very well in practice and usually only three to five iterations are needed to reach a convergence criterion. Since both α and β are unbounded, occasional data sets will yield very large values of $\hat{\alpha}$, $\hat{\beta}$. Typically large values of $\hat{\alpha}$ are rare, but values near zero are reasonably common. When this occurs, $\hat{\beta} = -\hat{\zeta}/\hat{\lambda}$ becomes very large. Thus, when these estimation procedures are implemented as computer programs, limit protection for $\hat{\alpha}$ and

$\hat{\beta}$ is necessary. For example, the LOGIST computer program (Wingersky, Barton, & Lord, 1982) sets the limits at $|\hat{\alpha}| < 4$, $|\hat{\beta}| < 10$. The problem of excessively large values of $\hat{\beta}$ is also a reason to use the ζ and λ parameterization. When the discrimination approaches zero, the intercept also approaches zero and the solution equations will converge.

One additional outcome of the maximum likelihood equations is that the asymptotic variances of the estimators and a χ^2 goodness-of-fit statistic can be obtained easily. The large sample variances of $\hat{\zeta}$, $\hat{\lambda}$ (Berkson, 1955b) are

$$S_{\hat{\lambda}}^2 = \frac{1}{\displaystyle\sum_{j=1}^{k} f_j W_j (\theta_j - \bar{\theta})^2} = S_{\hat{\alpha}}^2$$

$$S_{\hat{\zeta}}^2 = \frac{1}{\displaystyle\sum_{j=1}^{k} f_j W_j} + \bar{\theta}^2 S_{\hat{\lambda}}^2.$$

Since $\hat{\beta} = -\hat{\zeta}/\hat{\lambda}$ and $\hat{\alpha} = \hat{\lambda}$,

$$S_{\hat{\beta}}^2 = \frac{1}{\hat{\alpha}^2} \left[\frac{1}{\displaystyle\sum_{j=1}^{k} f_j W_j} + S_{\hat{\alpha}}^2 (\hat{\beta} - \bar{\theta})^2 \right].$$

Unfortunately, the small sample variances of $\hat{\alpha}$ and $\hat{\beta}$ are unknown. Baker (1962) performed a Monte Carlo study of the sampling distribution of $\hat{\alpha}$ and $\hat{\beta}$ for samples ranging from size 15 to 120 and nine different combinations of α and β. He found that for samples of 120, $\hat{\alpha}$ and $\hat{\beta}$ were approximately normally distributed with variances close to those yielded by the asymptotic formulas.

A χ^2 goodness-of-fit test between the normal ogive specified by the parameter estimates and the observed proportion of correct response can also be obtained. Garwood (1941) showed

$$\chi^2 = \sum_{j=1}^{k} f_j W_j v_j^2 \tag{2.16}$$

with $k - 2$ degrees of freedom. A degree of freedom is lost for each parameter estimated. Because all the terms in equation (2.16) are available in the estimation equations, a measure of the goodness of fit of the model to the data can be incorporated easily into the estimation process.

2.3 Maximum Likelihood Estimation: Logistic Model

Switching from a normal ogive to a logistic ogive model for an item's ICC results in a significant decrease in the computational demands of the maximum likelihood estimation procedure. Since the cumulative distribution of the logistic density has a closed form, that is, does not involve an integral, it can be computed easily. This computational advantage is the primary reason for using logistic ogive models for the ICC. Changing from the normal ogive model to the logistic ogive model for the item characteristic curve has no impact upon the framework of the maximum likelihood estimation procedures described above. The likelihood equation is the same and the basic Newton-Raphson procedure can be employed. However, differences arise in the first and second derivatives since a different ICC model is in use. Because of the underlying similarity, only an abbreviated version of the maximum likelihood estimation procedures will be presented for the two-parameter logistic ICC model.

The cumulative logistic distribution function is given by

$$P_i = P(\theta_i) = P(\zeta, \lambda, \theta_j) = \frac{1}{1 + e^{-(\zeta + \lambda \theta_j)}}. \tag{2.17}$$

Let $Z_j = \zeta + \lambda \theta_j$ be the logistic deviate, that is, the "logit." Some useful derivatives are

$$\frac{\partial P_j}{\partial \zeta} = P_j Q_j, \qquad \frac{\partial P_j}{\partial \lambda} = P_j Q_j \theta_j$$

and

$$\frac{\partial Q_j}{\partial \zeta} = -P_j Q_j, \qquad \frac{\partial Q_j}{\partial \lambda} = -P_j Q_j \theta_j.$$

These very simple expressions are a direct consequence of the closed form of the expression (2.17) for the cumulative logistic distribution function.

The likelihood function is

$$\text{Prob}(R) = \prod_{j=1}^{k} \frac{f_j!}{r_j!(f_j - r_j)!} P_j^{r_j} Q_j^{f_j - r_j}$$

and

$$L = \log \text{Prob}(R) = \text{constant} + \sum_{j=1}^{k} r_j \log P_j + \sum_{j=1}^{k} (f_j - r_j) \log Q_j. \tag{2.18}$$

As was the case for the normal ogive model, the first and second partial derivatives of L with respect to the parameters ζ and λ are needed. These are

$$L_1 = \frac{\partial L}{\partial \zeta} = \sum_{j=1}^{k} \frac{r_j}{P_j} P_j Q_j + \sum_{j=1}^{k} \frac{f_j - r_j}{Q_j}(-P_j Q_j)$$

$$= \sum_{j=1}^{k}[r_j Q_j - (f_j - r_j)P_j] = \sum_{j=1}^{k}[r_j Q_j - f_j P_j + r_j P_j]$$

$$= \sum_{j=1}^{k}(r_j - f_j P_j),$$

but

$$p_j = \frac{r_j}{f_j},$$

then substituting $(r_j - f_j P_j) = f_j(p_j - P_j)$ yields

$$L_1 = \sum_{j=1}^{k} f_j(p_j - P_j).$$

$$L_2 = \frac{\partial L}{\partial \lambda} = \sum_{j=1}^{k} \frac{r_j}{P_j} P_j Q_j \theta_j + \sum_{j=1}^{k} \frac{f_j - r_j}{Q_j}(-P_j Q_j \theta_j)$$

$$= \sum_{j=1}^{k}(r_j - f_j P_j)\theta_j$$

$$= \sum_{j=1}^{k} f_j(p_j - P_j)\theta_j.$$

$$L_{11} = \frac{\partial^2 L}{\partial \zeta^2} = \frac{\partial}{\partial \zeta}\left[\sum_{j=1}^{k}(r_j - f_j P_j)\right] = -\sum_{j=1}^{k} f_j P_j Q_j.$$

$$L_{22} = \frac{\partial^2 L}{\partial \lambda^2} = \frac{\partial}{\partial \lambda}\left[\sum_{j=1}^{k}(r_j - f_j P_j)\theta_j\right] = -\sum_{j=1}^{k} f_j P_j Q_j \theta_j^2.$$

$$L_{21} = \frac{\partial^2 L}{\partial \lambda \partial \zeta} = \frac{\partial}{\partial \lambda}\left[\sum_{j=1}^{k}(r_j - f_j P_j)\right] = -\sum_{j=1}^{k} f_j P_j Q_j \theta_j = L_{12}.$$

Using the logistic equivalent $W_j = P_j Q_j$ of the Urban-Müller weight, the partial derivatives can be written as:

$$L_1 = \sum_{j=1}^{k} f_j W_j \left(\frac{p_j - P_j}{P_j Q_j}\right)$$

$$L_2 = \sum_{j=1}^{k} f_j W_j \left(\frac{p_j - P_j}{P_j Q_j}\right) \theta_j$$

$$L_{11} = -\sum_{j=1}^{k} f_j W_j$$

$$L_{12} = L_{21} = -\sum_{j=1}^{k} f_j W_j \theta_j$$

$$L_{22} = -\sum_{j=1}^{k} f_j W_j \theta_j^2.$$

Using the same Taylor series approach as employed above, the solution equations are

$$-L_1 = L_{11} \Delta\hat\zeta + L_{12} \Delta\hat\lambda$$

and

$$-L_2 = L_{21} \Delta\hat\zeta + L_{22} \Delta\hat\lambda.$$

Since the second derivatives do not involve any observed terms, for example, $E(L_{11}) = L_{11}$, and the terms can be used as they are. Thus, for a two-parameter logistic ICC model, Fisher's method of scoring for parameters and the Newton-Raphson procedures are the same. Substituting for the derivatives, the solution equations become

$$\Delta\hat\zeta \left(-\sum_{j=1}^{k} f_j W_j\right) + \Delta\hat\lambda \left(-\sum_{j=1}^{k} f_j W_j \theta_j\right) = -\sum_{j=1}^{k} f_j W_j \left(\frac{p_j - P_j}{P_j Q_j}\right)$$

$$\Delta\hat\zeta \left(-\sum_{j=1}^{k} f_j W_j \theta_j\right) + \Delta\hat\lambda \left(-\sum_{j=1}^{k} f_j W_j \theta_j^2\right) = -\sum_{j=1}^{k} f_j W_j \left(\frac{p_j - P_j}{P_j Q_j}\right) \theta_j.$$

(2.19)

Expressing the iterative process for estimating ζ and λ in Newton-Raphson form.

$$\begin{bmatrix} \hat\zeta \\ \hat\lambda \end{bmatrix}_{t+1} = \begin{bmatrix} \hat\zeta \\ \hat\lambda \end{bmatrix}_{t} - \begin{bmatrix} -\sum_{j=1}^{k} f_j W_j & -\sum_{j=1}^{k} f_j W_j \theta_j \\ -\sum_{j=1}^{k} f_j W_j \theta_j & -\sum_{j=1}^{k} f_j W_j \theta_j^2 \end{bmatrix}_{t}^{-1} \begin{bmatrix} \sum_{j=1}^{k} f_j W_j \left(\frac{p_j - P_j}{P_j Q_j}\right) \\ \sum_{j=1}^{k} f_j W_j \left(\frac{p_j - P_j}{P_j Q_j}\right) \theta_j \end{bmatrix}$$

Let $v_j = (p_j - P_j)/P_j Q_j$, then the equation can be written as

$$\begin{bmatrix} \hat{\zeta} \\ \hat{\lambda} \end{bmatrix}_{t+1} = \begin{bmatrix} \hat{\zeta} \\ \hat{\lambda} \end{bmatrix}_{t} - \begin{bmatrix} -\sum\limits_{j=1}^{k} f_j W_j & -\sum\limits_{j=1}^{k} f_j W_j \theta_j \\ -\sum\limits_{j=1}^{k} f_j W_j \theta_j & -\sum\limits_{j=1}^{k} f_j W_j \theta_j^2 \end{bmatrix}_{t}^{-1} \begin{bmatrix} \sum\limits_{j=1}^{k} f_j W_j v_j \\ \sum\limits_{j=1}^{k} f_j W_j v_j \theta_j \end{bmatrix}_{t} .(2.20)$$

This equation is identical to equation (2.11) under the normal model except for the definitions of W_j and v_j. Consequently, the equations for $\Delta \hat{\zeta}_t$ and $\Delta \hat{\lambda}_t$ are also the same.

$$\Delta \hat{\lambda}_t = \frac{\sum\limits_{j=1}^{k} f_j W_j v_j \theta_j - \left[\sum\limits_{j=1}^{k} f_j W_j \theta_j \sum\limits_{j=1}^{k} f_j v_j / \sum\limits_{j=1}^{k} f_j W_j \right]}{\sum\limits_{j=1}^{k} f_j W_j \theta_j^2 - \left[\left(\sum\limits_{j=1}^{k} f_j W_j \theta_j \right)^2 / \sum\limits_{j=1}^{k} f_j W_j \right]}. \qquad (2.21)$$

Let

$$\bar{v} = \frac{\sum\limits_{j=1}^{k} f_j W_j v_j}{\sum\limits_{j=1}^{k} f_j W_j} \qquad \text{and} \qquad \bar{\theta} = \frac{\sum\limits_{j=1}^{k} f_j W_j \theta_j}{\sum\limits_{j=1}^{k} f_j W_j},$$

then

$$\Delta \hat{\zeta}_t = \frac{\bar{v} \sum\limits_{j=1}^{k} f_j W_j \theta_j^2 - \bar{\theta} \sum\limits_{j=1}^{k} f_j W_j v_j \theta_j}{\sum\limits_{j=1}^{k} f_j W_j \theta_j^2 - \left[\left(\sum\limits_{j=1}^{k} f_j W_j \theta_j \right)^2 / \sum\limits_{j=1}^{k} f_j W_j \right]} = \bar{v} - \Delta \hat{\lambda}_t \bar{\theta}. \qquad (2.22)$$

Again

$$\hat{\zeta}_{t+1} = \hat{\zeta}_t + \Delta \hat{\zeta}_t,$$

$$\hat{\lambda}_{t+1} = \hat{\lambda}_t + \Delta \hat{\lambda}_t,$$

and

$$\hat{Z}_{j,t+1} = \hat{\zeta}_{t+1} + \hat{\lambda}_{t+1} \theta_j.$$

$\hat{Z}_{j,t+1}$ becomes the "true" logit and the iterative process is continued until a convergence criterion is met.

Maxwell (1959) has provided the large sample variances of $\hat{\zeta}$, $\hat{\lambda}$, and $\hat{\beta}$ which are identical in form to those under a normal ogive ICC model:

$$S_{\hat{\lambda}}^2 = \frac{1}{\sum_{j=1}^{k} f_j W_j (\theta_j - \bar{\theta})^2} = S_{\hat{\alpha}}^2$$

$$S_{\hat{\zeta}}^2 = \frac{1}{\sum_{j=1}^{k} f_j W_j} + \bar{\theta}^2 S_{\hat{\lambda}}^2.$$

Since $\hat{\beta} = -\hat{\zeta}/\hat{\lambda}$ and $\hat{\alpha} = \hat{\lambda}$,

$$S_{\hat{\beta}}^2 = \frac{1}{\hat{\alpha}^2} \left[\frac{1}{\sum_{j=1}^{k} f_j W_j} + S_{\hat{\alpha}}^2 (\hat{\beta} - \bar{\theta})^2 \right].$$

When the θ_j are known, Berkson (1955a, p. 157) has shown that under the logistic model and binomial variation of the observed p_j, the terms $\sum_{j=1}^{k} f_j p_j$ and $\sum_{j=1}^{k} f_j p_j \theta_j$ are minimal sufficient statistics for ζ and λ, respectively. In addition, when the maximum likelihood estimators $\hat{\zeta}$, $\hat{\lambda}$ are finite, they are necessarily one-to-one functions of these sufficient statistics. After substituting for W_j and v_j and performing some algebraic manipulations, equations (2.21) and (2.22) can be written as

$$\Delta\hat{\lambda}_t = \frac{\sum_{j=1}^{k} f_j p_j \theta_j - \sum_{j=1}^{k} f_j P_j \theta_j - \dfrac{\sum_{j=1}^{k} f_j P_j Q_j \theta_j \left(\sum_{j=1}^{k} f_j p_j - \sum_{j=1}^{k} f_j P_j \right)}{\sum_{j=1}^{k} f_j W_j}}{\sum_{j=1}^{k} f_j P_j Q_j \theta_j^2 - \left[\left(\sum_{j=1}^{k} f_j P_j Q_j \theta_j \right)^2 \Big/ \sum_{j=1}^{k} f_j P_j Q_j \right]}$$

and

$$\Delta\hat{\zeta}_t = \frac{\sum_{j=1}^{k} f_j p_j - \sum_{j=1}^{k} f_j P_j - \Delta\hat{\lambda}_t \sum_{j=1}^{k} f_j P_j Q_j \theta_j}{\sum_{j=1}^{k} f_j P_j Q_j},$$

and the first terms in the numerators of these equations can be identified as the sufficient statistics. When the normal ogive model is employed, sufficient statistics for ζ and λ do not exist under maximum likelihood estimation

(Anscombe, 1956, p. 461). Unfortunately, under IRT the θ_j typically are unknown and these sufficient statistics are not available.

A common practice is to express the item discrimination parameter λ in numerical values corresponding to those under the normal ogive model. Presumably to maintain a consistent framework for interpretation. The transformation can be accomplished by replacing $\hat{\lambda}$ by $1.702\hat{\lambda}$ in the expression for the cumulative logistic distribution function (2.17) and the solution of the Newton-Raphson procedure will yield numerical values of $\hat{\zeta}$ and $\hat{\lambda}$ matching that of the normal model.

The expression for the χ^2 goodness-of-fit statistic is the same as equation (2.16), with

$$v_j = \frac{p_j - P_j}{P_j Q_j}$$

and the degrees of freedom remain at $k - 2$.

The maximum likelihood estimation of the item parameters under a two-parameter logistic ICC model can also be expressed in a regression format. If one lets z_j be the "working" logit and Z_j be the "true" logit, then, following Berkson (1955a),

$$z_j = Z_j + \left(\frac{p_j - P_j}{P_j Q_j} \right)$$

and

$$z_j - Z_j = \frac{p_j - P_j}{P_j Q_j}.$$

Substituting in the right hand side of equation (2.19) an alternative form of the solution equations can be obtained:

$$\Delta\hat{\zeta} \left(-\sum_{j=1}^{k} f_j W_j \right) + \Delta\hat{\lambda} \left(-\sum_{j=1}^{k} f_j W_j \theta_j \right) = -\sum_{j=1}^{k} f_j W_j (z_j - Z_j)$$

$$\Delta\hat{\zeta} \left(-\sum_{j=1}^{k} f_j W_j \theta_j \right) + \Delta\hat{\lambda} \left(-\sum_{j=1}^{k} f_j W_j \theta_j^2 \right) = -\sum_{j=1}^{k} f_j W_j (z_j - Z_j)\theta_j.$$

(2.23)

As was the case for the normal model, this alternative form is useful for comparative purposes, but will not be used below.

The implementation of the maximum likelihood estimation of the item parameters, under a two parameter ICC model, is illustrated in Appendix A. A computer program written in MICROSOFT QUICKBASIC is presented and an example data set analyzed. Enough detail of the computations is presented in the appendix to show the main features of the estimation process. The reader can use the program to obtain the full details of the solution of the estimation equations.

2.4 Influence of the Weighting Coefficients

The maximum likelihood estimation of the item parameters under both the normal and logistic models involves weighting the data points. In the case of the normal model, the weights are the Urban-Müller weights and, under the logistic model, they are $P_j Q_j$. These weights are shown in Table 2.1 for values of θ_j from -3 to $+3$ in steps of 0.5. Both sets of weights reach their maximum when $\theta_j = \beta$ and decrease as $|\theta_j - \beta|$ increases. The net effect is to give greater influence to data points near the value of β. Thus, the "best" estimate of an item's parameters are obtained when the item difficulty is matched to the mean ability of the group. It should be noted that the W_j term is always multiplied by an f_j term in the solution equations. As a result, larger group sizes will also increase the influence of data points in the estimation of item parameters. The values of α used in Table 2.1 correspond to a normal theory value of 1.0. If smaller values of α were involved the decrease in the value of the weighting coefficients will be less pronounced as the value of $|\theta_j - \beta|$ increases. The reverse holds when high values of α underlie the item response data. Thus, the maximum likelihood estimation process is influenced by the distribution of the subjects over the ability scale, the underlying value of the item discrimination parameter, and the difficulty of the item relative to the mean ability of the examinees. This does not contradict the basic property of the item parameters being invariant with respect to the ability of the group administered the item. It simply indicates that certain parts of the examinee distribution have a greater role in determining the values of the item parameter estimates.

Table 2.1. Weighting Coefficients Used in Maximum Likelihood Estimation Procedures

	Item Characteristic Curve Model				
	Normal			Logistic	
	($\alpha = 1.0$, $\beta = 0.0$)			($\alpha = 1.702$, $\beta = 0.0$)	
θ_j	P_j	h_j	W_j	P_j	W_j
-3.0	0.0013	0.0044	0.0146	0.0060	0.0060
-2.5	0.0062	0.0175	0.0498	0.0140	0.0138
-2.0	0.0228	0.0540	0.1311	0.0322	0.0311
-1.5	0.0668	0.1295	0.2691	0.0722	0.0670
-1.0	0.1587	0.2420	0.4386	0.1542	0.1304
-0.5	0.3085	0.3521	0.5810	0.2992	0.2097
0.0	0.5000	0.3989	0.6366	0.5000	0.2500
0.5	0.6915	0.3521	0.5810	0.7008	0.2097
1.0	0.8413	0.2420	0.4386	0.8458	0.1304
1.5	0.9332	0.1295	0.2691	0.9278	0.0670
2.0	0.9772	0.0540	0.1311	0.9678	0.0311
2.5	0.9938	0.0175	0.0498	0.9860	0.0138
3.0	0.9987	0.0044	0.0146	0.9940	0.0060

2.5 The Item Log-Likelihood Surface

Under a two-parameter logistic ICC model, the maximum likelihood estimation technique described above involves an item log-likelihood surface such as that shown in Figure 2.2. In this figure, the two lower axes are item difficulty β and item discrimination α while the vertical axis is the log-likelihood. The value 200 has been added to the value of the item log-likelihood to facilitate plotting. Under the iterative estimation procedure, the initial parameter estimates yield a value of the log-likelihood that places the solution somewhere on the surface. The increments to the item parameter estimates obtained in each iteration of the Newton-Raphson procedure moves the solution towards the maximum of the surface that is defined by the values of the item parameters. When convergence is reached the maximum value of the item log-likelihood is attained. As is the case with all such procedures, one runs the risk of finding a local rather than a global maximum. However, the existing research (Baker, 1988), suggests the item log-likelihood surface for a two-parameter ICC model has only a global maximum at the point defined by the item parameters.

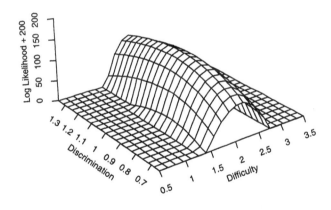

Fig. 2.2. Item log-likelihood surfaces for two-parameter model $\hat{\alpha}_i = 1.0$, $\hat{\beta}_i = 2.0$. Copyright 1988 by Applied Psychological Measurement, Inc., reproduced by permission.

It can be seen in Figure 2.2 that general form of the item log-likelihood surface is a ridge. Parallel to the difficulty axis, the surface is sharply peaked and the maximum with respect to difficulty would be found readily. Parallel to the discrimination axis, the top of this ridge is relatively flat suggesting that the maximum with respect to discrimination will be more difficult to locate. These characteristics of the item log-likelihood surface suggest that item difficulty would be easier to estimate than would discrimination. This agrees with what has been observed in practice.

2.6 Maximum Likelihood Estimation: Three-Parameter Model

Although the slope intercept parameterization of the ICC was employed for the two-parameter model, it will not be employed for the three-parameter model. Rather, the a, b, c form will be employed for the three-parameter model so that the current derivations can serve as the basis for the presentations in several of the later chapters. Because there is an additional parameter, a completely new set of Newton-Raphson equations must be developed that simultaneously estimate a, b, and c. While the three-parameter model can be based upon a normal ogive, where

$$P_j = c + (1-c)\Phi(Z_j)$$

and

$$Q_j = [1 - \Phi(Z_j)](1-c),$$

this model is rarely used in practice. However, the three-parameter model based upon the logistic ogive,

$$P_j = c + (1-c)\Psi(Z_j)$$

and

$$Q_j = [1 - \Psi(Z_j)](1-c),$$

is widely used and will be employed below. Let

$$\Psi(Z_j) = P_j^* = P^*(\theta_j) = \frac{1}{1 + e^{-a(\theta_j - b)}},$$

where P_j^* denotes a value based upon the two-parameter model. Some useful derivatives are as follows:

$$\frac{\partial P_j}{\partial a} = (\theta_j - b)(1-c)P_j^* Q_j^* = \frac{(\theta_j - b)Q_j(P_j - c)}{1-c}$$

$$\frac{\partial P_j}{\partial b} = -a(1-c)P_j^* Q_j^* = \frac{-aQ_j(P_j - c)}{1-c}$$

$$\frac{\partial P_j}{\partial c} = Q_j^* = \frac{Q_j}{1-c}.$$

Again, taking all response categories and ability score groups into account, the probability of the vector $R = (r_1, \ldots, r_k)$ is given by the likelihood function

$$\text{Prob}(R) = \prod_{j=1}^{k} \frac{f_j!}{r_j!(f_j - r_j)!} P_j^{r_j} Q_j^{f_j - r_j}$$

and the log-likelihood is

$$L = \log \text{Prob}(R) = \text{constant} + \sum_{j=1}^{k} r_j \log P_j + \sum_{j=1}^{k} (f_j - r_j) \log Q_j.$$

Now taking derivatives of the log-likelihood with respect to the parameters a, b, and c, the first derivatives are:

$$L_1 = \frac{\partial L}{\partial a} = \sum_{j=1}^{k} \frac{r_j}{P_j} \frac{\partial P_j}{\partial a} + \sum_{j=1}^{k} \frac{(f_j - r_j)}{Q_j} \frac{\partial Q_j}{\partial a}$$

$$= \sum_{j=1}^{k} \frac{r_j}{P_j} (\theta_j - b)(1 - c) P_j^* Q_j^* + \sum_{j=1}^{k} \frac{(f_j - r_j)}{Q_j} [-(\theta_j - b)(1 - c) P_j^* Q_j^*]$$

$$= \sum_{j=1}^{k} (\theta_j - b)(1 - c) P_j^* Q_j^* \left[\frac{r_j}{P_j} - \frac{(f_j - r_j)}{Q_j^*} \right]$$

$$= \sum_{j=1}^{k} (\theta_j - b)(1 - c) \frac{P_j^* Q_j^*}{P_j Q_j} [r_j Q_j - (f_j - r_j) P_j]$$

$$= \sum_{j=1}^{k} (\theta_j - b)(1 - c) \frac{P_j^* Q_j^*}{P_j Q_j} [r_j - f_j P_j].$$

Let $W_j = P_j^* Q_j^*$ and $r_j = f_j p_j$, where p_j is the observed proportion of correct response. Then

$$L_1 = \sum_{j=1}^{k} f_j (p_j - P_j)(\theta_i - b)(1 - c) W_j, \qquad (2.24)$$

but

$$(1 - c) W_j = \frac{P_j - c}{P_j (1 - c)},$$

then

$$L_1 = \sum_{j=1}^{k} f_j (p_j - P_j)(\theta_i - b) \left[\frac{P_j - c}{P_j (1 - c)} \right]$$

$$\sum_{j=1}^{k} f_j (p_j - P_j)(\theta_i - b) \left[\frac{P_j^*}{P_j} \right].$$

$$L_2 = \frac{\partial L}{\partial b} = \sum_{j=1}^{k} \frac{r_j}{P_j} \frac{\partial P_j}{\partial b} + \sum_{j=1}^{k} \frac{(f_j - r_j)}{Q_j} \frac{\partial Q_j}{\partial b}$$

$$= \sum_{j=1}^{k} \frac{r_j}{P_j} [-a(1-c)P_j^* Q_j^*] + \sum_{j=1}^{k} \frac{(f_j - r_j)}{Q_j} [-(-a)(1-c)P_j^* Q_j^*]$$

$$= -a \sum_{j=1}^{k} \frac{(1-c)P_j^* Q_j^*}{P_j Q_j} [r_j Q_j - (f_j - r_j)P_j]$$

$$= -a \sum_{j=1}^{k} \frac{(1-c)P_j^* Q_j^*}{P_j Q_j} [f_j p_j - f_j P_j]$$

$$= -a \sum_{j=1}^{k} f_j(p_j - P_j)(1-c)W_j. \tag{2.25}$$

Substituting for $(1-c)W_j$ yields

$$L_2 = -a \sum_{j=1}^{k} f_j(p_j - P_j) \frac{P_j - c}{P_j(1-c)} = -a \sum_{j=1}^{k} f_j(p_j - P_j) \left[\frac{P_j^*}{P_j}\right].$$

$$L_3 = \frac{\partial L}{\partial c} = \sum_{j=1}^{k} \frac{r_j}{P_j} \frac{\partial P_j}{\partial c} + \sum_{j=1}^{k} \frac{(f_j - r_j)}{Q_j} \frac{\partial Q_j}{\partial c}$$

$$= \sum_{j=1}^{k} \frac{r_j Q_j^*}{P_j} + \sum_{j=1}^{k} \frac{(f_j - r_j)}{Q_j}(-Q_j^*)$$

$$= \sum_{j=1}^{k} \frac{Q_j^*}{P_j Q_j} [r_j Q_j - (f_j - r_j)P_j] = \sum_{j=1}^{k} \frac{Q_j^*}{P_j Q_j} [f_j p_j - f_j P_j]$$

$$= \sum_{j=1}^{k} \frac{1}{(1-c)} \left[\frac{f_j p_j - f_j P_j}{P_j}\right] = \sum_{j=1}^{k} f_j(p_j - P_j) \left[\frac{1}{(1-c)P_j}\right].$$

Hence,

$$L_3 = \sum_{j=1}^{k} f_j(p_j - P_j) \frac{1}{(P_j - c)} \left[\frac{P_j^*}{P_j}\right]. \tag{2.26}$$

Some useful derivatives are:

$$\frac{\partial}{\partial a} \left[\frac{P_j - c}{P_j(1-c)}\right] = c(\theta_j - b) \left[\frac{P_j - c}{P_j(1-c)}\right] \left[\frac{Q_j}{P_j(1-c)}\right]$$

$$\frac{\partial}{\partial b} \left[\frac{P_j - c}{P_j(1-c)}\right] = -ac \left[\frac{P_j - c}{P_j(1-c)}\right] \left[\frac{Q_j}{P_j(1-c)}\right]$$

$$\frac{\partial}{\partial c} \left[\frac{P_j - c}{P_j(1-c)}\right] = - \left[\frac{P_j - c}{(1-c)^2}\right] \left[\frac{Q_j}{P_j^2}\right].$$

Note that $(P_j - c)/[P_j(1-c)] = P_j^* Q_j^* /[P_j Q_j(1-c)] = W_j(1-c)$. The second derivatives of the log-likelihood are:

$$
L_{11} = \frac{\partial^2 L}{\partial a^2} = \frac{\partial}{\partial a} L_1 = \frac{\partial}{\partial a} \left\{ \sum_{j=1}^{k} f_j(p_j - P_j)(\theta_j - b) \left[\frac{P_j - c}{P_j(1-c)} \right] \right\}
$$

$$
= \sum_{j=1}^{k} (\theta_j - b) \left\{ \left[\frac{P_j - c}{P_j(1-c)} \right] \frac{\partial}{\partial a}(f_j p_j - f_j P_j) \right.
$$

$$
\left. + (f_j p_j - f_j P_j) \frac{\partial}{\partial a} \left[\frac{P_j - c}{P_j(1-c)} \right] \right\}
$$

$$
= \sum_{j=1}^{k} (\theta_j - b) \left\{ \left[\frac{P_j - c}{P_j(1-c)} \right] \left[-f_j(\theta_j - b)Q_j \left(\frac{P_j - c}{1-c} \right) \right] \right.
$$

$$
\left. + (f_j p_j - f_j P_j) \left[c(\theta_j - b) \frac{(P_j - c)}{P_j(1-c)} \frac{Q_j}{P_j(1-c)} \right] \right\}
$$

$$
= \sum_{j=1}^{k} (\theta_j - b)^2 \left[\frac{P_j - c}{P_j(1-c)} \right]
$$

$$
\times \left[\frac{1}{P_j(1-c)} \right] [-f_j P_j Q_j (P_j - c) + (f_j p_j - f_j P_j) c Q_j].
$$

$$
L_{11} = \sum_{j=1}^{k} f_j (\theta_j - b)^2 \left[\frac{P_j - c}{P_j(1-c)} \right] \frac{Q_j}{P_j(1-c)} [-P_j^2 + c p_j]. \tag{2.27}
$$

$$
L_{22} = \frac{\partial L_2}{\partial b} = \frac{\partial}{\partial b} \left\{ -a \sum_{j=1}^{k} f_j(p_j - P_j) \left[\frac{P_j - c}{P_j(1-c)} \right] \right\}
$$

$$
= -\sum_{j=1}^{k} a \left\{ \left[\frac{P_j - c}{P_j(1-c)} \right] \left[\frac{-f_j[-aQ_j(P_j - c)]}{1-c} \right] \right.
$$

$$
\left. + (f_j p_j - f_j P_j) \left[-ac \frac{P_j - c}{P_j(1-c)} \frac{Q_j}{P_j(1-c)} \right] \right\}
$$

$$
= \sum_{j=1}^{k} a^2 \left[\frac{P_j - c}{P_j(1-c)} \right] \left[\frac{-f_j Q_j(P_j - c)}{1-c} + (f_j p_j - f_j P_j) \left(\frac{cQ_j}{P_j(1-c)} \right) \right].
$$

$$
L_{22} = a^2 \sum_{j=1}^{k} f_j \left[\frac{P_j - c}{P_j(1-c)} \right] \frac{Q_j}{P_j(1-c)} [-P_j^2 + p_j c]. \tag{2.28}
$$

$$
L_{33} = \frac{\partial}{\partial c} L_3 = \frac{\partial}{\partial c} \left\{ \sum_{j=1}^{k} f_j(p_j - P_j) \left[\frac{1}{(1-c)P_j} \right] \right\}
$$

$$= \sum_{j=1}^{k} \frac{1}{[P_j(1-c)]^2}$$

$$\times \left\{ P_j(1-c) \left(\frac{-f_j Q_j}{1-c} \right) - (f_j p_j - f_j P_j) \left[\frac{Q_j}{1-c} - \frac{cQ_j}{1-c} \right] \right\}$$

$$= - \sum_{j=1}^{k} \frac{f Q_j}{[P_j(1-c)]^2} (-P_j - p_j + P_j).$$

$$L_{33} = - \sum_{j=1}^{k} \frac{f_j p_j}{(1-c)^2} \left[\frac{Q_j}{P_j^2} \right]. \tag{2.29}$$

$$L_{12} = \frac{\partial}{\partial b} L_1 = L_{21}$$

$$= \sum_{j=1}^{k} \left\{ (\theta_j - b) \frac{\partial}{\partial b} \left[\frac{P_j - c}{P_j(1-c)} (f_j p_j - f_j P_j) \right] + \right.$$

$$\left. \left[\frac{P_j - c}{P_j(1-c)} (f_j p_j - f_j P_j) \right] \frac{\partial}{\partial b} (\theta_j - b) \right\}$$

$$= \sum_{j=1}^{k} (\theta_j - b) \left\{ \left[\frac{P_j - c}{P_j(1-c)} \right] \left[\frac{f_j a Q_j (P_j - c)}{(1-c)} \right] \right.$$

$$+ (f_j p_j - f_j P_j) \left[\frac{-ac(P_j - c)}{P_j(1-c)} \frac{Q_j}{P_j(1-c)} \right]$$

$$\left. - (f_j p_j - f_j P_j) \frac{(P_j - c)}{P_j(1-c)} \right\}$$

$$= \sum_{j=1}^{k} (\theta_j - b) \left\{ \left[\frac{P_j - c}{P_j(1-c)} \right] \left[\frac{f_j a Q_j (P_j - c)}{(1-c)} \right] \right.$$

$$\left. - (f_j p_j - f_j P_j) \left[\frac{ac(P_j - c)}{P_j(1-c)} \frac{Q_j}{P_j(1-c)} + \frac{(P_j - c)}{P_j(1-c)} \right] \right\}$$

$$L_{12} = \sum_{j=1}^{k} f_j (\theta_j - b) \left\{ \left[\frac{a(P_j - c)^2}{(1-c)^2} \right] \left[\frac{Q_j}{P_j} \right] \right.$$

$$\left. - (p_j - P_j) \left[\frac{ac(P_j - c)}{P_j(1-c)} \frac{Q_j}{P_j(1-c)} + \frac{(P_j - c)}{P_j(1-c)} \right] \right\}.$$

$$L_{13} = \frac{\partial}{\partial c} L_1 = \frac{\partial}{\partial c} \left\{ \sum_{j=1}^{k} f_j (\theta_j - b) \frac{(P_j - c)}{P_j(1-c)} (p_j - P_j) \right\}$$

$$= \sum_{j=1}^{k} f_j(\theta_j - b) \left\{ \frac{P_j - c}{P_j(1-c)} \left[\frac{-Q_j}{(1-c)} \right] + (p_j - P_j) \left[\frac{-(P_j - c)}{(1-c)^2} \frac{Q_j}{P_j^2} \right] \right\}$$

$$= -\sum_{j=1}^{k} f_j(\theta_j - b) \left\{ \left[\frac{(P_j - c)}{(1-c)^2} \right] \frac{Q_j}{P_j} + (p_j - P_j) \left[\frac{(P_j - c)}{(1-c)^2} \frac{Q_j}{P_j^2} \right] \right\}.$$

$$L_{23} = \frac{\partial}{\partial c} L_2 = \frac{\partial}{\partial c} \left\{ -a \sum_{j=1}^{k} f_j \left[\frac{(P_j - c)}{P_j(1-c)} \right] (p_j - P_j) \right\}$$

$$= -a \sum_{j=1}^{k} f_j \left\{ \frac{P_j - c}{P_j(1-c)} \left[\frac{-Q_j}{(1-c)} \right] + (p_j - P_j) \left[\frac{-(P_j - c)}{(1-c)^2} \frac{Q_j}{P_j^2} \right] \right\}$$

$$= a \sum_{j=1}^{k} \left\{ f_j \left[\frac{(P_j - c)}{(1-c)^2} \right] \frac{Q_j}{P_j} + f_j(p_j - P_j) \left[\frac{(P_j - c)}{(1-c)^2} \frac{Q_j}{P_j^2} \right] \right\}.$$

As was the case with the maximum likelihood estimation of the item parameters under the normal ICC model, the second derivatives contain observed values. Therefore it is necessary again to replace the observed p_j values by their expectations P_j in the second derivatives. Since this was not necessary in the two-parameter case, it is probable that, when the θ_j are known, sufficient estimates for the regression parameters do not exist under the three-parameter model. The expectations of the second derivatives are:

$$E(L_{11}) = \Lambda_{11} = \sum_{j=1}^{k} f_j(\theta_j - b)^2 \left[\frac{P_j - c}{P_j(1-c)} \right] \frac{Q_j}{P_j(1-c)} [-P_j^2 + cP_j]$$

$$= -\sum_{j=1}^{k} f_j(\theta_j - b)^2 \left[\frac{P_j - c}{P_j(1-c)} \right]^2 \frac{Q_j}{P_j} = -\sum_{j=1}^{k} f_j(\theta_j - b)^2 P_j Q_j \left[\frac{P_j^*}{P_j} \right]^2$$

$$E(L_{22}) = \Lambda_{22} = a^2 \sum_{j=1}^{k} f_j \left[\frac{P_j - c}{P_j(1-c)} \right] \frac{Q_j}{P_j(1-c)} (-P_j^2 + P_j c)$$

$$= -a^2 \sum_{j=1}^{k} f_j \left[\frac{P_j - c}{P_j(1-c)} \right]^2 \frac{Q_j}{P_j} = -a^2 \sum_{j=1}^{k} f_j P_j Q_j \left[\frac{P_j^*}{P_j} \right]$$

$$E(L_{33}) = \Lambda_{33} = -\sum_{j=1}^{k} \frac{f_j P_j}{(1-c)^2} \left[\frac{Q_j}{P_j^2} \right] = -\sum_{j=1}^{k} f_j \frac{Q_j}{(1-c)^2} \left[\frac{1}{P_j} \right] =$$

$$= -\sum_{j=1}^{k} \frac{f_j}{(1-c)^2} \left[\frac{Q_j}{P_j} \right] = -\sum_{j=1}^{k} f_j \frac{Q_j}{(1-c)} \frac{1}{P_j - c} \left[\frac{P_j^*}{P_j} \right]$$

$$E(L_{12}) = E(L_{21}) = \Lambda_{12} = \Lambda_{21}$$

$$= \sum_{j=1}^{k} f_j(\theta_j - b) \left\{ a \left[\frac{(P_j - c)}{(1 - c)} \right]^2 \left[\frac{Q_j}{P_j} \right] \right.$$

$$\left. - (P_j - P_j) \left[\frac{ac(P_j - c)}{P_j(1 - c)} \frac{Q_j}{P_j(1 - c)} + \frac{(P_j - c)}{P_j(1 - c)} \right] \right\}$$

and the second term is eliminated; thus,

$$\Lambda_{12} = \sum_{j=1}^{k} a f_j(\theta_j - b) \left[\frac{(P_j - c)}{(1 - c)} \right]^2 \left[\frac{Q_j}{P_j} \right] = \sum_{j=1}^{k} a f_j(\theta_j - b) P_j Q_j \left[\frac{P_j^*}{P_j} \right].$$

$$E(L_{13}) = \Lambda_{13}$$

$$= -\sum_{j=1}^{k} f_j(\theta_j - b) \left\{ \left[\frac{(P_j - c)}{(1 - c)^2} \right] \frac{Q_j}{P_j} + (P_j - P_j) \left[\frac{(P_j - c)}{(1 - c)^2} \frac{Q_j}{P_j^2} \right] \right\}$$

and the second term is eliminated, yielding

$$\Lambda_{13} = -\sum_{j=1}^{k} f_j(\theta_j - b) \frac{(P_j - c)}{(1 - c)^2} \frac{Q_j}{P_j} = -\sum_{j=1}^{k} f_j(\theta_j - b) \frac{Q_j}{1 - c} \left[\frac{P_j^*}{P_j} \right].$$

$$L_{23} = \Lambda_{23} = \sum_{j=1}^{k} a \left\{ f_j \left[\frac{(P_j - c)}{(1 - c)^2} \right] \frac{Q_j}{P_j} + f_j(P_j - P_j) \left[\frac{(P_j - c)}{(1 - c)^2} \frac{Q_j}{P_j^2} \right] \right\}$$

$$= \sum_{j=1}^{k} a f_j \left[\frac{(P_j - c)}{(1 - c)^2} \right] \left[\frac{Q_j}{P_j} \right] = \sum_{j=1}^{k} a f_j \left[\frac{Q_j}{1 - c} \right] \left[\frac{P_j^*}{P_j} \right].$$

The simultaneous estimation of a, b, and c under the three-parameter model can be achieved using the iterative Newton-Raphson procedure again:

$$\begin{bmatrix} \hat{a} \\ \hat{b} \\ \hat{c} \end{bmatrix}_{t+1} = \begin{bmatrix} \hat{a} \\ \hat{b} \\ \hat{c} \end{bmatrix}_t + \begin{bmatrix} \Delta\hat{a} \\ \Delta\hat{b} \\ \Delta\hat{c} \end{bmatrix}_t ,$$

substituting for

$$\begin{bmatrix} \Delta\hat{a} \\ \Delta\hat{b} \\ \Delta\hat{c} \end{bmatrix}_t ,$$

yields

$$\begin{bmatrix} \hat{a} \\ \hat{b} \\ \hat{c} \end{bmatrix}_{t+1} = \begin{bmatrix} \hat{a} \\ \hat{b} \\ \hat{c} \end{bmatrix}_t - \begin{bmatrix} \Lambda_{11} & \Lambda_{12} & \Lambda_{13} \\ \Lambda_{21} & \Lambda_{22} & \Lambda_{23} \\ \Lambda_{31} & \Lambda_{32} & \Lambda_{33} \end{bmatrix}_t^{-1} \begin{bmatrix} L_1 \\ L_2 \\ L_3 \end{bmatrix}_t .$$

Because of the algebraic labor involved, the explicit formulas for $\Delta\hat{a}$, $\Delta\hat{b}$, $\Delta\hat{c}$ will not be presented. However, computer program implementation is straightforward using matrix algebra.

The mathematics of the maximum likelihood estimation under the three-parameter model does not yield as simple expressions as did the two-parameter model. The first and second derivatives with respect to a and b had the same basic form under both models but under the three-parameter model a "compensation" term $[P_j^*/P_j]$ was appended as a multiplier. This term reflects the fact that the three-parameter ICC had a lower asymptote. As a result, the ICC under the three-parameter model will always yield the larger P_j value and the ratio will be less than unity. One consequence of this is that the standard errors of the item difficulty and discrimination estimates under the three-parameter model will be larger than those obtained under a two-parameter model. In addition, there was a row and column in the information matrix due to the inclusion of the c parameter. When this matrix is inverted, the terms corresponding to the a and b parameters will be impacted by those due to c. It should be noted that although the mathematics presented above appears to be straightforward, the simultaneous estimation of the three-parameters is difficult to implement. The Newton-Raphson equations do not converge readily and achieving a solution depends upon initial estimates that are very close to the actual values. Also, the increments in the estimates must be tightly bounded and headed in the proper direction. These properties are associated with the form of log-likelihood surface associated with a three-parameter model. Because the surface exists in a four dimensional space, it is difficult to depict. Baker (1988) employed the following scheme to graph the item log-likelihood surface shown in Figure 2.3. The log-likelihood surface was computed over a range of values for the item difficulty b and discrimination a parameters for a fixed value of c. The value of c was increased and another log-likelihood surface computed. This process was repeated for three values of c. The three log-likelihood surfaces could be plotted on a common set of coordinates and the maximums might be connected with a heavy line. The overall maximum of the three-parameter likelihood surface will be on this line. It should be noted that this line is relatively flat suggesting that the maximum would be relatively hard to find with respect to the parameter c. A suggestion that agrees with what is observed empirically for this model. One of the problems with a flat ridge is that it is difficult to find the maximum. If the iterative procedure yields large increments to the parameters, the next solution will be positioned a considerable distance from the true maximum. This can lead to "hunting" on the part of the process and finding the maximum is difficult. Conversely, if the increments are small, the solution will "creep" up to the maximum and the overall process will converge slowly. Thus, the implementor of the estimation process needs to write computer programs that find a suitable balance between these two situations.

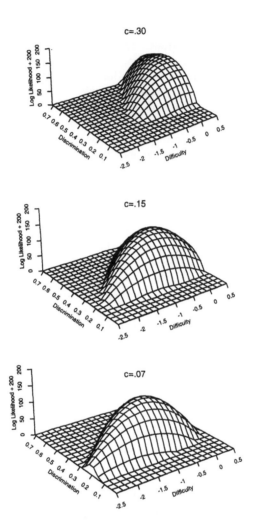

Fig. 2.3. Item log-likelihood surfaces for the three-parameter model $\hat{\alpha}_i = .36$, $\hat{\beta}_i = -.5$, $\hat{c}_i = .07$, .15, and .30. Copyright 1988 by Applied Psychological Measurement, Inc., reproduced by permission.

2.7 Minimum χ^2 and Minimum Transform χ^2 Estimations

Maximum likelihood is only one of several schemes for deriving estimates of parameters that meet the usual criteria of being unbiased, consistent, efficient, and sufficient. Berkson (1955a) advocated using procedures that minimize a χ^2 criterion rather than maximizing a likelihood function. The starting point of these procedures is the Pearson χ^2 which is defined as

$$\chi^2 = \sum_{j=1}^{n} \frac{(O_j - E_j)^2}{E_j},$$

where k is the number of categories, O_j is the observed frequency, and E_j is the expected frequency at ability level θ_j. Now, the dichotomously scored item response data for a single item can be arranged in the following $k \times 2$ contingency table (see Table 2.3).

Table 2.2. Dichotomously Scored Item Response Data for a Single Item

Ability Score	Correct	Wrong	No. of Subjects
θ_1	r_1	$f_1 - r_1$	f_1
θ_2	r_2	$f_2 - r_2$	f_2
θ_3	r_3	$f_3 - r_3$	f_3
.	.	.	.
.	.	.	.
.	.	.	.
θ_j	r_j	$f_j - r_j$	f_j
.	.	.	.
.	.	.	.
.	.	.	.
θ_k	r_k	$f_k - r_k$	f_k
			$\sum_j f_j = N$

At a given ability score level θ_j the observed proportion of correct response is

$$p_j = \frac{r_j}{f_j} \quad \text{and} \quad r_j = f_j p_j.$$

The expected frequency is $f_j P_j$ where P_j is obtained from the item characteristic curve evaluated at $Z_j = \zeta + \lambda \theta_j$. The observed proportion of incorrect response is

$$q_j = \frac{f_j - r_j}{f_j}$$

and $f_j - r_j = f_j q_j$. The expected frequency of incorrect response is $f_j Q_j$, where $Q_j = 1 - P_j$. Thus, the χ^2 term for the row of the table corresponding to θ_j is

$$\chi_j^2 = \frac{(f_j p_j - f_j P_j)^2}{f_j P_j} + \frac{(f_j q_j - f_j Q_j)^2}{f_j Q_j} = \frac{f_j}{P_j Q_j}(p_j - P_j)^2.$$

Now summing over all ability score levels the overall χ^2 is

$$\chi_j = \sum_{j=1}^{k} \frac{f_j}{P_j Q_j}(p_j - P_j)^2.$$

Using the same logic that was used in equation (2.4), let

$$\left(\frac{p_j - P_j}{P_j Q_j}\right)^2 = (z_j - Z_j)^2,$$

then $(p_j - P_j)^2 = (P_j Q_j)^2 (z_j - Z_j)^2$; an alternative form is

$$\chi^2 = \sum_{j=1}^{k} f_j P_j Q_j (z_j - Z_j)^2.$$

2.7.1 Minimum χ^2 Estimation

Under the minimum χ^2 estimation procedure, the derivatives of χ^2 with respect to ζ and λ are taken and set equal to zero. The resulting normal equations (Berkson, 1955a) are:

$$\sum_{j=1}^{k} f_j \left(\frac{P_j q_j - Q_j p_j}{P_j Q_j}\right)(p_j - P_j) = 0$$

$$\sum_{j=1}^{k} f_j \left(\frac{P_j q_j - Q_j p_j}{P_j Q_j}\right)(p_j - P_j)\theta_j = 0.$$

(2.30)

However, since the P_j and Q_j are functions of the parameters to be estimated and the P_j are not linear in the parameters, the minimum χ^2 estimation procedure is iterative as was the maximum likelihood procedure. Thus, we shall not pursue minimum χ^2 estimation further.

2.7.2 Minimum Transform χ^2 Estimation

Berkson (1944) proposed a non-iterative procedure which he designated as minimum transform χ^2 estimation. Minimum normit χ^2 estimation was used with the normal ogive model and minimum logit χ^2 was used with the logistic ogive model. The "logit" χ^2 was defined as follows:

$$\chi^2(\text{logit}) = \sum_{j=1}^{k} f_j p_j q_j (l_j - \hat{l}_j)^2,$$

(2.31)

where $l_j = \log(p_j/q_j)$ is the observed logit and $\hat{l}_j = \hat{\zeta} + \hat{\lambda}\theta_j$ represents the estimated value of the logit. It is important to notice that the p_j, q_j in equation (2.31) are the observed values. Taking derivatives with respect to ζ and λ and setting them equal to zero, the parameter estimates minimizing the χ^2 function can be found. The resulting normal equations are:

$$\sum_{j=1}^{k} f_j p_j q_j (l_j - \hat{l}_j) = 0$$
$$\sum_{j=1}^{k} f_j p_j q_j (l_j - \hat{l}_j) \theta_j = 0. \tag{2.32}$$

Substituting for $\hat{l}$, multiplying through, and reorganizing yield the normal equations:

$$\hat{\zeta} \sum_{j=1}^{k} f_j p_j q_j + \hat{\lambda} \sum_{j=1}^{k} f_j p_j q_j \theta_j = \sum_{j=1}^{k} f_j p_j q_j l_j$$
$$\hat{\zeta} \sum_{j=1}^{k} f_j p_j q_j \theta_j + \hat{\lambda} \sum_{j=1}^{k} f_j p_j q_j \theta_j^2 = \sum_{j=1}^{k} f_j p_j q_j l_j \theta_j. \tag{2.33}$$

Letting $w_j = p_j q_j$ and writing (2.33) in matrix notation,

$$D = \begin{bmatrix} \sum_{j=1}^{k} f_j w_j & \sum_{j=1}^{k} f_j w_j \theta_j \\ \sum_{j=1}^{k} f_j w_j \theta_j & \sum_{j=1}^{k} f_j w_j \theta_j^2 \end{bmatrix}$$

$$Y = \begin{bmatrix} \sum_{j=1}^{k} f_j w_j l_j \\ \sum_{j=1}^{k} f_j w_j l_j \theta_j \end{bmatrix}$$

$$X = \begin{bmatrix} \hat{\zeta} \\ \hat{\lambda} \end{bmatrix},$$

then $Y = DX$ and

$$X = D^{-1}Y$$

$$\begin{bmatrix} \hat{\zeta} \\ \hat{\lambda} \end{bmatrix} = \begin{bmatrix} \sum_{j=1}^{k} f_j w_j & \sum_{j=1}^{k} f_j w_j \theta_j \\ \sum_{j=1}^{k} f_j w_j \theta_j & \sum_{j=1}^{k} f_j w_j \theta_j^2 \end{bmatrix}^{-1} \begin{bmatrix} \sum_{j=1}^{k} f_j w_j l_j \\ \sum_{j=1}^{k} f_j w_j l_j \theta_j \end{bmatrix}. \tag{2.34}$$

The basic form of the matrix and the vectors is identical to that found earlier; see equation (2.11). Thus the solutions can be written directly

$$\hat{\zeta} = \frac{\displaystyle\sum_{j=1}^{k} f_j w_j \theta_j^2 \sum_{j=1}^{k} f_j w_j l_j - \sum_{j=1}^{k} f_j w_j \theta_j \sum_{j=1}^{k} f_j w_j l_j \theta_j}{\displaystyle\sum_{j=1}^{k} f_j w_j \sum_{j=1}^{k} f_j w_j \theta_j^2 - \left(\sum_{j=1}^{k} f_j w_j \theta_j\right)^2}$$

$$= \frac{\displaystyle\bar{l}\sum_{j=1}^{k} f_j w_j \theta_j^2 - \bar{\theta}\sum_{j=1}^{k} f_j w_j l_j \theta_j}{\displaystyle\sum_{j=1}^{k} f_j w_j \theta_j^2 - \left(\sum_{j=1}^{k} f_j w_j \theta_j\right)^2 \Big/ \left(\sum_{j=1}^{k} f_j w_j\right)}$$

(2.35)

$$\hat{\lambda} = \frac{\displaystyle\sum_{j=1}^{k} f_j w_j \sum_{j=1}^{k} f_j w_j l_j \theta_j - \sum_{j=1}^{k} f_j w_j \theta_j \sum_{j=1}^{k} f_j w_j l_j}{\displaystyle\sum_{j=1}^{k} f_j w_j \sum_{j=1}^{k} f_j w_j \theta_j^2 - \left(\sum_{j=1}^{k} f_j w_j \theta_j\right)^2}$$

$$= \frac{\displaystyle\sum_{j=1}^{k} f_j w_j l_j \theta_j - \left(\sum_{j=1}^{k} f_j w_j \theta_j \sum_{j=1}^{k} f_j w_j l_j\right) \Big/ \left(\sum_{j=1}^{k} f_j w_j\right)}{\displaystyle\sum_{j=1}^{k} f_j w_j \theta_j^2 - \left(\sum_{j=1}^{k} f_j w_j \theta_j\right)^2 \Big/ \left(\sum_{j=1}^{k} f_j w_j\right)}.$$

(2.36)

Although the form of equations (2.35) and (2.36) is identical to equations (2.22) and (2.21), respectively, there are important differences. First, under maximum likelihood the W_j are expressed in terms of "true" values and minimum logit χ^2 the w_j are functions of the observations. Second, the observed logits l_j are functions of the observations. Finally, equations (2.35) and (2.36) yield $\hat{\zeta}$ and $\hat{\lambda}$ directly, while equations (2.21) and (2.22) yield $\Delta\hat{\zeta}_t$ and $\Delta\hat{\lambda}_t$; thus, $\hat{\zeta}$ and $\hat{\lambda}$ must be obtained iteratively via the Newton-Raphson procedure under maximum likelihood estimation.

The minimum logit (normit) χ^2 estimation procedure has computational advantages due to its non-iterative nature. Berkson (1955b) for the normal ogive model has shown that the parameter estimates obtained are very close to those obtained under maximum likelihood with respect to bias and sampling error. The minimum logit χ^2 estimation procedure has been implemented in an IRT computer program (Baker, 1986a). The empirical results of an item parameter recovery study (Baker, 1987b) showed that the obtained item parameter estimates are comparable to those yielded via maximum likelihood.

The major disadvantage of this estimation procedure lies in the use of the observed logit $l_j = \log(p_j/q_j)$. As is often the case among low ability subjects p_j can be zero and among high ability subjects p_j can be unity. As a result the observed logit is infinite. Berkson (1955a) proposed using the $1/2n$ rule. When $p_j = 0$, it is replaced by $1/2f_j$ and when $p_j = 1$, it is replaced by $1 - (1/2f_j)$. In situations where a small number of ability score levels are used and an item is very easy or very hard, values of $p_j = 0$ or 1 can occur. In such cases, the $1/2n$ rule will have the effect of reducing the value of $\hat{\lambda}$. However, in most item analysis data, a relatively large number of examinees are found at each ability score level and the chances of an observed p_j being exactly 0 or 1 are reasonably small. Thus, the $1/2n$ rule would have only a minor impact upon the obtained values of $\hat{\zeta}$ and $\hat{\lambda}$. Research reported by Kim, Baker, and Subkoviak (1989) suggests that the $1/2n$ rule works well, but for small samples, a $1/4n$ rule may be more appropriate as it can cope with ability groups having only one or two examinees.

While the minimum logit χ^2 estimation procedure is not widely used in practice, one can take advantage of the good agreement between the item parameter estimates it yields and the maximum likelihood estimates. Baker and Martin (1969) used the minimum logit χ^2 estimation procedure as the first stage in a computer program for maximum likelihood estimation to provide the initial estimates of the item parameters under a two-parameter ICC model.

2.8 Summary

The maximum likelihood procedure for estimating the parameters of a single item has been presented for the situation where the test item has been dichotomously scored for examinees having known ability levels. These maximum likelihood estimation procedures, "the bioassay solution," will be at the core of the item parameter estimation techniques presented in Chapters 4–7. Because the same essential mathematics appears within so many different procedures, the elements of the Newton-Raphson/Fisher equations have been presented here in considerable detail. Having done so at this point will simplify later discussions and eliminate much of the repetition inherent in the item parameter estimation techniques presented in subsequent chapters. One characteristic of the IRT literature is that several different parameterizations of the ICC are employed. The slope intercept form is has been used here in conjunction with the two-parameter ICC models since its origins are in the bioassay literature. In addition, many of the computer program implementations of the maximum likelihood estimation procedures employ this form. Under the three-parameter ICC model, the a, b, c parameterization has been used to provide a direct linkage to the IRT literature.

Berkson, (1949, 1955a) has shown that, for dichotomously scored responses, the several item characteristic curve models and estimation procedures share a common form of the normal equations. This common form can be expressed in two ways. First, all the normal equations have the form

$$\sum_{j=1}^{k} W_j(p_j - P_j) = 0,$$

where W_j has absorbed terms such as f_j, h_j, P_j, Q_j, and θ_j. Berkson (1949) observed, when the p_j are binomially distributed, the normal equations impose the requirement that the weighted average of the observed proportions of correct response equal the weighted average of the estimated proportions. Thus,

$$\sum_{j=1}^{k} W_j p_j = \sum_{j=1}^{k} W_j P_j.$$

Second, an alternative form of the normal equations involving the "working deviate" was also presented above. This form is somewhat easier to use when contrasting the various models and estimation procedures. The normal equations are

$$\sum_{j=1}^{k} f_j W_j(z_j - Z_j) = 0,$$

where z_j is the working deviate based upon the observed data and Z_j is the "true" deviate based upon the values of $\hat{\zeta}$ and $\hat{\lambda}$ at any iteration in the solution process. The various situations of interest vary only in the definitions of W_j, z_j, and Z_j. Thus, in all of the estimation procedures derived above, the fitting of a cumulative distribution function to the observed item response data can be reduced to the weighted fitting of a linear regression line. This was accomplished by using either a normal deviate transformation of the proportion of correct response or a logistic deviate transformation of the odds ratio. The maximum likelihood estimation of the item parameters was implemented via a Taylor series expansion in terms of increments $\Delta\hat{\zeta}$ and $\Delta\hat{\lambda}$ to estimate the parameters ζ and λ via the iterative Newton-Raphson or Fisher's method of scoring procedures. A nice feature of the mathematics of the item parameter estimation procedures is that large sample (asymptotic) variances of the parameter estimates as well as a χ^2 goodness-of-fit test are obtained from the terms computed in the estimation equations. This serves to remind one that all of the mathematics associated here with item parameter estimation falls into the realm of asymptotic statistics. As a consequence, the techniques are best suited to large data sets. As will be seen in Chapter 4, a large body of literature deals with the determination of the appropriate numbers of items and examinees needed to obtain suitable item parameter estimates under the various ICC models.

The ICC is defined as the functional relation between the probability of correct response and the ability scale the frequency distribution of the examinees over the ability scale does not enter into this definition. However, in the estimation of the item parameters the f_j, W_j terms appear in all elements of the Newton-Raphson/Fisher equations. As a result, there is an interplay between the weighting coefficients that depend upon the ICC and this frequency distribution. A consequence of this interplay is that certain regions of the ability scale will play a larger role than others in the estimation of the item parameters. In particular, when the item difficulty is matched to the mean of the examinee ability distribution, the central portion of the distribution essentially determines the values of the item parameter estimates.

Inspection of the elements of the Fisher's method of scoring equations for the two and three-parameter models reveals that the common elements differ only in one respect. Under a three-parameter ICC, the two-parameter element is multiplied by a compensation term involving $[P_j^*/P_j]$. This term adjusts for the existence of a lower bound on the probability of correct response under the three-parameter that is not present in the two-parameter model. When $c = 0$, this term becomes unity. But estimating the parameters of a three-parameter ICC having $c = 0$ is not the same as a two-parameter solution. The information matrix of the former still has a row and column corresponding to the presence of a c parameter that is not present in the two-parameter solution. Experience has shown that the estimation of the item parameters under a two- parameter ICC model is quite robust and the Newton-Raphson/Fisher equations nearly always converge. In sharp contrast, the estimation of item parameters under a three-parameter ICC model are not very robust. It takes a considerable effort on the part of the software implementer to develop a scheme where the solution equations will generally converge. The consequences of this lack of robustness will examined in Chapter 4.

3. Maximum Likelihood Estimation of Examinee Ability

3.1 Introduction

Within the context of a test theory, a major goal is to obtain a measure of the ability of each examinee administered a educational test or psychological instrument. For a given examinee, this measure will be the maximum likelihood estimate of their unknown ability based upon their responses to the n items of the test and the values of the parameters of these items. In order to accomplish this estimation, three assumptions are made: First, the values of the parameters of the n dichotomously scored test items are known; Second, the examinees are independent objects and ability can be estimated on an examinee by examinee basis; Third, all n items in the test are modeled by ICC's of the same family. The maximum likelihood procedure for estimating an examinee's ability constitutes the second of the two basic building blocks of test analysis under IRT. While the overall logic of the maximum likelihood estimation of ability is the same under all ICC models, the mathematical details will differ as a function of the model. In the present chapter, these will be shown for the two- and three-parameter models.

3.2 Maximum Likelihood Estimation of Ability

A given examinee responds to the n items of a test and the responses are dichotomously scored, $u_{ij} = 0, 1$, where i designates the item $i = 1, \ldots, n$ and j designates the examinee $j = 1, \ldots, N$, yielding a vector of item responses of length n denoted by $U_j = (u_{1j}, u_{2j}, u_{3j}, \ldots, u_{nj}|\theta_j)$. Under the local independence assumption, the u_{ij} are statistically independent. Thus, the probability of the vector of item responses for a given examinee is given by the likelihood function

$$\text{Prob}(U_j|\theta_j) = \prod_{i=1}^{n} P_i^{u_{ij}}(\theta_j) Q_i^{1-u_{ij}}(\theta_j). \tag{3.1}$$

To simplify the notation, let $P_i(\theta_j) = P_{ij}$ and $Q_i(\theta_j) = Q_{ij}$, then

$$\text{Prob}(U_j|\theta_j) = \prod_{i=1}^{n} P_{ij}^{u_{ij}} Q_{ij}^{1-u_{ij}}. \tag{3.2}$$

Taking the natural logarithm of the likelihood function yields

$$L = \log \text{Prob}(U_j|\theta_j) = \sum_{i=1}^{n}[u_{ij} \log P_{ij} + (1 - u_{ij}) \log Q_{ij}]. \qquad (3.3)$$

Now, the P_{ij} are functions of the item characteristic curve and the deviate Z_{ij} (either normal or logistic) will be expressed in terms of the parameters ζ_i and λ_i, hence, $Z_{ij} = \zeta_i + \lambda_i\theta_j$. This parameterization is used here to maintain notational consistency with the previous chapter. Since the parameters for all n items are assumed to be known and are the "true values," only derivatives of the log-likelihood with respect to a given examinee will need to be taken:

$$\frac{\partial L}{\partial \theta_j} = \sum_{i=1}^{n} u_{ij} \frac{1}{P_{ij}} \frac{\partial P_{ij}}{\partial \theta_j} + \sum_{i=1}^{n}(1 - u_{ij}) \frac{1}{Q_{ij}} \frac{\partial Q_{ij}}{\partial \theta_j}. \qquad (3.4)$$

The derivatives of P_{ij} and Q_{ij} with respect to the ability parameter will be dependent upon the particular item characteristic curve model employed. Hence, for purposes of a generalized presentation these derivatives will be left in their definitional form.

As was the case for single items, the Newton-Raphson technique will be used to obtain the estimates of an ability parameter via an iterative procedure. Thus, the second order partial derivatives of the likelihood function with respect to the ability parameter also will be needed. Now, for a given examinee a Newton-Raphson equation can be established to be solved iteratively for the maximum likelihood estimate of ability. This equation is as follows:

$$\left[\hat{\theta}_j\right]_{t+1} = \left[\hat{\theta}_j\right]_t - \left[\frac{\partial^2 L}{\partial \theta_j^2}\right]_t^{-1} \left[\frac{\partial L}{\partial \theta_j}\right]_t. \qquad (3.5)$$

When this Newton-Raphson procedure has been performed, an ability estimator θ_j for the examinee is obtained. In the paragraphs below, the Newton-Raphson equations for ability parameter estimation are presented under each of item characteristic curve models of interest.

3.2.1 Normal Model

As was the case of item parameter estimation, the normal ogive model will be used initially to introduce ability estimation. Again this is done in order to link elements within the estimation process to important IRT concepts discussed later in the chapter. From equation (3.5) it can be seen that both the first and second derivatives of the log-likelihood function, equation (3.3), with respect to θ_j will be needed. The first order derivatives are

$$\frac{\partial P_{ij}}{\partial \theta_j} = \lambda_i h_i(\theta_j) \quad \text{and} \quad \frac{\partial Q_{ij}}{\partial \theta_j} = -\lambda_i h_i(\theta_j).$$

Let $h_i(\theta_j) = h_{ij}$; then,

$$\frac{\partial L}{\partial \theta_j} = \sum_{i=1}^{n} u_{ij} \frac{1}{P_{ij}} (\lambda_i h_{ij}) + \sum_{i=1}^{n} (1 - u_{ij}) \frac{1}{Q_{ij}} (-\lambda_i h_{ij})$$

$$= \sum_{i=1}^{n} \lambda_i h_{ij} \left(\frac{u_{ij} - P_{ij}}{P_{ij} Q_{ij}} \right).$$

Let $W_{ij} = h_{ij}^2 / P_{ij} Q_{ij}$, where W_{ij} are the Urban-Müller weights, then

$$\frac{\partial L}{\partial \theta_j} = \sum_{i=1}^{n} \lambda_i W_{ij} \left(\frac{u_{ij} - P_{ij}}{h_{ij}} \right). \tag{3.6}$$

The second order derivative with respect to θ_j is

$$\frac{\partial^2 L}{\partial \theta_j^2} = \frac{\partial}{\partial \theta_j} \left[\sum_{i=1}^{n} \lambda_i W_{ij} \left(\frac{u_{ij} - P_{ij}}{h_{ij}} \right) \right]$$

$$= \sum_{i=1}^{n} \frac{\lambda_i^2 h_{ij}}{P_{ij} Q_{ij}} \left[(u_{ij} - P_{ij}) \left(-Z_{ij} - \frac{h_{ij}}{P_{ij}} + \frac{h_{ij}}{Q_{ij}} \right) - h_{ij} \right]$$

$$= \sum_{i=1}^{n} \frac{\lambda_i^2 h_{ij}}{P_{ij} Q_{ij}} \left[(u_{ij} - P_{ij}) \left(-Z_{ij} - \frac{h_{ij}}{P_{ij}} + \frac{h_{ij}}{Q_{ij}} \right) \right] - \sum_{i=1}^{n} \frac{\lambda_i^2 h_{ij}^2}{P_{ij} Q_{ij}}.$$

As is the usual maximum likelihood practice, the observed u_{ij} are replaced by their expectations, P_{ij}, based upon the "true" item parameters. This causes the first term to vanish and then

$$E \left(\frac{\partial^2 L}{\partial \theta_j^2} \right) = - \sum_{i=1}^{n} \lambda_i^2 W_{ij}. \tag{3.7}$$

After substituting $E(\partial^2 L / \partial \theta_j^2)$ for for $\partial^2 L / \partial \theta_j^2$ in equation (3.5), the Fisher scoring equation becomes

$$\left[\hat{\theta}_j \right]_{t+1} = \left[\hat{\theta}_j \right]_t - \left[-\sum_{j=1}^{n} \lambda_i^2 W_{ij} \right]_t^{-1} \left[\sum_{i=1}^{n} \lambda_i W_{ij} \left(\frac{u_{ij} - P_{ij}}{h_{ij}} \right) \right]_t$$

$$\left[\hat{\theta}_j \right]_{t+1} = \left[\hat{\theta}_j \right]_t - \left[\frac{\sum_{i=1}^{n} \lambda_i W_{ij} \left(\frac{u_{ij} - P_{ij}}{h_{ij}} \right)}{-\sum_{i=1}^{n} \lambda_i^2 W_{ij}} \right]_t, \tag{3.8}$$

where W_{ij}, P_{ij}, and h_{ij} are evaluated using the item's "true" parameter values and the minus sign has been factored out of the inverse. The large sample variance of θ_j will be given by

$$S_{\hat{\theta}_j}^2 = \frac{1}{-E(\partial^2 L / \partial \theta_j^2)} = \frac{1}{\sum\limits_{j=1}^{n} \lambda_i^2 W_{ij}} = \frac{1}{\sum\limits_{j=1}^{n} \alpha_i^2 W_{ij}} \tag{3.9}$$

and the standard error will be

$$\text{SE}_{\hat{\theta}_j} = \sqrt{S^2_{\hat{\theta}_j}}.$$

3.2.2 Logistic Model

As indicated in Chapter 2, the two-parameter logistic ogive ICC model is preferred over the normal ogive model for computational reasons. Consequently, the equations for the maximum likelihood estimation of an examinees ability will be derived below using the this logistic model. Again, the first and second derivatives of the log-likelihood with respect to ability will be needed. Some useful derivatives with respect to θ_j are

$$\frac{\partial P_{ij}}{\partial \theta_j} = \lambda_i P_{ij} Q_{ij} \quad \text{and} \quad \frac{\partial Q_{ij}}{\partial \theta_j} = -\lambda_i P_{ij} Q_{ij}.$$

Substituting these derivatives in equation (3.4) yields the first derivative of the log-likelihood with respect to θ_j, and it is

$$\begin{aligned}
\frac{\partial L}{\partial \theta_j} &= \sum_{i=1}^{n} u_{ij} \frac{1}{P_{ij}} (\lambda_i P_{ij} Q_{ij}) + \sum_{i=1}^{n} (1 - u_{ij}) \frac{1}{Q_{ij}} (-\lambda_i P_{ij} Q_{ij}) \\
&= \sum_{i=1}^{n} \lambda_i (u_{ij} - P_{ij}) \\
&= \sum_{i=1}^{n} \lambda_i u_{ij} - \sum_{i=1}^{n} \lambda_i P_{ij}.
\end{aligned} \tag{3.10}$$

It is interesting to note that the first term on the right in the final form of equation (3.10) is the minimal sufficient statistic for the estimate of ability under a logistic model (Birnbaum, 1968). Letting $W_{ij} = P_{ij} Q_{ij}$,

$$\frac{\partial L}{\partial \theta_j} = \sum_{i=1}^{n} \lambda_i W_{ij} \left(\frac{u_{ij} - P_{ij}}{P_{ij} Q_{ij}} \right). \tag{3.11}$$

The second order derivative of the log-likelihood function with respect to θ_j is

$$\begin{aligned}
\frac{\partial^2 L}{\partial \theta_j^2} &= \frac{\partial}{\partial \theta_j} \left[\sum_{i=1}^{n} \lambda_i (u_{ij} - P_{ij}) \right] \\
&= \frac{\partial}{\partial \theta_j} \left[\sum_{i=1}^{n} \lambda_i u_{ij} \right] - \frac{\partial}{\partial \theta_j} \left[\sum_{i=1}^{n} \lambda_i P_{ij} \right] \\
&= -\sum_{i=1}^{n} \lambda_i^2 P_{ij} Q_{ij} = -\sum_{i=1}^{n} \lambda_i^2 W_{ij}.
\end{aligned} \tag{3.12}$$

As was the case when estimating item parameters, the second derivative of the log-likelihood did not contain observed data. Thus, the Newton-Raphson and Fisher's method of scoring for parameters approaches are the same. Substituting (3.12) for the $\partial^2 L/\partial\theta_j^2$ in equation (3.5) yields

$$\left[\hat{\theta}_j\right]_{t+1} = \left[\hat{\theta}_j\right]_t - \left[-\sum_{j=1}^{n}\lambda_i^2 W_{ij}\right]_t^{-1}\left[\sum_{i=1}^{n}\lambda_i W_{ij}\left(\frac{u_{ij}-P_{ij}}{P_{ij}Q_{ij}}\right)\right]_t$$

$$\left[\hat{\theta}_j\right]_{t+1} = \left[\hat{\theta}_j\right]_t - \left[\frac{\displaystyle\sum_{i=1}^{n}\lambda_i W_{ij}\left(\frac{u_{ij}-P_{ij}}{P_{ij}Q_{ij}}\right)}{-\displaystyle\sum_{i=1}^{n}\lambda_i^2 W_{ij}}\right]_t, \tag{3.13}$$

which at a given stage would be solved iteratively for the value of $\hat{\theta}_j$ for each examinee. The large sample variance of $\hat{\theta}_j$ will be given by

$$S_{\hat{\theta}_j}^2 = \frac{1}{-E(\partial^2 L/\partial\theta_j^2)} = \frac{1}{\displaystyle\sum_{j=1}^{n}\lambda_i^2 W_{ij}} = \frac{1}{\displaystyle\sum_{j=1}^{n}\alpha_i^2 W_{ij}} \tag{3.14}$$

and the standard error will be

$$SE_{\hat{\theta}_j} = \sqrt{S_{\hat{\theta}_j}^2}.$$

The implementation of the maximum likelihood estimation of ability, under a two parameter ICC model is illustrated in Appendix B. A computer program written in MICROSOFT QUICKBASIC is presented and an examinee's item response vector is used in the estimation of their ability. The computational aspects of the estimation process are reported to illustrate the details of the estimation process.

3.2.3 Birnbaum's Three-Parameter (Logistic) Model

The final ICC model to be employed in the estimation of an examinee's ability is Birnbaum's three-parameter (logistic) model. The necessary derivatives with respect to θ_j are:

$$\frac{\partial P_{ij}}{\partial\theta_j} = (1-c_i)P_{ij}^* Q_{ij}^* a_i$$

but, as was shown in Chapter 2,

$$Q_{ij}^* = \frac{Q_{ij}}{1-c_i} \quad \text{and} \quad P_{ij}^* = \frac{P_{ij}-c_i}{1-c_i},$$

then

$$\frac{\partial P_{ij}}{\partial\theta_j} = (1-c_i)\left(\frac{P_{ij}-c_i}{1-c_i}\right)\left(\frac{Q_{ij}}{1-c_i}\right)a_i = a_i Q_{ij}\left(\frac{P_{ij}-c_i}{1-c_i}\right).$$

Now,

$$\frac{\partial L}{\partial \theta_j} = \sum_{i=1}^{n} u_{ij} \frac{1}{P_{ij}} \left[a_i Q_{ij} \left(\frac{P_{ij} - c_i}{1 - c_i} \right) \right] + \sum_{i=1}^{n} (1 - u_{ij}) \frac{1}{Q_{ij}} \left[-a_i Q_{ij} \left(\frac{P_{ij} - c_i}{1 - c_i} \right) \right]$$

$$= \sum_{i=1}^{n} a_i \left(\frac{P_{ij} - c_i}{1 - c_i} \right) \left[u_{ij} \frac{Q_{ij}}{P_{ij}} - (1 - u_{ij}) \right]$$

$$= \sum_{i=1}^{n} a_i \left(\frac{P_{ij} - c_i}{1 - c_i} \right) \left(\frac{u_{ij} - P_{ij}}{P_{ij}} \right)$$

$$= \sum_{i=1}^{n} a_i W_{ij} \left(\frac{u_{ij} - P_i}{P_{ij} Q_{ij}} \right) \left[\frac{(P_{ij} - c_i)}{(1 - c_i)} \frac{1}{P_{ij}} \right]$$

$$= \sum_{i=1}^{n} a_i W_{ij} \left(\frac{u_{ij} - P_i}{P_{ij} Q_{ij}} \right) \left[\frac{P_{ij}^*}{P_{ij}} \right]$$

$$\frac{\partial^2 L}{\partial \theta_j^2} = \frac{\partial^2}{\partial \theta_j} \left[\sum_{i=1}^{n} a_i \left(\frac{P_{ij} - c_i}{1 - c_i} \right) \left(\frac{u_{ij} - P_{ij}}{P_{ij}} \right) \right] = \frac{\partial^2}{\partial \theta_j} \left[\sum_{i=1}^{n} a_i P_{ij}^* \left(\frac{u_{ij} - P_{ij}}{P_{ij}} \right) \right]$$

$$= \sum_{i=1}^{n} a_i \left\{ P_{ij}^* \left[\frac{-P_{ij}(1 - c_i) P_{ij}^* Q_{ij}^* a_i - (u_{ij} - P_{ij})(1 - c_i) P_{ij}^* Q_{ij}^* a_i}{P_{ij}^2} \right] \right.$$

$$\left. + \left(\frac{u_{ij} - P_{ij}}{P_{ij}} \right) (P_{ij}^* Q_{ij}^* a_i) \right\}$$

$$= \sum_{i=1}^{n} a_i^2 P_{ij}^* Q_{ij}^* \left\{ \left[\frac{-(1 - c_i) P_{ij} P_{ij}^* - (u_{ij} - P_{ij})(1 - c_i) P_{ij}^*}{P_{ij}^2} \right] \right.$$

$$\left. + \left(\frac{u_{ij} - P_{ij}}{P_{ij}} \right) \right\}$$

$$= \sum_{i=1}^{n} a_i^2 P_{ij}^* Q_{ij}^* \left(\frac{u_{ij} c_i - P_{ij}^2}{P_{ij}^2} \right)$$

$$= \sum_{i=1}^{n} a_i^2 \left(\frac{P_{ij} - c_i}{1 - c_i} \right) \left(\frac{Q_{ij}}{1 - c_i} \right) \left(\frac{u_{ij} c_i - P_{ij}^2}{P_{ij}^2} \right)$$

$$= \sum_{i=1}^{n} a_i^2 \frac{(P_{ij} - c_i)}{(1 - c_i)^2} \frac{Q_{ij}}{P_{ij}} \left(\frac{u_{ij} c_i - P_{ij}^2}{P_{ij}} \right)$$

$$= \sum_{i=1}^{n} a_i^2 W_{ij} \left[\frac{(P_{ij} - c_i)}{(1 - c_i)^2} \frac{1}{P_{ij}^2} \left(\frac{u_{ij} c_i - P_{ij}^2}{P_{ij}} \right) \right].$$

Again, this second derivative contains observed values. Hence, it is necessary to replace the u_{ij} by its expected value P_{ij}, yielding

$$E\left(\frac{\partial^2 L}{\partial\theta_j^2}\right) = \Lambda_{\theta\theta} = \sum_{i=1}^{n} a_i^2 W_{ij} \left[\frac{(P_{ij} - c_i)}{(1 - c_i)^2} \frac{1}{P_{ij}^2}(c_i - P_{ij})\right]$$

$$= -\sum_{i=1}^{n} a_i^2 W_{ij} \left[\frac{P_{ij}^*}{P_{ij}}\right]^2.$$

Substituting $E(\partial^2 L/\partial\theta_j^2)$ and $\partial L/\partial\theta_j$ in equation (3.5), the Fisher scoring equation for estimating ability is obtained:

$$\left[\hat{\theta}_j\right]_{t+1} = \left[\hat{\theta}_j\right]_t - \left[\frac{\sum_{j=1}^{n} a_i W_{ij}[(u_{ij} - P_{ij})/P_{ij}Q_{ij}][P_{ij}^*/P_{ij}]}{-\sum_{i=1}^{n} a_i^2 W_{ij}[P_{ij}^*/P_{ij}]^2}\right]_t. \qquad (3.15)$$

The basic terms in the Fisher equations for the ability parameters are the same under the two- and three-parameter logistic ICC models. In the case of the three-parameter model each basic term is multiplied by a factor involving P_{ij}^* and P_{ij} that compensates the terms for "guessing." It is also worth noting that algebraically most terms were expressed in terms of P_{ij}, Q_{ij} whereas the item parameters are terms in the logistic functions for P_{ij}^*. This was done to simplify the equations algebraically and P_{ij} can easily be obtained from P_{ij}^* since $P_{ij} = c_i + (1 - c_i)P_{ij}^*$.

The large sample variance of $\hat{\theta}_j$ will be given by

$$S_{\hat{\theta}_j}^2 = \frac{1}{-E(\partial^2 L/\partial\theta_j^2)} = \frac{1}{\sum_{j=1}^{n} a_i^2 W_{ij}\left[\frac{P_{ij}^*}{P_{ij}}\right]^2} \qquad (3.16)$$

and the standard error will be

$$SE_{\hat{\theta}_j} = \sqrt{S_{\hat{\theta}_j}^2}.$$

3.3 Information Functions

The sections above have shown how to obtain a maximum likelihood estimate of an examinee's unknown ability. It is also of interest to know something about the sampling distribution of these estimates about the examinee's underlying ability level. Within IRT, the main interest is in the sampling variability of these estimates. Equations (3.9), (3.14), and (3.16) define these variances under the ICC models of interest. The terms in all three equations are available in the corresponding Newton-Raphson equations (3.8), (3.13), and (3.15). An interesting feature of these three equations is that they do not depend upon the examinee's responses to the items. They depend only on the values of the parameters of the n items. In the present case, these are known and the variances can be computed easily. Although, such variances and the standard errors are of interest when a given examinees ability is estimated, it is also of interest to examine them over the whole ability scale. In IRT, this is accomplished through the item and test information functions which reflect how well the individual items and the test as a whole estimate ability over the ability scale.

While the original derivations of the test and item information functions (Birnbaum 1957, 1968, Chapters 17, 20) were based upon formula test scores within a classification framework, this approach is very convoluted both conceptually and mathematically. As a result some of Birnbaum's results will be taken and put in a more understandable and useful framework. When the original work was done, a means for obtaining maximum likelihood estimates of an examinee's ability was not fully developed. However, at the present time such estimates are easily obtained via computer programs such as LO-GIST (Wingersky, Patrick, & Lord, 1999; Wood, Wingersky, & Lord, 1976) and BILOG (Mislevy & Bock, 1982a, 1986, 1989). The availability of such estimates allows the whole issue of observed test scores and formula scoring to be avoided since ability estimates can be obtained directly.

One can conceptualize a conditional sampling distribution of ability estimates $\hat{\theta}$ about a common underlying ability level θ. Cramér (1946, p. 500) has shown that a maximum likelihood estimator, such as $\hat{\theta}$, has a normal asymptotic distribution with mean θ and variance $\sigma^2 = 1/I(\theta)$, where $I(\theta)$ is the amount of information and $\sqrt{1/I(\theta)}$ is the standard error. Some additional mathematical conditions are necessary, but the usual item characteristic curve models meet them (Samejima, 1977). The amount of information is a concept due to Sir R. A. Fisher and is the reciprocal of the variance of an estimate. In the present context, the variance of interest is that of the conditional distribution of the $\hat{\theta}$ at a given ability level, that is, $\sigma^2_{\hat{\theta}|\theta}$. Thus, the larger this variance the less precise the estimate of θ and the less information one has as to an examinee's unknown ability level. Birnbaum (1968, Chapters 17, 20) has defined the test information function as

$$I(\theta) = \sum_{i=1}^{n} \frac{[P_i'(\theta)]^2}{P_i(\theta)Q_i(\theta)}, \qquad (3.17)$$

where $P_i(\theta)$ is obtained by evaluating the item characteristic curve model at θ, and $P_i'(\theta) = \partial P_i/\partial\theta$. Birnbaum also has shown that this expression is the upper bound upon the amount of information yielded by any possible test scoring formula, thus, providing another advantage to using the maximum likelihood estimate of ability. It should be noted that this expression involves only the ability level θ and the item characteristic curves of the n items in the test. Since the right hand side is a sum, it can be decomposed into the contribution of each item to the amount of test information. The amount of information contributed by an individual item is given by

$$I_i(\theta) = \frac{[P_i'(\theta)]^2}{P_i(\theta)Q_i(\theta)}.$$ (3.18)

Given this, the test information function can be rewritten as follows:

$$I(\theta) = \sum_{j=1}^{n} I_i(\theta) = \sum_{j=1}^{n} \frac{[P_i'(\theta)]^2}{P_i(\theta)Q_i(\theta)}.$$ (3.19)

From a test theory point of view, defining the test information function in terms of the variance of the conditional distribution of the maximum likelihood estimates of ability is a crucial concept. It takes the test information function out of the realm of formula scoring and into a modern statistical framework. It also provides a vehicle for ready substantive interpretation of the meaning of the amount of information. The greater the amount of information at a given ability level, the closer the maximum likelihood estimates of ability will be clustered around the true but unknown ability level and, hence, the estimate is more precise.

Although the concept of the amount of test information is due to Birnbaum (1968), it has several antecedents in the IRT literature. Lord (1952, p. 21) developed an index for the discrimination power of a test at a given ability level. This index was the ratio of the slope of the test characteristic curve to the standard deviation of the observed test scores at the given ability level. This index is the $\sqrt{1/I(\theta, X)}$ when the scoring formula is $X = \sum_{i=1}^{n} u_i$, that is, the unweighted case and a normal ogive model was used for the item characteristic curves. In a subsequent paper, Lord (1953) derived the standard error of the maximum likelihood estimate of $\hat{\theta}$ of θ as well as the large sample confidence interval for the estimate of a given ability level. These earlier results are technically equivalent to Birnbaum's; however, Lord did not extract the test information function concept from them.

With this background, the next sections will provide the technical details of the information functions beginning at the item level.

3.3.1 Item Information Function

In equation (3.19), it was shown that the test information is simply the sum over items of the amount of item information at the ability level of interest. Even though one's primary interest is in the test information function, introduction of the mathematics of the item information function before that of the test information function makes for a better presentation sequence.

The explicit mathematical expression for the amount of item information depends upon the model employed for the item characteristic curve and its parameters. The normal ogive model will be used to explore the expression for the amount of item information in greater detail. Under this model, the item information function is defined as

$$I_i(\theta) = \frac{[P_i'(\theta)]^2}{P_i(\theta)Q_i(\theta)},$$ (3.20)

where $P_i'(\theta) = \partial P_i(\theta)/\partial\theta = \alpha_i h_i(\theta)$. Recall that $\alpha_i = \lambda_i$ and $h_i(\theta)$ is the ordinate of the unit normal density corresponding to $P_i(\theta)$. Then,

$$I_i(\theta) = \alpha_i^2 \left\{ \frac{[h_i(\theta)]^2}{P_i(\theta)Q_i(\theta)} \right\}.$$

The composition of $I_i(\theta)$ under the normal model for the item characteristic curve is very revealing. The term within the braces is identically the Urban-Müller weight involved in the weighted least squares and maximum likelihood approaches to estimating the item and ability parameters. When fitting the normal ogive model to the observed proportions of correct response to an item, the Urban-Müller weights, multiplied by f_j, were the reciprocal of the large sample variances of the normal deviate transformations of the observed proportions of correct response at a given ability level. The smaller this variance, the greater the contribution of a given ability level to the estimation of the item parameters. Consequently, the reciprocal of this variance, the Urban-Müller weight, is an information measure. The term preceding the braces is the square of the item discrimination parameter, which is also a measure of information since $\alpha_i^2 = 1/\sigma_i^2$, where σ_i^2 is the variance of the normal ogive serving as the item characteristic curve. Overall, Birnbaum's item information function is the product of the amount of information provided two components: first, the amount of information provided by the item's capacity to discriminate among ability levels in the neighborhood of θ and measured by σ_i^2; second, an amount of information that depends upon the probability of correct response at a given ability level and measured by the Urban-Müller weights. This particular fractionalization of the item information function reveals a close connection between the amount of information and the item parameter estimation procedures.

The item information function can also be fractionalized in a somewhat different manner to relate it to Birnbaum's "locally best" weights. Equation (3.20) can be rewritten as

$$I_i(\theta) = P_i'(\theta) \left[\frac{P_i'(\theta)}{P_i(\theta)Q_i(\theta)} \right] \tag{3.21}$$

and the term within the bracket is the "locally best" weight. The $P_i'(\theta)$ is the rate of change of the item characteristic curve at θ. Thus, the item information function is a weighted rate of change in discrimination. This fractionalization reflects Birnbaum's classification approach to the information function.

Since the amount of information can be determined at each ability level, as shown in Table 3.1, one can plot $I_i(\theta)$ as a function of ability. Figure 3.1 depicts an item information function based upon the normal model with $\alpha_i = .9$, $\beta_i = 0$. This item information curve is symmetrical about $\beta_i = 0$ and is bell-shaped indicating that the extreme ability levels provide less information.

Table 3.1. Item Information Function Under the Normal Model ($\alpha_i = 0.9$, $\beta_i = 0.0$)

θ	Z_i	$P_i(\theta)$	$h_i(\theta)$	$I_i(\theta)$
-3.0	-2.70	0.0035	0.0104	0.0255
-2.5	-2.25	0.0122	0.0317	0.0676
-2.0	-1.80	0.0359	0.0790	0.1458
-1.5	-1.35	0.0885	0.1604	0.2583
-1.0	-0.90	0.1841	0.2661	0.3819
-0.5	-0.45	0.3264	0.3605	0.4789
0.0	0.00	0.5000	0.3989	0.5157
0.5	0.45	0.6736	0.3605	0.4789
1.0	0.90	0.8159	0.2661	0.3819
1.5	1.35	0.9115	0.1604	0.2583
2.0	1.80	0.9641	0.0790	0.1458
2.5	2.25	0.9878	0.0317	0.0676
3.0	2.70	0.9965	0.0104	0.0255

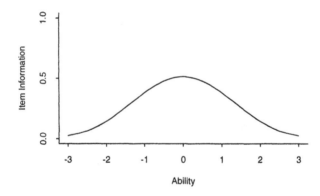

Fig. 3.1. Item information curve, normal model, $\alpha_i = 0.9$, $\beta_i = 0.0$.

The expressions for the item information functions under the remaining item characteristic curve models are:

- Two-parameter logistic

$$I_i(\theta) = \frac{[P_i'(\theta)]^2}{P_i(\theta)Q_i(\theta)} = \alpha_i^2 P_i(\theta)Q_i(\theta).$$ (3.22)

- Rasch logistic

$$I_i(\theta) = P_i(\theta)Q_i(\theta)$$ (3.23)

since $\alpha_i = 1$.

- Birnbaum three-parameter (logistic)

$$I_i(\theta) = a_i^2 \frac{Q_i(\theta)}{P_i(\theta)} \left[\frac{P_i(\theta) - c_i}{1 - c_i} \right]^2 = a_i^2 P_i(\theta) Q_i(\theta) \left[\frac{P_i^*(\theta)}{P_i(\theta)} \right]^2. \tag{3.24}$$

In all three of these expressions, the term $P_i(\theta)Q_i(\theta)$ appears. It is the logistic counterpart of the Urban-Müller weights in the item parameter estimation procedures (see Table 2.2). Thus, the basic structure of the item information function under the logistic and normal models is the same. The expression for the item information function under a three-parameter model involves an extra term resulting from the "guessing" parameter. This term decreases the amount of information by the square of the ratio of the probability that a person of the given ability will answer the item correctly under a two-parameter model to that under a three-parameter model. This term compensates for the lifting of the lower asymptote of the logistic ogive under the three-parameter model. It should be noted that this term is the same as the "compensation" term appearing in the information matrix of the Fisher scoring equation used to estimate the item parameters under a three-parameter model.

One property of the item information function of considerable practical interest is the maximum amount of information obtainable. When the two-parameter normal and logistic models for the item characteristic curve are used, the maximum occurs when $\alpha_i(\theta - \beta_i) = 0$, that is, when $\theta = \beta_i$. The maximum under these models (Birnbaum, 1968, p. 461) is:

(a) Normal

$$\max[I_i(\theta)] = \alpha_i^2 \frac{[.3989]^2}{(.5)(.5)} = .64\alpha_i^2, \tag{3.25}$$

where .3989 is the ordinate of the normal density when $Z = 0$.

(b) Logistic

$$\max[I_i(\theta)] = \alpha_i^2(.5)(.5) = .25\alpha_i^2. \tag{3.26}$$

(c) Rasch

$$\max[I_i(\theta)] = (.5)(.5) = .25. \tag{3.27}$$

It should be noted that the item discrimination parameter under the logistic model was not rescaled in equations (3.26) and (3.27). To put the amount of information for the logistic models on a common basis with the normal model, it must be multiplied by $(1.702)^2 = 2.8968$. When this is done the maximum under the logistic model is $.724\alpha_i^2$ which is slightly higher than the normal model.

Under all three of these models, the item information curve always will be bell-shaped with its maximum at $\theta = \beta_i$. In all cases, the amount of information can be quite small for ability levels that deviate considerably from β_i. This indicates that the estimation of ability is better when the difficulty of the item, β_i, is matched to the examinee's ability.

The maximum of the information function under Birnbaum's three-parameter (logistic) model will not be obtained at $\theta = b_i$ because its median does not occur at b_i. Birnbaum (1968, p. 464) derived the following expression for the ability level at which the maximum is obtained:

$$\theta_{\max} = b_i + \frac{1}{a_i} \log \left(\frac{1 + \sqrt{1 + 8c_i}}{2} \right). \tag{3.28}$$

Since the maximum amount of information under Birnbaum's three-parameter (logistic) model is a function of the ability level $\theta_{\max}$ and the guessing parameter c_i, a simple expression for its maximum cannot be obtained but it can be found by plotting the item information function. Figure 3.2, based upon the data of Table 3.2, presents an item information curve with parameters $a_i = 0.8$, $b_i = 0.0$, and $c_i = 0.2$. It can be seen that the maximum amount of information was obtained at an ability level above b_i, occurring at $\theta_{\max} = .334$ which is the value yielded by equation (3.28) in the example. The maximum amount of information obtained at $\theta_{\max}$ was:

$$I_i(\theta) = .8^2 (.2265)(.8672)^2$$
$$= .64(.2265)(.7521)$$
$$= .1090$$

Again, if comparability with the normal model is desired, a_i in equation (3.28) should be replaced by $1.702a_i$. Yielding .3126 for the maximum amount of information in the present example.

Table 3.2. Item Information Function for Birnbaum's Three-Parameter (Logistic) Model ($a_i = 0.8$, $b_i = 0.0$, $c_i = 0.20$)

θ	Z_i	$P_i^*(\theta)$	$P_i(\theta)$	$P_i^*(\theta)/P_i(\theta)$	$I_i(\theta)$
-3.0	-2.40	0.0832	0.2665	0.3120	0.0122
-2.5	-2.00	0.1192	0.2954	0.4036	0.0217
-2.0	-1.60	0.1680	0.3344	0.5024	0.0359
-1.5	-1.20	0.2315	0.3852	0.6010	0.0547
-1.0	-0.80	0.3100	0.4480	0.6920	0.0758
-0.5	-0.40	0.4013	0.5210	0.7702	0.0947
0.0	0.00	0.5000	0.6000	0.8333	0.1067
0.5	0.40	0.5987	0.6790	0.8818	0.1085
1.0	0.80	0.6900	0.7520	0.9175	0.1005
1.5	1.20	0.7685	0.8148	0.9432	0.0859
2.0	1.60	0.8320	0.8656	0.9612	0.0688
2.5	2.00	0.8808	0.9046	0.9736	0.0523
3.0	2.40	0.9168	0.9335	0.9822	0.0383

When the "guessing" parameter, c_i is set to zero, the "compensation" term yields a value of unity. Thus, the information function reduces to that for the two-parameter logistic ICC model. As a result the information function for the two-parameter logistic model is the upper bound for the infor-

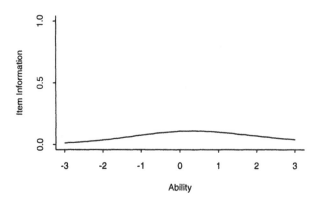

Fig. 3.2. Item information curve for Birnbaum's three-parameter (logistic) model, $a_i = 0.8$, $b_i = 0.0$, $c_i = 0.2$.

mation function for Birnbaum's three-parameter (logistic) model given the same values of α_i, β_i.

3.3.2 Samejima's Approach to the Item Information Function

Although the formulas for the amount of item information derived by Birnbaum (1968) are correct, his method is based upon procedures aimed at classifying examinees. As a result, his approach obscures some concepts that are of fundamental importance to IRT. A more generalized approach to the item information function has been developed by Samejima (1969, 1972, 1977) and will be presented below.

A dichotomously scored item has two item response categories, correct and incorrect, and an examinee's response can be assigned to either category. Samejima assumes that both response categories provide information about the estimation of the examinee's ability. Let $I_k(\theta)$ denote the amount of information yielded by response category k, where $k = 1, 2$. The information function for an item response category was given as

$$I_k(\theta) = -\frac{\partial^2 \log P_k(\theta)}{\partial \theta^2} = -\frac{\partial}{\partial \theta}\left[\frac{P_k'(\theta)}{P_k(\theta)}\right] = \frac{[P_k'(\theta)]^2 - P_k(\theta)P_k''(\theta)}{[P_k(\theta)]^2}, \quad (3.29)$$

where $P_k''(\theta) = \partial P_k'(\theta)/\partial \theta$ and $P_k'(\theta) = \partial P_k(\theta)/\partial \theta$. In the case of a binary item, however, an examinee with ability θ has probability $P_1(\theta)$ of choosing the correct response and $P_2(\theta) = Q(\theta) = 1 - P_1(\theta)$ of choosing the incorrect response. As a result the contribution of the correct response category to the amount of item information is $I_1(\theta)P_1(\theta)$. The contribution of the incorrect response is $I_2(\theta)Q(\theta) = I_2(\theta)P_2(\theta)$. Now in general, the amount of information share of item response category k, $k = 1, 2$, is given by

$$I_k(\theta)P_k(\theta) \tag{3.30}$$

and the amount of item information is

$$I_i(\theta) = \sum_{k=1}^{2} I_{ik}(\theta)P_{ik}(\theta) = I_{i1}(\theta)P_{i1}(\theta) + I_{i2}(\theta)P_{i2}(\theta). \tag{3.31}$$

From this equation, it can be seen that the amount of item information at a given value of θ is composed of two shares, one contributed by the correct response category and the other contributed by the incorrect response category. This is a very important concept within the IRT and is often overlooked in the literature. Substituting equation (3.29) in equation (3.31), the expression for the item information function of a binary item can be obtained:

$$I_i(\theta) = \sum_{k=1}^{2} \left\{ \frac{[P'_{ik}(\theta)]^2 - P_{ik}(\theta)P''_{ik}(\theta)}{[P_{ik}(\theta)]^2} \right\} P_{ik}(\theta) \tag{3.32}$$

$$= \sum_{k=1}^{2} \left\{ \frac{[P'_{ik}(\theta)]^2}{P_{ik}(\theta)} - P''_{ik}(\theta) \right\},$$

and letting $P_i(\theta) = P_{i1}(\theta)$ and $Q_i(\theta) = P_{i2}(\theta)$ yields

$$I_i(\theta) = \left\{ \frac{[P'_i(\theta)]^2 - P_i(\theta)P''_i(\theta)}{[P_i(\theta)]^2} \right\} P_i(\theta) + \left\{ \frac{[-P'_i(\theta)]^2 + Q_i(\theta)P''_i(\theta)}{[Q_i(\theta)]^2} \right\} Q_i(\theta)$$

$$= \frac{[P'_i(\theta)]^2 Q_i(\theta) - P_i(\theta)Q_i(\theta)P''_i(\theta) + [-P'_i(\theta)]^2 P_i(\theta) + P_i(\theta)Q_i(\theta)P''_i(\theta)}{P_i(\theta)Q_i(\theta)}$$

$$= \frac{[P'_i(\theta)]^2 [P_i(\theta) + Q_i(\theta)]}{P_i(\theta)Q_i(\theta)}$$

$$I_i(\theta) = \frac{[P'_i(\theta)]^2}{P_i(\theta)Q_i(\theta)}.$$

This result if identical to the item information function expression due to Birnbaum (1968) for a dichotomously scored item. The problem with Birnbaum's derivation based upon classification procedures was that it obscures the contribution, that is, the amount of information share, of each of the two item response categories to the amount of item information at a given ability level. Conceptually, it is important to recognize that all response categories of an item contribute to the amount of item information, not just the correct response.

Samejima (1973) provides a caution regarding Birnbaum's three-parameter model. She indicates that, under this model, the amount of item response category information can be negative under certain circumstances even though the overall amount of item information will always be positive.

3.3.3 Test Information Function

As was indicted above, the test information function provides a measure of how precisely the n items are estimating ability at any point along the ability scale. This function plays a role within IRT analogous to that of reliability in classical test theory. However, it has a distinct advantage over the latter as it provides a measure of precision at each ability level of interest rather than a global measure. Equation (3.17) shows that the test information is simply the sum of the n item informations at a given ability level. Thus, expressions for the test information function under the various item characteristic curve models can be obtained using the results given above for the amount of the item information. The formulas are as follows:

(a) Normal

$$I(\theta) = \sum_{i=1}^{n} \frac{\alpha_i^2 [h_i(\theta)]^2}{P_i(\theta)Q_i(\theta)} \tag{3.33}$$

(b) Logistic

$$I(\theta) = \sum_{i=1}^{n} \alpha_i^2 P_i(\theta)Q_i(\theta) \tag{3.34}$$

(c) Rasch

$$I(\theta) = \sum_{i=1}^{n} P_i(\theta)Q_i(\theta) \tag{3.35}$$

(d) Birnbaum's three-parameter (logistic)

$$I(\theta) = \sum_{i=1}^{n} a_i^2 P_i(\theta)Q_i(\theta) \left[\frac{P_i^*(\theta)}{P_i(\theta)} \right]^2 \tag{3.36}$$

It is also important to recognize that these formulas are actually composed of the amount of information shares of the item response categories. Consequently, the definition of the test information function is actually

$$I(\theta) = \sum_{i=1}^{n} \sum_{k=1}^{m_i} I_{ik}(\theta) P_{ik}(\theta), \tag{3.37}$$

where m_i is the number of item response categories for item i. This definition is more general than that due to Birnbaum and encompasses other item response schemes such as ordered response and nominal response cases.

As was the case for items, the amount of information for a test $I(\theta)$ can be plotted as a function of θ. The resulting curve is known as the test information function, and the shape of the test information function will depend upon the mix of values of the parameters of the items in the test. When all of the items in the test are equivalent and share common values of

α_i, β_i, the amount of information in the test is simply n times the amount of information in an item. In this case, the test information function is simply an elevated version of the item information function. However, the common situation is one where each item parameter has a distribution of values. To illustrate a test information function, a hypothetical test of 60 items was created in which the β_i were normally distributed with mean zero and unit variance and the α_i were normally distributed with mean one and variance .2. These item parameter values were randomly paired to define an item. The resulting test information curve is plotted in Figure 3.3. The curve shows that ability is measured best from $-.5$ to $+.5$ on the ability scale and the precision decreases towards the extreme ability values. When constructing tests, a common practice is to define a "target test information function" that depends upon the goals of the test and then select items from an item pool such that the target is matched (see Lord, 1980).

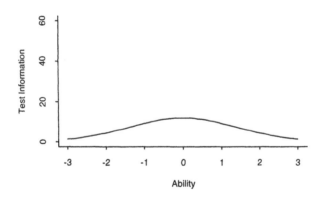

Fig. 3.3. Test information curve for a 60 item test.

It is also of some historical interest to show the test information function for the type of tests employed by Lawley (1943) in his classic paper on IRT. He used a test in which all of the items had a discrimination parameter of the same value. Using Lawley's results, Birnbaum (1968, p. 467) was able to express the test information function in terms of the item parameters of a test, assuming that the location parameters are normally distributed with

$$\bar{\beta} = \frac{\sum_{i=1}^{n} \beta_i}{n} \quad \text{and} \quad \sigma_\beta^2 = \frac{1}{n}\sum_{i=1}^{n}(\beta_i - \bar{\beta})^2.$$

Then, using a logistic model for the ICC, Birnbaum showed the following:

$$I(\theta) = \frac{n\alpha}{\sigma}\phi\left[\frac{\theta - \bar{\beta}}{\sigma}\right],$$ (3.38)

where $\sigma^2 = \sigma_\beta^2 + \alpha^{-2}$, ϕ is the normal density, $\bar{\beta}$ is the mean difficulty of the items in the test, and α is the fixed common value of the discrimination index. Thus for this special case, the test information function has the general form of a normal density multiplied by the factor $n\alpha/\sigma$.

3.4 Summary

The estimation of a single examinee's ability was based upon his/her vector of responses to the n binary items of the test and the known values of the item parameters. The mathematical details of the solution equations for the ability estimate varies as a function of the ICC model used for the test's items. In the case of the two-parameter logistic model, the estimation process was formulated as an iterative Newton-Raphson procedure. In the case of the normal model and the Birnbaum three-parameter (logistic) models, the second derivative with respect to ability was replaced by it's expectation to obtain an information matrix, and Fisher's "method of scoring for the parameters" was actually the basic solution method. Given an examinee's ability estimate, it is also of interest to have some measure of how "precise" an estimate it is. At a given ability level, a sampling distribution of the maximum likelihood estimates of that ability level exists. The Newton-Raphson/Fisher scoring equations yield an estimate of the variance of this conditional distribution; thus, the standard error of the examinee's ability estimate is available for this purpose. Since the item parameters are assumed to be known, the variance of this conditional distribution of ability estimates depends only upon these item parameters, and it can be computed for all points along the ability scale. This variance can be computed on the basis of a single item or for all items in the test. The former is the basis for the item information function and the latter for test information functions. These functions show the relationship between the amount of information, that is, precision, and ability over the complete ability scale. In a rough sense, the test information function is the IRT substitute for reliability under classical test theory. The conceptual advantage of the information function is that it indicates how well ability is being estimated at every ability level. Whereas, the classical test theory reliability coefficient is simply an overall measure of consistency of measurement.

Although it is beyond the scope of the present book, several uses of the information function should be mentioned as they relate to how well an examinee's ability is being estimated. Within computerized adaptive testing (Wainer, 1990), the standard error desired for a given examinee's ability estimate can be established a priori. As the sequence of items is administered to the examinee, the standard error is computed after each item. When this computed value is equal or less than the a priori value, the testing is complete. The advantage of this stopping rule is that the ability of all examinees tested can be estimated to the same level of precision even though they were administered different numbers and sequences of items. A second application is in automated test construction via mathematical programming. Although employed by Yen (1983) at an earlier date, the initial methodological paper is due to Theunissen (1985). Under this approach, a "target" test information function is specified over an ability range of interest, and items are selected from a precalibrated item pool via integer or linear programming procedures, the goal being to select those items whose test information function will match that of the target. Again, the underlying rationale is one of constructing a test that will have a desired precision of estimation over the range of ability of interest. Baker, Cohen, and Barmish (1988) have examined the characteristics of the tests constructed via this technique.

The maximum likelihood procedures for the estimation of an examinee's ability is the second of the two basic building blocks underlying IRT test analysis procedures for dichotomously scored items. In Chapters 4–7, these two building blocks will be incorporated in various ways to yield procedures that estimate both the item and examinee parameters for a set of test results.

4. Procedures for Estimating Both Ability and Item Parameters

4.1 Introduction

The typical educational or psychological test usually contains many items and is administered to a group of examinees. When such a test is analyzed, it is of interest to be able to estimate the parameters of the test items and the ability parameters of the examinees. A major factor in the practical application of IRT has been the development of maximum likelihood estimation procedures for these two sets of parameters and the implementation of these procedures in the form of computer programs. The development of these procedures was a rather slow process. Lord (1953), in a very prophetic article, presented the basic framework of the simultaneous estimation of the item and ability parameters via maximum likelihood using a normal ogive model for the item characteristic curves. However, the modern digital computer was not yet available and the computational aspects were not practical. Birnbaum (1968, pp. 420–422) presented a simplified paradigm for jointly estimating the item and ability parameters that has become the basis for many computer implementations of these maximum likelihood estimation procedures. The first actual implementation of this paradigm was due to Lord (1968), who described a computer program which evolved into the widely used LOGIST program (Wingersky, Barton, & Lord, 1982). It was also implemented by Wright and Panachapakesan (1969) in a computer program for the Rasch model that became the BICAL program (Wright & Mead, 1978). Because of the computational demands of these estimation procedures, IRT is not practical without such computer programs. Yet, to wisely select and use such programs one must have an understanding of the basic estimation procedures they implement. Consequently, the present chapter has two goals: first, to present the statistical logic of the Birnbaum paradigm and, second, to explore the domain of parameter estimation under this paradigm from a technical point of view. There has been a great deal of work on various facets of such parameter estimation, and much of it is of interest because it deals with issues of importance to the application of the theory.

4.2 Joint Maximum Likelihood Estimation: The Birnbaum Paradigm

In Chapter 2, the maximum likelihood estimation of the parameters of a single item, when the ability scores were known, was presented. In Chapter 3, the maximum likelihood estimation of an examinee's ability, when the item parameters are known, was presented. Since these procedures are the two basic building blocks of the joint maximum likelihood estimation (JMLE) paradigm, the algebraic detail of those chapters will not be repeated here, and the presentation will be at a more general level. Once the paradigm used for the JMLE of the item and ability parameters is in place, the second goal can be approached. With this caveat, the maximum likelihood procedures for the joint estimation of item parameters, under a two-parameter ICC model, and of examinee ability are presented below.

Under typical testing conditions, a sample of N examinees are drawn at random from a population of examinees possessing the underlying trait, that is, the ability of interest. No assumption as to the distribution of examinees over the ability continuum is necessary (Lord & Novick, 1968, p. 375). These examinees respond to the n items of the test, and the responses are dichotomously scored, $u_{ij} = 0, 1$, where i designates the item, $i = 1, \ldots, n$, and j designates the examinee, $j = 1, \ldots, N$. For each examinee there will be a vector of item responses of length n denoted by $(u_{1j}, \ldots, u_{nj} | \theta_j)$. Under the local independence assumption, the u_{ij} are statistically independent for all persons having the same ability. There will be one such vector for each examinee, hence, there will be N vectors. The resulting n by N matrix of item responses is denoted by $U = [u_{ij}]$ and θ is the vector of the N examinee ability scores $(\theta_1, \ldots, \theta_N)$. Thus, the probability of the $n \times N$ matrix of item responses is given by the likelihood function

$$\text{Prob}(U|\theta) = \prod_{j=1}^{N} \prod_{i=1}^{n} P_i^{u_{ij}}(\theta_j) Q_i^{1-u_{ij}}(\theta_j). \tag{4.1}$$

To simplify the notation, let $P_i(\theta_j) = P_{ij}$ and $Q_i(\theta_j) = Q_{ij}$; then

$$\text{Prob}(U|\theta) = \prod_{j=1}^{N} \prod_{i=1}^{n} P_i^{u_{ij}} Q_i^{1-u_{ij}}. \tag{4.2}$$

Taking the logarithm of the likelihood function yields

$$L = \log \text{Prob}(U|\theta) = \sum_{j=1}^{N} \sum_{i=1}^{n} [u_{ij} \log P_{ij} + (1 - u_{ij}) \log Q_{ij}]. \tag{4.3}$$

Now, the P_{ij} are functions of the item characteristic curve, and the deviate Z_{ij} (either normal or logistic) will be expressed in terms of the linear item parameters ζ_i and λ_i, $Z_{ij} = \zeta_i + \lambda_i \theta_j$. Since the parameters for all n items and the ability scores for all N examinees are unknown, it will be necessary to take the derivatives of L with respect to these parameters, equate them to zero, and solve the $2n + N$ simultaneous equations to obtain the maximum likelihood estimates of the unknown parameters. Taking the derivatives with respect to the intercept and slope parameters,

$$\frac{\partial L}{\partial \zeta_i} = \sum_{j=1}^{N} u_{ij} \frac{1}{P_{ij}} \frac{\partial P_{ij}}{\partial \zeta_i} + \sum_{j=1}^{N} (1 - u_{ij}) \frac{1}{Q_{ij}} \frac{\partial Q_{ij}}{\partial \zeta_i}, \tag{4.4}$$

$$\frac{\partial L}{\partial \lambda_i} = \sum_{j=1}^{N} u_{ij} \frac{1}{P_{ij}} \frac{\partial P_{ij}}{\partial \lambda_i} + \sum_{j=1}^{N} (1 - u_{ij}) \frac{1}{Q_{ij}} \frac{\partial Q_{ij}}{\partial \lambda_i}, \tag{4.5}$$

and there will be n pairs of such derivatives. The summation in these equations is over the individual examinees, as is the case in the LOGIST program, rather than over groups of examinees, as was the case in Chapter 2.

Taking the derivative of the log likelihood with respect to the ability parameters yields

$$\frac{\partial L}{\partial \theta_i} = \sum_{j=1}^{N} u_{ij} \frac{1}{P_{ij}} \frac{\partial P_{ij}}{\partial \theta_i} + \sum_{j=1}^{N} (1 - u_{ij}) \frac{1}{Q_{ij}} \frac{\partial Q_{ij}}{\partial \theta_i}, \tag{4.6}$$

and there will be one such derivative for each of the N ability parameters. The derivatives of P_{ij} and Q_{ij} with respect to the parameters will be dependent upon the particular item characteristic curve model employed. Hence, for purposes of a generalized presentation these derivatives will be left in their definitional forms.

The iterative Newton-Raphson technique will be used to obtain the estimates of the item and ability parameters. Thus, the second order partial derivatives of the likelihood function with respect to the parameters will be needed. Again, these can be represented symbolically. The necessary derivatives are

$$\frac{\partial^2 L}{\partial \zeta_i^2}, \quad \frac{\partial^2 L}{\partial \lambda_i^2}, \quad \frac{\partial^2 L}{\partial \zeta_i \lambda_i}, \quad \frac{\partial^2 L}{\partial \theta_j^2}, \quad \frac{\partial^2 L}{\partial \theta_j \zeta_i}, \quad \text{and} \quad \frac{\partial^2 L}{\partial \theta_j \lambda_i}.$$

The Newton-Raphson equation will be of the following form:

$$A_{t+1} = A_t - B_t^{-1} F_t, \tag{4.7}$$

where A is the column vector of item and ability parameter estimates of length $2n+N$, B is the matrix of second order partial derivatives of dimension $(2n+N) \times (2n+N)$, F is the column vector of the first order derivatives of the likelihood function and is of length $2n + N$, and t indexes the iteration. The full representation of the Newton-Raphson equation is presented in Figure 4.1.

Since the dimensionality of all the terms in equation (4.7) are of order $2n + N$, when the number of items or number of subjects is reasonably large,

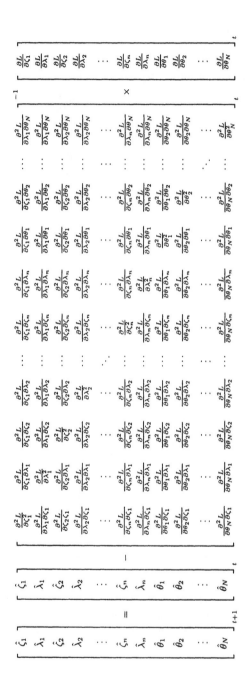

Fig. 4.1. Full Newton-Raphson equation.

matrices and vectors of considerable size result. These are beyond the capabilities of most digital computers, so that ways to reduce the dimensionality must be found. Three assumptions are made to obtain equations that are amenable to efficient solution. First, since each examinee is drawn at random from the population, they are independent objects. Thus, the cross derivatives between pairs of examinees will have expectations of zero, and the corresponding terms can be eliminated. In the lower right-hand $N \times N$ submatrix of B, terms will appear only on the diagonal. Second, there is no reason to believe that there should be any covariation between an individual examinee and either of the parameters of a given item. Thus, examinees and items can be assumed to be independent, and the corresponding cross derivatives can be eliminated. As a result, the upper right-hand $2n \times N$ submatrix and the lower left-hand $N \times 2n$ submatrix of the B matrix will be zero filled. These two assumptions have greatly simplified the internal structure of equation (4.7), but have not reduced its dimensionality. The simplified equations are shown in Figure 4.2.

In the equation shown in Figure 4.2, there is a $2n \times 2n$ submatrix in the upper left-hand corner of matrix B consisting of the derivatives corresponding to all pairwise combinations of the $2n$ item parameters. In the lower right-hand corner there is an $N \times N$ diagonal submatrix containing one term for each examinee. From the structure of the B matrix it can be seen that the items and the examinees have been orthogonalized. But, a $2n \times 2n$ submatrix is associated with the item parameters. Again when n is large (e.g., $n > 50$), this presents computational difficulties. The final simplification assumes that the items are independent. Thus, the corresponding second order derivatives of the likelihood function can be removed, leaving only the 2×2 submatrices for each item. It should be noted that within an item there will be covariation between an item's parameters, and consequently, there will be cross derivative terms in the Hessian or information matrix. The Newton-Raphson equation resulting from these three simplifying assumptions is shown in Figure 4.3.

The structure of the B matrix is now very sparse, containing 2×2 submatrices along the diagonal of the upper left-hand $2n \times 2n$ submatrix. There is one 2×2 submatrix for each item and the estimation of the item parameters across items has been orthogonalized. Inspection of the derivatives inside each 2×2 submatrix reveals that they are identically the same second order derivatives of the likelihood function as those for a single item presented in Chapter 2. In addition, there is one term for each of the N examinees. If the equation in Table 4.3 were solved, it would still involve the inversion of a $2n + N$ matrix, even if it is a very sparse matrix. What Birnbaum did was to take advantage of the fact that both items and examinees had been orthogonalized in this equation.

It is at this point where the Birnbaum (1968, p. 420) paradigm comes into play. He proposed using a "back and forth" two-stage procedure to solve for the estimates of the $2n + N$ parameters. In the first stage, the item pa-

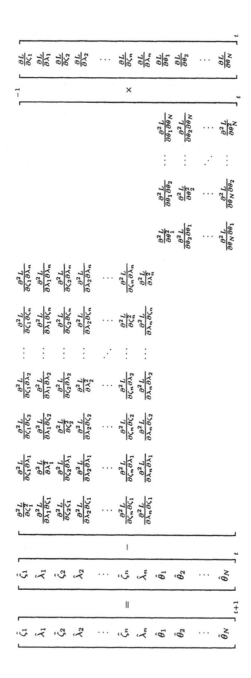

Fig. 4.2. Newton-Raphson equation under assumption of item by examinee independence.

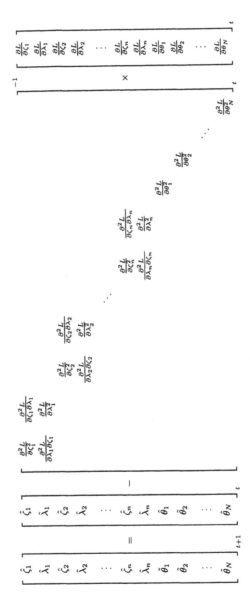

Fig. 4.3. Newton-Raphson equation under assumption of independence of items and examinees.

rameters are estimated assuming the examinee's abilities are known. In the second stage, the examinee's abilities are estimated assuming that the item parameters are known. To initiate the process, a rough estimate of each examinee's ability is obtained. The standardized raw test score is commonly used as the initial "known" value of an examinee's ability, for example, see Wood, Wingersky, and Lord (1976). Now, if each item is considered separately, the single item, known ability situation presented in Chapter 2 holds. Then, after having replaced the second derivatives by their expectations, there is a separate Newton-Raphson or Fisher scoring equation for each item, as shown below, to be solved iteratively for the item parameters $\hat{\zeta}_i$, $\hat{\lambda}_i$:

$$\begin{bmatrix} \hat{\zeta}_i \\ \hat{\lambda}_i \end{bmatrix}_{t+1} = \begin{bmatrix} \hat{\zeta}_i \\ \hat{\lambda}_i \end{bmatrix}_t - \begin{bmatrix} E\left(\frac{\partial^2 L}{\partial \zeta_i^2}\right) & E\left(\frac{\partial^2 L}{\partial \zeta_i \lambda_i}\right) \\ E\left(\frac{\partial^2 L}{\partial \lambda_i \zeta_i}\right) & E\left(\frac{\partial^2 L}{\partial \lambda_i^2}\right) \end{bmatrix}_t^{-1} \begin{bmatrix} \frac{\partial L}{\partial \zeta_i} \\ \frac{\partial L}{\partial \lambda_i} \end{bmatrix}_t . \qquad (4.8)$$

One such set of equations is solved individually for each of the n items. At this point a set of $2n$ item parameter estimates is available. In the second stage, these parameter estimates can be treated as the "true" item parameters, and the situation of Chapter 3 for estimating an examinee's ability holds. Now, for each of the N examinees, a Newton-Raphson equation can be established to be solved for the maximum likelihood estimates of ability. Again, the observed data in the second derivatives are replaced by their expectations under the Fisher scoring method. These equations are as follows:

$$[\hat{\theta}_j]_{t+1} = [\hat{\theta}_j]_t - \left[E\left(\frac{\partial^2 L}{\partial \theta_j^2}\right)\right]_t^{-1} \left[\frac{\partial L}{\partial \theta_j}\right]_t . \qquad (4.9)$$

After this estimation procedure has been performed for each examinee, a vector of maximum likelihood estimates $\hat{\theta}_j$ of length N is obtained.

At this point a single cycle of the Birnbaum paradigm has been completed, and the initial crude estimate of each examinee's ability has been replaced by the second stage set of maximum likelihood estimates $\hat{\theta}_j$. In addition, the first stage set of maximum likelihood estimates of the parameters ζ_i, λ_i are available for each of the n items. However, the metric of the ability estimates $\hat{\theta}_j$ is unique only up to a linear transformation (Lord & Novick, 1968, p. 366; Wood, Wingersky, & Lord, 1976). In IRT, this is known as the "identification problem." Since the location and scale of this metric is unknown, there are only $N - 2$ independent $\hat{\theta}_j$, even though N values were obtained. Thus, it is necessary to create values for the location and scale of the obtained ability metric, that is, "anchor" the $\hat{\theta}_j$. How this is done is somewhat arbitrary and a number of different schemes are available. In the LOGIST program, the anchoring of the metric is accomplished by computing the mean and variance of the examinee's $\hat{\theta}_j$ values and making a standard score transformation $\hat{\theta}_j^* = (\hat{\theta}_j - \bar{\hat{\theta}})/S_{\hat{\theta}}$ at the end of each cycle. After making the transformation, Wood, Wingersky, and Lord (1976) also adjust the item parameters for this scaling using the following equations:

$$\hat{b}_i^* = (\hat{b}_i - \bar{\hat{\theta}})/S_{\hat{\theta}}, \quad \hat{a}_i^* = S_{\hat{\theta}}\hat{a}_i.$$

An overall convergence criterion is needed to determine when a sufficient number of cycles have been performed and a final set of item and ability parameter estimates have been obtained. Such a criterion is arbitrary. Wood, Wingersky, and Lord (1976) use the likelihood function itself to determine convergence. This requires using the $\hat{\zeta}_i$ and $\hat{\lambda}_i$ as well as the $\hat{\theta}_j$ to numerically evaluate the likelihood function [equation (4.3)] at the end of each cycle. The difference in the likelihoods between two successive cycles then must be less than a specified value to terminate the process.

At the beginning of the second cycle, the rescaled maximum likelihood estimates $\hat{\theta}_j^*$ are considered to be the "known" ability scores of the N examinees. Then, in the first stage, the item parameters are estimated via the Newton-Raphson procedure for each item. Next, the new item parameter estimates are treated as the "true" values and new ability estimates computed for each examinee. With the new set of $\hat{\theta}_j$, $\hat{\zeta}_i$, $\hat{\lambda}_i$ is in hand, the convergence criterion is evaluated. If the criterion is met, $\hat{\zeta}_i$, $\hat{\lambda}_i$ can be adjusted for the final rescaling of the $\hat{\theta}_j$. If not, the procedures for a cycle are repeated until convergence is reached or the process is arbitrarily terminated if more than a specified number of cycles have been performed. It should be noted that, if the identification problem is not resolved, the item and ability parameter estimates will "drift" off towards infinity as the number of cycles increases and convergence will never be achieved.

The Birnbaum JMLE paradigm has several important features of interest. First, there is an overall iterative process that is performed in stages. Within the first stage, the items are treated as being independent, and a Newton-Raphson procedure, as described in Chapter 2, is performed for each item with the ability scores assumed to be known. Within each second stage, the examinees are considered independent, and a Newton-Raphson procedure, as described in Chapter 3, is performed for each examinee with the item parameters assumed to be known. Second, as Sanathanan (1974) and Mislevy and Bock (1984) have pointed out, the estimation procedure is like a "fixed effects" ANOVA in that items and examinees have been selected a priori. Because of this, the parameters of the n items and the N examinee parameters are the only ones of interest, and no inferences can be made beyond the test items and examinees actually used. Third, it can be employed under one-, two-, and three-parameter ICC models. Fourth, the anchoring procedure employed plays an important role in determining the metric of the obtained parameter estimates. Fifth, because the equations solved for the item parameter estimates are the same as those for a single item, the formulas for the large sample variances of the item parameter estimates will be the same those of Chapter 2. Also, the χ^2 statistic for the goodness of fit of the item characteristic curve specified by $\hat{\zeta}_i$, $\hat{\lambda}_i$ and the data remains the same. Sixth, the procedure does not actually estimate the item and ability parameters simultaneously. Due to the several independence assumptions made and the "back and forth" manner with which the two stages are employed within a cycle, the paradigm simply jointly estimates them; hence, the JMLE label.

In parallel with the theoretical developments there has been an evolutionary series of computer programs based on the JMLE paradigm. These programs (Wingersky, Lees, Lennon, & Lord, 1969; Lees, Wingersky, & Lord, 1972; Wingersky & Lord, 1973) that have eventuated in the program known as LOGIST (Wood, Wingersky, & Lord, 1976; Wingersky, Barton, & Lord, 1982; Wingersky, Patrick, & Lord, 1999), which is widely distributed. The LOGIST program is a generalized program capable of using a one-, two-, or three-parameter (logistic) ICC model, with emphasis upon the latter. Any empirical evaluation of the JMLE procedure and the three ICC models is intimately interwoven with their implementation in the LOGIST computer program used to analyze the data sets.

4.2.1 Some Additional Facets of the Birnbaum Paradigm

Although the JMLE paradigm is straightforward, there are a number of technical and practical considerations involved in its implementation and application. In the present section, a number of these considerations are discussed in order to provide a context for the use of this paradigm and the interpretation of the results it yields.

Use of Grouping in the Joint Estimation Paradigm. The equations used to estimate the parameters of a single item in Chapter 2 and those used for an item within this chapter are basically the same. They differ only in respect to how the summation over ability is handled. In the single-item, known-ability-level situation of Chapter 2, it was assumed that a group of f_g subjects existed at each ability level (n.b., subscript g is now used). Of these, r_g responded correctly to the item, and $r_g = f_g p_g$. The likelihood function was a product over the G ability levels

$$\text{Prob}(U|\theta) = \prod_{g=1}^{G} P_i^{r_g}(\theta_g) Q_i^{f_g - r_g}(\theta_g).$$

The likelihood equations to be solved for the item parameter estimates under a logistic model are:

$$\sum_{g=1}^{G} f_g W_{ig} \left(\frac{p_{ig} - P_{ig}}{P_{ig} Q_{ig}} \right) = 0$$

$$\sum_{g=1}^{G} f_g W_{ig} \left(\frac{p_{ig} - P_{ig}}{P_{ig} Q_{ig}} \right) \theta_g = 0. \tag{4.10}$$

In these equations, the sums were over the G ability levels, and the observed proportion of correct response appears in the term within the parentheses.

When grouping is not employed, the likelihood equations solved for the item parameters under the JMLE procedure described above and a logistic model are:

$$\sum_{j=1}^{N} W_{ij} \left(\frac{u_{ij} - P_{ij}}{P_{ij}Q_{ij}} \right) = 0$$

$$\sum_{j=1}^{N} W_{ij} \left(\frac{u_{ij} - P_{ij}}{P_{ij}Q_{ij}} \right) \theta_j = 0.$$

$$(4.11)$$

The summation in these equations is over the number of examinees and the item response u_{ij} appears in the term in the parentheses. One of the problems associated with these latter equations is that one is fitting the item characteristic curve to zero-one data. Instead of the observed proportions of correct response following an ogive form, the zeros and ones scatter themselves above and below the ogive. In the aggregate, the item responses will yield the appropriate item characteristic curve. However, one is making the fitting process more difficult than it need be since getting a good fit using $(u_{ij} - P_{ij})/P_{ij}Q_{ij}$ is more tedious than using $(p_{ig} - P_{ig})/P_{ig}Q_{ig}$. In the former, there will generally be a large discrepancy to be averaged out over all examinees, while the latter will generally show a reasonably small discrepancy. The result is that the Newton-Raphson procedure should converge more rapidly when grouping and equations of the form of equation (4.10) are used. Bock (1972) recognized this in another context, and Kolakowski and Bock (1973a) took advantage of it in their NORMOG program. Bock (1972) suggested the ability scale be divided into intervals where these intervals were to be small enough that a local independence assumption held for those ability levels in the neighborhood of each θ_g. When this is done, the summations in the terms of the Newton-Raphson equations for items are over the G intervals rather than the N examinees. Also the first derivatives will involve $(p_{ig} - P_{ig})$ rather than $(u_{ij} - P_{ij})$. The Newton-Raphson equations for the estimation of each examinee's ability are unchanged. The early LOGIST program did not group the examinees when estimating item parameters. Wood, Wingersky, and Lord (1976) reported that LOGIST usually takes between 10 and 15 cycles of the joint estimation procedure to reach a suitable convergence criterion. Kolakowski and Bock (1973a) reported that NORMOG usually converged in four to six cycles. While the absence of the estimation of the "guessing" parameter c_i in NORMOG accounts for most of the difference, grouping undoubtedly also contributes. These authors also used the grouping procedure in the LOGOG program (Kolakowski & Bock, 1973b) since it greatly reduces the amount of computation within a stage. Under the non-grouped procedure the P_{ij} value must be computed for each examinee, while in the grouped case it is only computed once for the f_g examinees in the group. In the absence of other compelling arguments, it would appear that grouping the examinees into ability intervals for item parameter estimation purposes results in a more efficient computing procedure. The version

5.2 of the LOGIST program (Wingersky, Barton, & Lord, 1982) provides for grouping of examinees as an analysis option.

Bounding the Estimates. Items which are either missed by all examinees or answered correctly by all examinees lead to values of $\hat{\lambda}_i$ that are infinite and cannot be computed. This is a common problem with maximum likelihood estimates and was the basis for the well-known Berkson-Fisher controversy (Berkson, 1954). Also, if an examinee answers all items correctly or incorrectly, their ability estimate will go to positive or negative infinity, respectively. Common practice is to remove such items and examinees from the data set via data screening procedures (see Wood, Wingersky, & Lord 1976; Wright & Panachapakesan, 1969). Preferably, this screening is done for these obviously extreme cases before the first stage is initiated. Similar problems can occur with items that only a few persons either answer correctly or miss or with examinees who either answer only a few items correctly or incorrectly. A more common case is one in which an item has low or no discrimination, and this causes the item difficulty parameter to move towards infinity. In addition, it is not possible to predict whether deviate parameter estimates will be produced by a given data set. In LOGIST, upper and lower bounds are placed upon the values of $\hat{\alpha}_i$, $\hat{\beta}_i$, and $\hat{\theta}_i$. When these bounds are breached, remedial procedures should be performed, such as substituting a fixed value for the parameter for the remaining cycles or the offending item or examinee can be removed at that point. Each of the available computer programs implementing the Birnbaum paradigm differs with respect to how it copes with this problem. Wright (1977a, p. 103) severely criticizes the two- and three-parameter item characteristic curve models for needing to have such bounds set. In fact, Wright and Panachapakesan (1969) used similar screening and bound setting procedures in the BICAL program as even under the Rasch model an occasional item difficulty can go off towards infinity.

The Heywood Case. When the two- and three-parameter ICC models are employed in the JMLE procedure, a phenomenon occurs in certain data sets: discrimination estimates for one or more items can become very large which, in turn, results in large values of the ability estimates for examinees answering those items correctly. In successive cycles, both the discriminations and the ability estimates go towards infinity, and the overall solution diverges, that is, "blows up." Wingersky (1983) indicated that in the LOGIST program upper limits on the values of the discrimination estimates must be imposed to handle such data sets. The problem does not occur under a one-parameter logistic model as all of the discrimination indices are fixed at unity. Much of Wright's (1977a) criticism of the two- and three-parameter models is based upon this phenomenon. He stated that it always happens under these models and, therefore, it is impossible to estimate the discrimination parameter. However, there is no evidence that it occurs in all data sets. Recently, it has been recognized that this problem is the Heywood case of factor analysis

(Swaminathan & Gifford, 1985). Therefore, the discussion will digress for a bit to relate factor analysis to item parameter estimation.

Lawley's (1944) second paper on test theory contained an approach to the simultaneous estimation of the item discrimination indices for all n items in a test. Prior to this article, the basic theory of item analysis was to estimate the parameters of each item separately. To accomplish the simultaneous estimation of the item discrimination parameters α_i, he developed maximum likelihood factor analysis. Using this procedure, he factor analyzed the matrix of interitem tetrachoric correlation coefficients. The item discrimination indices α_i were shown to be the item's loading on the first common factor. In addition, Henrysson (1962) has shown that these factor loadings are highly correlated with the biserial correlation between the dichotomized item response variable and the total test score. Tucker (1946) has shown that this biserial correlation is directly related to the discrimination parameter of IRT under an assumption of bivariate normality [see equations (1.9) and (1.10)]. Lawley (1944) did not estimate the item difficulties; however, later authors (Tucker, 1946; Lord, 1952) used the normal deviate corresponding to the proportion of subjects in the total group answering the items correctly as the "standard" item difficulty measure. This factor analytically based approach to the simultaneous estimate of the classical item parameters π, ρ_{bis} has been called the "heuristic approach" by Bock and Lieberman (1970). Lawley's 1944 paper is also important in the field of factor analysis as it introduced maximum likelihood factor analysis.

Mislevy (1986a) indicted that, under the common factor model, the Heywood case occurs when one or more unique variances takes a value of zero. He indicated that under maximum likelihood factor analysis, the unique variances do not appear as parameters to be estimated, hence their values are implied through the values of the item discrimination indices. As a result, the Heywood case becomes apparent when one or more discrimination estimates becomes infinite.

Missing Data. As presented above, the JMLE procedure makes no allowances for missing data. In most practical testing situations, students will omit occasional items, and slow workers may not finish the test. Common practice is to simply score such items as incorrect, however, which is improper. In addition, items omitted and items not reached involve quite different psychological processes. An item may not be reached because the test was speeded or the student didn't reach it within a reasonable time. In the case of items not reached, the LOGIST program estimates an examinee's ability without using the items not reached. However, the fact that an examinee omitted an item contains information, indicating the examinee did not know the correct answer; hence, under the three-parameter model, the examinee's probability of correct response was c_i. Lord (1974a) developed a joint estimation procedure in which omitted items were taken into account. He replaced the likelihood function [equation (4.2)] by a new function,

$$\text{Prob}(V|\theta) = \prod_{j=1}^{N}\prod_{i=1}^{n} P_{ij}^{v_{ij}} Q_{ij}^{1-v_{ij}}, \qquad\qquad (4.12)$$

where V is the matrix $[v_{ij}]$ and $v_{ij} = 1$, 0, or c_i, depending upon whether the jth examinee's response to the ith item was correct, wrong, or omitted. Using essentially the same mathematical procedures as employed in the maximum likelihood estimation procedures described above, Newton-Raphson equations were established for the items and ability parameters. Lord (1974a) states that these parameter estimates are not maximum likelihood estimates. They are close relatives, however, and if there are no omits in the complete data set, the estimation equations reduce to those for maximum likelihood. In fact, in later work (Wood, Wingersky, & Lord, 1976), the estimates obtained in both cases are referred to as the maximum likelihood estimates. If the omitted responses are filled in at random, then the usual likelihood function [equation (4.2)] can be employed. What interested Lord (1974a) was determining which of these procedures was preferable. Item parameter and ability estimates were obtained via the new procedure for four different data sets involving large numbers of examinees and items (see Lord, 1974a, for the exact definition of these data sets). Two of the tests were analyzed via the usual JMLE procedures with the omits randomly filled in by 1s or 0s. Lord found very high correlations between the item parameter estimates, .995, .996 and .990 for $\hat{a}_i$, $\hat{b}_i$, $\hat{c}_i$, respectively, as estimated by the new method and the usual method with random fill-in of omitted responses. Using the same procedure as employed in Lord (1970), the frequency distribution of estimated ability $\hat{\theta}$ under the new procedure was compared with that obtained via a transformation of the true score distribution. Again, very good agreement was obtained. Finally, Lord was able to show analytically that the new method had a smaller asymptotic variance of $\hat{\theta}$ than does the maximum likelihood joint estimation procedure when filled- in data is used.

The Obtained Ability Metric. Although an underlying ability scale metric is postulated within IRT, recovering this metric from item response data is a difficult task. A fundamental problem with the JMLE paradigm is that the obtained metric is unique only up to a linear transformation. The choice of an anchoring procedure to handle this indeterminacy is arbitrary and depends upon the item characteristic curve model as well as the implementer's point of view. Both the origin and unit of measurement must be fixed in order to anchor the ability scale metric. In the LOGIST computer program this is accomplished by standardizing the ability scores of the N examinees. Thus, the midpoint of the ability metric is zero and the unit of measurement is $1\sigma_{\hat{\theta}}$. In the present section, this standard score metric will be called the LOGIST metric and denoted by θ_N. The theoretical underlying ability metric will be denoted by θ_T and referred to as the theta metric. It has a midpoint of zero and a unit of measurement of one. Since the present interest is in determining if the LOGIST computer program recovers the θ_T metric, a means is needed for evaluating the obtained results. This will be accomplished by transforming the ability scores in the LOGIST metric to the θ_T metric. A transformation has been given by Loyd and Hoover (1980) that is based upon matching the item characteristic curve of a given item expressed in two different metrics. When matched

$$P(\theta_1) = P(\theta_2)$$

and

$$\alpha_1(\theta_1 - \beta_1) = \alpha_2(\theta_2 - \beta_2)$$

holds for all equivalent pairs of θ_1, θ_2. Solving for θ_1 yields

$$\theta_1 = \frac{\alpha_2}{\alpha_1}\theta_2 + \left[\beta_1 - \frac{\alpha_2}{\alpha_1}\beta_2\right] \tag{4.13}$$

which holds when parameters are employed.

Although this equation holds for individual items, the average values of α_i, β_i over the n items would be used when dealing with a whole test. Thus, the basic metric transformation equation to be used with test results is

$$\theta_1 = \frac{\bar{\alpha}_2}{\bar{\alpha}_1}\theta_2 + \left[\bar{\beta}_1 - \frac{\bar{\alpha}_2}{\bar{\alpha}_1}\bar{\beta}_2\right] = A\theta_2 + K. \tag{4.14}$$

This equation will be used below to examine the agreement of the LOGIST metric and the underlying θ_T metric.

Data Set 1. The first data set was designed to illustrate the impact of the LOGIST anchoring procedure on the metric recovery. Simulated item response data for a population of 1100 examinees whose ability scores in the θ_T metric are known will be used. The known examinee ability scores were uniformly distributed over the ability scale from -2.5 to 2.5. There were 100 examinees at each of 11 score points spaced .5 units apart. The frequency distribution of these ability scores had a mean of zero and a standard deviation of 1.581. The simulated test consisted of 20 equivalent items ($\alpha_T = .5$, $\beta_T = 0.0$) based upon a normal ogive model for the common item characteristic curve. Given these specifications, the GENIRV computer program (Baker, 1986b) was used to generate the dichotomously scored item responses of the 1100 examinees to the 20 items. The use of simulated data here is advantageous since the underlying parameter values are known. In the LOGIST analysis of this data, 33 examinees who had raw scores of 0 or 20 were removed leaving a group size of 1077 in which $\sigma_{\theta_T} = 1.558$. The summary statistics for the parameter estimates were:

$$\bar{\hat{\beta}}_N = -.0182 \quad S_{\hat{\beta}_N} = .034 \quad \bar{\hat{\alpha}}_N = .858$$

$$S_{\hat{\alpha}_N} = .077 \quad \bar{\hat{\theta}}_N = 0.0 \quad S_{\hat{\theta}_N} = 1.00.$$

Clearly, something is amiss in value of $\bar{\hat{\alpha}}$, which does not agree with the underlying value of .5. The discrepancy is due to the LOGIST anchoring procedure, in which the ability score estimates were standardized using $\sigma_{\hat{\theta}_N}$. The effect of this is to compress the ability scale and thus elevate the obtained value of $\bar{\hat{\alpha}}_N$. The value of $\bar{\alpha}_T$ can be obtained from $\bar{\alpha}_N$ via

$$\bar{\alpha}_T = \bar{\alpha}_N \left(\frac{\sigma_{\theta_N}}{\sigma_{\theta_T}} \right). \tag{4.15}$$

Using the obtained value of $\bar{\hat{\alpha}}_N$, $\sigma_{\theta_T} = 1.558$, and $\sigma_{\theta_N} = 1.0$ yields

$$\bar{\hat{\alpha}}_T = \bar{\hat{\alpha}}_N \left(\frac{1.0}{1.558} \right) = 0.551$$

which is close to the underlying value of $\bar{\alpha}_T$. Thus, in this data set, the unit of measurement yielded by LOGIST encompasses too many units of the underlying θ_T metric, and this fact will need to be taken into account when transforming the θ_N metric.

The summary statistics of the obtained ability estimates based upon the 1077 examinees can be transformed to the θ_T metric using the following values in equation (4.14):

$$\theta_1 = \bar{\hat{\theta}}_T \qquad \sigma_{\theta_T} = 1.558 \qquad \alpha_1 = \bar{\alpha}_T = .5 \qquad \beta_1 = \bar{\beta}_T = 0$$
$$\theta_2 = \bar{\hat{\theta}}_N \qquad \sigma_{\hat{\theta}_T} = 1.0 \qquad \alpha_2 = \bar{\alpha}_N = .858 \qquad \beta_2 = \bar{\beta}_N = -.0182$$

and the estimated mean is

$$\bar{\hat{\theta}}_T = \frac{.858}{.5}(0) + \left[0 - \frac{.858}{.5}(-.0182) \right] = .031.$$

The standard deviation obtained via equation (4.15) is

$$S_{\hat{\theta}_T} = \frac{.858}{.5}(1.0) = 1.716.$$

The transformed mean is a reasonably close estimate of the underlying value of .011, and the standard deviation overestimates the underlying value of 1.558 by about 10%.

Data Set 2. The JMLE procedure can only ascertain the relative locations of $\hat{\theta}$ and $\hat{\beta}$ rather than their absolute locations on the theta metric. A second simulated data set will be used to illustrate this phenomena. There were 1100 examinees whose true ability scores were normally distributed over the θ_T scale with a mean of $\bar{\theta}_T = -.5$ and unit variance. The simulated test consisted of 40 items based upon a normal ogive model for the item characteristic curve. The random item discrimination parameters α_{T_i} were uniformly distributed over the range .19 to .39 with $\bar{\alpha}_T = .27$. The random item difficulty parameters β_{T_i} were normally distributed with $\bar{\beta}_T = .464$ and $\sigma_{\beta_T} = 1.084$. Again the GENIRV program was used to generate the binary item responses of the 1100 examinees based upon the 40 items defined by random pairing of the α_{T_i} and β_{T_i}. The data generation specifications yielded an absolute difference of .964 theta units between $\bar{\theta}_T$ and $\bar{\beta}_T$. The generated data was analyzed via the LOGIST computer program and the results were $\bar{\hat{\theta}}_N = -.016$,

$S_{\hat{\theta}_N} = 1.052$, $\bar{\hat{\alpha}}_N = .321$, $S_{\hat{\alpha}_N} = .083$, and $\hat{\beta} = .830$, $S_{\hat{\beta}_N} = 1.000$. Since LOGIST anchors its metric at $\bar{\hat{\theta}}_N = 0.0$, the true value of $\bar{\theta}_T = -.5$ was not recovered. However, the difference $\bar{\hat{\theta}} - \bar{\hat{\beta}} = -.846$ indicates $\bar{\hat{\beta}}_N$ is greater than $\bar{\hat{\theta}}_N$, but the difference is also an underestimate of the true value. Rewriting equation (4.14) slightly yields

$$\bar{\theta}_T - \bar{\beta}_T = \frac{\bar{\hat{\alpha}}_N}{\bar{\alpha}_T}(\bar{\hat{\theta}}_N - \bar{\hat{\beta}}_N). \tag{4.16}$$

Substituting the following values

$$|\bar{\theta}_T - \bar{\beta}_T| = .964 \qquad |\bar{\hat{\theta}}_N - \bar{\hat{\beta}}_N| = |-.846|$$
$$\bar{\alpha}_T = .270 \qquad \bar{\hat{\alpha}}_N = .321$$

yields

$$|\bar{\theta}_T - \bar{\beta}_T| = \frac{.321}{.270}|-.846| = 1.0058$$

which is reasonably close to the true absolute difference of .964, given that there are estimation errors in the obtained results.

The above shows that the metric yielded by the LOGIST implementation of the JMLE paradigm is a function of the underlying ability metric. It should be clear, however, that the anchoring procedure based upon standardizing the ability estimates has the unfortunate property of making the obtained ability metric dependent upon the variability of the frequency distribution of the examinees' ability scores. When $\sigma_{\theta_T} \neq 1$, the anchoring procedure standardizes the ability estimates so that $\sigma_{\hat{\theta}_N} = 1$, the consequence being that the numerical values of the item and ability parameter estimates do not agree with the underlying values. In addition, data set 2 confirmed that the joint estimation paradigm recovers the relative, not the absolute, location of $\bar{\theta}$ and $\bar{\beta}$ on the underlying ability scale. In the present examples, the underlying values of the item and ability parameters were known, and the transformation to the underlying metric was direct. Yet, in applied situations, these are not available, and the user of test results needs to be aware of the impact of the anchoring procedures upon the obtained metric. When the LOGIST-style anchoring procedure is used, the ability estimates will have mean zero and unit variance, and the item parameters will reflect the difficulty and discrimination of those items for the group of examinees at hand. If the test is administered to another group, the same situation prevails. The two sets of results are not directly comparable, however, since no linkage has been provided. Both ability scales will have mean zero and unit variance, yet the groups may differ considerably in true ability. Fortunately, the metric information is contained in the item discrimination and difficulty parameter estimates. Using an equating procedure such as the characteristic curve method due to Stocking and Lord (1983), one can put the results of one group into the metric of another if there are common items in the two tests. Once a common metric has been established, direct comparisons of the values of the parameter estimates across groups and items can be made. Baker, Al-Karni, and Al-Dosary (1991) have implemented this equating procedure in a FORTRAN computer program for the IBM PC series of micro-computers.

The Consistency Issue. The desirable properties of maximum likelihood estimators, such as consistency, efficiency, and sufficiency, are achieved when a finite set of parameters are estimated via a large set of observations. In the context of the JMLE procedure, the property of consistency has not been proved analytically for the two- and three-parameter ICC models. In simple terms, a consistent estimator is one in which $P(\hat{\mu} \to \mu) \to 1$ as $N \to \infty$, that is, the estimator approaches the parameter with probability approaching 1 as N increases. The basic problem in the present situation is that, as the number of observations increases, so does the number of parameters to be estimated. Fortunately, in most situations the number of items in a test is fixed; hence, the parameters of only the n items need be estimated. However, increasing the sample size results in a one-to-one increase in the number of ability parameters to be estimated. In situations similar to this, Neyman and Scott (1948) distinguished between structural parameters and incidental parameters. Where these two types of parameters were defined by Andersen (1970) in the following fashion:

> ... let X_1, X_2, X_3, ..., X_j be independently distributed random variables and let the distribution of X_j depend upon the parameters τ and θ_j. The τ is the same regardless of j, while θ_j changes with j. The τ is a structural parameter and the θ_j are the incidental parameters.

In the present context, the item parameters α_i, β_i or ζ_i, λ_i are the structural parameters. The examinees' abilities θ_j are the incidental parameters since there is one such parameter for each examinee. Neyman and Scott (1948) showed that, in the presence of incidental parameters, the maximum likelihood estimates of the structural parameters need not be consistent. They also indicated that the property of efficiency may not hold, even when the estimates of the structural parameters were consistent. Since the JMLE procedure is exactly the situation dealt with by Neyman and Scott, the consistency of the item parameter estimates are suspect. There are, however, situations in which the maximum likelihood estimators of the item parameters will be consistent. Kiefer and Wolfowitz (1956) have shown that, if the incidental parameters are independently distributed chance variables with a common unknown distribution function, then consistent estimates of the structural parameters can be obtained. One additional result in the Kiefer and Wolfowitz (1956) work was that the maximum likelihood estimation $\hat{F}_0$ of the distribution function F_0 of the incidental parameters converges to F_0 at every point of continuity with probability one (also, see Mislevy, 1984). Neyman and Scott (1948) devised a procedure for coping with the deficiencies of maximum likelihood estimates of the structural parameters in the presence of incidental parameters. This procedure has become known as conditional maximum likelihood estimation. Andersen (1970, 1972, 1973a, 1977) has employed this approach in the estimation of item parameters. Basically, minimal sufficient statistics are substituted for the incidental parameters. Since such statistics are independent of the incidental parameters, the estimates of the structural (item) parameters are also independent of the incidental parameters. Thus, the conditional maximum likelihood estimation procedure could be used to estimate the item parameters. Since the only ICC model for which conditional maximum likelihood estimation is appropriate is the one-parameter logistic ICC (Rasch) model, an extended discussion will be deferred until Chapter 5. However, the reader should be reminded that the consistency issue remains an important problem when maximum likelihood estimation is used. The marginal maximum likelihood procedures of Chapter 6 were developed specifically to resolve this issue.

Although all the statistical properties of the item parameter estimates yielded by the JMLE paradigm have not been demonstrated analytically for the three-parameter model, empirical evidence suggests that consistency may conform to the theoretical expectations. Swaminathan and Gifford (1983) performed a simulation study examining the consistency of the item parameter estimates under a three-parameter model as yielded by the LOGIST program. Tests of 10, 15, and 20 items and sample sizes of 50, 200, and 1000 examinees were used. Swaminathan and Gifford found that, as the number of items and the sample size increased, the regression lines for the relation of the difficulty and discrimination estimates to their underlying parameter values differed only slightly from the theoretical 45 degree line. Thus, the empirical data suggested that the estimates of item difficulty and discrimination in the three-parameter model are consistent. Results for the guessing parameter were not reported.

4.2.2 Quality of the Parameter Estimates

Item Parameter Recovery Studies. A number of studies have examined the degree to which the JMLE paradigm can recover the underlying item parameters. Typically, these studies used simulated data, so that the estimates and the true values can be related. Lord (1975a) used simulated data for a 90-item test and 2995 examinees. As is usual in these studies, the item parameter values under a three-parameter (logistic) model were matched to those of an existing test. The correlation between the discrimination estimates and their parameters was .920. When c was overestimated, a also tended to be overestimated. Item difficulty was overestimated for larger absolute values of b, $|b| > 3$, and slightly underestimated for medium values of b. The correlation between the difficulty estimates and the parameter values was $r = .988$. In the case of both a and b, the estimates were scattered rather tightly about the 45 degree theoretical relationship line. The estimates of the c parameter were widely scattered about the corresponding parameter values and generally underestimated the parameter values.

Hulin, Lissak, and Drasgow (1982) investigated the recovery of item parameters under the two- and three-parameter models. Using the root mean square of the difference (RMSD) between the recovered and empirical ICCs as the criterion, they found that minimum test lengths and sample sizes depended upon the model. Under a two-parameter model, 30 items and 500 examinees gave satisfactory results in terms of RMSD, while the three-parameter model required 60 items and 1000 examinees. They also found that a tradeoff between test length and sample size gave comparable results, at least for the data used. The correlation between the estimates and the parameter values was also higher for the two-parameter model than for the three-parameter model. When $N > 500$, the correlation of $\hat{\alpha}$ and α was roughly .9 for the two-parameter model and .5 for the three-parameter model. Under a two-parameter model, $r_{\hat{\beta}\beta}$ was greater than .94 for all sample sizes and test lengths. Under a three-parameter model, $r_{\hat{\beta}\beta}$ was greater than .94 only when $N \geq 1000$ and $n > 30$ items.

Bias of the Item Parameter Estimates. Lord (1983) derived expressions for the asymptotic bias of the maximum likelihood estimators of the item parameters under a three-parameter model. The bias expressions were for a single item, assuming known examinee ability parameters. To investigate the characteristics of the asymptotic biases, Lord used simulated data having parameter values roughly equal to the values of a, b, c, and θ_j yielded by 2995 examinees on a 90-item verbal Scholastic Aptitude Test. In the case of item difficulty, easy and medium difficulty items had a negative bias and only difficult items were positively biased. The bias in the estimates of item discrimination was always positive. The bias in the estimate of c was negative for all items. In general, if an item's parameter estimates had large standard errors, the biases were also large. However, the magnitude of an estimator's bias was typically was about .1 of its standard error and very seldom was greater that .2 of its standard error. Lord concluded that because the standard errors are inversely proportional to the sample size, when N is large, the numerical value of the bias is probably negligible. These results, however, are for a restricted set of conditions. Swaminathan and Gifford (1983) also examined the bias of the item parameters yielded by LOGIST under the three-parameter model. They found that, for 20 replications of a 20-item test and a sample of 200 examinees, the LOGIST program yielded an over estimate of small values of a and accurate estimates of large values of a. The overall bias of the discrimination estimates was small. In the case of item difficulty, they found that LOGIST yielded a slight underestimate of negative values of b and close estimates of large positive values of b. The bias in the estimate of c was generally negative.

Standard Errors of the Item Parameter Estimates. The characteristics of the standard errors of the item parameter estimates are also an important issue. The large sample variances of the item parameter estimates are given by the inverse of the information matrix used in the Fisher scoring procedure. Thus, they are readily available as a byproduct of the item parameter estimation process. The values obtained in practice are those yielded by the final cycle of the JMLE procedure when the overall convergence criterion has been met.

Thissen and Wainer (1982) investigated the asymptotic standard errors of the item parameters for the one-, two-, and three-parameter ICC models under the assumption that the abilities of the examinees were known and normally distributed $N(0, 1)$. Tables of the minimum asymptotic standard errors were reported for combinations of parameter values under each of the three models. Plots of the standard error of $\hat{b}$ for all three models showed a concave surface that increased at the extremes of the ability scale.

An interesting set of results was given by the two-parameter and the three-parameter models when $c = 0$. Even though the numerical values of a and b would be the same, the information matrices are not. The three-parameter matrix still has a row and column corresponding to the c parameter. When an item was easy and had low discrimination, the standard errors under the two-parameter model were roughly .09 of those reported for the three-parameter model. Clearly, the two-parameter model and the three-parameter model with $c = 0$ are not the same with respect to the standard errors of the item parameter estimates. The asymptotic standard errors for item difficulty under the one-parameter model were consistently smaller than those obtained for the other two models. In particular, the increase in standard error with the departure of item difficulty from zero was much less pronounced.

Based upon these results, Thissen and Wainer felt that the three-parameter model was inferior to the other two models. The standard errors of the item parameters under a three-parameter model were acceptable only in the middle of the ranges of b and a with low values of c. Thissen and Wainer concluded:

> . . . the use of an unrestricted maximum likelihood estimation for the three-parameter model either yields results too inexact to be of any practical use, or requires samples of such enormous size so as to make them prohibitively expensive. (p. 403)

One of the limitations of the equations for the asymptotic standard errors of the item parameters used above is that they were based upon a single item and known ability. Lord and Wingersky (1985) derived a method for computing the asymptotic sampling variance-covariance matrix under JMLE when all parameters are unknown. They employed this method to study the asymptotic standard errors of the three-parameter model (Wingersky & Lord, 1984). Parameter sets for simulated tests of 45 and 90 items were established in which the parameter values of 15 items were replicated. Two ability distributions, bell shaped and uniform were used with each of the two sample sizes, $N = 1500$ and $N = 6000$. Only the standard errors of the basal 15 items were reported. The scale used required that the mean of the difficulty parameters of certain selected items be 0 (the origin) and that the difference between two such sets of selected items be 1 (the scale unit). This metric was called the "capital" scale. The standard errors were reported for the item parameter estimates a, b, c in this metric.

Lord and Wingersky (1985) concluded that, for the values of n and N employed, the standard errors of the item parameter estimates a, b, c varied inversely as the square root of N but were only slightly affected by changes in the number of items. Using a rectangular ability distribution yielded smaller standard errors for the item parameters than did doubling the number of items under a bell-shaped ability distribution. For low a's and for c's from items with $b - 2/a$ less than 1, the standard errors computed with a rectangular distribution of ability were nearly as low as the standard errors computed with a bell-shaped distribution and quadruple the number of examinees. This concurs with Baker's (1967) conclusion that, for item parameter estimation purposes, the uniform distribution of examinees over the ability scale is preferred.

De Gruijter (1985) derived equations for the asymptotic standard errors of the item difficulty under the Rasch model. He noted that the size of the standard errors yielded by these formulas and those due to Wingersky and Lord (1984) depend strongly upon the restrictions introduced to solve the identification problem.

Estimation of the c Parameter Under a Three-Parameter Model. One recurring theme in the IRT literature concerns the problems associated with the parameter c of the three-parameter model. For example, Kolen (1981) found for three tests that 92%, 53%, and 39% of the cs were not successfully estimated. Lord (1980) and many other authors have suggested that the c parameter cannot be estimated unless a considerable lower tail to the ICC is present in the range of ability scores employed. In addition, very few studies have plotted the proportions of correct response to an item as a function of ability in order to show that a three-parameter model is needed. McKinley and Reckase (1980) plotted such data for 50 items taken from the Iowa Tests of Educational Development and a sample of 1,999 examinees. Only 8 of the 50 items showed a lower tail, suggesting that a three-parameter model was needed. The data were analyzed using the LOGIST program and the fitted ICC's were graphed. For these 8 items, the scatter of the empirical proportions of correct response about the lower tail of the ICC was much larger than elsewhere along the curve and did not appear to be binomially distributed. These data suggest that c would be poorly estimated even for these items.

In order to explore the properties of the three-parameter model and LOGIST in depth, Lord (1975a) performed an extensive evaluation using artificial data. He used the values of the parameters estimates from an actual data set to drive the generation of the artificial data via a computer program due to Hambleton and Rovinelli (1973). The results of analyzing the artificial data via LOGIST were then compared with the same analysis of the actual data. A number of results of interest were obtained, some of which had also been observed in the earlier studies. The estimation of c, the lower asymptote, has been uniformly troublesome. Lord states:

> When items are easy, there is often no way to estimate c from the data, since even low ability examinees have a good chance of getting the right answer. ... Left to itself the maximum likelihood procedures would produce unacceptable values of $\hat{c}_i$.

Lord also found that LOGIST tended to underestimate the true value of c_i for reasons that were not apparent. Wood, Wingersky, and Lord (1976) also reported that many of the estimates of c are poorly determined and that any movement of their values during the joint estimation process must be severely limited. The index $b - 2/a$ (Lord, 1975a) identifies the point on the ability scale at which the ICC is within .03 of its lower asymptote (c). When the criterion for the $b - 2/a$ index is not met, the LOGIST program sets the value of c_i for the item to $\tilde{c}$, the mean of the values of c_i for those items where c_i was successfully estimated. Lord (1975a) concluded that c_i cannot be estimated if the index $(\hat{b}_i - 2/\hat{a}_i)$ is too negative. The LOGIST manual (Wingersky, Barton, & Lord, 1982) suggests an index value of -3.5 for samples of 2000 to 3000 and a value of -2.5 for smaller samples. Since the study due to Lord (1975a) involved 2995 examinees and 90 items, one would conclude that the estimation of c would degenerate markedly for small numbers of examinees. Thus, with smaller samples, $N \leq 1000$, the two-parameter model is probably a better model in practice from the point of view of item parameter estimation.

The difficulties associated with the estimation of c_i also carry over into the estimation of ζ_i and λ_i and, hence, of the item parameters a_i, b_i. Inspection of the terms in the Newton-Raphson equation for item parameter estimation given in Chapter 2 reveals terms for the cross derivatives $c_i a_i$ and $c_i b_i$. In addition, all derivatives involve the terms $(P_{ij} - c_i)$ and $(1 - c_i)$. As a result, when c_i is poorly estimated there will be an impact upon the estimation of the remaining item parameters. The ability estimates will be indirectly affected when the item parameters are in error. They will also be directly affected by the $(P_{ij} - c_i)$ and $(1 - c_i)$ terms that appear in the Newton-Raphson equation (3.15) for ability estimation. In his study, Lord (1975a) found that the item difficulty parameter b_i was estimated very accurately. This parameter is the mean of the logistic ogive, and means are usually accurately measured by large samples. Lord (1975a) reported a tendency to underestimate low values of a_i and to overestimate high values. The estimation of ability revealed that low ability levels $\theta < -3$ are estimated with large sampling errors. However, except for the extremes, $|\theta_j| > 3$, the agreement of the estimated and actual ability levels of the 2995 examinees was good. Despite these results, one should pay close attention when the three-parameter model is used. Examination of LOGIST results for a variety of data sets has shown that it uses the value of $\tilde{c}$ rather than an estimated value in a large proportion of the items. The $b_i - 2/a_i$ index is employed to determine when the value of $\tilde{c}$ is to be used and it appears to be a rather conservative criterion that favors items near the mean ability having high discrimination. As a result, very easy items or those with low discrimination will generally employ $\tilde{c}$ as the value of c_i.

Relatively little attention has been paid to why the estimation of the parameters of the three-parameter model is fraught with so much difficulty. Analytical results due to Holland (1990) suggest that a unidimensional test can only support two parameters per item. Thus, it may be that the three-parameter ICC model is over-parameterized. A basic mathematical problem is that items often yield information matrices under this model that are ill-conditioned. In addition, the Newton-Raphson/Fisher scoring procedure moves over the log-likelihood surface in a quadratic fashion and it is possible for the increments to the item parameter estimates to improperly position the estimates on the surface. When this happens, the iterative process diverges rather than converges. Householder (1953) showed that a parameter estimate must be within a certain neighborhood of the parameter in order for the Newton-Raphson procedure to converge. Kale (1962) extended these results to maximum likelihood estimation of multiple parameters under Newton-Raphson and Fisher's method of scoring. He also showed that the initial estimates must be consistent estimators if the iterative estimation procedure is to converge.

The problems in finding the maximum of both the overall and item log-likelihoods were recognized early by the authors of the LOGIST computer program. Constraints were imposed upon the size of the increments to the parameter estimates within each iteration of the maximum likelihood estimation procedures as well as the setting of upper and lower bounds on the parameter estimates themselves. Within stages, these constraints prevent large movements of the parameter estimates. To facilitate finding the overall maximum, a set of steps were established across the two stages of the JMLE procedure where some parameters were held fixed while others were estimated (Wingersky, 1983). In the first step, θ and b were estimated stagewise while the remaining parameters were held fixed. In step 2, θ was fixed while a, b, c, and $\tilde{c}$ were estimated. Step 3 was the same as step 1. In step 4, θ and value of $\tilde{c}$ were held fixed while a, b, and c were estimated. Finding the maximum with respect to b in the first step will locate the ridge in the item log-likelihood surface such as shown in Figure 2.3. Once on the ridge, holding θ fixed and estimating all the item parameters will move the solution along the surface. Then, only estimating b in step three repositions the solution on the ridge once again. Step 4 moves the solution over the surface. These four steps are repeated until the overall log-likelihood is maximized. This four step scheme coupled with the upper and lower bounds and the limits placed upon the size of the increments to the parameter estimates was designed to keep the JMLE procedure moving towards a maximum. Although Lord's writings do not appear to mention plotting the likelihood surfaces, it is clear he understood their characteristics. The plots of the item log likelihoods due to Baker (1988) tend to confirm that the procedures implemented in the LOGIST program are those that would facilitate finding the maxima of the item log-likelihoods and, hence, that of the overall log-likelihood. In contrast to the situation for item parameters, the estimation of an examinee's ability does not seem to involve any significant problems. When an examinee's ability estimate becomes deviant or the estimation procedure fails, it can usually be traced to an unusual pattern of responses to items having atypical values of their item parameter estimates.

4.3 Summary

Birnbaum's JMLE paradigm has been the basis for the estimation of item and ability parameters in a number of IRT test analysis computer programs. For the two-parameter model, under the independence assumptions invoked, this paradigm reduces the solution of $2n + N$ simultaneous equations to the solution of n sets of equations for the parameters of the items in a test and the solution of N individual equations for the ability parameters. In each stage of the paradigm, an iterative Newton-Raphson/Fisher scoring procedure is used to obtain the parameter estimates. The overall paradigm is also iterative since it alternately estimates item and then ability parameters in a "back and forth" manner until a convergence criterion is met. Since the ability scale metric of the parameter estimates is unique only up to a linear transformation, the "identification problem," it is necessary to fix the origin and scale of this metric. Because of this, it is necessary to impose a metric anchoring procedure to remove the indeterminacy. The anchoring procedure used in the LOGIST program results in the origin and unit of measurement being dependent upon the frequency distribution of the examinees over the underlying scale. Thus, there is no guarantee that the metric yielded by two different groups for even the same test will be in a common metric other than one having $S_{\hat{\theta}}$ as its unit of measurement. As a result, some care must be exercised when interpreting the results for different data sets analyzed via the JMLE procedure. To make proper comparisons, the several sets of results must be equated to a common metric using one of the available IRT equating procedures such as that due to Stocking and Lord (1983).

In order to implement the JMLE paradigm in the LOGIST computer program, a large number of ad hoc constraints and a rather convoluted estimation sequence were incorporated to prevent failure of the paradigm in the face of certain data characteristics. This, in turn, has led to computer run times that can be rather long and expensive for large data sets. The marriage of the three-parameter ICC model and the JMLE procedure has not been a happy one. When Birnbaum's three-parameter ICC model is employed, the estimation of the item parameters can be very difficult in some data sets. The Newton-Raphson equations for items can quickly diverge and, hence, the parameters cannot be estimated. For successful estimation, the initial estimates of the item parameters a, b, and c must be reasonably close to the underlying values and the increments to the estimates controlled very closely from iteration to iteration. This is in sharp contrast to the two-parameter models for which only rough estimates of ζ_i and λ_i are needed and the Newton-Raphson procedure usually converges very rapidly. The inclusion of the "guessing" parameter does considerable violence to the mathematics underlying the estimation process that is not apparent in the resultant equations. In particular, the information matrices for such items are often ill-conditioned, and technical difficulties arise in inversion of the matrix which effects the estimation of the item parameters. Perhaps the most telling result is the large number of items in which the c parameter is not estimated and only the value of $\tilde{c}$ is reported as the value of c_i. As a result, the user of LOGIST needs to be aware of how the JMLE paradigm has been implemented and what the various analysis and control options accomplish. Despite these caveats, experience and some strong empirical evidence indicates that the JMLE paradigm is effective and yields reasonable estimates of the unknown parameters.

The mathematics of the JMLE procedures embedded in the joint estimation paradigm are based upon asymptotic results. Thus, the variances of the estimators are large sample variances and the goodness-of-fit statistics are asymptotically distributed as χ^2. As a result, the JMLE procedure works best for large groups of examinees and long tests, say, 1000 examinees and 60 items. Certain problems, in particular the lack of proof of consistent estimators, are associated with maximum likelihood estimation in the present context. The item parameters are structural parameters, the ability parameters are incidental parameters, and large numbers of incidental parameters affect the consistency of the estimates of the structural parameters. It was the "consistency issue" of the JMLE paradigm that led to the development of the marginal maximum likelihood estimation procedures of Chapter 6.

Despite the recognized problems of the JMLE procedure, the estimates yielded by its implementation in the LOGIST program have been the de-facto standard against which all other procedures are compared. While the LOGIST program (Wingersky, Patrick, & Lord, 1999) is still used, the estimation techniques described in Chapters 6 and 7 can cope with many, but not all, of the problems inherent in the JMLE procedure. Thus, over the long term, the JMLE approach will be supplanted by the implementations of these newer techniques. The interesting observation is that even these newer techniques employ the same basic building blocks of Chapters 2 and 3 as well as the independence assumptions of the JMLE paradigm.

5. The Rasch Model

5.1 Introduction

Of all the facets of IRT, the Rasch model for dichotomously scored items is probably the most widely recognized by practitioners. This is due in no small part to the efforts of Professor Benjamin D. Wright of the University of Chicago. As a result of the work of Wright and his followers, this Rasch model has received considerable attention in applied settings. Unfortunately, the early literature dealing with this Rasch model was somewhat disjoint from the main body of IRT literature. The original work due to Rasch (1960, 1961, 1966a, 1966b) was within the framework of a strictly probabilistic approach to item analysis and made no reference to any of the IRT literature. The 1960 book by Rasch has been reissued with a foreword and afterword by Wright (Rasch, 1980). Since the original work, a number of extensions and variations of the several Rasch models have appeared, and the article by Masters and Wright (1984) organizes five different Rasch measurement models within a coherent IRT framework. In the present chapter, the Rasch model label will refer only the case of dichotomously scored items. The widely known work due to Wright (Wright, 1968; Wright & Panchapakesan, 1969) dealt with parameter estimation procedures and properties of this model. The book by Wright and Stone (1979) provides an excellent discussion of how to use the Rasch model in test analysis and test construction. Work by Andersen (1972, 1973a, 1973b) has dealt with conditional maximum likelihood estimation procedures. Although there was a certain degree of continuity within the early Rasch model literature, it lacked an integration with the bulk of the work on IRT. To this end, this chapter attempts to present the Rasch model for dichotomously scored items in a coherent fashion. The historically separate development makes this somewhat difficult and, as a result, a certain amount of presentation that parallels earlier chapters will be necessary. The notation of the preceding chapters will be used in an attempt to link the Rasch model somewhat more closely to previous work. Some care must be exercised because the terms represented by a common symbol may have subtle differences in meaning.

With this introduction, let us turn our attention to the basics of the Rasch model. The presentation will be structured as follows: the probabilistic approach due to Rasch will be reproduced rather closely to present the basic logic and mathematics. The estimation of the item parameters via conditional maximum likelihood will be presented because it is the natural procedure to be used with the model. Then, the joint maximum likelihood estimation (JMLE) procedure of Chapter 4 will be applied to the Rasch model. At this point, the presentation of the theoretical aspects of the model will be complete. It will be followed by a discussion of the features of the model and research related to its application.

5.2 The Rasch Model

In 1960, the Danish mathematician Georg Rasch published a small book in which he approached test analysis from a probabilistic frame of reference. Three different models were presented, one for the number of misreadings in oral text, one for the speed of reading, and one for the items of a test. It is this latter model, presented in Chapters 5, 6, and 10 of his book, that has become known as the Rasch Model. The mathematical presentation of the model is based on the two-dimensional data matrix obtained by administering an n item test to N examinees. The responses to the items are dichotomously scored, $u_{ij} = 0$ or 1, where i designates the item, $i = 1, 2, \ldots, n$, and j designates the examinee, $j = 1, 2, \ldots, N$. For each examinee, there will be a column vector (u_{ij}) of item responses of length n. There will be one such vector for each examinee; hence, there will be N such vectors. Let $U_{ij} = [u_{ij}]$ denote the $n \times N$ matrix formed by these item response vectors. The vector of column marginal totals $(u_{.j})$ contains the total test scores r_j of the examinees. The vector of row marginal totals $(u_{i.})$ contains the item scores s_i. The basic data matrix is shown in Table 5.1.

Table 5.1. Data Matrix Under the Rasch Model

		Examinee						
		1	2	$\cdots$	j	$\cdots$	N	
	1	u_{11}	u_{12}	$\cdots$	u_{1j}	$\cdots$	u_{1N}	$u_{1.} = s_1$
	2	u_{21}	u_{22}	$\cdots$	u_{2j}	$\cdots$	u_{2N}	$u_{2.} = s_2$
	$\cdot$	$\cdot$	$\cdot$	$\cdots$	$\cdot$	$\cdots$	$\cdot$	$\cdot$
	$\cdot$	$\cdot$	$\cdot$	$\cdots$	$\cdot$	$\cdots$	$\cdot$	$\cdot$
Item	i	u_{i1}	u_{i2}	$\cdots$	u_{ij}	$\cdots$	u_{iN}	$u_{i.} = s_i$
	$\cdot$	$\cdot$	$\cdot$	$\cdots$	$\cdot$	$\cdots$	$\cdot$	$\cdot$
	$\cdot$	$\cdot$	$\cdot$	$\cdots$	$\cdot$	$\cdots$	$\cdot$	$\cdot$
	n	u_{n1}	u_{n2}	$\cdots$	u_{nj}	$\cdots$	u_{nN}	$u_{n.} = s_n$
		$u_{.1} = r_1$	$u_{.2} = r_2$	$\cdots$	$u_{.j} = r_j$	$\cdots$	$u_{.N} = r_N$	

In this context, Rasch (1960, Chapter 5) employed two symbols: η_j, called the ability of examinee j $(0 < \eta_j < \infty)$, and δ_i, called the difficulty of item i ($0 < \delta_i < \infty$). He wanted these symbols to represent properties of the examinees and the items such that if examinee 1 was twice as able as examinee 2, then $\eta_1 = 2\eta_2$. Similarly, if item 1 was twice as difficult as item 2, then $\delta_1 = 2\delta_2$. When these are true, the ratio of η_1/δ_1 should be the same as the ratio η_2/δ_2. In general terms, the multiplicative factor 2 could be any real valued constant K, and then

$$\frac{\eta_1}{\delta_1} = \frac{K\eta_2}{K\delta_2} = \frac{\eta_2}{\delta_2}.$$

Thus, one should be able to establish similar ratios using arbitrary sets of people and items as long as the same value of K was involved. As a result, it was necessary to impart meaning simultaneously to the terms ability and item difficulty and, hence to, the symbols η_j, δ_i. This was done by considering the ratio $\xi_{ij} = \eta_j/\delta_i$, where ξ_{ij} is referred to as the situational parameter. It should be noted that the situational parameter can also be defined as $\xi_{ij} = \eta_j\epsilon_i$, where ϵ_i is an item easiness parameter. Hence, $\delta_i = 1/\epsilon_i$ is an item difficulty parameter and serves as a location parameter, but it is not the same item difficulty that was employed in the preceding chapters. As now defined, ξ_{ij} is a ratio, and the immediate problem is to find a function of ξ_{ij} that takes on only values from 0 to 1 as ξ_{ij} goes from 0 to ∞. Rasch (1960) selected the simplest such function,

$$f(\xi_{ij}) = \frac{\xi_{ij}}{1 + \xi_{ij}}.$$

Then,

$$P(u_{ij} = 1|\xi_{ij}) = \frac{\xi_{ij}}{1 + \xi_{ij}}$$

and

$$P(u_{ij} = 0|\xi_{ij}) = 1 - \frac{\xi_{ij}}{1 + \xi_{ij}} = \frac{1}{1 + \xi_{ij}}.$$

Substituting $\xi_{ij} = \eta_j/\delta_i$ yields

$$P(u_{ij} = 1|\eta_j, \delta_i) = \frac{\eta_j/\delta_i}{1 + (\eta_j/\delta_i)} = \frac{\eta_j}{\delta_i + \eta_j} \tag{5.1}$$

and

$$P(u_{ij} = 0|\eta_j, \delta_i) = \frac{1}{1 + (\eta_j/\delta_i)} = \frac{\delta_i}{\delta_i + \eta_j}, \tag{5.2}$$

and the general term for the probability of response is given by

$$P(u_{ij}|\eta_j, \delta_i) = \frac{(\eta_j/\delta_i)^{u_{ij}}}{1 + (\eta_j/\delta_i)}. \tag{5.3}$$

The situational parameter ξ_{ij} can also be defined in terms of betting odds, that is, the ratio of the probability of correct response to the probability of incorrect response:

$$\xi_{ij} = \frac{P(u_{ij} = 1|\eta_j, \delta_i)}{P(u_{ij} = 0|\eta_j, \delta_i)} = \frac{\eta_j/(\delta_i + \eta_j)}{\delta_i/(\delta_i + \eta_j)} = \frac{\eta_j}{\delta_i}. \tag{5.4}$$

Thus, if $P(u_{ij} = 1) = .67$ and $P(u_{ij} = 0) = .33$, then the odds are 2 to 1 that an examinee with ability η_j will answer an item with difficulty δ_i correctly. The odds concept has a long and honored position within mathematics and was a rather natural choice for Rasch.

Since the situational parameter ξ_{ij} is the ratio of the ability parameter η_j to the item difficulty parameter δ_i, there is an indeterminacy in its definition. Ratios are unique up to a multiplication of the numerators and denominators by the same constant. Hence, if one multiplies η_j and δ_i by the same constant, the ratio is unchanged since

$$\xi_{ij} = \frac{K\eta_j}{K\delta_i} = \frac{\eta_j}{\delta_i}$$

as long as $K > 0$ holds. For the present, this indeterminacy can be ignored. When one examines the procedures for estimating η_j and δ_i, however, the indeterminacy must be resolved in some way. This is the same "identification problem" encountered in the joint estimation procedures of Chapter 4.

Equation 5.1 relates the probability of correctly answering a given item and the ability scale. Thus, it defines an item characteristic curve that depends on only a single parameter δ_i. A plot of this function for $0 < \eta_j < \infty$ and $\delta_i = 1.822$ is shown in Figure 5.1 based on the values in Table 5.2. The form of the curve does not look like that of any of the previous ICC models. In addition, the ability scale has a range of $0 < \eta < \infty$ with an arbitrary unit of measurement. Because Rasch started from a probabilistic frame of reference and did not attempt to put his work in the context of previous test theory, his modeling and mathematics are quite different from those of previous chapters.

Table 5.2. Item Characteristic Curve Under the Rasch Model ($\delta_i = 1.822$, $0 < \eta_j < 21$)

η_j	ξ_{ij}	$P(u_{ij} = 1)$	Log Ability
0.0498	0.0273	0.0266	−3.0
0.0821	0.0451	0.0431	−2.5
0.1353	0.0743	0.0691	−2.0
0.2231	0.1224	0.1091	−1.5
0.3679	0.2019	0.1680	−1.0
0.6065	0.3329	0.2497	−0.5
1.0000	0.5488	0.3544	0.0
1.6487	0.9049	0.4750	0.5
2.7183	1.4919	0.5987	1.0
4.4817	2.4598	0.7110	1.5
7.3891	4.0555	0.8022	2.0
12.1825	6.6863	0.8699	2.5
20.0855	11.0239	0.9168	3.0

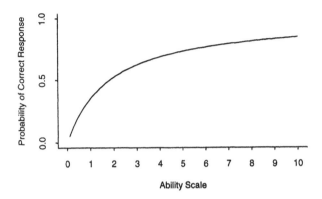

Fig. 5.1. Item characteristic curve under the Rasch model, $\delta_i = 1.822$.

It will be useful to digress for a bit to show how the Rasch item characteristic curve can be brought into alignment with existing models used in IRT. A logarithmic transformation can be applied to both the ability scale and the item difficulty scale as follows:

$$\theta_j = \log \eta_j \qquad -\infty < \theta_j < \infty$$

$$\beta_i = \log \delta_j \qquad -\infty < \beta_j < \infty$$

also,

$$\beta_i = -\log \epsilon_i.$$

And the inverse transformation yields

$$\eta_j = e^{\theta_j},$$

$$\delta_i = e^{\beta_i},$$

and

$$\epsilon_i = e^{-\beta_i}.$$

Then, from Equation 5.3,

$$P(u_{ij}|\theta_j, \beta_i) = \frac{(e^{\theta_j} e^{-\beta_i})^{u_{ij}}}{1 + e^{\theta_j} e^{-\beta_i}} = \frac{e^{(\theta_j - \beta_i)u_{ij}}}{1 + e^{(\theta_j - \beta_i)}} \tag{5.5}$$

and

$$P(u_{ij} = 1|\theta_j, \beta_i) = \frac{e^{(\theta_j - \beta_i)}}{1 + e^{(\theta_j - \beta_i)}} = \frac{1}{1 + e^{-(\theta_j - \beta_i)}} \tag{5.6}$$

and

$$P(u_{ij} = 0|\theta_j, \beta_i) = \frac{1}{1 + e^{(\theta_j - \beta_i)}} = \frac{e^{-(\theta_j - \beta_i)}}{1 + e^{-(\theta_j - \beta_i)}}. \tag{5.7}$$

Equation 5.6 reveals familiar ground as the probability of correct response is simply a two-parameter logistic model with the item discrimination parameter $\alpha_i = 1$ and a difficulty parameter β_i. The ability scale is now centered at $\theta_j = 0$ and is expressed in terms of the logarithm of the original arbitrary unit of measurement. The nonlinear nature of the logarithmic transformation is such that it stretches and compresses the item characteristic curve of Figure 5.2 in such a way as to yield the familiar logistic model for the one-parameter item characteristic curve. This particular representation of the Rasch model also highlights the role of the metric of ability scale. If one plots the probability of correct response from Table 5.1 against the natural logarithm of the ability scale, a logistic ogive with parameters $\beta_i = 0.6$ and $\alpha_i = 1.0$ results as shown in Figure 5.3. Thus, a change in the ability metric resulted in obtaining an item characteristic curve whose form is in keeping with previous models. From a practical point of view, it is important to recognize that the item characteristic curve yielded by this transformation is the two-parameter logistic model with $\alpha_i = 1$. It is commonly stated (see Whitely & Dawis, 1974) that the Rasch model requires items having equal values of the item discrimination index. When the common value of $\alpha_i \neq 1$, however, the unit of measurement of ability scale must be adjusted to yield $\alpha_i = 1$ for all items. This adjustment in the ability scale metric would not be obvious to someone analyzing a data set. It would become apparent when one equates the item difficulty parameters of two tests with differing common levels of α_i, say, $\alpha_i = 1$ and $\alpha_i \neq 1$. Thissen (1982) has developed an IRT model and estimation procedures in which all items share a common α, not necessarily equal to unity, but have separate difficulty parameters.

Another interesting insight into the mathematics of the Rasch model is provided by the odds ratio. Recall that the situational parameter could be defined as

$$\xi_{ij} = \frac{\eta_j}{\delta_i} = \frac{P(u_{ij} = 1|\theta_j, \beta_i)}{P(u_{ij} = 0|\theta_j, \beta_i)}.$$

Substituting Equations (5.6) and (5.7) for the numerator and denominator, respectively,

$$\xi_{ij} = \frac{1/[1 + e^{-(\theta_j - \beta_i)}]}{e^{-(\theta_j - \beta_i)}/[1 + e^{-(\theta_j - \beta_i)}]} = e^{(\theta_j - \beta_i)}.$$

Now, taking the natural logarithm of the odds yields

$$\log \xi_{ij} = \theta_j - \beta_i$$

which is linear in the parameters of the overall Rasch model and is the logistic deviate.

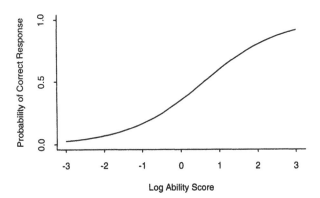

Fig. 5.2. Rasch item characteristic curve obtained after logarithmic transformation, $\alpha_i = 1.0$, $\beta_i = 0.6$.

5.3 Separation of Parameters

With this connection between IRT and Rasch's item characteristic curve shown, let us return to the presentation of the overall model as developed by Rasch, which will result in a unique set of mathematics in a separate notation. Once the developments due to Rasch have been shown, the work of Andersen (1973a, 1973b) and of Wright and Douglas (1977b) will be used to present the results of interest in a more familiar mathematical framework. The reason for this rather redundant approach is to make explicit what Rasch did and to facilitate distinguishing it from subsequent work.

Before proceeding, the assumptions underlying Rasch's work are:

1. The probability of correct response of examinee j to a dichotomously scored item i is given by $P(u_{ij} = 1|\eta_j, \delta_i) = \xi_{ij}/(1 + \xi_{ij})$, where ξ_{ij} is the situational parameter.
2. Given the values of the parameters, all answers are stochastically independent (Rasch, 1966a, p. 50).

These two assumptions have a number of implications that need to be clarified. If assumption 1 holds, then a single latent trait is involved which will be called ability. The second assumption includes the local independence assumption of item response theory where, for examinees having a given ability η_j, responses to the n items are independent. It also includes local independence of the responses of the examinees to a given item. As will become apparent in the subsequent mathematics, the Rasch model treats persons (examinees) and items as independent under assumption 2. This is also apparent in the logarithm of the odds ratio. Although a given examinee's answers to the set of items are stochastically independent, variation in ability levels of the examinees will lead to interitem correlations as has been the case with previous IRT models. It is implicit in assumption 2 that parameters are dealt with in a manner analogous to a two-way fixed effects analysis of variance with no interaction. As a result, the only parameters of interest are those corresponding to the N examinees and the n items actually at hand (Sanathanan, 1974). This analogy with fixed effects ANOVA is important in all of IRT.

Since Rasch assumed local independence, the probability of obtaining a given pattern of item responses when a single examinee of ability η_j responds to n items of difficulty δ_i is given by

$$P(u_{1j}, u_{2j}, \ldots, u_{nj} | \eta_j, \delta_1, \delta_2, \ldots, \delta_n)$$

$$= P(u_{1j} | \eta_j, \delta_1) P(u_{2j} | \eta_j, \delta_2) \cdots P(u_{nj} | \eta_j, \delta_n) \tag{5.8}$$

but, from Equation 5.3 and for a single item,

$$P(u_{ij} | \eta_j, \delta_i) = \frac{(\eta_j/\delta_i)^{u_{ij}}}{1 + (\eta_j/\delta_i)}.$$

The mathematics can be considerably simplified if, instead of item difficulty δ_i, item easiness ϵ_i is used, where $\epsilon_i = 1/\delta_i$; then,

$$P(u_{ij} | \eta_j, \epsilon_i) = \frac{(\eta_j \epsilon_i)^{u_{ij}}}{1 + (\eta_j \epsilon_i)}. \tag{5.9}$$

Substituting in Equation 5.8 yields

$$P(u_{1j}, u_{2j}, \ldots, u_{nj} | \eta_j, \epsilon_1, \epsilon_2, \ldots, \epsilon_n) = \frac{\prod_{i=1}^{n} \eta_j^{u_{ij}} \epsilon_i^{u_{ij}}}{\prod_{i=1}^{n} (1 + \eta_j \epsilon_i)}, \tag{5.10}$$

but η_j is a constant with respect to items and can be factored out as $\eta_j^{u_{.j}}$, where $u_{.j}$ is the sum of the examinee's response to the n items. Because the items are dichotomously scored, $u_{.j}$ is the examinee's raw test score. Then,

$$P(u_{1j}, u_{2j}, \ldots, u_{nj} | \eta_j, \epsilon_1, \epsilon_2, \ldots, \epsilon_n) = \frac{\eta_j^{u_{.j}} \prod_{i=1}^{n} \epsilon_i^{u_{ij}}}{\prod_{i=1}^{n} (1 + \eta_j \epsilon_i)}. \tag{5.11}$$

Now, this general formula for the probability of a response vector can be particularized to the case in which the sum of the number of correct responses is a specified test score, say, r. The formula will be developed via induction, beginning with a score of zero. In this case, there is only one possible item response vector, the null vector. When this is the case, the probability of response to a single item is given by

$$P(u_{ij} = 0 | \eta_j, \epsilon_i) = \frac{(\eta_j \epsilon_i)^0}{1 + \eta_j \epsilon_i} = \frac{1}{1 + \eta_j \epsilon_i},$$

and the probability of the item response vector is

$$P(u_{.j} = 0 | \eta_j, \epsilon_1, \epsilon_2, \ldots, \epsilon_n) = \frac{1}{\prod\limits_{i=1}^{n}(1 + \eta_j \epsilon_i)}.$$

To simplify the notation, let

$$d(\eta_j) = \prod_{i=1}^{n}(1 + \eta_j \epsilon_i)$$

and

$$P(u_{.j} = 0 | \eta_j, \epsilon_1, \epsilon_2, \ldots, \epsilon_n) = \frac{1}{d(\eta_j)}.$$

When $r = 1$, there are n different possible item response vectors each with $n - 1$ zeros and one correct response. The probabilities are

$$P(u_{1j} = 1 | \eta_j, \epsilon_1, \epsilon_2, \ldots, \epsilon_n) = \frac{\eta_j \epsilon_1}{\prod\limits_{i=1}^{n}(1 + \eta_j \epsilon_i)} = \frac{\eta_j \epsilon_1}{d(\eta_j)}$$

$$P(u_{2j} = 1 | \eta_j, \epsilon_1, \epsilon_2, \ldots, \epsilon_n) = \frac{\eta_j \epsilon_2}{\prod\limits_{i=1}^{n}(1 + \eta_j \epsilon_i)} = \frac{\eta_j \epsilon_2}{d(\eta_j)}$$

$$\vdots$$

$$P(u_{nj} = 1 | \eta_j, \epsilon_1, \epsilon_2, \ldots, \epsilon_n) = \frac{\eta_j \epsilon_n}{\prod\limits_{i=1}^{n}(1 + \eta_j \epsilon_i)} = \frac{\eta_j \epsilon_n}{d(\eta_j)}.$$

Then the probability of a test score of $r = 1$ is the sum of the probabilities of the n different item response vectors,

$$P(u_{.j} = 1 | \eta_j, \epsilon_1, \epsilon_2, \ldots, \epsilon_n) = \sum_{i=1}^{n} \frac{\eta_j \epsilon_i}{d(\eta_j)} = \frac{\eta_j \sum\limits_{i=1}^{n} \epsilon_i}{d(\eta_j)}.$$

A test score of $r = 2$ can be obtained from n items in $_nC_2$ or $\binom{n}{2}$ different ways and the individual item response vectors will have probabilities of

$$P(u_{1j} = 1, u_{2j} = 1 | \eta_j, \epsilon_1, \epsilon_2, \ldots, \epsilon_n) = \frac{\eta_j^2 \epsilon_1 \epsilon_2}{d(\eta_j)}$$

$$P(u_{1j} = 1, u_{3j} = 1|\eta_j, \epsilon_1, \epsilon_2, \ldots, \epsilon_n) = \frac{\eta_j^2 \epsilon_1 \epsilon_3}{d(\eta_j)}$$

$$\vdots$$

$$P(u_{(n-1)j} = 1, u_{nj} = 1|\eta_j, \epsilon_1, \epsilon_2, \ldots, \epsilon_n) = \frac{\eta_j^2 \epsilon_{n-1} \epsilon_n}{d(\eta_j)}.$$

The sum of these probabilities is the probability of obtaining a test score of 2:

$$P(u_{.j} = 2|\eta_j, \epsilon_1, \epsilon_2, \ldots, \epsilon_n) = \frac{\eta_j^2 (\epsilon_1 \epsilon_2 + \epsilon_1 \epsilon_3 + \cdots + \epsilon_{n-1} \epsilon_n)}{d(\eta_j)},$$

where the sum in the parentheses includes all $_nC_2$ possible combinations of item easiness parameters. Now, in the general case, any value of r can be obtained in $_nC_r$ ways, and the sum of the probabilities of the individual item response vectors is the probability of $u_{.j} = r$, which is given by

$$P(u_{.j} = r|\eta_j, \epsilon_1, \epsilon_2, \ldots, \epsilon_n) = \frac{\eta_j^r \gamma_r}{d(\eta_j)}, \tag{5.12}$$

where

$$\gamma_r = (\epsilon_1 \epsilon_2 \cdots \epsilon_r) + (\epsilon_2 \epsilon_3 \cdots \epsilon_{r+1}) + \cdots + (\epsilon_{n-r} \epsilon_{n-r+1} \cdots \epsilon_n)$$

and the terms in the parentheses contain the individual combinations obtained when $_nC_r$ enumerated for each value of r (see Table 5.3).

Table 5.3. Symmetric Functions

r	γ_r	Terms	Number of terms
0	γ_0	1	1
1	γ_1	$\epsilon_1 + \epsilon_2 + \epsilon_3 + \cdots + \epsilon_n$	n
2	γ_2	$\epsilon_1 \epsilon_2 + \epsilon_1 \epsilon_3 + \cdots + \epsilon_{n-1} \epsilon_n$	$_nC_2$
3	γ_3	$\epsilon_1 \epsilon_2 \epsilon_3 + \epsilon_1 \epsilon_2 \epsilon_4 + \cdots + \epsilon_{n-2} \epsilon_{n-1} \epsilon_n$	$_nC_3$
$\vdots$	$\vdots$	$\vdots$	$\vdots$
r	γ_r	$(\epsilon_1 \epsilon_2 \epsilon_3 \epsilon_4 \cdots \epsilon_r) + (\epsilon_1 \epsilon_2 \epsilon_3 \cdots \epsilon_{r-1} \epsilon_{r+1}) + \cdots + (\epsilon_{n-r} \epsilon_{n-r+1} \cdots \epsilon_n)$	$_nC_r$
$\vdots$	$\vdots$	$\vdots$	$\vdots$
n	γ_n	$\prod_{i=1}^{n} \epsilon_i$	1

Expressing γ_r in more general notation

$$\gamma_r = \gamma(r, \epsilon_1, \epsilon_2, \ldots, \epsilon_n) = \sum_{(u_{ij})} \left(\prod_{i=1}^{n} \epsilon_i^{u_{ij}} \right), \tag{5.13}$$

where $\sum_{(u_{ij})}^{r}$ indicates that the number of terms in the sum is a function of $_nC_r$ for a given value of r, and (u_{ij}) denotes an item response vector for person j that yields a test score of r. Then Equation 5.12 becomes

$$P(u_{.j} = r|\eta_j, \epsilon_1, \epsilon_2, \ldots, \epsilon_n) = \frac{\eta_j^r}{d(\eta_j)} \sum_{(u_{ij})}^{r} \left(\prod_{i=1}^{n} \epsilon_i^{u_{ij}} \right). \tag{5.14}$$

Since the scores $r = 0, 1, \ldots, n$ exhaust all the possible test scores at any given ability level, the following holds:

$$\sum_{r=0}^{n} P(u_{.j} = r|\eta_j, \epsilon_1, \epsilon_2, \ldots, \epsilon_n) = 1$$

and, from Equation 5.12,

$$\sum_{r=0}^{n} \frac{\eta_j^r \gamma_r}{d(\eta_j)} = 1$$

but $d(\eta_j)$ does not depend on r. Then,

$$\sum_{r=0}^{n} \eta_j^r \gamma_r = d(\eta_j),$$

but

$$d(\eta_j) = \prod_{i=1}^{n} (1 + \eta_j \epsilon_i)$$

and

$$\sum_{r=0}^{n} \eta_j^r \gamma_r = \prod_{i=1}^{n} (1 + \eta_j \epsilon_i).$$

Notice that the product on the right is over the number of items and the summation on the left is over the possible test scores, assuming dichotomous scoring. Recall, in the present case of a single person, that the η_j is a constant; consequently, the γ_r are the coefficients, expressed in terms of ϵ_i, of the powers of η_j in the algebraic expansion of the product in $d(\eta_j)$. The γ_r are the elementary symmetric functions of $\epsilon_1, \epsilon_2, \ldots, \epsilon_n$ (see David, Kendall, & Barton, 1966). For elaboration, see Table 5.2. The expansion of $d(\eta_j)$ can be written as

$$\prod_{i=1}^{n} (1 + \eta_j \epsilon_i) = 1 + \gamma_1 \eta_j^1 + \gamma_2 \eta_j^2 + \gamma_3 \eta_j^3 + \cdots + \gamma_n \eta_j^n = \sum_{r=0}^{n} \gamma_r \eta_j^r.$$

Two comments are in order at this point. First, the symmetric functions are very difficult to evaluate. They involve continuing products of order r, and it is difficult to maintain numerical accuracy when r obtains even modest values. Second, although Rasch has approached the topic from the point of

view of combinatorial mathematics, there is a distinct parallelism with the
distributional approach taken by Birnbaum (1968, Chapter 17). Recall that,
under previous item characteristic curve models, a conditional distribution of
the item response variable u_{ij} existed at each ability level. Depending on what
form was assumed for these conditional distributions, particular weighting
coefficients were obtained in the fitting of the item characteristic curve. When
a normal ogive model was used for the item characteristic curve and the
conditional distribution followed the binomial, the Urban-Müller weights were
obtained. Rather than assume a particular distribution of test scores at a
given ability level, Rasch used combinatorial mathematics to obtain the exact
discrete distribution of r. Although the approach is different, the result is the
same in that the form of the conditional distribution has been defined. In the
subsequent derivations, Rasch used the individual probabilities associated
with a value of r rather than the form of the distribution.

Once the probability of obtaining an item response pattern whose sum of
correct responses is a particular value of r has been developed, the essential
feature of the Rasch model can be demonstrated. This is accomplished via
the conditional probability of a pattern of item responses (u_{ij}), given that
their sum is r and person j is the responder. This conditional probability is

$$P((u_{ij})|u_{.j} = r, \epsilon_1, \epsilon_2, \ldots, \epsilon_n) = \frac{P((u_{ij})|\eta_j, \epsilon_1, \epsilon_2, \ldots, \epsilon_n)}{P(u_{.j} = r|\eta_j, \epsilon_1, \epsilon_2, \ldots, \epsilon_n)}$$

$$= \frac{\eta_j^r \prod_{i=1}^{n} \epsilon_i^{u_{ij}} / d(n_j)}{\eta_j^r \gamma_r / d(\eta_j)} = \frac{\prod_{i=1}^{n} \epsilon_i^{u_{ij}}}{\sum_{(u_{ij})}^{r} \left(\prod_{i=1}^{n} \epsilon_i^{u_{ij}} \right)}. \tag{5.15}$$

The interesting aspect of this conditional probability is that it does not involve the ability parameter η_j of the particular person responding to the
items. What it indicates is that, given a particular test score, the probability
of a particular pattern of item responses depends only on the item easiness
parameters ϵ_i and which of the n items were answered correctly, that is, on
the u_{ij}. The direct implication is that, for a group of examinees having different ability levels but sharing a common raw test score, the probability of
obtaining a pattern of item responses containing r correct responses does not
depend on the examinee's ability level. Thus, it follows "that once $u_{.j}$ has
been obtained, any extra information about which of the items were answered
correctly is, according to our model, useless as a source of inference about η_j"
(Rasch, 1966a, p. 98). Therefore, $u_{.j}$ meets the requirements of a sufficient
statistic (see Mood, 1950, p. 151) and is an estimator of the unknown parameter η_j. However, the statistic $u_{.j}$ is only relatively sufficient since it depends
on knowing the values of the item parameters ϵ_i. Since these are not known,
the relative sufficiency is not particularly useful in the present context. Thus,
to achieve his goal of a sufficient statistic, Rasch needs to improve the level

from relatively sufficient to sufficient and, it is hoped, to minimally sufficient. That this can be accomplished is clear since, for the two-parameter model, Birnbaum (1968, Chapter 18) has shown that

$$T_j = \sum_{i=1}^{n} \alpha_i u_{ij}$$

is a minimal sufficient estimator for the ability η_j. But, under the Rasch model, $\alpha_i = 1$ for all items, and

$$T_j = \sum_{i=1}^{n} u_{ij} = r_j$$

is simply a raw test score. Thus, under the Rasch model, the raw test score should be the minimal sufficient estimator of ability. It is also clear that, under the two-parameter logistic model, even though several item response patterns may have the same number of items correct, they may yield different values of T_j and hence different ability estimates because of the presence of the α_i. Under the Rasch model, all such patterns yield a single value of the estimator since all $\alpha_i = 1$, even though persons of differing ability may have yielded the particular response patterns having a common value of r. This confirms Rasch's (1966a, p. 98) statement about $u_{.j}$. One could replicate the logic of the above presentation with the roles of examinees and items reversed and find a relatively sufficient estimator for item easiness.

Rasch (1960, p. 178) recognized the distinction between structural and incidental parameters in an informal way but did not cite the work of Neyman and Scott (1948). Consequently, he did not pursue this differentiation in detail. Rather, the sufficiency of the estimators of both ability and item easiness was proved en route to what Rasch termed the "separation of parameters." He first considered the special case in which all N subjects obtained the same test score $u_{.j} = r$ and let $(u_{ij}) = (u_{1j}, u_{2j}, \ldots, u_{nj})$ be a set of item responses for person j. The Equation 5.15 can be written as

$$P((u_{ij})|u_{.j} = r, \epsilon_1, \epsilon_2, \ldots, \epsilon_n) = \frac{\prod_{i=1}^{n} \epsilon_i^{u_{ij}}}{\gamma_r}. \tag{5.16}$$

Under the basic assumption of the independence of item responses of the N persons, the joint probability of the full set of observed item responses, the $n \times N$ matrix $[u_{ij}]$ is the product of equations of the form (5.16) for the individual patterns. The resultant product is

$$P([u_{ij}]|u_{.j} = r, \epsilon_1, \epsilon_2, \ldots, \epsilon_n) = \frac{\epsilon_1^{u_{1.}} \cdot \epsilon_2^{u_{2.}} \cdots \epsilon_n^{u_{n.}}}{\gamma_r^N}, \tag{5.17}$$

where $u_{i.} = \sum_{j=1}^{N} u_{ij} = s_i$, which is the number of times the item was answered correctly, that is, the item score and ϵ_1 denotes that the easiness of item 1, etc., was used over all examinees. At this point, Rasch has shown that, for the special case of all persons having the same test score, the conditional probability of the $n \times N$ matrix of item responses is a function only of the item easiness parameters. It is also the basis for Wright's (1977b, p. 219) comment, "In fact a set of items may be calibrated even when all persons in the calibration sample have earned one and the same score." It is important to note that the exponents of the ϵ_i in Equation 5.17 are the item scores s_i. Again, it is clear that the marginal totals of the item response matrix play an important role while the item response vectors are of interest only insofar as they contribute to the marginal totals. While this special case reveals the general trend of Rasch's approach, what remains is to generalize the mathematics to encompass the situation in which many different test scores appear in the results. Rasch (1960, 1966a, 1966b) called this final step "the separation of parameters."

Consider the responses of N persons with parameters $\eta_1, \eta_2, \ldots, \eta_N$ to n items with parameters $\epsilon_1, \epsilon_2, \ldots, \epsilon_n$, where the response of person j to item i is given by u_{ij}. Recall that the probability of response given η_j and ϵ_i is

$$P(u_{ij}|\eta_j, \epsilon_i) = \frac{(\eta_j \epsilon_i)^{u_{ij}}}{1 + \eta_j \epsilon_i}. \tag{5.18}$$

Under the basic assumption that the item responses are stochastically independent (i.e., assumption 2) the joint probability of the $n \times N$ item responses is

$$
\begin{aligned}
P([u_{ij}]|(\eta_j), (\epsilon_i)) &= \prod_{j=1}^{N} \prod_{i=1}^{n} P(u_{ij}|\eta_j, \epsilon_i) \\
&= \frac{\displaystyle\prod_{j=1}^{N} \prod_{i=1}^{n} (\eta_j \epsilon_i)^{u_{ij}}}{\displaystyle\prod_{j=1}^{N} \prod_{i=1}^{n} (1 + \eta_j \epsilon_i)} \\
&= \frac{\displaystyle\prod_{j=1}^{N} \prod_{i=1}^{n} (\eta_j \epsilon_i)^{u_{ij}}}{d(\Xi, E)},
\end{aligned} \tag{5.19}
$$

where Ξ and E are the vectors (η_j) and (ϵ_i), respectively, and

$$d(\Xi, E) = \prod_{j=1}^{N} \prod_{i=1}^{n} (1 + \eta_j \epsilon_i).$$

Equation 5.19 can be rewritten by taking advantage of the fact that η_j is a constant relative to subscript i and ϵ_i is a constant relative to subscript j to obtain

$$P([u_{ij}]|(\eta_j),(\epsilon_i)) = \frac{\prod\limits_{j=1}^{N} \eta_j^{u_{.j}} \prod\limits_{i=1}^{n} \epsilon_i^{u_{i.}}}{d(\Xi, E)} = \frac{\prod\limits_{j=1}^{N} \eta_j^{r_j} \prod\limits_{i=1}^{n} \epsilon_i^{s_i}}{d(\Xi, E)}. \qquad (5.20)$$

Written in this form, the exponents of the parameters are the corresponding marginals of the item response matrix. Consequently, one needs to derive the probability that the marginal vectors $u_{.1}, u_{.2}, \ldots, u_{.N}$ and $u_{1.}, u_{2.}, \ldots, u_{n.}$ take on values $r_1, r_2, \ldots, r_j$ and $s_1, s_2, \ldots, s_n$, respectively. Generalizing on the approach taken earlier in which all the possible item response vectors (u_{ij}) that yielded the same test score were used, now all possible item response matrices $[u_{ij}]$ that yield a particular set of marginal totals are needed. There will be many different such matrices, but the probability of each is the same and is given by Equation 5.20. As a result, an enumeration of the different matrices is not needed; hence, only a notation for the number is needed. Let the following notation represent the total number

$$[rs],$$

where $r = r_1, r_2, \ldots, r_N$ and $s = s_1, s_2, \ldots, s_n$, and multiplication is not implied in the notation. Taking advantage of the fact that the η_j are constant with respect to the subscript i and that the ϵ_i are constant with respect to subscript j, the joint probability of the set of marginal totals may be written as

$$P((u_{.j} = r_j), (u_{i.} = s_i)|(\eta_j), (\epsilon_i)) = \frac{\prod\limits_{j=1}^{N} \eta_j^{r_j} \prod\limits_{i=1}^{n} \epsilon_i^{s_i}}{d(\Xi, E)}[rs]. \qquad (5.21)$$

The joint probability distribution of the row totals $u_{i.}$ and column totals $u_{.j}$ contains just as many parameters, the η_j and ϵ_i, as observables. Now, when Equation 5.20 is divided by Equation 5.21, the probability of obtaining the item response matrix conditional on a particular set of row and column marginal totals is given by

$$P([u_{ij}]|(u_{.j} = r_j), (u_{i.} = s_i)) = \frac{1}{[rs]}. \qquad (5.22)$$

The interesting result is that this probability of the set of observations is independent of all parameters; however, the row and column marginal totals must be known. Because of this, Rasch (1966b, p. 101) concluded:

> Therefore, once the totals have been recorded, any further statement as regards which of the items were answered correctly by which persons is, according to our model, useless as a source of information about the parameters . . . Thus, the row totals and the column totals are not only suitable for estimating the parameters: They imply every possible statement about the parameters that can be made on the basis of the observations.

Consequently, the r_j are sufficient estimators of the η_j, and the s_i are sufficient estimators of the ϵ_i. In addition, they will be minimally sufficient (see Birnbaum, 1968, Chapter 18; Andersen, 1970). It should be noted here that the sufficiency of the item scores as estimates of the item easiness parameters was accomplished somewhat incidentally here. The result provided by Equation 5.22 can also be used as the basis for a goodness-of-fit test, "controlling the model" in Rasch's terminology, in a manner similar to that used in a two dimensional contingency table. Although, Rasch has now shown that the test scores and the item scores are sufficient estimators of the corresponding ability and item parameters, he has not yet demonstrated what he called the "separation of parameters."

Since Equation 5.21 involves both sets of parameters, it could be used to establish a system of simultaneous equations to solve for both sets of parameters. But, it was shown, in the case of a single person or where all persons had the same test score, the probability of a given vector of item responses, conditional on a particular test score, was independent of the ability parameters of the persons. Hence, a generalization of this characteristic to the full item response matrix would appear to be possible. To accomplish this generalization, the distribution of the column marginal totals irrespective of the values of the row marginal totals is needed. This can be done by summing Equation 5.21 over all possible combinations of $s_1, s_2, \ldots, s_n$, the row marginal totals. Let

$$\gamma(E) = \sum_{(s_i)}^{s} [rs] \prod_{i=1}^{n} \epsilon_i^{s_i},$$

where $\sum_{(s_i)}^{s}$ indicates summation over all item vectors yielding an item score of s_i as well as over all possible values of s. Then, the probability of the vector column marginal totals $(u_{.j})$ taking on a particular set of values (r_j) is

$$P((u_{.j} = r_j)|(\eta_j), (\epsilon_i)) = \frac{\gamma(E) \prod_{j=1}^{N} \eta_j^{r_j}}{d(\Xi, E)}. \tag{5.23}$$

As was the case in Equation 5.15 earlier, if the ϵ_i were known, Equation 5.23 could be used to estimate the η_j from the column marginals.

A parallel derivation can be made to obtain the distribution of the row marginal totals irrespective of the column marginal totals. This can be done by summing Equation 5.21 over all possible configurations of $r_1, r_2, \ldots, r_n$. Following the previous case, let

$$\gamma(\Xi) = \sum_{(r_j)}^{r} [rs] \prod_{j=1}^{N} \eta_j^{r_j},$$

where $\sum_{(r_j)}^{r}$ indicates summation over all item response vectors yielding a test score of r_j as well as over all possible values of r. Using this notation, the probability that the vector of row marginals $(u_{i.})$ will take on a particular set of values (s_i) is given by

$$P((u_{i.} = s_i)|(\eta_j), (\epsilon_i)) = \frac{\gamma(\Xi) \prod\limits_{i=1}^{n} \epsilon_i^{s_i}}{d(\Xi, E)}. \qquad (5.24)$$

Thus, if the ability parameters were known, the ϵ_i could be estimated, and the s_i are relatively sufficient statistics for the ϵ_i.

The consequence of Equations 5.23 and 5.24 is that each set of parameters can be estimated if the other is known, which incidentally suggests the JMLE procedure of Chapter 4. Even though Rasch had studied under Sir R. A. Fisher, he did not pursue a maximum likelihood estimation line of inquiry. Instead, he overcame the impasse by obtaining the conditional probability of $(u_{.j} = r_j)$, given that $(u_{i.} = s_i)$. This is accomplished by dividing Equation 5.21 of the joint probability of $u_{.j} = r_j$ and $u_{i.} = s_i$ by Equation 5.24 of the probability of $u_{i.} = s_i$. Then the conditional probability is

$$P((u_{.j} = r_j)|(u_{i.} = s_i), (\eta_j)) = \frac{[rs] \prod\limits_{j=1}^{N} \eta_j^{r_j}}{\gamma(\Xi)}, \qquad (5.25)$$

which does not involve the item easiness parameters. Similarly, the conditional probability of $(u_{i.} = s_i)$, given that $(u_{.j} = r_j)$, is obtained by dividing Equation 5.21 by Equation 5.23, yielding

$$P((u_{i.} = s_i)|(u_{.j} = r_j), (\epsilon_i)) = \frac{[rs] \prod\limits_{i=1}^{n} \epsilon_i^{s_i}}{\gamma(E)}, \qquad (5.26)$$

which does not involve the ability parameters.

Thus, Rasch has achieved his goal of separation of parameters since Equation 5.25 does not involve the item parameters and Equation 5.26 does not involve the ability parameters. The importance of achieving this goal is that these expressions for the conditional probabilities can be used as the basis for estimation procedures in which the two sets of parameters can be estimated separately. Rasch (1960, pp. 178–181) outlined an estimation procedure but subsequently indicated (Rasch, 1966a) that it was not practical. The basic problem is the presence of the symmetric functions $\gamma(E)$ and $\gamma(\Xi)$, which involve continuing products. He also presented procedures for examining the goodness of fit of the data and the model. However, it remained for other workers to develop estimation procedures for the Rasch model.

Several features of the above presentation are worthy of comment. First, the actual Rasch model deals with modeling the $n \times N$ item response matrix from a purely probabilistic point of view. Rasch did not approach this modeling from an IRT point of view. In the applied literature, there is a tendency to consider the logistic version of the function

$$\xi_{ij} = \frac{\eta_j/\delta_i}{1 + \eta_j/\delta_i}$$

as the Rasch model. While this function involves both parameters, focusing narrowly on it ignores the broader definition of the overall Rasch model. Second, in all the mathematics above, the ability parameters were considered on an individual person basis. All of the products or sums involving the subscript j had an upper limit of N, the group size. This is important in that, although Rasch recognized the distinction between incidental and structural parameters, in his mathematics the incidental parameters, the η_j, were treated as structural parameters. Once separation of parameters was achieved, it was clear that one set of parameters could be estimated by replacing the other with its minimal sufficient statistic. Yet, from a mathematical point of view, all N ability parameters would need to be estimated. However, Rasch (1960, Chapter 6) recognized that all persons having the same test score would obtain the same ability estimate, and this could be used to advantage in the estimation process. The estimation procedure developed by Rasch, which has been called the log method, is described in detail by Wright and Panchapakesan (1969). Finally, in Equation 5.26, the probability statement on the left side does not contain the η_j. Rasch (1960, p. 178) states, "We may eliminate the parameters $\eta_1, \eta_2, \ldots, \eta_N$, having, as it were, replaced them by the observed marginals $r_1, r_2, \ldots, r_N$." The "as it were" is important here because, on the right-hand side, the terms involving the η_j simply canceled out. Thus, replacement of the η_j by the r_j on the left-hand side of Equation 5.26 is due to a conditional probability having been obtained. Clarification of this issue is complicated by the notational inconsistency of Rasch's writings. In two articles, Rasch (1966a, 1966b) wrote the conditional probability as

$$P((u_{i.} = s_i)|(u_{.j} = r_j), (\eta_j), (\epsilon_i))$$

indicating that the marginal totals $(u_{i.})$ were conditional on $(u_{.j} = r_j)$, (η_j), and (ϵ_i). But in his earlier work (Rasch, 1960, Equation 5.22), the notation for the (η_j) was not present. From a purely theoretical point of view, the conditional probability depends on the ability parameters, but the mathematical expression does not involve them on the right. A similar situation exists in regard to Equation 5.25.

Although Equations 5.25 and 5.26 apply to the total item response matrix, Equation 5.25 can be applied to any subgroup of the total group of N subjects that have responded to the n items. Consequently, the ability parameters of the persons in the subgroups can be evaluated without regard to the ability parameters of the remainder of the total group. Equation 5.25 shows that this can be done independently of the parameters of the n items. This feature of the Rasch model must be interpreted very carefully as there are a number of unstated assumptions. First, it assumes that the persons in the subgroups belong to the same total group, that is, are drawn (not necessarily at random) from a population of persons possessing the latent trait. Second, as Whitely and Dawis (1974) point out, it assumes that the items behave the same in all such subgroups; that is, the item parameters are group invariant. Similarly, if all N subjects respond to a subset of items, the corresponding item parameters can be estimated without regard to the parameter values of the remaining items. Equation 5.26 shows that this can be done independently of the ability parameters. Now, these two features of the Rasch model are simply a reflection of a property of all IRT models; namely, the item characteristic curve is a functional relationship between the probability of correct response and the latent trait or ability continuum. In this regard, the Rasch model is no different than other models. These two features are also a reflection of the fundamental assumption that subjects and items are independent (see Whitely & Dawis, 1974, p. 168).

5.4 Specific Objectivity

Given these two features of the model, Rasch (1966b, p. 104–105) defined what he called "specific objectivity" in the following manner:

> The comparison of any two subjects can be carried out in such a way that no parameters are involved other than those of the two subjects . . . Similarly, any two stimuli can be compared independently of all other parameters than those of the two stimuli, the parameters of all other stimuli as well as the parameters of the subjects having been replaced with observable numbers. It is suggested that comparisons carried out under such circumstances be designated as specific objective.

In addition, he indicated (Rasch, 1960, p. 155; 1961, p. 325) that specific objectivity also involved "checks on the model which are independent of all parameters, relying only on the observations." What Rasch is essentially saying is that, using his model, a metric for the latent trait continuum can be established, the ability and item parameters will be expressed in this metric, and that this metric can be determined from the observable data, namely, the item-by-person response matrix. Once this metric is established, interperson or interitem comparisons are possible. In addition, these comparisons do not depend on the form of the population or item distributions. Rasch (1960, p. 108) recognized that there was an in determinacy in this metric and suggested ways for coping with it (the identification problem of Chapter 4).

It is somewhat difficult to establish exactly what was meant by "specific objectivity" from Rasch. Fortunately, Scheiblechner (1977) has provided a lucid explanation, which will be followed closely below. For scientific statements to be "objective," two conditions must hold. First, the statement must be independent of individual components; that is, they must not be situationally dependent, for example dependent on who scored the test items.

Second, the statements must be independent of "their specific coordinate system." This means that the truth of the statement must hold under a class of transformations on the underlying measurement system, for example, under a linear transformation. Rasch (1961) indicates that the class is not chosen arbitrarily. In his definition of specific objectivity, Rasch (1966b) dealt with comparisons rather than statements or measurements since a comparison is the most elementary case. Scheiblechner (1977) then states ". . . a comparison is specifically objective if it depends exclusively on properties residing in the objects and is invariant with respect to the instruments by means of which the comparison is made." Wright (1968) called this "item-free person calibration." Thus, the comparison of the ability levels of two examinees should be invariant with respect to which subsets of items are used to measure their abilities. When one has precalibrated items, this is a characteristic of all IRT models. According to Rasch (1960, 1961), the necessary and sufficient conditions for specifically objective comparisons are:

1. Completeness of the parameter space
2. Solvability
3. Continuity and differentiability with respect to the parameters

The first condition indicates that the ability and item difficulty parameters are the only parameters involved in defining the probability of a correct response. The second condition requires that all inverse functions of the probability function, that is, the item characteristic curve function, must exist. Scheiblechner (1977) states that these two conditions imply that the relation between two examinees must be the same as the relation between the two instruments to which they react equivalently, that is, the principle of equivalent reactions. This principle was used by Rasch when he established his model. The basic starting point was the ratio $\eta_1/\delta_1 = K\eta_2/K\delta_2$, which represented the probability of correct response to items of different difficulty by examinees of different ability, where K was the multiplicative factor defined by $\eta_1/\eta_2 = \delta_1/\delta_2 = K$. Finally, specific objectivity requires that inferences on the structural parameters must not depend on the incidental parameters.

Thus, comparison of the difficulty of two items should not depend on the ability levels of the groups used to measure the difficulty of the items. One must be a bit careful here since the issue of structural and incidental parameters has not gone completely away. Early in the presentation of Rasch's approach, there was an ability parameter for each examinee (see Equation 5.1) and since, as $N \to \infty$, the number of parameters approaches ∞, the ability parameters are incidental parameters. Later in the presentation, identification of the individual examinees was abandoned, and attention turned to raw score groups (see Equation 5.12). Eventually, it was shown that all persons having the same raw score will obtain the same estimated ability (see Equation 5.15). This was a consequence of the $\alpha_i = 1$ in the item characteristic curve model. Thus, under the Rasch model, one cannot, in fact, obtain a unique ability estimate for each examinee, and the number of ability parameters does not increase as sample size increases. The powerful result is that minimum sufficient statistics exist for this reduced set of ability parameters.

Underlying the whole issue of specific objectivity is a basic property of all IRT models, namely, that they establish a common metric for the item and ability parameters. Because of this, it is clear that specific objectivity is essentially a restatement of the invariance principle common to all IRT models. Some controversy has arisen with regard to the concept of specific objectivity (see Whitely & Dawis, 1974; Wright, 1977b; Whitely, 1977). Unfortunately, the discussions were not narrowly focused on specific objectivity and became a comparison of IRT and classical test theory. Upon examining the Rasch model from a classical theory point of view, Whitely and Dawis (1974) concluded:

> ... applying the Rasch model to typical trait data does not necessarily yield objective measurement, since some of the claimed advantages of applying the model depend directly upon the characteristics of the item pool rather than the model. For an item pool to fully possess objective measurement, a set of rigorous conditions must be met.

Wright (1977b), in his rejoinder, indicated that this basic conclusion is correct, but that, when these conditions are met or at least closely approximated, "specific objectivity will be attained."

5.5 Conditional Maximum Likelihood Estimation Procedures

Rasch (1960, 1966a) indicated that the expressions for the conditional probabilities (Equations 5.25 and 5.26) could be used for estimation purposes. He provided a rather general approach (Rasch, 1960, Chapter 10) and one implementation (Rasch, 1960, Chapter 6), but neither were satisfactory in practice. A maximum likelihood approach to parameter estimation based on conditional probabilities was developed by Andersen (1972, 1973a). Because of the problem of structural and incidental parameters, this conditional maximum likelihood estimation procedure is applied only to estimating the item difficulty parameters. Once the item pool is calibrated, the ability estimate corresponding to each raw score can easily be obtained. In the preceding section, the work due to Rasch was presented in the same fashion as it appeared in the literature. It is now necessary to shift to the logistic form of Rasch's item characteristic curve. This is done to bring the results back into the common mathematical framework used earlier and to align the Rasch model more closely with the bulk of IRT. In doing so, Andersen's (1972) general approach, particularized to the case of dichotomous response (Andersen, 1973a; Wright & Douglas, 1977b), will be followed.

When the free response of examinee j to item i is dichotomously scored, under Rasch's model, the probability of response is given by

$$P(u_{ij}|\theta_j, \beta_i) = \frac{e^{u_{ij}(\theta_j - \beta_i)}}{1 + e^{(\theta_j - \beta_i)}}. \tag{5.27}$$

When the item is answered correctly, $u_{ij} = 1$, and

$$P(u_{ij} = 1|\theta_j, \beta_i) = \frac{e^{(\theta_j - \beta_i)}}{1 + e^{(\theta_j - \beta_i)}} = \frac{1}{1 + e^{-(\theta_j - \beta_i)}} \tag{5.28}$$

which is called the one-parameter logistic ICC model. Note also that the logarithmic transformations of the ability and item difficulty parameters (see Equation 5.5) are used. Under the assumption of local independence, the probability of person j yielding item response vector (u_{ij}) when responding to the n items of the test is given by

$$
\begin{aligned}
P((u_{ij})|\theta_j, (\beta_i)) &= \prod_{i=1}^{n} P(u_{ij}|\theta_j, \beta_i) = \prod_{i=1}^{n} \frac{\exp[u_{ij}(\theta_j - \beta_i)]}{1 + e^{(\theta_j - \beta_i)}} \\
&= \frac{\exp\left(\sum_{i=1}^{n} u_{ij}\theta_j - \sum_{i=1}^{n} u_{ij}\beta_i\right)}{\prod_{i=1}^{n}\left[1 + e^{(\theta_j - \beta_i)}\right]} \\
&= \frac{\exp\left(r_j\theta_j - \sum_{i=1}^{n} u_{ij}\beta_i\right)}{\prod_{i=1}^{n}\left[1 + e^{(\theta_j - \beta_i)}\right]}
\end{aligned} \tag{5.29}
$$

which is analogous to Equations 5.8–5.11. As was the case in section 5.3, it is necessary to find the probability of a given test score r_j. A person of ability θ_j can obtain a given test score r_j in $_nC_{r_j}$ different ways and the sum of the probabilities of these individual ways is the probability that $u_{.j} = r_j$. Thus,

$$
\begin{aligned}
P(u_{.j} = r_j|\theta_j, (\beta_i)) &= \sum_{(u_{ij})}^{r_j} P((u_{ij})|\theta_j, (\beta_i)) \\
&= \frac{e^{r_j\theta_j} \sum_{(u_{ij})}^{r_j} \exp\left(-\sum_{i=1}^{n} u_{ij}\beta_i\right)}{\prod_{i=1}^{n}\left[1 + e^{(\theta_j - \beta_i)}\right]} \\
&= \frac{e^{r_j\theta_j}\gamma_{r_j}}{\prod_{i=1}^{n}\left[1 + e^{(\theta_j - \beta_i)}\right]},
\end{aligned} \tag{5.30}
$$

where γ_{r_j} is the elementary symmetric function. Since the logarithmic transformations of the parameters have been taken, the expressions for the γ_{r_j} will reflect the transformation as follows:

$$\gamma_{r_j} = \sum_{(u_{ij})}^{r_j} \exp\left(-\sum_{i=1}^{n} u_{ij}\beta_i\right) = \gamma(r_j; \beta_1, \beta_2, \ldots, \beta_n).$$

Letting $d(\theta_j) = \prod_{i=1}^{n}[1 + e^{(\theta_j - \beta_i)}]$ yields the expression for the probability of person j having score r_j

$$P(u_{.j} = r_j | \theta_j, (\beta_i)) = \frac{e^{r_j\theta_j}\gamma_{r_j}}{d(\theta_j)} \tag{5.31}$$

which is analogous to Equation 5.12. Then the conditional probability of the item response vector (u_{ij}), given that person j has test score r_j, is obtained by dividing Equation 5.29 by Equation 5.31, resulting in

$$
\begin{aligned}
P((u_{ij})|u_{.j} = r_j) &= \frac{P((u_{ij})|\theta_j, (\beta_i))}{P(u_{.j} = r_j|\theta_j, (\beta_i))} \\
&= \frac{\left[\exp\left(r_j\theta_j - \sum_{i=1}^{n} u_{ij}\beta_i\right)\right] / d(\theta_j)}{\left[e^{r_j\theta_j}\gamma_{r_j}\right] / d(\theta_j)} \\
&= \frac{\exp\left(-\sum_{i=1}^{n} u_{ij}\beta_i\right)}{\gamma(r_j; \beta_1, \beta_2, \ldots, \beta_n)}
\end{aligned}
\tag{5.32}
$$

which is analogous to Equation 5.15. Notice here that r_j is used in place of the person's ability parameter θ_j; hence, this probability is not a function of the person's ability level. Thus, the probability of the item response vector is conditional on the test score and the set of item parameters (β_i). If N persons have responded to the set of items and the responses from different persons are assumed independent, the conditional probability distribution of the matrix of item responses, given the values of the raw scores $r_1, r_2, \ldots, r_N$ of the N persons, is

$$P([u_{ij}]|r_1, r_2, \ldots, r_N, \beta_1, \beta_2, \ldots, \beta_n) = \frac{\exp\left(-\sum_{j=1}^{N}\sum_{i=1}^{n} u_{ij}\beta_i\right)}{\prod_{j=1}^{N} \gamma(r_j; \beta_1, \beta_2, \ldots, \beta_n)} \tag{5.33}$$

which is analogous to Equation 5.26. Now Equation 5.33 can be used as the likelihood function for the estimation of the item parameters $\beta_1, \beta_2, \ldots, \beta_n$. At this point, Anderson (1973a) takes advantage of the fact that the possible values of the test score r (i.e., $0, 1, 2, \ldots, n$) is usually less than the number of persons possessing these scores. As a result, the denominator of Equation 5.33 can be rewritten as

$$\prod_{j=1}^{N} \gamma(r_j; \beta_1, \beta_2, \ldots, \beta_n) = \prod_{r=0}^{n} [\gamma(r; \beta_1, \beta_2, \ldots, \beta_n)]^{f_r},$$

where f_r is the number of persons having score r. Letting $s_i = \sum_{j=1}^{N} u_{ij}$ be the item score and $\beta \equiv (\beta_i) = (\beta_1, \beta_2, \ldots, \beta_n)$, Equation 5.33 can be written as

$$l = \frac{\exp\left(-\sum_{i=1}^{n} s_i \beta_i\right)}{\prod_{r=0}^{n} [\gamma(r, \beta)]^{f_r}}. \tag{5.34}$$

To obtain maximum likelihood estimates of the item parameters, it is necessary to take derivatives of the likelihood with respect to these parameters. In order to do this, it is necessary to be able to take derivatives of the symmetric functions. Andersen (1973a) provides the following results:

$$\gamma(r, \beta) = \sum_{(u_{ij})}^{r} \exp\left(-\sum_{i=1}^{n} u_{ij}\beta_i\right), \tag{5.35}$$

where j is fixed for a given person. Take the derivative of Equation 5.35 with respect to an arbitrary item parameter, say, β_h, yields

$$\frac{\partial}{\partial \beta_h}\gamma(r, \beta) = -\sum_{(u_{ij})}^{r} u_{hj} \exp\left(-\sum_{i=1}^{n} u_{ij}\beta_i\right)$$

$$= -e^{-\beta_h} \sum_{(u_{ij})}^{r-1} \exp\left(-\sum_{i=1}^{n} u_{ij}\beta_i\right),$$

where i cannot equal h. Letting

$$\gamma(r-1, \beta(h)) = \sum_{(u_{ij})}^{r-1} \exp\left(-\sum_{i=1}^{n} u_{ij}\beta_i\right),$$

where $i \neq h$, and $\beta(h)$ denotes that item h has been removed from the vector of item parameters. Then,

$$\frac{\partial}{\partial \beta_h}\gamma(r, \beta) = -e^{-\beta_h}\gamma(r-1, \beta(h)) = -e^{-\beta_h} \sum_{(u_{ij})}^{r-1} \exp\left(-\sum_{i=1}^{n} u_{ij}\beta_i\right). \tag{5.36}$$

The second derivative of the symmetric functions with respect to the item parameters will also be needed. Let h, k denote an arbitrary pair of items, and then a typical member of the set of n^2 second-order partial derivatives will be

$$\frac{\partial^2}{\partial\beta_h\partial\beta_k}\gamma(r,\beta) = \begin{cases} e^{-\beta_h}\gamma(r-1,\beta(h)) & \text{for } h = k \\ e^{-\beta_h-\beta_k}\gamma(r-2,\beta(h,k)) & \text{for } h \neq k, \end{cases} \tag{5.37}$$

where $\beta(h,k)$ denotes that items h and k have been removed from the vector of item parameters.

With these derivatives at hand, the ordinary maximum likelihood procedures can be used. Since the raw score of zero and n yield infinite maximum likelihood estimators, the range of possible values of r is restricted to $r = 1, 2, \ldots, (n-1)$. From Equation 5.34, the log likelihood is

$$L = \log l = -\sum_{i=1}^{n} s_i\beta_i - \sum_{r=1}^{n-1} f_r \log \gamma(r,\beta). \tag{5.38}$$

There will be n first derivatives with respect to the item parameters of the following general form:

$$\frac{\partial L}{\partial\beta_h} = -s_h - \sum_{r=1}^{n-1} f_r \frac{\partial}{\partial\beta_h} \log \gamma(r,\beta) \qquad \text{for } h = 1, 2, \ldots, n.$$

Substituting from Equation 5.36 yields

$$L_h = \frac{\partial L}{\partial\beta_h} = -s_h + \sum_{r=1}^{n-1} f_r \frac{e^{-\beta_h}\gamma(r-1,\beta(h))}{\gamma(r,\beta)} = 0 \tag{5.39}$$

for $h = 1, 2, \ldots, n$. Normally, n of these equations would be solved simultaneously for the estimates of the β_h.

Typically, the set of n equations defined by Equation 5.39 is not solved directly, but the iterative Newton-Raphson procedure is used. In order to employ this technique, the n^2 second-order derivatives of the log-likelihood are also needed. There will be n derivatives, where $h = k$, of the general form (Wright & Douglas, 1977b),

$$\begin{aligned} L_{hh} &= \frac{\partial^2}{\partial\beta_h^2} = \frac{1}{\gamma(r,\beta)}\frac{\partial^2}{\partial\beta_h^2}\gamma(r,\beta) - \frac{1}{\gamma^2(r,\beta)}\left[\frac{\partial\gamma(r,\beta)}{\partial\beta_h}\right]^2 \\ &= \frac{e^{-\beta_h}\gamma(r-1,\beta(h))}{\gamma(r,\beta)} - \left[\frac{e^{-\beta_h}\gamma(r-1,\beta(h))}{\gamma(r,\beta)}\right]^2. \end{aligned} \tag{5.40}$$

There will be $n(n-1)$ second-order derivatives, in which $h \neq k$, having the general form (Wright & Douglas, 1977b),

$$\begin{aligned} L_{hk} &= \frac{\partial^2}{\partial\beta_h\partial\beta_k} \\ &= \frac{1}{\gamma(r,\beta)}\frac{\partial^2}{\partial\beta_h\partial\beta_k}\gamma(r,\beta) - \frac{1}{\gamma^2(r,\beta)}\left[\frac{\partial\gamma(r,\beta)}{\partial\beta_h}\right]\left[\frac{\partial\gamma(r,\beta)}{\partial\beta_k}\right] \\ &= \frac{e^{-\beta_h-\beta_k}\gamma(r-2,\beta(h,k))}{\gamma(r,\beta)} - \left[\frac{e^{-\beta_h}\gamma(r-1,\beta(h))}{\gamma(r,\beta)}\right]\left[\frac{e^{-\beta_k}\gamma(r-1,\beta(k))}{\gamma(r,\beta)}\right]. \end{aligned} \tag{5.41}$$

It should be noted that, in both Equations 5.40 and 5.41, the approximation

$$\frac{\partial^2 f(x_i)}{\partial x_i \partial x_i'} = \left[\frac{\partial f(x_i)}{\partial x_i}\right] \times \left[\frac{\partial f(x_i)}{\partial x_i'}\right]$$

has been used to simplify the mathematics.

Given these derivatives, the Newton-Raphson equation can be established that will be solved iteratively for the conditional estimates of the item difficulty parameters,

$$
\begin{bmatrix} \hat{\beta}_1 \\ \hat{\beta}_1 \\ \vdots \\ \hat{\beta}_n \end{bmatrix}_{t+1}
=
\begin{bmatrix} \hat{\beta}_1 \\ \hat{\beta}_1 \\ \vdots \\ \hat{\beta}_n \end{bmatrix}_{t}
-
\begin{bmatrix} \hat{L}_{11} & \hat{L}_{12} & \cdots & \hat{L}_{1n} \\ \hat{L}_{21} & \hat{L}_{22} & \cdots & \hat{L}_{2n} \\ \vdots & \vdots & \ddots & \vdots \\ \hat{L}_{n1} & \hat{L}_{n2} & \cdots & \hat{L}_{nn} \end{bmatrix}_{t}^{-1}
\times
\begin{bmatrix} \hat{L}_1 \\ \hat{L}_2 \\ \vdots \\ \hat{L}_n \end{bmatrix}_{t}.
\tag{5.42}
$$

As was the case with previous item parameter estimation procedures, there is an indeterminacy in the definition of the underlying metric. Andersen (1973a, 1977) imposed the linear constraint $\sum_{i=1}^{n} \beta_i = 0$. Wright and Douglas (1977b) implemented this by subtracting the derivatives for one item from those of the remaining items, thus eliminating one item, say, item n from the estimation process. The resulting derivatives were

$$L_h' = L_h - L_n$$

for $h = 1, 2, \ldots, n - 1$, and

$$L_{hk}' = L_{hk} - L_{hn} - L_{nk} + L_{nn}$$

for $h = 1, 2, \ldots, n-1$ and $k = 1, 2, \ldots, n-1$. The resulting Newton-Raphson equation is

$$
\begin{bmatrix} \hat{\beta}_1 \\ \hat{\beta}_1 \\ \vdots \\ \hat{\beta}_{n-1} \end{bmatrix}_{t+1}
=
\begin{bmatrix} \hat{\beta}_1 \\ \hat{\beta}_1 \\ \vdots \\ \hat{\beta}_{n-1} \end{bmatrix}_{t}
-
\begin{bmatrix} \hat{L}_{11}' & \hat{L}_{12}' & \cdots & \hat{L}_{1(n-1)}' \\ \hat{L}_{21}' & \hat{L}_{22}' & \cdots & \hat{L}_{2(n-1)}' \\ \vdots & \vdots & \ddots & \vdots \\ \hat{L}_{(n-1)1}' & \hat{L}_{(n-1)2}' & \cdots & \hat{L}_{(n-1)(n-1)}' \end{bmatrix}_{t}^{-1}
\times
\begin{bmatrix} \hat{L}_1' \\ \hat{L}_2' \\ \vdots \\ \hat{L}_{n-1}' \end{bmatrix}_{t}.
\tag{5.43}
$$

Wright and Douglas (1977b) have implemented the estimation process via the following steps:

1. Obtain initial estimates of the item difficulty parameters using

$$\hat{\beta}_i^{(0)} = \log\left(\frac{N - s_i}{s_i}\right) - \frac{\sum_{i=1}^{n} \log[(N - s_i)/s_i]}{n}.$$

2. Using the current set of $\hat{\beta}_i$, evaluate the symmetric functions, $\gamma(r, \beta)$, $\gamma(r - 1, \beta(h))$, and $\gamma(r - 2, \beta(h, k))$.
3. Using these values, evaluate the first and second derivatives of the log likelihood function, $\hat{L}_h$, $\hat{L}_{hn}$, and $\hat{L}_{hk}$.

4. Reduce the number of item parameters to be estimated by adjusting the derivatives to $\hat{L}'_h$, $\hat{L}'_{hn}$, and $\hat{L}'_{hk}$, where $h = 1, 2, \ldots, n - 1$ and $k = 1, 2, \ldots, n - 1$.

5. Solve the Newton-Raphson Equation 5.43 iteratively to obtain improved estimates of the $n - 1$ item parameters. Set $\hat{\beta}_n = -\sum_{i=1}^{n-1} \hat{\beta}_i$.

6. Repeat steps 2–5 until the convergence criterion,

$$\sum_{i=1}^{n-1} \frac{(\hat{\beta}_i^{(t+1)} - \hat{\beta}_i^{(t)})^2}{n} < 0.001,$$

is met.

Because this estimation process is conditional on the sufficient statistics r for ability, it yields only item parameter estimates. Wright and Douglas (1977b) indicate that this procedure, called FCON for fully conditional, is extremely sensitive to round-off errors in the calculation of the symmetric functions. As a result, they concluded that it was not practical for more than 10–15 items. Part of the problem is that all n item difficulty parameters are being estimated simultaneously and the matrix of second-order partial derivatives includes the $\hat{L}_{hk}$ terms. If one assumes that the items are independent, the off-diagonal terms can be deleted. This reduces the matrix to a diagonal matrix and also eliminates the matrix inversion calculations that are also prone to round-off error accumulation. The resulting procedure has been called ICON, incomplete conditional estimation, by Wright and Douglas (1977b). This procedure delays the round-off error problem until 20 to 30 items are employed, but Gustafsson (1980a) reports that convergence of the overall estimation process is rather slow. These authors provide some evidence (Wright & Douglas, 1975) that the two procedures provide nearly identical results where both can be employed. As a result, they recommend the ICON procedure for conditional estimation of the item difficulty parameters. However, ICON is also limited to a rather small number of items.

The limiting aspect of the conditional maximum likelihood estimation procedure is the computation of the elementary symmetric functions on a digital computer. Gustafsson (1980a) states that a symmetric function of order r consists of the sum of $_nC_r$ products, each of which consists of r terms. When $n = 50$ and $r = 25$, the symmetric function is defined as a sum of about 1.264×10^{14} terms, each of which is a product of 25 terms. Thus, the computational demands are very large when the symmetric functions are computed directly. As a result, considerable effort has been devoted to finding efficient computing procedures that can handle tests with a practical number of items. Recursive formulas exist that make the rapid computation of the symmetric functions possible. The algorithm for computing the symmetric functions implemented in the FORTRAN computer program due to Fischer and Forman (1972) can handle somewhat more than 40 items and executes on an IBM PC microcomputer. Using recursive formulas for the difference algorithm, and the summation algorithm due to Fischer (1974, pp. 242–243), Gustafsson (1980a) developed an improved algorithm. The new one combined the speed of the difference algorithm and the numerical accuracy of the summation algorithm. He reported that it was is both reasonably fast and could handle up to 80 items and, under some conditions, up to 100 items. Supporting evidence as to the accuracy of this algorithm is limited (see Gustafsson, 1980a; Wainer, Morgan, & Gustafsson, 1980). However, it does appear that the symmetric functions can be computed efficiently for up to 80 items, and this encompasses most tests of interest. Sanathanan (1974) reported a procedure attributed to Scheiblechner (1971) for the iterative solution of the conditional likelihood equations that avoided the computation of the symmetric functions. While several computer programs for the conditional maximum likelihood estimation of item parameters are available in Europe (Gustafsson, 1977, Fischer & Forman, 1972), they have not been widely circulated in the United States. The lack of a readily available computer program is a significant deficiency of the conditional maximum likelihood estimation approach. If the approach is to become widely used, it must be implemented in a user-friendly and well-supported computer program.

In the discussion above, only the item difficulty parameter estimates were obtained via conditional maximum likelihood estimation. Conceptually, one can turn the approach around and obtain ability estimates conditional on the sufficient statistics for the item parameters. Wainer, Morgan, and Gustafsson (1980) report that this approach is fraught with both theoretical and computational problems. However, if the number of persons is large relative to the number of items, the ability estimation equations reduce to those obtained under the JMLE procedure. The usual practice is to treat the obtained item parameter estimates as the known item parameters and then compute the ability estimates using maximum likelihood as in Chapter 3.

5.6 Application of the JMLE Procedure to the Rasch Model

Although Rasch took a rather different approach to deriving his model, the item characteristic curve involved was a member of the logistic family. Consequently, the JMLE procedure of Chapter 4 can be employed. Wright and Panchapakesan (1969) particularized this procedure to the one-parameter logistic model and called it "unconditional maximum likelihood estimation" (UCON) since the likelihood function did not employ one of Rasch's conditional probability equations.

As was the case with the conditional maximum likelihood procedure, the logistic form of the item characteristic curve will be used. The probability of an item response is given by

$$P(u_{ij}|\theta_j, \beta_i) = \frac{\exp[u_{ij}(\theta_j - \beta_i)]}{1 + e^{(\theta_j - \beta_i)}}. \tag{5.44}$$

Under assumption 2 of the Rasch model, the likelihood of the $n \times N$ matrix of dichotomously scored item responses is

$$l = \prod_{j=1}^{N} \prod_{i=1}^{n} P(u_{ij}|\theta_j, \beta_i) = \frac{\exp\left[\sum_{j=1}^{N}\sum_{i=1}^{n} u_{ij}(\theta_j - \beta_i)\right]}{\prod_{j=1}^{N}\prod_{i=1}^{n}\left[1 + e^{(\theta_j - \beta_i)}\right]}. \tag{5.45}$$

Then the log likelihood is given by

$$L = \log l = \sum_{j=1}^{N}\sum_{i=1}^{n} u_{ij}(\theta_j - \beta_i) - \sum_{j=1}^{N}\sum_{i=1}^{n} \log\left[1 + e^{(\theta_j - \beta_i)}\right]. \tag{5.46}$$

Distributing the summation signs in the first term on the right yields

$$L = \sum_{j=1}^{N}\sum_{i=1}^{n} u_{ij}\theta_j - \sum_{j=1}^{N}\sum_{i=1}^{n} u_{ij}\beta_i - \sum_{j=1}^{N}\sum_{i=1}^{n} \log\left[1 + e^{(\theta_j - \beta_i)}\right]. \tag{5.47}$$

The first two terms on the right can be simplified. The first term can be written $\sum_{j=1}^{N} \theta_j \sum_{i=1}^{n} u_{ij}$, but, for a given person j, $\sum_{i=1}^{n} u_{ij}$ is simply the test score r_j. The second term can be written $\sum_{i=1}^{n} \beta_i \sum_{j=1}^{N} u_{ij}$, but $\sum_{j=1}^{N} u_{ij}$ is the item score s_i for a given item.

The log likelihood then can be written as

$$L = \sum_{j=1}^{N} r_j\theta_j - \sum_{i=1}^{n} s_i\beta_i - \sum_{j=1}^{N}\sum_{i=1}^{n} \log\left[1 + e^{(\theta_j - \beta_i)}\right]. \tag{5.48}$$

In order to use the iterative Newton-Raphson procedure for both the item parameter and ability parameter estimation process, first and second derivatives of the likelihood function with respect to the θ_j and β_i are needed. These will be identical to those of Chapter 2 for the two-parameter logistic with $\alpha_i = 1$ in all cases and can be written by inspection from the previous results, yielding

$$\frac{\partial L}{\partial \theta_j} = L_j = r_j - \sum_{i=1}^{n} P_{ij} \tag{5.49}$$

for $j = 1, 2, \ldots, N$ and

$$\frac{\partial L}{\partial \beta_i} = L_i = -s_i + \sum_{j=1}^{N} P_{ij} \tag{5.50}$$

for $i = 1, 2, \ldots, n$, where $P_{ij} = \{1 + \exp[-(\theta_j - \beta_i)]\}^{-1}$. It is interesting to note that the first derivative with respect to ability involves the difference between the person's test score and their true score. Conversely, the first derivative with respect to item difficulty involves the difference between the item score and, so to speak, a true item score. The second derivatives are

$$\frac{\partial^2 L}{\partial \theta_j^2} = -\sum_{i=1}^{n} P_{ij}(1 - P_{ij}) = -\sum_{i=1}^{n} P_{ij}Q_{ij} \tag{5.51}$$

and

$$\frac{\partial^2 L}{\partial \beta_i^2} = -\sum_{j=1}^{N} P_{ij}(1 - P_{ij}) = -\sum_{j=1}^{N} P_{ij}Q_{ij}. \tag{5.52}$$

Since, under assumption 2, people and items are independent, the cross derivative $\partial^2 L/\partial \theta_j \partial \beta_i = 0$ for all pairs of i and j.

Although one can proceed directly from Equations 5.49–5.52, to obtain the parameter estimates, some adjustments can be made as a result of the characteristics of the item response data under the Rasch model. Since the items are dichotomously scored, only integer test scores are possible, and a number of different item response patterns, specifically $_nC_r$, can yield the same test score. In addition, it was shown above that all persons having the same test score will obtain the same estimated ability, regardless of the true value of their ability parameter. Following Rasch (1960), Wright and Panchapakesan (1969) restructured the basic data matrix. All persons having a common test score were pooled into a subgroup, and the person dimension was reduced from length N to $n - 1$. In addition, the subgroups having test scores of zero and n were discarded since they would lead to infinite maximum likelihood estimators of the corresponding ability parameters. The resulting data matrix is now of size $n \times (n-1)$, with the number of persons f_{ig} having test score r_g $(g = 1, 2, \ldots, n-1)$ answering item i correctly as the cell entry. The column marginal totals are the number of persons $f_{.g}$ with a particular test score. The row marginal totals are the number of persons $f_{i.}$ correctly answering an item which is the same as s_i. The resulting data matrix is shown below.

Test Score

Item		1	2	$\cdots$	g	$\cdots$	$n-1$	
	1	f_{11}	f_{12}	$\cdots$	f_{1g}	$\cdots$	$f_{1,n-1}$	$f_{1.} = s_1$
	2	f_{21}	f_{22}	$\cdots$	f_{2g}	$\cdots$	$f_{2,n-1}$	$f_{2.} = s_2$
	$\vdots$	$\vdots$	$\vdots$	$\ddots$	$\vdots$	$\ddots$	$\vdots$	$\vdots$
	i	f_{i1}	f_{i2}	$\cdots$	f_{ig}	$\cdots$	$f_{i,n-1}$	$f_{i.} = s_i$
	$\vdots$	$\vdots$	$\vdots$	$\ddots$	$\vdots$	$\ddots$	$\vdots$	$\vdots$
	n	f_{n1}	f_{n2}	$\cdots$	f_{ng}	$\cdots$	$f_{n,n-1}$	$f_{n.} = s_n$
		$f_{.1}$	$f_{.2}$	$\cdots$	$f_{.g}$	$\cdots$	$f_{.,n-1}$	

Although, it has not been done so above, the data matrix is also edited to remove items that are answered correctly by all or none of the examinees (see

Wright & Douglas, 1977a). Such items also lead to infinite maximum likelihood estimates for their difficulty parameters. The editing was not incorporated here in order to keep the notation as simple as possible. Restructuring
the data matrix also influences the derivative of the likelihood equation with
respect to the item parameters but not with respect to the ability parameters.

Taking into account the grouping of the examinees on the basis of their
raw scores, the item characteristic curve model becomes

$$P(u_{ig} = 1|\theta_g, \beta_i) = \frac{e^{(\theta_g - \beta_i)}}{1 + e^{(\theta_g - \beta_i)}}, \tag{5.53}$$

where θ_g is the ability level associated with raw score r_g.

It should be noted that there will be a distribution of ability for those
persons obtaining a given raw score. Under the Rasch model, all such persons
will receive the same estimated ability $\hat{\theta}_g$. Thus, θ_g is the ability level that
$\hat{\theta}_g$ estimates.

At this point, it is useful to switch from parametric terms to estimates
since, in the maximum likelihood procedures, the true P_{ig} are replaced by
values obtained if the current parameter estimates were considered to be the
true values. Then,

$$\hat{P}_{ig} = \frac{e^{(\hat{\theta}_g - \hat{\beta}_i)}}{1 + e^{(\hat{\theta}_g - \hat{\beta}_i)}}.$$

Since estimates are involved, advantage can be taken of the relationship

$$\sum_{j=1}^{N} \hat{P}_{ij} = \sum_{g=1}^{n-1} f_{.g} \hat{P}_{ig}$$

which assumes that persons with scores of 0 or n have not been included in
the summations. All this equation says is that the estimated item score is the
same before and after grouping of the ability dimension.

With these two changes (grouping and shifting to estimates), the first
derivatives of the likelihood become

$$\hat{L}_g = \frac{\partial L}{\partial \theta_g} = r_g - \sum_{i=1}^{n} \hat{P}_{ig} \tag{5.54}$$

and

$$\hat{L}_i = \frac{\partial L}{\partial \beta_i} = -s_i - \sum_{g=1}^{n-1} f_{.g} \hat{P}_{ig}. \tag{5.55}$$

The second derivatives become

$$\hat{L}_{gg} = -\sum_{i=1}^{n-1} \hat{P}_{ig} \hat{Q}_{ig} \tag{5.56}$$

and

$$\hat{L}_{ii} = -\sum_{i=1}^{n-1} f_{.g}\hat{P}_{ig}\hat{Q}_{ig}. \tag{5.57}$$

Under maximum likelihood procedures, a Newton-Raphson equation would be established to estimate the vectors of parameters (θ_g) and (β_i) simultaneously. The equation would be

$$\begin{bmatrix} (\hat{\theta}_g) \\ (\hat{\beta}_i) \end{bmatrix}_{t+1} = \begin{bmatrix} (\hat{\theta}_g) \\ (\hat{\beta}_i) \end{bmatrix}_t - \begin{bmatrix} [\hat{L}_{gg}] & [\hat{L}_{gi}] \\ [\hat{L}_{ig}] & [\hat{L}_{ii}] \end{bmatrix}_t^{-1} \times \begin{bmatrix} (\hat{L}_g) \\ (\hat{L}_i) \end{bmatrix}_t, \tag{5.58}$$

where t is the iteration index and the terms in the matrix are submatrices of second derivatives over all possible pairs of the parameters. Since persons and items are considered independent, the submatrices denoted by $[\hat{L}_{gi}]$ and $[\hat{L}_{ig}]$ will contain zeros, and the full matrix reduces to a diagonal matrix of two submatrices. This same assumption and subsequent reduction was employed in the JMLE procedure of Chapter 4. The result is that the ability parameters and the item parameters can be estimated separately. Now, under the joint estimation procedure, one set of parameters is considered to be known when the other is being estimated. The set considered to be known is alternated until convergence of both sets of parameter estimates is achieved (the well-known Birnbaum paradigm). Wright and Panchapakesan (1969) used this same paradigm. Thus, there will be a separate Newton-Raphson equation for each set of parameters. In the case of the ability parameters,

$$\begin{bmatrix} \hat{\theta}_1 \\ \hat{\theta}_2 \\ \vdots \\ \hat{\theta}_g \\ \vdots \\ \hat{\theta}_{n-1} \end{bmatrix}_{t+1} = \begin{bmatrix} \hat{\theta}_1 \\ \hat{\theta}_2 \\ \vdots \\ \hat{\theta}_g \\ \vdots \\ \hat{\theta}_{n-1} \end{bmatrix}_t -$$

$$\begin{bmatrix} \hat{L}_{11} & \hat{L}_{12} & \cdots & \hat{L}_{1g} & \cdots & \hat{L}_{1,n-1} \\ \hat{L}_{21} & \hat{L}_{22} & \cdots & \hat{L}_{2g} & \cdots & \hat{L}_{2,n-1} \\ \vdots & \vdots & \ddots & \vdots & \ddots & \vdots \\ \hat{L}_{g1} & \hat{L}_{g2} & \cdots & \hat{L}_{gg} & \cdots & \hat{L}_{g,n-1} \\ \vdots & \vdots & \ddots & \vdots & \ddots & \vdots \\ \hat{L}_{n-1,1} & \hat{L}_{n-1,2} & \cdots & \hat{L}_{n-1,g} & \cdots & \hat{L}_{n-1,n-1} \end{bmatrix}_t^{-1} \times \begin{bmatrix} \hat{L}_1 \\ \hat{L}_2 \\ \vdots \\ \hat{L}_g \\ \vdots \\ \hat{L}_{n-1} \end{bmatrix}_t. \tag{5.59}$$

Since persons are considered to be independent, the off-diagonal terms in the matrix of second derivatives will be zero, and the matrix reduces to a diagonal matrix. The inverse of this matrix will simply be a diagonal matrix of the reciprocals of the derivatives. The algebra of Equation 5.59 is such that $n - 1$ independent equations are obtained, one for each θ_g, which can be solved independently of the rest. The equations will have the following general form:

$$\hat{\theta}_g^{(t+1)} = \hat{\theta}_g^{(t)} - \left(\frac{r_g - \sum\limits_{i=1}^{n} \hat{P}_{ig}}{-\sum\limits_{i=1}^{n} \hat{P}_{ig}\hat{Q}_{ig}} \right)^{(t)} \qquad g = 1, 2, \ldots, n-1. \qquad (5.60)$$

It is interesting to note that the last term is $\Delta\theta_g^{(t)}$ and is in the form of a standard score. In Equation 5.60, r_g is the observed test score, $\sum_i^n = \hat{P}_{ig}$ is the estimated true score, and the denominator $\sum_{i=1}^{n} \hat{P}_{ig}\hat{Q}_{ig}$ is the pooled estimate of the true score variance. These latter two terms treat the current values of $\hat{\theta}_g^{(t)}$ and $(\hat{\beta}_i)^{(t)}$ as the "true" values of these parameters. What is happening during the iterative Newton-Raphson procedure is that the ability parameter is being adjusted until the standardized difference between the observed test score and the estimated true score is minimized.

The Newton-Raphson equation for the item parameters is

$$
\begin{bmatrix} \hat{\beta}_1 \\ \hat{\beta}_2 \\ \vdots \\ \hat{\beta}_g \\ \vdots \\ \hat{\beta}_n \end{bmatrix}_{t+1}
=
\begin{bmatrix} \hat{\beta}_1 \\ \hat{\beta}_2 \\ \vdots \\ \hat{\beta}_i \\ \vdots \\ \hat{\beta}_n \end{bmatrix}_{t}
-
\begin{bmatrix} \hat{L}_{11} & \hat{L}_{12} & \cdots & \hat{L}_{1i} & \cdots & \hat{L}_{1n} \\ \hat{L}_{21} & \hat{L}_{22} & \cdots & \hat{L}_{2i} & \cdots & \hat{L}_{2n} \\ \vdots & \vdots & \ddots & \vdots & \ddots & \vdots \\ \hat{L}_{i1} & \hat{L}_{i2} & \cdots & \hat{L}_{ii} & \cdots & \hat{L}_{in} \\ \vdots & \vdots & \ddots & \vdots & \ddots & \vdots \\ \hat{L}_{n1} & \hat{L}_{n2} & \cdots & \hat{L}_{ni} & \cdots & \hat{L}_{nn} \end{bmatrix}_{t}^{-1}
\times
\begin{bmatrix} \hat{L}_1 \\ \hat{L}_2 \\ \vdots \\ \hat{L}_i \\ \vdots \\ \hat{L}_n \end{bmatrix}_{t}.
\qquad (5.61)
$$

Again, the items are considered to be independent, the off-diagonal second derivatives are zero, and the matrix becomes a diagonal matrix. Thus, Equation 5.61 reduces to a set of n independent equations, one for each item, that can be solved separately. The general form of these equations is

$$\hat{\beta}_i^{(t+1)} = \hat{\beta}_i^{(t)} - \left(\frac{-s_i + \sum\limits_{g=1}^{n-1} f_{.g}\hat{P}_{ig}}{-\sum\limits_{g=1}^{n-1} f_{.g}\hat{P}_{ig}\hat{Q}_{ig}} \right)^{(t)} \qquad i = 1, 2, \ldots, n. \qquad (5.62)$$

Note that the $\Delta\beta_i^{(t)}$ term is also in the form of a standard score.

The only remaining problem is how to cope with the indeterminacy of the parameters, that is, the identification problem. Rasch (1960) suggested selecting a given person as the "standard person" and arbitrarily setting their parameter to 1 or, alternatively, selecting an arbitrary item as the "standard item" and setting its difficulty value to 1. Since log $\log \eta_j$ and $\log \delta_i$ are being used here, the equivalent procedure would be to set the standard to 0.

The anchoring procedure employed (see Wright & Douglas, 1977a; Wright & Mead, 1978; Wright & Stone, 1979) is to standardize the item difficulties in a particular manner. This is done via the following equation,

$$\hat{\beta}_i^* = \frac{\hat{\beta}_i - E(\hat{\beta}_i)}{\sigma_\beta}, \tag{5.63}$$

where

$$E(\hat{\beta}_i) = \frac{\sum_{i=1}^{n} \hat{\beta}_i}{n}$$

and σ_β is set to an a priori value of unity. The result is to center the estimated item difficulties on a value of zero, and the origin of the obtained metric has been shifted. It can be shown easily that anchoring the $\hat{\beta}_i$ at zero also anchors the metric of the ability estimates at the same point. Before anchoring $E(\hat{\beta}_i) = 0.5$, $\alpha_i = 1.0$, $\theta_g = -0.5$, $Z_{ig} = -1.0$, and $P_i(\theta_g) = .2689$. Now anchoring the $E(\hat{\beta}_i)$ at zero yields the logit $Z_{ig}^* = 1(\theta_g^* - 0)$ and, in order to retain the proper value of $P_i(\theta_g)$ is retained, $\theta_g^* = -1.0$. Thus, the ability scale metric has been shifted by the same amount. It should also be noted that anchoring the $\hat{\beta}_i$ at zero means that the median of the test characteristic curve will correspond to a raw score of $n/2$ or one-half the raw score range after editing, as well as to a value of zero for $\hat{\theta}_g$.

5.6.1 Implementation of the JMLE paradigm

With these preliminaries in hand, the two-stage iterative maximum likelihood procedure for estimation of the ability and item parameters under the Rasch model is identical to that used with the previous ICC models (see Chapter 4). Only the minor details will differ. To make this explicit, the two-stage paradigm due to Wright and Panchapakesan (1969), detailed by Wright and Douglas (1975, 1977b) and Wright and Stone (1979), and extracted from the actual FORTRAN code for the BICAL computer program (Wright & Mead, 1978) is presented below. An abridged version of the program, written in MICROSOFT QUICKBASIC, is given in Appendix C to illustrate both the JMLE paradigm and parameter estimation under the Rasch model.

1. Preliminary estimates of the $\hat{\theta}_g$ and $\hat{\beta}_i$ are obtained using a technique called PROX due to Cohen (1979). It assumes that the examinees are normally distributed over the log ability scale.

   a) The basal set of item parameter estimates are defined using

   $$\hat{\beta}_i^{(0)} = \log\left(\frac{N - s_i}{s_i}\right) - \frac{\sum_{i=1}^{n} \log\left[(N - s_i)/s_i\right]}{n},$$

where for each item, $\hat{\beta}_i^{(0)}$ is the maximum likelihood estimate of β_i for a sample of N persons having the same ability level (Wright & Douglas, 1977a, p. 284). These basal values are anchored at a value of zero. The basal item parameter estimates also can be related to the log odds. $(N - s_i)/N$ is the sample probability q_i of getting item i wrong, while s_i/N is the sample probability p_i of getting the item correct. Then,

$$\frac{q_i}{p_i} = \frac{(N - s_i)/N}{s_i/N} = \frac{N - s_i}{s_i}$$

is the odds of getting item i wrong. Thus, the basal estimates of β_i is a function of the logarithm of the odds ratio. Wright and Stone (1979) call it the sample score logit of an item.

b) The basal set of $\hat{\theta}_g$ are defined as

$$\hat{\theta}_g^{(0)} = \log \left(\frac{r_g}{n - r_g} \right),$$

where, for each score, $\hat{\theta}_g^{(0)}$ is the maximum likelihood estimate of θ_g for a test of n equivalent items centered at a score of zero (Wright & Douglas, 1977a, p. 284). Again, a log odds interpretation is possible. $(n - r_g)/n$ is the sample probability of not getting score r_g, while r_g/n is the sample probability of getting score r_g. Then,

$$\frac{r_g/n}{(n - r_g)/n} = \frac{r_g}{n - r_g}$$

is the odds of getting score r_g. Thus, the basal estimate of θ_g is the logarithm of an odds. Wright and Stone (1979, p. 22) refer to it as the item score logit of a person.

c) These basal estimates are then jointly rescaled, taking the variances of the $\hat{\beta}_i^{(0)}$ and $\hat{\theta}_g^{(0)}$ into account. This rescaling sets the means of both sets of parameter estimates to zero. The resulting $\hat{\beta}_i$ and $\hat{\theta}_g$ are then considered the initial parameter estimates for the two-stage iterative joint estimation paradigm.

2. The Newton-Raphson procedure, Equation 5.62, is iterated up to 10 times or until $|\hat{\beta}_i^{(t+1)} - \hat{\beta}_i^{(t)}| < .05$ for each of the n items. The result is a set of improved item parameter estimates.

3. The improved $\hat{\beta}_i$ are anchored at zero by subtracting $\bar{\hat{\beta}}$ from each estimate.

4. Treating the improved and reanchored set of $\hat{\beta}_i$ as the true parameter values, the Newton-Raphson procedure, Equation 5.58, is performed separately for each raw score. The iterative process is performed until either five iterations have been completed or the convergence criterion $|\hat{\theta}_g^{(t+1)} - \hat{\theta}_g^{(t)}| < .05$ is met. The result is an improved set of ability parameter estimates.

5. Steps 2–4 are repeated until the overall convergence criterion

$$\frac{\sum_{i=1}^{n} \left| \hat{\beta}_i^{(y+1)} - \hat{\beta}_i^{(y)} \right|}{n} < .025$$

is met, where y indexes the stage in the overall paradigm. It should be noted that this criterion does not agree with the

$$\frac{\sum_{i=1}^{n} \left| \hat{\beta}_i^{(y+1)} - \hat{\beta}_i^{(y)} \right|}{n} < .0001$$

reported by Wright and Stone (1979, p. 62) or Wright and Douglas (1977a, p. 285).

6. The final $\hat{\beta}_i$ are corrected for bias;

$$\hat{\beta}_i = \hat{\beta}_i \left(\frac{n-1}{n} \right).$$

7. An additional stage of ability estimation is then performed using these correct item parameters in the Newton-Raphson procedure (Equation 5.58). The ability estimates yielded by this additional stage are then corrected for bias as follows:

$$\hat{\theta}_g = \hat{\theta}_g \left(\frac{n-2}{n-1} \right).$$

8. The asymptotic estimates of the standard errors of each item difficulty estimates can be found via the negative reciprocal of the second derivatives of the log-likelihood used in the Newton-Raphson procedure, Equation 5.62:

$$SE(\hat{\beta}_i) = \frac{1}{\sqrt{\sum_{g=1}^{n-1} f_{.g} \hat{P}_{ig} \hat{Q}_{ig}}}.$$

The standard error of the ability estimates from Equation 5.60 is

$$SE(\hat{\theta}_g) = \frac{1}{\sqrt{\sum_{i=1}^{n} \hat{P}_{ig} \hat{Q}_{ig}}}$$

and the test information function is

$$I(\hat{\theta}_g) = \sum_{i=1}^{n} \hat{P}_{ig} \hat{Q}_{ig}$$

which is identical to that for the two-parameter logistic when all $\alpha_i = 1$.

Other than obtaining the initial estimates and the anchoring scheme, this two-stage procedure is the same as Birnbaum's JMLE procedure. Under previous applications, there was a set of two or three simultaneous equations to be solved iteratively at each stage for the item parameters α_i, β_i, and possibly c_i. Since $\alpha_i = 1$ for the Rasch model and there is no allowance for guessing, these reduce to a single equation for the β_i of each item. In addition, the equation for the estimation of ability under the two-parameter model reduces identically to that of the Rasch model (Equation 5.59) when α_i is set equal to unity. Grouping the subjects by raw score groups under the Rasch model reduces the number of ability parameters to be estimated from N to $n - 1$. This reduction is a bit misleading, however, as there are still N ability parameters involved. Because of the restriction of $\alpha_i = 1$, the Rasch model is precluded from yielding N unique ability estimates. The best it can do is produce $n - 1$ ability estimates, one for each raw score except 0 and n. A natural inclination is to assume the true ability levels associated with the subgroup of persons having a common raw score are homogeneous. It is quite easy to show (see Birnbaum, 1968, Chapter 17) that, for all ability levels, a probability exists that a person of ability θ_i, will obtain any one of the possible raw scores. Conversely, for a given raw score, there will be a distribution of true ability levels for the persons in a subgroup, all of whom share a common raw test score. What the Rasch model has done is trade off the differentiation among persons obtaining the same raw score for simplicity of the item characteristic curve model. Under the Rasch model a large number $_nC_r$ of item response patterns yields a single value of the ability estimate. One can conceive of a person of low true ability, say, $\theta_j = -1$, and a person of high true ability, say, $\theta_j = 1$, both getting a raw score of r. The Rasch model would give both persons the same estimated ability. In contrast, the two- and three-parameter models would yield different ability estimates for these two examinees as long as they answered a different set of items correctly and the item discrimination parameters differed. Thus, the two- and three-parameter models offer a finer degree of differentiation among examinee's estimated ability at the cost of more complex models and somewhat more involved computational procedures. The lack of differentiation among persons having the same raw score has a certain appeal to the practitioner. It is intuitively appealing to have all persons with the same raw score obtain the same estimated ability (also, see Wright & Stone, 1979, Chapter 5).

5.6.2 Bias of the Parameter Estimates

Wright and Douglas (1975, 1977a) reported the item parameters yielded by the UCON (JMLE) procedure were biased relative to those yielded by the conditional maximum likelihood procedure. The magnitude of the bias was determined by comparing the estimation equation for $\hat{\beta}_i$ under the UCON procedure with that under the conditional procedure. In the latter case, the item of interest was removed from the expression for the ability estimates used to evaluate the symmetric functions. Wright and Douglas were able to show algebraically that, if after rescaling the item difficulty estimates to mean zero, they are multiplied by the factor $(n - 1)/n$, the bias in the $\hat{\beta}_i$ was removed. They reported a small Monte Carlo type of study for tests of different lengths and item difficulty distributions. The study confirmed that the correction factor yielded values of $\hat{\beta}_i$ that agreed very closely with those obtained via conditional estimation. Wright and Douglas (1977a, p. 289) concluded, "Thus, we found the corrected UCON procedure to be indistinguishable in results from the conditional procedure." The latter was the approximate (i.e., PROX) calibration procedure. The validity of this bias correction factor has been questioned by Jansen, van den Wollenberg, and Wierda (1988), who indicated that the simulation study of Wright and Douglas (1977a, 1977b) was based on two assumptions that led to the $(n - 1)/n$ bias correction term only in special cases. In particular, the distribution of the item difficulties must be symmetrically distributed about zero. When a small number of items were involved and the item difficulties were asymmetrically distributed, the bias was not removed. Jansen and co-workers concluded that, when the JMLE procedure was used to estimate the parameters of under a Rasch model, the nature of the bias was more complicated that could be adjusted for by a simple multiplicative factor. In a subsequent paper (van den Wollenberg, Wierda, & Jansen, 1988), they performed a simulation study to investigate the effect of the $(n - 1)/n$ bias correction factor. When a small number of items $n < 10$ was employed, the JMLE item difficulty estimates were biased relative to those from conditional maximum likelihood (CML). In addition, the range of values of the underlying item difficulty parameters had an impact on the bias. They concluded (p. 311), "the bias is dependent not only on the number of items, but also on seemingly irrelevant characteristics of the data set such as range and skewness of the item parameter distributions." In his rejoinder to these two papers, Wright (1988) pointed out that these results were for a very small number of items $n \leq 10$, which is not a realistic test size. In addition, the van den Wollenberg, Wierda, and Jansen (1988) results showed that, for 10 items, the amount of bias was very small. Wright argued that, while the algebraic results showed that the bias correction factor was more complex than $(n - 1)/n$, the empirical results show that, for a reasonable number of items $n \geq 20$, this factor works well.

Since relative bias is a function of the standard, a bias relative to the conditional procedure is of limited interest. Of greater concern is the bias of the item difficulty estimates relative to their true values. To this end, Wright and Douglas (1977a) used known item parameters to generate item response data for samples of size 500. Fifteen replications of each combination of test length and item difficulty parameter dispersion were used. The mean absolute difference in $|\hat{\beta}_i - \beta_i|$ ranged from .08 to .12, with 9 of the 12 values being .08 or .09. Although the bias correction brings the item parameter estimates into line with those of the conditional estimation procedure, it does not bring them identical with the true values. With only 15 replications, the results are only indicative; however, the root mean squares of the biases were reasonably small, .03, in 9 of the 12 cases.

Most of the literature due to Wright and his colleagues places considerable emphasis on correcting the bias in the item parameter estimates. Yet the BICAL program (Wright & Mead, 1978) also applies a correction factor $(n - 2)/(n - 1)$ to the ability estimates yielded by the joint estimation procedure. The origin of this factor is not presented. For example, Wright and Douglas (1977a) and Wright and Stone (1979) make no mention of this correction factor.

5.7 Measuring the Goodness of Fit of the Rasch Model

One of the criteria for evaluating any model is how well it replicates the observed data. The closer the fit between what the model predicts and what is observed, the better the model. Since the estimation procedures reduce what is basically an $n+N$ parameter situation to an $n+(n-1)$ parameter situation, the goodness of fit of this simplified model is of considerable interest. In his writings, Rasch (1960, 1961, 1966a, 1966b) referred to the goodness-of-fit issue as "controlling the model." A number of different goodness-of-fit techniques have been developed for the Rasch model, and several of them will be described below.

5.7.1 Chi-square Tests for Goodness of Fit

Wright and Panchapakesan (1969) presented a procedure based on the observed and expected cell frequencies in the item by raw score data matrix, such as shown on page 146. They defined the standard score

$$z_{ig} = \frac{f_{ig} - E(f_{ig})}{\sqrt{\text{Var}(f_{ig})}}, \tag{5.64}$$

where f_{ig} is the observed number of persons with raw score g that correctly answer item i. They assumed z_{ig} was asymptotically normally distributed with mean zero and unit variance. The distribution of each cell frequency was assumed to be binomial with parameters P_{ig} and $f_{.g}$, where P_{ig} was obtained from Equation 5.5 and evaluated for a particular θ_g and β_i combination and $f_{.g}$ is the number of persons having a given raw score. Following the binomial, the expectation of f_{ig} is

$$E(f_{ig}) = f_{.g}P_{ig} = f_{.g}\left(\frac{1}{1 + e^{-(\theta_g - \beta_i)}}\right)$$

and the variance is

$$\text{Var}(f_{ig}) = f_{.g}P_{ig}Q_{ig}.$$

Since the true values of the parameters are unknown, the P_{ig} and Q_{ig} are calculated using the $\hat{\theta}_g$ and $\hat{\beta}_i$. Consequently, the estimated standard residual is given by

$$\hat{z}_{ig} = \frac{f_{ig} - f_{.g}\hat{P}_{ig}}{\sqrt{f_{.g}\hat{P}_{ig}\hat{Q}_{ig}}}. \tag{5.65}$$

There will be such a standardized residual for each cell in the $n \times (n-1)$ data matrix. The overall goodness of fit of the Rasch model to the data can be indexed by

$$\chi^2 = \sum_{i=1}^{n}\sum_{g=1}^{G} \hat{z}_{ig}^2 \qquad \text{with } (n-1)(G-1) \text{ degrees of freedom,} \tag{5.66}$$

where G is the number of score groups where $f_{.g} \neq 0$. In the case in which all scores from 1 to $n-1$ are represented, $G = n-1$, and there are $(n-1)(n-2)$ degrees of freedom. If a significant χ^2 was obtained, one would proceed as if the Rasch model was not appropriate for the data set.

Although a measure of the overall fit is of some use, it is of limited value when a lack of fit is indicated. Thus, in later work (Wright & Mead, 1978; Wright & Stone, 1979), the goodness of fit is examined in greater detail. For dichotomously scored items, the estimated standardized residual for the response of person j to item i is given by

$$z_{ij} = (2u_{ij} - 1)\exp\left[\frac{(2u_{ij} - 1)(\hat{\beta}_i - \hat{\theta}_j)}{2}\right]$$

and

$$z_{ij}^2 = \exp\left[(2u_{ij} - 1)(\hat{\beta}_i - \hat{\theta}_j)\right]. \tag{5.67}$$

Wright and Stone (1979) use this residual in a number of different ways. For an individual person, one can use their vector of item responses to obtain an index of the goodness of fit. Let

$$v_j = \frac{\sum_{i=1}^{n} \hat{z}_{ij}^2}{n-1}.$$

Then,

$$t_j = [\log(v_j) + v_j - 1]\frac{\sqrt{n-1}}{\sqrt{8}}$$

is approximately normally distributed with mean zero, and unit variance. Similarly for items, let

$$v_j = \frac{\sum_{j=1}^{N} \hat{z}_{ij}^2}{N-1},$$

where N is the sample size. Then,

$$t_i = [\log(v_i) + v_i - 1]\,\frac{\sqrt{N-1}}{\sqrt{8}}$$

is approximately normally distributed with mean zero and unit variance, that is, $t_i \simeq N(0,1)$. These large sample statistics are used to identify persons or items for which the model is inappropriate. For a comparison of several types of χ^2-related goodness-of-fit indices that could be used with the Rasch model, the reader is referred to Yen (1981). Glas (1988) has developed two techniques for indexing the overall goodness of fit to the Rasch model based on employing a marginal distribution of examinees and on a conditional distribution using the sufficient statistics for ability. Both of the goodness-of-fit measures were asymptotically distributed as χ^2. A rather ingenious approach to analyzing the fit of the Rasch model is due to Wright, Mead, and Draba (1976), who performed an analysis of variance on the residuals obtained after fitting the Rasch model. This enables one to look at the main effects of items, raw scores as well as the item-raw score interaction of the residuals.

The use of these goodness-of-fit indices leads to a second level of editing used to assist in obtaining an overall fit of the model. An iterative editing procedure is employed by which items, persons, or both are removed from the data set when their t statistics are significant. Once removed, the joint estimation process is performed again, and new goodness-of-fit indices are calculated (Wright & Stone, 1979, Chapter 4, provide an example of this scheme). Presumably, the editing is done until the overall χ^2 statistic (Equation 5.65) becomes non-significant. Since a lack of fit can occur when the $\alpha_i = 1$ assumption is violated, it is of interest to examine the item discrimination indices. However, the Rasch model and the estimation procedures do not involve such indices. Consequently, Wright and Mead (1975, p. 10) provide approximations to the discrimination indices based on data available in the estimation process. They estimate the item discrimination parameter α_i of the two-parameter logistic model as follows:

$$\hat{\alpha}_i - 1 = \frac{-\displaystyle\sum_{g=1}^{n-1}\hat{\beta}_{ig}\hat{\theta}_g}{\displaystyle\sum_{g=1}^{n-1}\hat{\beta}_{ig}^2}, \tag{5.68}$$

where $\sum_{g=1}^{n-1}\hat{\beta}_{ig} = 0$ and $\hat{\beta}_{ig}$ is the estimate of the item difficulty parameter based only on persons in raw score group g. Equation 5.68 is the regression of item difficulty for a score group on the ability estimate of the group. This rough estimate is used to identify items not following the one-parameter logistic model. Again, such items could be eliminated to obtain a better overall fit of the Rasch model. Andersen (1973a) has provided a means for obtaining estimates of the item discrimination parameters when conditional maximum likelihood estimation has been employed.

5.7.2 Likelihood Ratio Tests for Goodness of Fit

When maximum likelihood estimation procedures are used, an alternative overall goodness-of-fit measure is the likelihood ratio. With this approach, one computes the maximum of the likelihood function under two different configurations and obtains the ratio of these two maxima. Andersen (1973b) used this technique to test the fit of the Rasch model when the item parameters were estimated via conditional maximum likelihood estimation. The two configurations established were the overall conditional approach, in which the item parameters were estimated using the total group of examinees, and the restricted conditional approach, in which the item parameters were estimated successively for each raw score group. The likelihood ratio (LR) was given by

$$\text{LR} = \frac{l(\hat{\beta}_i)}{\sum_{g=1}^{n-1} l_g(\hat{\beta}_{ig})}, \tag{5.69}$$

where $(\hat{\beta}_i)$ is the vector of item difficulty parameter estimates under the overall approach; $(\hat{\beta}_{ig})$ is the vector of item difficulty parameter estimates within a score group; $l(\hat{\beta}_i)$ is the likelihood function under the overall approach; and $l_g(\hat{\beta}_{ig})$ is the likelihood function under the restricted approach. When the Rasch model holds, the item difficulty parameters are the same across all raw score groups, and LR should be close to unity. If it is not the case, then LR will be less than unity. Under the likelihood ratio approach, the test statistic is

$$z = -2\log(\text{LR}) = 2\sum_{g=1}^{n-1} \log l_g(\hat{\beta}_{ig}) - 2\log l(\hat{\beta}_i). \tag{5.70}$$

This statistic is asymptotically distributed as χ^2 with $(n-1)(n-2)$ degrees of freedom. A significant χ^2 value indicates the item difficulty parameters differ across the raw score groups and that the Rasch model does not hold. Andersen (1973b, p. 135) indicated the goodness-of-fit procedure should be powerful against alternatives to the Rasch model, such as the two- and three-parameter logistic models. However, the formulas for calculating the power of the test against such alternatives have not been derived. Gustafsson (1980b) has indicated that the likelihood ratio approach can be applied to other groupings of examinees, such as gender, school or other demographic variables. In such cases, the test would be one of unidimensionality, reflecting the lack of group invariance of the item parameters, that is, specific objectivity. However, the grouping should not be related to performance differences on the test, in which case the test would also be sensitive to discrimination differences. Gustafsson also presented a test for unidimensionality when the items are grouped a priori. These likelihood tests are global measures of how all the items of a test fit the Rasch model. Lord (1975b) found the approach to be suspect when the three-parameter logistic model was employed. The basic problem was that the distribution of $-2\log(\text{LR})$ did not follow the χ^2 distribution. Whether the problem will be evident when the one-parameter logistic is used has not been investigated. Molenaar (1983) has presented a number of different procedures for a detailed analysis of the fit of items and the data set as a whole to the Rash model. Kelderman (1984) formulated the Rasch model as a quasi-loglinear model and presented a likelihood ratio test of the goodness of fit. Although the literature contains a wide variety of goodness-of-fit tests for the Rasch model, these tests tend not to become incorporated within the available computer programs for the Rasch model. We would wish that, at some point, one or more of these techniques will become the de facto standard tests.

5.8 The Rasch Model and Additive Conjoint Measurement

Two similar papers (Brogden, 1977; Perline, Wright, & Wainer, 1977) dealt with the relationship between the Rasch model and additive conjoint measurement (Luce & Tukey, 1964). Although the former is a stochastic model and the latter is a deterministic model, some interesting parallels exist and, under certain assumptions, the Rasch model is a special case of additive conjoint measurement. In both approaches, a two-dimensional table (raw scores by items) is established, with the proportion of correct response as the cell entry. Under conjoint measurement, the goal is to find a monotonic transformation of the dependent variable P_{ij}, measured on an ordinal scale, such that it can be represented as a sum of row and column effects. This is equivalent to determining whether the row column interaction can be removed by a monotonic rescaling of the dependent variable. Note the similarity to a two-way, fixed-effects ANOVA. In the case of the Rasch model, it is assumed that the necessary transformation is the one-parameter logistic function. The procedures of additive conjoint measurement are aimed at finding the transformation, although its mathematical form remains unknown. Brogden (1977) shows that, if the P_{ij} can be considered an ordinal variable, then additive conjoint measurement is possible if a transformation $T(P_{ij})$ exists such that

$$T(P_{ij}) = \theta_j - \beta_i, \tag{5.71}$$

where T is an order-preserving transformation and the θ_j and β_i are constants for given persons and items. Such a transformation exists if, for all i and j, the following hold:

1. Items and people can each be sufficiently finely graded.

2. The so-called cancellation axiom (Brogden, 1977, p. 638) holds.
3. If the item characteristic curve is the one-parameter logistic model.

Then the necessary transformation T is the one parameter logistic, and the items and persons are measured on an interval scale with a common unit. Perline, Wright, and Wainer (1977) applied the Rasch model and the procedures of additive conjoint measurement to a number of data sets. When the Rasch model fit the data, both the ability and item difficulty parameter estimates yielded by a Rasch model computer program (Wright & Mead, 1978) were the same linear function of the corresponding estimates as under additive conjoint measurement. Thus, in this case, the Rasch model has been shown to be a special case of additive conjoint measurement. While this relationship is of limited practicality, it is of interest as it links the Rasch model and, hence, IRT to a well-developed axiomatic theory of measurement. For additional discussions of the relationship between scaling and the Rasch model see Andrich (1978b) and Jansen (1984).

5.9 Research Related to the Rasch Model

Since the Rasch model employs a single item parameter, it is of interest to compare its applicability to item response data with that of other logistic models. This comparison was approached rather tangentially by Hambleton and Traub (1973). These authors compared the distribution of the sufficient estimators of ability, $t_j = \sum_{i=1}^{n} u_{ij}$ and $t_j = \sum_{i=1}^{n} \alpha_i u_{ij}$ under the Rasch and two-parameter logistic models, respectively. Hambleton and Trauba, assuming that the examinees were normally distributed over the ability scale, used a χ^2 statistic to compare the observed number of examinees in a score group with the number expected in the distribution of t. This statistic was computed for three different tests of scholastic ability. They found that the two-parameter logistic model yielded a lower χ^2 statistic for all three tests. The number of items in the three tests were 20, 45, and 80, and the differences in the χ^2 statistics between models were an inverse function of the number of items. This result agreed with an earlier conjecture due to Birnbaum (1968, p. 492) that inferences as to an examinee's ability under these two models will become quite similar as the number of items in the test increases, the assumption being that the distribution of item discrimination parameters will be such that their average value will outweigh the effect of their variability.

Dinero and Hartel (1977) studied the effect on the Rasch model of varying item discrimination parameters via a Monte Carlo study. They established a set of 16 tests of 30 items, each of which varied with respect to the form of the distribution (uniform, normal, skewed) of the α_i and the variability of the α_i. All distributions had a mean $\alpha_i = 1.0$. The responses of the examinees to each of these 16 tests were then simulated. The evaluation criterion used was the mean square error of the item difficulty parameter (β_i) and the maximum discrepancy between the observed and actual value. The same two statistics were computed for the ability estimates corresponding to the raw scores. Whereas there was some tendency for these statistics to increase as the variability of the α_i increased, the form of the distribution had a dramatic impact. When the α_i were uniformly distributed with $\mu_{\alpha_i} = 1$, $\sigma_{\alpha_i} = .05$, .10, .15, or .25, the mean square error of both the item difficulty and ability estimates were roughly 10–100 times larger than when the distributions of the α_i were normal or skewed. The maximum misfits showed a similar pattern. Dinero and Hartel concluded that the Rasch model was robust to variability in the item discrimination parameters, except when these parameters were uniformly distributed. These results should be considered tentative because of the extremely small sample sizes employed.

Van de Vijiver (1986) performed a simulation study to investigate the robustness of the Rasch model estimates obtained via conditional maximum likelihood using the computer program due to Vehelst, Glas, and van der Sluis (1984). Data sets were created for the following number of items and examinees: $n = 10$ and $N = 25$; $n = 25$ and $N = 50$; $n = 25$ and $N = 100$; $n = 25$ and $N = 500$; and $n = 50$ and $N = 500$. The data were first generated under the Rasch model and then under varying degrees of heterogeniety of the item discrimination indices and of guessing. Within each possible situation, 25 replications were generated and analyzed. Three criteria were used to evaluate the results: correlation of the parameter estimates with the generating parameters, bias of the estimates, and the root mean square (RMS) of the difference between the estimate and the generating parameter. When the data was generated under the Rasch model, the $n = 25$ and $N = 25$ data sets yielded correlations of the item and ability estimates with the generating parameters that were all greater than .90. When the assumptions of the Rasch model were not met, the data showed that these correlations were not very sensitive to heterogeniety of the item discriminations. However, the presence of guessing reduced the correlations by about .15. There did not appear to be any interactive effect on the correlations of the combination of heterogeniety of discrimination and guessing. Heterogeniety of the item discriminations did not appear to have much effect on the bias and RMS of the item difficulty estimates. However, the presence of guessing considerably increased both the bias and RMS measures for item difficulty. The effect of these two factors on the ability estimates was similar to those for the item difficulties. Van de Vijiver concluded that the CML estimation procedure was quite robust with respect to heterogeniety of the item discriminations underlying the generated item response data. In direct contrast, he stated that these procedures were not robust with respect to the presence of guessing. Because of this, it was recommended that measures of the overall fit of the Rasch model should be used to determine whether the assumptions of the model are met in the data.

Yen (1981) performed a simulation study aimed at determining how one should choose an ICC model. She generated item response data for 36 item tests and 1000 examinees under one-, two- and three-parameter logistic ICC models. These data sets were then analyzed under each of the three ICC models. When the ICC model used to generate the data matched that used to analyze the data, there was a strong linear relationship between the parameter estimates and their underlying values for both items and examinees. When the data were generated via a three-parameter model and analyzed under the Rasch model, there was a fairly linear relationship of the ability estimates with those yielded under a three parameter model. Yen attributed this to the values of the item parameters used to generate the data. It was also found that, when the Rasch model was used with multiple-choice items, the goodness of fit was less than that found under two- and three-parameter models. As a consequence, Yen recommended analyzing item response data under all three models and ascertaining which model yielded the best average goodness-of-fit index.

From these studies, it is clear that using items not meeting the Rasch model requirements of $\alpha_i = 1$, with no guessing, has an impact on the parameter estimates. This illustrates a point made by Wright and Panchapakesan (1969) and Wright (1977a) that the choice of models and test construction procedures are not independent. Since it is extremely difficult for an item writer to produce items with a priori values of $\alpha_i = 1$, one needs to create item pools that can be screened for items meeting the model selected. In the case of the Rasch model, at a minimum, this requires that one limit the variability of the α_i as much as possible and avoid a uniform distribution of the α_i. Certain Rasch model computer programs, such as CALFIT (see Wright & Mead, 1975), perform such screening as part of the estimation process. However, a better initial procedure would be to follow Yen's (1981) recommendation to use a program such as LOGIST or BILOG to estimate the item parameters under all three logistic ICC models and then to select the items that fit the Rasch model. In a second stage, the resulting test could be analyzed using one of the Rasch model programs. Much of the past controversy surrounding the Rasch model stemmed from a lack of appreciation for the interdependence of models and test construction procedures. Wright (1977a, 1977b) clearly understands this issue, but others often fail to recognize its importance.

5.10 Summary

The original work due to Rasch approached test analysis from the point of view of a mathematician rather than that of a psychometrician. In choosing to consider test analysis as a problem in probability theory, Rasch ignored the existing literature on both IRT and classical test theory. However, his probabilistic approach was also based on the underlying trait, independence—independence of items, independence of persons, joint independence of items and people, and local independence. In addition, once the ability and item parameters were subjected to logarithmic transformations, the two-parameter logistic model, with α_i set to unity, was obtained. What is unique to Rasch's probabilistic approach is what he has called "separation of parameters," which made it clear that minimal sufficient statistics in the form of item scores and test scores existed for the item and ability parameters. In the case of the ability parameters, the raw scores are sufficient statistics is a direct consequence of the restriction that all item discrimination parameters are set to unity, which makes the ICC a member of the exponential family (Haberman, 1977). Because of the unit value of the discrimination parameter, the Rasch model cannot distinguish among persons having different patterns of item response, but having the same raw score. That is, all item response patterns having the same number of correct responses yield the same estimated ability, a characteristic that is attractive to practitioners. On the other hand, the two- and three-parameters models will yield a separate ability estimate for each pattern of item responses. Whether this difference is important depends on one's theoretical and practical orientations.

The Rasch approach has also led to the development of conditional maximum likelihood estimation procedures for the item difficulty parameters. Since the β_i are structural parameters and the θ_j are incidental parameters, it is advantageous to get rid of the latter in the item parameter estimation process. Since the random variable defined as the raw score is a minimum sufficient statistic, the conditional maximum likelihood estimation equations employ these raw scores. This estimation process is based on the same assumption of an itemized test (Ross & Lumsden, 1968) as other IRT models are since the raw scores depend on the number of items in the test. What is advantageous about the conditional maximum likelihood estimators is that the item difficulty parameters can be estimated via the Newton-Raphson technique without using a joint estimation process also involving the ability parameters. The major drawback to the conditional maximum likelihood estimation process is a computational one. Calculation of the symmetric functions with a reasonable degree of accuracy is very difficult. Recent work has increased the number of items that can be used to a level that encompasses most instruments, but data relating to the overall adequacy of these procedures is limited.

The application of the Birnbaum paradigm to the estimation of the item and ability parameters yields a JMLE procedure for the Rasch model. It has been implemented in the BICAL computer program and is widely used in applied settings. The item parameter estimates produced by this program have not been studied as extensively as those yielded by LOGIST for the other ICC models. However, the available results show that the JMLE procedure produces parameter estimates with acceptable properties. In particular, the combination of the JMLE procedure and the Rasch model appears to work well for data sets that are considerably smaller than those required for adequate results under the two- and three-parameter models. The available evidence also indicates that the numerical values of the parameter estimates yielded by via JMLE do not differ much from those obtained via conditional maximum likelihood. Thus, the choice of which approach to use rests on computational and goodness-of-fit procedure considerations.

6. Parameter Estimation via MMLE and an EM Algorithm

6.1 Introduction

[1]Since it was formulated in 1968 by Dr. Allan Birnbaum, the primary approach to item parameter estimation has been the joint maximum likelihood estimation (JMLE) paradigm. A distinguishing characteristic of this paradigm is that examinee abilities are unknown and, hence, must be estimated along with the item parameters. As discussed in Chapter 4, the item parameters are the "structural" parameters, which are fixed in number by the size of the test. The ability parameters of the examinees are the "incidental" parameters, the number of which depends on the sample size. From a theoretical point of view, this paradigm has an inherent problem first recognized in another context by Neyman and Scott (1948) (see also Little & Rubin, 1983). Neyman and Scott showed that, when structural parameters are estimated simultaneously with the incidental parameters, the maximum likelihood estimates of the former need not be consistent as sample size increases. If sufficient statistics are available for the incidental parameters, the conditional maximum likelihood estimation procedure (Andersen, 1972) can be established for the consistent estimation of the structural parameters. As explained in Chapter 5, such conditional estimation can be established only for the one-parameter (Rasch) model. As a result, an estimation procedure for two- and three-parameter IRT models that avoids the problem of inconsistent estimates of structural parameters has considerable value. The basic paper in this regard was due to Bock and Lieberman (1970), who developed a marginal maximum likelihood procedure for estimating item parameters. Unfortunately, the Bock and Lieberman approach posed a formidable computational task and was practical for only very short tests. A subsequent reformulation of this marginal maximum likelihood estimation (MMLE) approach by Bock and Aitkin (1981) has resulted in a procedure that is both theoretically acceptable and computationally feasible. Their reformulation, under certain conditions, is an instance of an expectation-maximization (EM) algorithm. As a result, MMLE/EM will be used to identify the Bock and Atkin

---

[1] This chapter is a modified version of the paper by M. R. Harwell, F. B. Baker, and M. Zwarts (1988). Item parameter estimation via marginal maximum likelihood and an EM algorithm: A didactic. *Journal of Educational Statistics, 13*, 243–271 (used with permission).

procedure for estimating item parameters (see Section 6.4.1 for a definition of EM).

6.2 Item Parameter Estimation via Marginal Maximum Likelihood

Under the MMLE approach to item parameter estimation, it is assumed that the examinees represent a random sample from a population where ability is distributed according to a density function $g(\theta|\tau)$, where τ is the vector containing the parameters of the examinee population ability distribution. The situation now corresponds to a mixed-effects ANOVA model, with items considered to be a fixed effect and abilities a random effect. The essence of the Bock and Lieberman (1970) solution is to integrate over the ability distribution, thus removing the random nuisance (ability) parameters from the likelihood function. Hence, the item parameters are estimated in the marginal distribution, and the item parameter estimation is freed from its dependence on the estimation of each examinee's ability though not from its dependence on the ability distribution. This produces consistent estimates of item parameters for samples of any size since increasing sample size does not require the estimation of additional examinee parameters.

Although MMLE is not considered a Bayesian estimation technique, the marginalization process incorporates Bayes' theorem in its mathematics. Thus, it is necessary to discuss briefly the theorem and its role. In the MMLE approach, the point estimator of ability for a subject, say, $\hat{\theta}_j$, is replaced by a distribution that allocates the data of subject j across the ability scale in proportion to the probability of their being at any given point along the ability scale. Thus, instead of a single-valued $\hat{\theta}_j$ for an examinee, the probability is found of an examinee having each possible θ_j value along the ability scale conditional on the item response vector $\mathbf{u}_j$, the item parameters in $\boldsymbol{\xi}$, and the population distribution of ability.[2] In Bayesian statistics, this probability is known as the posterior probability. Since ability is assumed to be continuous, a plot of this probability as a function of ability results in a smooth curve. This curve represents a summary of the information about θ contained in the observed item response data for examinee j. Each examinee's item response vector would generate such a curve. The posterior probability employed here is crucial in the MMLE process because it is the vehicle for freeing the item parameter estimation process from its dependence on the estimation of each examinee's ability as is the case under the JMLE procedure. This is achieved by employing the following form of Bayes' theorem to compute the entire posterior ability distribution:

---

[2] It should be noted that the symbol $\boldsymbol{\xi}$ used here is not the situational parameter of Chapter 5 but is the vector of item parameters for the n items of the test.

$$P(\theta_j|\mathbf{u}_j, \boldsymbol{\tau}, \boldsymbol{\xi}) = \frac{P(\mathbf{u}_j|\theta_j, \boldsymbol{\xi})g(\theta_j|\boldsymbol{\tau})}{\displaystyle\int P(\mathbf{u}_j|\theta_j, \boldsymbol{\xi})g(\theta_j|\boldsymbol{\tau})d\theta_j}. \tag{6.1}$$

Assuming local independence, the first term in the numerator is given by

$$P(\mathbf{u}_j|\theta_j, \boldsymbol{\xi}) = \prod_{i=1}^{n} P_i(\theta_j)^{u_{ij}} Q_i(\theta_j)^{1-u_{ij}} \tag{6.2}$$

which is a likelihood function. It is the probability of the examinee's item response vector $\mathbf{u}_j$ conditional on the examinee's ability θ_j and the item parameters in $\boldsymbol{\xi}$.

The second term in the numerator, $g(\theta_j|\boldsymbol{\tau})$, is the probability density function of ability in the population of examinees. Since the form of this distribution is not known before the data is collected, one makes a judgement as to its form. This judgement is based on one's knowledge of the distribution of ability for the test and examinees of interest; often, its form will be the unit normal, but it could have other forms. This distribution, with parameter vector $\boldsymbol{\tau}$, defines the relative probability of the θ_j values. In Bayesian statistics, it is known as the prior distribution and, in the present context, as the prior ability distribution. A necessary assumption is that the prior distribution of ability is the same for all examinees. Thus, the product in the numerator of Equation 6.1, $P(\mathbf{u}_j|\theta_j, \boldsymbol{\xi})g(\theta_j|\boldsymbol{\tau})$, is the joint distribution of $\mathbf{u}_j$ and θ_j.

In the denominator of Equation 6.1, we have

$$\int P(\mathbf{u}_j|\theta_j, \boldsymbol{\xi})g(\theta_j|\boldsymbol{\tau})d\theta_j.$$

Because the integration is with respect to the nuisance parameter θ_j, this expression is the marginal (unconditional) probability of the item response vector $\mathbf{u}_j$ with respect to the item parameters and the population ability density.

In Equation 6.1, the posterior distribution of ability has been defined via an application of Bayes' theorem. This posterior probability distribution combines the information from the prior distribution of ability and that in the likelihood function. What the prior distribution essentially does is to distribute the data, in the item response vectors, across the ability scale in proportion to the posterior probability of the examinees being at each point on the scale. The net result is that the expected number of examinees at each point along the ability scale is determined. As will be seen below, this result plays a crucial role in MMLE.

6.3 The Bock and Lieberman Solution

In this presentation the Bock and Lieberman (1970) marginal maximum likelihood solution, both its similarities to and differences from the bioassay solution of Chapter 2 and the JMLE solution of Chapter 4 will be emphasized.

Although Bock and Lieberman used a normal ogive ICC model in their article, the three-parameter logistic ICC model with parameterization a_i, b_i, c_i will be used here. To simplify the notation, let $P(\mathbf{u}_j) = \int P(\mathbf{u}_j|\theta_j, \boldsymbol{\xi})g(\theta_j|\tau)d\theta_j$. Then, following Bock and Lieberman (1970), the marginal likelihood function is

$$L = \prod_{j=1}^{N} P(\mathbf{u}_j). \tag{6.3}$$

The logarithm of L is

$$\log L = \sum_{j=1}^{N} \log P(\mathbf{u}_j) \tag{6.4}$$

and, to find the marginal likelihood equation for a_i, take

$$\frac{\partial}{\partial a_i}(\log L) = 0.$$

Then,

$$\frac{\partial}{\partial a_i}(\log L) = \sum_{j=1}^{N} \frac{\partial}{\partial a_i}[\log P(\mathbf{u}_j)]$$

$$= \sum_{j=1}^{N} [P(\mathbf{u}_j)]^{-1} \frac{\partial}{\partial a_i}\left[\int P(\mathbf{u}_j|\theta_j, \boldsymbol{\xi})g(\theta_j|\tau)d\theta_j\right].$$

For convenience, the j subscript on θ is dropped in subsequent expressions because θ_j can be seen as a random subject sampled from a population. In addition, by interchanging the differentiation and integration (see Kendall & Stuart, 1979, p. 10),

$$\frac{\partial}{\partial a_i}(\log L) = \sum_{j=1}^{N} [P(\mathbf{u}_j)]^{-1} \int \frac{\partial}{\partial a_i}[P(\mathbf{u}_j|\theta, \boldsymbol{\xi})]\, g(\theta|\tau)d\theta.$$

Using the relation

$$\frac{\partial}{\partial a_i}[P(\mathbf{u}_j|\theta, \boldsymbol{\xi})] = \frac{\partial}{\partial a_i}[\log P(\mathbf{u}_j|\theta, \boldsymbol{\xi})]\, P(\mathbf{u}_j|\theta, \boldsymbol{\xi}),$$

result in

$$= \sum_{j=1}^{N} [P(\mathbf{u}_j)]^{-1} \int \frac{\partial}{\partial a_i}[\log P(\mathbf{u}_j|\theta, \boldsymbol{\xi})]\, P(\mathbf{u}_j|\theta, \boldsymbol{\xi})g(\theta|\tau)d\theta$$

$$= \sum_{j=1}^{N} \int \frac{\partial}{\partial a_i}[\log P(\mathbf{u}_j|\theta, \boldsymbol{\xi})]\left[\frac{P(\mathbf{u}_j|\theta, \boldsymbol{\xi})g(\theta|\tau)}{P(\mathbf{u}_j)}\right]d\theta.$$

Where the second bracketed term is defined by Equation 6.1, replacing this term yields

$$\frac{\partial}{\partial a_i} (\log L) = \sum_{j=1}^{N} \int \frac{\partial}{\partial a_i} \left[\log P(\mathbf{u}_j|\theta, \boldsymbol{\xi})\right] \left[P(\theta|\mathbf{u}_j, \boldsymbol{\xi}, \boldsymbol{\tau})\right] d\theta. \tag{6.5}$$

Replacing $P(\mathbf{u}_j|\theta, \boldsymbol{\xi})$ in the first bracketed term of Equation 6.5 with the likelihood expression given in Equation 6.2 yields

$$\frac{\partial}{\partial a_i} (\log L) = \sum_{j=1}^{N} \int \frac{\partial}{\partial a_i} \left[\log \prod_{i=1}^{n} P_i(\theta)^{u_{ij}} Q_i(\theta)^{1-u_{ij}}\right] \times$$
$$[P(\theta|\mathbf{u}_j, \boldsymbol{\xi}, \boldsymbol{\tau})] d\theta$$
$$= \sum_{j=1}^{N} \int \left[\prod_{i=1}^{n} P_i(\theta)^{u_{ij}} Q_i(\theta)^{1-u_{ij}}\right]^{-1} \times$$
$$\frac{\partial}{\partial a_i} \left[\prod_{i=1}^{n} P_i(\theta)^{u_{ij}} Q_i(\theta)^{1-u_{ij}}\right] [P(\theta|\mathbf{u}_j, \boldsymbol{\xi}, \boldsymbol{\tau})] d\theta. \tag{6.6}$$

We need

$$\frac{\partial}{\partial a_i} \left[\prod_{i=1}^{n} P_i(\theta)^{u_{ij}} Q_i(\theta)^{1-u_{ij}}\right]$$
$$= \left[\prod_{h \neq i}^{n} P_h(\theta)^{u_{hj}} Q_h(\theta)^{1-u_{hj}}\right] \frac{\partial}{\partial a_i} \left[P_i(\theta)^{u_{ij}} Q_i(\theta)^{1-u_{ij}}\right], \tag{6.7}$$

where

$$\frac{\partial}{\partial a_i} \left[P_i(\theta)^{u_{ij}} Q_i(\theta)^{1-u_{ij}}\right]$$
$$= \frac{\partial}{\partial a_i} \left[P_i(\theta)^{u_{ij}}\right] Q_i(\theta)^{1-u_{ij}} + P_i(\theta)^{u_{ij}} \frac{\partial}{\partial a_i} \left[Q_i(\theta)^{1-u_{ij}}\right]$$

$$= u_{ij} P_i(\theta)^{u_{ij}-1} \left[\frac{\partial P_i(\theta)}{\partial a_i}\right] Q_i(\theta)^{1-u_{ij}} +$$
$$P_i(\theta)^{u_{ij}} (1 - u_{ij}) Q_i(\theta)^{1-u_{ij}-1} \left[\frac{\partial Q_i(\theta)}{\partial a_i}\right].$$

Using the relationship $\partial Q_i(\theta)/\partial a_i = -\partial P_i(\theta)/\partial a_i$ and gathering terms,

$$= \left[\frac{\partial P_i(\theta)}{\partial a_i}\right] \left[P_i(\theta)^{u_{ij}} Q_i(\theta)^{1-u_{ij}}\right] \left[\frac{u_{ij}}{P_i(\theta)} - \frac{1 - u_{ij}}{Q_i(\theta)}\right]$$
$$= \left[\frac{\partial P_i(\theta)}{\partial a_i}\right] \left[P_i(\theta)^{u_{ij}} Q_i(\theta)^{1-u_{ij}}\right] \left[\frac{u_{ij} - P_i(\theta)}{P_i(\theta)Q_i(\theta)}\right].$$

From the above results,

$$\frac{\partial}{\partial a_i}\left[\prod_{i=1}^{n} P_i(\theta)^{u_{ij}} Q_i(\theta)^{1-u_{ij}}\right]$$

$$=\left[\prod_{i=1}^{n} P_i(\theta)^{u_{ij}} Q_i(\theta)^{1-u_{ij}}\right]\left[\frac{\partial P_i(\theta)}{\partial a_i}\right]\left[\frac{u_{ij}-P_i(\theta)}{P_i(\theta)Q_i(\theta)}\right].$$

Note that the product is now over $h \neq i$ rather than i in Equation 6.7. This results from the dropping of the $P_i(\theta)$ and $Q_i(\theta)$ terms when taking the derivative with respect to a_i. Letting $v = a_i(\theta - b_i)$, note that

$$\frac{\partial P_i(\theta)}{\partial a_i} = (1-c_i)\left[e^v(-1)(1+e^v)^{-2}e^v(\theta-b_i)+(1+e^v)^{-1}e^v(\theta-b_i)\right]$$

$$= (1-c_i)(\theta-b_i)P_i^*(\theta)Q_i^*(\theta) \tag{6.8}$$

$$= K.$$

Continuing,

$$\frac{\partial}{\partial a_i}\left[\prod_{i=1}^{n} P_i(\theta)^{u_{ij}} Q_i(\theta)^{1-u_{ij}}\right]$$

$$=\left[\prod_{h\neq i}^{n} P_h(\theta)^{u_{hj}} Q_h(\theta)^{1-u_{hj}}\right]\times$$

$$\left[P_i(\theta)^{u_{ij}}\frac{\partial}{\partial a_i}\left[Q_i(\theta)^{1-u_{ij}}\right]+Q_i(\theta)^{1-u_{ij}}\frac{\partial}{\partial a_i}\left[P_i(\theta)^{u_{ij}}\right]\right] \tag{6.9}$$

$$=\left[\prod_{h\neq i}^{n} P_h(\theta)^{u_{hj}} Q_h(\theta)^{1-u_{hj}}\right]\times$$

$$\left[P_i(\theta)^{u_{ij}}(1-u_{ij})Q_i(\theta)^{1-u_{ij}-1}(-K)+Q_i(\theta)^{1-u_{ij}}u_{ij}P_i(\theta)^{u_{ij}-1}(K)\right].$$

Gathering terms,

$$= K\left[\prod_{h\neq i}^{n} P_h(\theta)^{u_{hj}} Q_h(\theta)^{1-u_{hj}}\right]\times$$

$$\left[Q_i(\theta)^{1-u_{ij}}u_{ij}P_i(\theta)^{u_{ij}-1}-P_i(\theta)^{u_{ij}}(1-u_{ij})Q_i(\theta)^{1-u_{ij}-1}\right].$$

The second bracketed term is +1 if $u_{ij} = 1$ and -1 when $u_{ij} = 0$; hence, it can be written as $(-1)^{u_{ij}+1}$. Then,

$$\frac{\partial}{\partial a_i}\left[\prod_{i=1}^{n} P_i(\theta)^{u_{ij}} Q_i(\theta)^{1-u_{ij}}\right]$$

$$= (-1)^{(u_{ij}+1)}(1 - c_i) \left[\prod_{h \neq i}^{n} P_h(\theta)^{u_{hj}} Q_h(\theta)^{1-u_{hj}} \right] \times$$

$$\left[e^v(-1)(1 + e^v)^{-2} e^v(\theta - b_i) + (1 + e^v)^{-1} e^v(\theta - b_i) \right]$$

$$= (-1)^{(u_{ij}+1)}(1 - c_i) \left[\prod_{h \neq i}^{n} P_h(\theta)^{u_{hj}} Q_h(\theta)^{1-u_{hj}} \right] \times$$

$$(\theta - b_i) P_i^*(\theta) Q_i^*(\theta).$$

Substituting the above expression into Equation 6.6 and returning the j subscript yields

$$\frac{\partial}{\partial a_i} (\log L) = \sum_{j=1}^{N} (-1)^{(u_{ij}+1)}(1 - c_i) \int (\theta_j - b_i) \left[\prod_{i=1}^{n} P_i(\theta_j)^{u_{ij}} Q_i(\theta_j)^{1-u_{ij}} \right]^{-1}$$

$$\times \left[\prod_{h \neq i}^{n} P_h(\theta_j)^{u_{hj}} Q_h(\theta_j)^{1-u_{hj}} \right] P_i^*(\theta_j) Q_i^*(\theta_j) \left[P(\theta_j | \mathbf{u}_j, \boldsymbol{\xi}, \boldsymbol{\tau}) \right] d\theta_j. \quad (6.10)$$

This is the marginal likelihood equation for a_i for the three-parameter logistic model. If the numerator and denominator are multiplied by $\left[P_i(\theta_j)^{u_{ij}} Q_i(\theta_j)^{1-u_{ij}} \right]$, the product operator in the bracketed term over $h \neq i$ reverts to a term over i. If, in the numerator, one then collects the $(-1)^{(u_{ij}+1)}$ and also takes the product operator over i, Equation 6.10 can be written as follows:

$$\frac{\partial}{\partial a_i} (\log L) = (1 - c_i) \sum_{j=1}^{N} \int [u_{ij} - P_i(\theta_j)] W_{ij}(\theta_j - b_i) \left[P(\theta_j | \mathbf{u}_j, \boldsymbol{\xi}, \boldsymbol{\tau}) \right] d\theta_j, (6.11)$$

where

$$W_{ij} = \frac{P_i^*(\theta_j) Q_i^*(\theta_j)}{P_i(\theta_j) Q_i(\theta_j)}.$$

It should be noted that this W_{ij} is the same as W_j in Equation 2.24. Understanding the transition from Equations 6.10 to 6.11 is facilitated by the following relationship:

$$\sum_{i=1}^{n} (-1)^{u_{ij}+1} \frac{1}{P_i(\theta_j)^{u_{ij}} Q_i(\theta_j)^{1-u_{ij}}} = \sum_{i=1}^{n} [u_{ij} - P_i(\theta_j)] \frac{1}{P_i(\theta_j) Q_i(\theta_j)}.$$

Suppose that $u_{ij} = 1$; then,

$$(-1)^2 \frac{1}{P_i(\theta_j)} = \frac{1 - P_i(\theta_j)}{P_i(\theta_j) Q_i(\theta_j)} = \frac{1}{P_i(\theta_j)}$$

and, when $u_{ij} = 0$,

$$(-1)^1 \frac{1}{Q_i(\theta_j)} = \frac{0 - P_i(\theta_j)}{P_i(\theta_j) Q_i(\theta_j)} = \frac{1}{Q_i(\theta_j)}.$$

Expression 6.11 illustrates that the marginal likelihood equation for a_i of Bock and Lieberman has a form similar to the unmarginalized likelihood Equation 2.23 employed in both the bioassay and JMLE solutions but multiplied by $[P(\theta_j|\mathbf{u}_j, \boldsymbol{\xi}, \boldsymbol{\tau})]$ and integrated over the posterior distribution.

The likelihood equations for b_i and c_i are

$$\frac{\partial}{\partial b_i}(\log L) = -a_i(1-c_i)\sum_{j=1}^{N}\int [u_{ij} - P_i(\theta_j)]\,W_{ij}\,[P(\theta_j|\mathbf{u}_j, \boldsymbol{\xi}, \boldsymbol{\tau})]\,d\theta_j = 0 \qquad (6.12)$$

and

$$\frac{\partial}{\partial c_i}(\log L) = (1-c_i)^{-1}\sum_{j=1}^{N}\int \left[\frac{u_{ij} - P_i(\theta_j)}{P_i(\theta_j)}\right][P(\theta_j|\mathbf{u}_j, \boldsymbol{\xi}, \boldsymbol{\tau})]\,d\theta_j = 0. \qquad (6.13)$$

6.3.1 Quadrature Distributions

Before continuing, let us detour from the general discussion and note that the previous equations involved an integral. Since, in a digital computer, such integrals are difficult to evaluate, a means must be found for approximating the integral. Fortunately, the technique known as Hermite-Gauss quadrature can be employed for approximating such integrals. If $g(\theta|\boldsymbol{\tau})$ is a continuous distribution with finite moments, it can be approximated to any desired degree of accuracy by a discrete distribution over a finite number of points (i.e., by a histogram).

Under the quadrature approximation approach, the problem of finding the sum of the area under the continuous curve is replaced by the simpler problem of finding the sum of the areas of a finite number of rectangles that approximate the area under the curve as depicted in Figure 6.1. The midpoint of each rectangle on the ability scale, X_k ($k = 1, 2, \ldots, q$), is called a "node." Each node has an associated weight $A(X_k)$ that takes into account the height of the density function $g(\theta|\boldsymbol{\tau})$ in the neighborhood of X_k and the width of the rectangles. The values of X_k and $A(X_k)$ are found by solving a set of equations that involve the continuous distribution to be approximated and the specified number of nodes (see Hildebrand, 1956, pp. 327–330). Stroud and Secrest (1966) give tables of X_k and the corresponding weights $A(X_k)$ for approximating the Gaussian error curve but not the unit normal distribution. Bock and Lieberman (1970) indicated that, to approximate the latter distribution, the tabled values of X_k are multiplied by $\sqrt{2}$ and the tabled values of $A(X_k)$ are divided by $\sqrt{\pi}$. It is not necessary for $g(\theta|\boldsymbol{\tau})$ to be normal in form; in general, it can be empirically defined (Mislevy & Bock, 1982a). Finally, note the similarity between the q values of X_k and the notion of known ability levels θ_j. This relationship is vital in establishing the link between the marginal solution of Bock and Lieberman (1970) and the (unmarginalized) bioassay solution of Chapter 2.

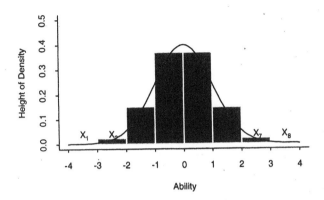

Fig. 6.1. Area under a curve.

For a random subject (i.e., without subscript j) sampled from a population with $g(\theta|\boldsymbol{\tau})$, substituting the quadrature approximation for the integral in Equation 6.11 results in

$$a_i: \quad (1-c_i)\sum_{j=1}^{N}\sum_{k=1}^{q}[u_{ij} - P_i(X_k)]\, W_{ik}(X_k-b_i)\,[P(X_k|\mathbf{u}_j,\boldsymbol{\xi},\boldsymbol{\tau})] = 0, \quad (6.14)$$

where $W_{ik} = P_i^*(X_k)Q_i^*(X_k)/P_i(X_k)Q_i(X_k)$ is the weighting term and the θ have been replaced by X_k. The quadrature form of the Bock and Lieberman MMLE equation for the difficulty parameter b_i can be written by inspection from Equation 6.14

$$b_i: \quad -a_i(1-c_i)\sum_{j=1}^{N}\sum_{k=1}^{q}[u_{ij} - P_i(X_k)]\, W_{ik}\,[P(X_k|\mathbf{u}_j,\boldsymbol{\xi},\boldsymbol{\tau})] = 0 \quad (6.15)$$

and that for c_i can be written from Equation 6.13

$$c_i: \quad (1-c_i)^{-1}\sum_{j=1}^{N}\sum_{k=1}^{q}\left[\frac{u_{ij} - P_i(X_k)}{P_i(X_k)}\right][P(X_k|\mathbf{u}_j,\boldsymbol{\xi},\boldsymbol{\tau})] = 0. \quad (6.16)$$

Then, expressions 6.14–6.16 represent the quadrature form of the marginal likelihood equations for the three-parameter logistic model. Under the Bock and Lieberman (1970) approach, the Newton-Raphson technique would be employed to estimate the $3n$ item parameters of a test simultaneously. As was the case in JMLE, the observed values of the u_{ij} in the second derivatives of the log-likelihood function were replaced by their expected values.

Thus, Fisher's method of scoring was actually used. However, the method is computationally unattractive because it requires the inversion of a $3n \times 3n$

information matrix. Because of this, the estimation procedures were limited to a very small number of items. In addition, the item parameter estimates are not "sample-free" since the method requires the population distribution of ability that is sampled to be known in advance.

6.4 The Bock and Aitkin Solution

A reformulation of the Bock and Lieberman likelihood equations by Bock and Aitkin (1981) yielded a solution that is computationally feasible and, under the assumption that the population distribution is known or is concurrently estimated with the correct specification, produces consistent item parameter estimates. As was the case in the Birnbaum JMLE paradigm, Bock and Aitkin assumed that the items are independent, that the examinees are independent, and that items and examinees are independent. Consequently, the item parameters can be estimated on a one item at a time basis, with the examinee's ability estimated on a per-examinee basis. The reformulation results in likelihood equations for an item that have the same general form as the bioassay likelihood equations presented in Chapter 2. To link the MMLE and bioassay solutions more formally, we begin in the likelihood equation for a_i (Equation 6.14) by multiplying through by the $P(X_k|\mathbf{u}_j, \boldsymbol{\xi}, \boldsymbol{\tau})$ term and distributing the summation over the j operator. This results in

$$a_i : \quad (1 - c_i) \sum_{k=1}^{q} (X_k - b_i) \times$$

$$\left[\sum_{j=1}^{N} u_{ij} P(X_k|\mathbf{u}_j, \boldsymbol{\xi}, \boldsymbol{\tau}) - P_i(X_k) \sum_{j=1}^{N} P(X_k|\mathbf{u}_j, \boldsymbol{\xi}, \boldsymbol{\tau}) \right] W_{ik} = 0. \qquad (6.17)$$

Putting Equation 6.1 in its quadrature form yields

$$P(X_k|\mathbf{u}_j, \boldsymbol{\xi}, \boldsymbol{\tau}) = \frac{\prod_{i=1}^{n} P_i(X_k)^{u_{ij}} Q_i(X_k)^{1-u_{ij}} A(X_k)}{\sum_{k=1}^{q} \prod_{i=1}^{n} P_i(X_k)^{u_{ij}} Q_i(X_k)^{1-u_{ij}} A(X_k)}, \qquad (6.18)$$

which is the posterior probability of an examinee having ability level X_k, and

$$P_i(X_k) = c_i + (1 - c_i) \frac{\exp[a_i(X_k - b_i)]}{1 + \exp[a_i(X_k - b_i)]}.$$

Hence, $Q_i(X_k) = 1 - P_i(X_k)$. Then, from the second element within the bracketed term of Equation 6.17, the following term can be defined:

$$\bar{f}_{ik} = \sum_{j=1}^{N} P(X_k | \mathbf{u}_j, \boldsymbol{\xi}, \boldsymbol{\tau})$$

$$= \sum_{j=1}^{N} \left[\frac{\prod_{i=1}^{n} P_i(X_k)^{u_{ij}} Q_i(X_k)^{1-u_{ij}} A(X_k)}{\sum_{k=1}^{q} \prod_{i=1}^{n} P_i(X_k)^{u_{ij}} Q_i(X_k)^{1-u_{ij}} A(X_k)} \right]. \qquad (6.19)$$

Note that this is simply the sum over the N examinees of the posterior probability of an examinee having an ability of X_k. The first element within the bracketed term of Equation 6.17 yields

$$\bar{r}_{ik} = \sum_{j=1}^{N} u_{ij} P(X_k | \mathbf{u}_j, \boldsymbol{\xi}, \boldsymbol{\tau})$$

$$= \sum_{j=1}^{N} \left[\frac{\prod_{i=1}^{n} u_{ij} P_i(X_k)^{u_{ij}} Q_i(X_k)^{1-u_{ij}} A(X_k)}{\sum_{k=1}^{q} \prod_{i=1}^{n} P_i(X_k)^{u_{ij}} Q_i(X_k)^{1-u_{ij}} A(X_k)} \right]. \qquad (6.20)$$

Following Bock and Aitkin, let

$$L(X_k) = \prod_{i=1}^{n} P_i(X_k)^{u_{ij}} Q_i(X_k)^{1-u_{ij}} \qquad (6.21)$$

represent the quadrature form of the conditional probability of $\mathbf{u}_j$, given $\theta = X_k$ and the item parameters. Then the quadrature form of expression 6.18 can be written

$$P(X_k | \mathbf{u}_j, \boldsymbol{\xi}, \boldsymbol{\tau}) = \frac{L(X_k) A(X_k)}{\sum_{k=1}^{q} L(X_k) A(X_k)}. \qquad (6.22)$$

The expressions for $\bar{f}_{ik}$ and $\bar{r}_{ik}$ then can be written more succinctly as

$$\bar{f}_{ik} = \sum_{j=1}^{N} \left[\frac{L(X_k) A(X_k)}{\sum_{k=1}^{q} L(X_k) A(X_k)} \right] \qquad (6.23)$$

and

$$\bar{r}_{ik} = \sum_{j=1}^{N} \left[\frac{u_{ij} L(X_k) A(X_k)}{\sum_{k=1}^{q} L(X_k) A(X_k)} \right]. \qquad (6.24)$$

Because of their conceptual importance, the $\bar{r}_{ik}$ and $\bar{f}_{ik}$ terms need to be examined a bit more closely. The $\bar{f}_{ik}$ are the number of examinees in a population of size N expected to have ability score X_k. This results from distributing an examinee's data over the q quadrature nodes in proportion to the posterior probabilities of their being at each node. The $\bar{r}_{ik}$ are the number of examinees in the population at ability X_k expected to respond correctly to the item. Note that Equations 6.23 and 6.24 are the same except for the u_{ij} that appears in Equation 6.24. Since this term is either 1 or 0, only the elements corresponding to correct responses are accumulated; hence, the frequency of correct response at X_k is obtained via Equation 6.24.

With the $\bar{r}_{ik}$ and $\bar{f}_{ik}$ quantities in hand, the quadrature form of the marginal likelihood equations can be written as follows:

$$a_i : \qquad (1 - c_i) \sum_{k=1}^{q} (X_k - b_i) \left[\bar{r}_{ik} - \bar{f}_{ik} P_i(X_k) \right] W_{ik} = 0 \qquad (6.25)$$

$$b_i : \qquad -a_i(1 - c_i) \sum_{k=1}^{q} \left[\bar{r}_{ik} - \bar{f}_{ik} P_i(X_k) \right] W_{ik} = 0 \qquad (6.26)$$

$$c_i : \qquad (1 - c_i)^{-1} \sum_{k=1}^{q} \left[\frac{\bar{r}_{ik} - \bar{f}_{ik} P_i(X_k)}{P_i(X_k)} \right] = 0. \qquad (6.27)$$

These equations are identical in form to those presented for the bioassay solution in Chapter 2 (Equations 2.24–2.26) and, in fact, are the likelihood equations for a probit analysis for item i in which X_k is the independent variable. The difference is that, in the bioassay solution, the total number of examinees f_{ik} responding at an ability level (θ_k) and the number exhibiting the correct response r_{ik} are known from the data. In MMLE solution, these quantities are unknown, and the expected number of examinees responding to item i and the expected number of correct responses are substituted for these unknown quantities at each quadrature node (X_k). In the IRT literature $\bar{f}_{ik}$ and $\bar{r}_{ik}$ are known as the "artificial data" since they are created via Equations 6.23 and 6.24.

As was the case with the bioassay solution, the item parameters of an item will be estimated using the iterative Newton-Raphson/Fisher procedure. Since the structure of the second and cross derivatives of the item parameters are the same as those derived in Chapter 2, they will not be repeated here. Typically, initial estimates of the item parameters based upon Equations 1.9 and 1.10 are used to start the iterative process. Within each iteration, adjustments are made to the item parameter estimates until a suitable convergence criterion is met. In the item parameter estimation stage of the JMLE procedure, the f_{ik} and r_{ik} are known. However, under the MMLE approach, the values of $\bar{f}_{ik}$ and $\bar{r}_{ik}$ depend on the values of the item parameter estimates. But, Equations 6.25–6.27 do not make any provision for this dependence.

Consequently, a paradigm needs to be established in which both the "artificial data" and the item parameter estimates can be obtained iteratively. The approach taken by Bock and Aitkin (1981) was to employ the EM algorithm due to Dempster, Laird, and Rubin (1977).

6.4.1 Some Background on the EM Algorithm

In general, the EM algorithm is an iterative procedure for finding maximum likelihood estimates of parameters of probability models in the presence of unobserved random variables. The E stands for the expectation step and the M for the maximization step. In the typical IRT case, we wish to find estimates of item parameters in the presence of an unobservable random variable (θ). To make inferences about θ, some observable representation based on the item responses ($\mathbf{U}$) is used. When the EM algorithm is conceived as an extension of the missing information principle of Orchard and Woodbury (1972), ($\mathbf{U}, \theta$) is considered to be the unobserved (complete) data, and $\mathbf{U}$ is the observed (incomplete) data. Let $f(\mathbf{U}, \theta|\boldsymbol{\xi})$ represent the joint probability density function of the complete data ($\mathbf{U}, \theta$), where $\boldsymbol{\xi}$ represents the item parameters to be estimated. Given the matrix of provisional item parameters at the pth cycle, $\boldsymbol{\xi}^{p+1}$ is computed by maximizing the posterior expectation,

$$E\left[\log f(\mathbf{U}, \theta|\boldsymbol{\xi})|\mathbf{U}, \boldsymbol{\xi}^p\right], \tag{6.28}$$

with respect to $\boldsymbol{\xi}$. This process is repeated until a convergence criterion is satisfied. The two general steps of the algorithm are:

E-step: Compute $E\left[\log f(\mathbf{U}, \theta|\boldsymbol{\xi})|\mathbf{U}, \boldsymbol{\xi}^p\right]$.
M-step: Choose $\boldsymbol{\xi}^{p+1}$ such that the posterior expectation is maximized.

There are three forms of the EM algorithm, distinguished by particular restrictions placed on the probability model, that is, the relation between the IRT (probability) model and the exponential family of distributions. It is important to emphasize that the probability models being evaluated for membership in the exponential family are IRT models, characterized by the joint distribution of the model and $g(\theta|\boldsymbol{\tau})$.

If the probability model is a member of the regular exponential family of distributions, simple sufficient statistics for θ exist, and the algorithm reduces to taking the expectation of the sufficient statistic, conditional on the observed data and the provisional parameter estimates (E-step), substituting these conditional expectations into the maximization or M-step, and carrying out usual maximum likelihood estimation. In IRT, only the one-parameter (Rasch) model is a member of the exponential family. However, an EM algorithm can be employed even if the probability model does not follow any kind of exponential family.

For its application to IRT, note that, for the two- and three-parameter logistic IRT models, the distribution of $f(\mathbf{U}, \theta|\boldsymbol{\xi})$ is unknown and not a member of the exponential family and, hence, the sufficient statistics are not available. As a substitute, the expected values (i.e., posterior expectations) of the $\log f(\mathbf{U}, \theta|\boldsymbol{\xi})$, conditional on some observed representation of θ, are taken, and these quantities treated as if they were known (E-step). These expected values are then used to find the item parameter estimates maximizing the log of the likelihood function (M-step) employing maximum likelihood methods. Details of this procedure for the Bock and Aitkin solution follow below.

We begin with the assumption that ability is restricted to a finite set of values θ_k, with probabilities $\vartheta_1, \ldots, \vartheta_k, \ldots, \vartheta_q$. Also, let $\mathbf{f} = (f_{i1}, \ldots, f_{ik}, \ldots, f_{iq})'$ represent the number of examinees at each of the q ability values for the ith item and $\mathbf{r} = (r_{i1}, \ldots, r_{ik}, \ldots, r_{iq})'$ the number of correct responses at each of the q ability values for the ith item. If the abilities of N examinees are randomly sampled from the above distribution, the joint probability that the $f_{i1}, \ldots, f_{ik}, \ldots, f_{iq}$ examinees will have ability levels $\theta_1, \ldots, \theta_k, \ldots, \theta_q$ is given by the multinomial function

$$\left[\frac{f!}{f_{i1}! \cdots f_{ik}! \cdots f_{iq}!}\right] \prod_{k=1}^{q} \vartheta_k^{f_{ik}}. \tag{6.29}$$

Given f_{ik} and θ_k, the probability of the response vector $\mathbf{r}$ is

$$\prod_{i=1}^{n} \binom{f_{ik}}{r_{ik}} P_i(\theta_k)^{r_{ik}} Q_i(\theta_k)^{f_{ik}-r_{ik}} \tag{6.30}$$

and the joint probability of the $\mathbf{f}$ and $\mathbf{r}$ vectors is

$$\prod_{k=1}^{q} \prod_{i=1}^{n} \binom{f_{ik}}{r_{ik}} P_i(\theta_k)^{r_{ik}} Q_i(\theta_k)^{f_{ik}-r_{ik}} \left[\frac{f!}{f_{i1}! \cdots f_{ik}! \cdots f_{iq}!}\right] \prod_{k=1}^{q} \vartheta_k^{f_{ik}}. \tag{6.31}$$

From the factorization criterion (see Halmos & Savage, 1949), it can be seen that $(\mathbf{f}, \mathbf{r})$ is the sufficient statistic for the complete data $(\mathbf{U}, \theta)$.

Ignoring the constant terms, the log-likelihood function for the complete data can be written in a form similar to that of the bioassay solution:

$$\log L = \sum_{k=1}^{q} \sum_{i=1}^{n} \left[r_{ik} \log P_i(\theta_k) + (f_{ik} - r_{ik}) \log Q_i(\theta_k) + \sum_{k=1}^{q} f_{ik} \log \vartheta_k \right] . \tag{6.32}$$

Here $(\mathbf{f}, \mathbf{r})$ is unobserved but, by taking the posterior expectation of the log-likelihood function, given $\boldsymbol{\xi}$, we obtain

$$E(\log L) = \sum_{k=1}^{q} \sum_{i=1}^{n} \left\{ E(r_{ik}|\mathbf{U}, \boldsymbol{\xi}) \log P_i(\theta_k) + \right.$$

$$\left. E\left[(f_{ik} - r_{ik})|\mathbf{U}, \boldsymbol{\xi}\right] \log Q_i(\theta_k) + \sum_{k=1}^{q} E(f_{ik}|\mathbf{U}) \log \vartheta_k \right\} . \tag{6.33}$$

The last sum in Equation 6.33 can be ignored since it does not depend on the $\boldsymbol{\xi}$. The remaining terms are just like the sum of log-likelihood functions encountered in the bioassay solution when $(\mathbf{f}, \mathbf{r})$ is replaced by $E(\mathbf{f}, \mathbf{r} | \mathbf{U}, \boldsymbol{\xi})$, which are finite-dimensional sufficient statistics. Maximizing Equation 6.33 is equivalent to maximizing the E-step expression in Equation 6.28, and, for a given $g(\theta | \tau)$ and IRT model, provides item parameter estimates that solve the marginal maximum likelihood Equations 6.25–6.27 (Bock & Aitkin, 1981). Since the items are assumed to be independent, the cross second derivatives of different items are zero in the M-step, and the maximization of $E(\log L)$ is carried out for each item singly.

Using EM in the Item Parameter Estimation Process. The Bock and Aitkin solution for the marginal maximum likelihood estimation of a given item's parameters was implemented via an EM algorithm as follows:

1. The expectation (E) step:
    a) Use the quadrature form given in expression 6.21 and provisional item parameter estimates to compute the likelihood of each examinee's item score vector at each of the q quadrature nodes.
    b) Use expression 6.22 and the quadrature weights $A(X_k)$ at each of the q nodes to compute the posterior probability that the ability of the jth examinee is X_k.
    c) Use Equations 6.23 and 6.24 to generate $\bar{f}_{ik}$ and $\bar{r}_{ik}$, the expected number of examinees attempting item i and the expected number of correct responses for that item at each of the q ability nodes.
2. The maximization (M) step: Solve the likelihood Equations 6.25–6.27 for the item parameter estimates using $\bar{r}_{ik}$ and $\bar{f}_{ik}$, the "artificial data." Because these values depend on the $P(X_k | \mathbf{u}_j, \boldsymbol{\xi}, \tau)$ terms which, in turn, depend on the unknown item parameters, the likelihood equations are implicit and must be solved iteratively (e.g., via the Taylor series and Newton-Raphson/Fisher procedure).
3. If the overall likelihood is unchanged from the previous cycle, the item estimation process has converged and the process is terminated. Otherwise, steps 1 and 2 are repeated.

Maximizing the marginal likelihood equations is, with respect to the resulting item parameter estimates, equivalent to maximizing the posterior expectation of expression 6.33. The expected number of attempts $(\bar{f}_{ik})$ and the expected number of correct responses at each ability level $(\bar{r}_{ik})$, computed in step 3, serve as E values. The likelihood function is then maximized (M-step) in step b using these values, which are based on the incomplete data pretending that they are based on the complete data (i.e., as if they were known). This leads to an iterative process in which the E- and M-steps are repeated until the convergence criterion is met. It should be noted that, because the two- and three-parameter IRT models are not members of the exponential family, convergence under an EM framework is not guaranteed. However, reported empirical work (e.g., Mislevy & Bock, 1985) suggests that the method does, in general, converge although a large number of cycles may be necessary.

Implementation of the MMLE/EM Approach. The MMLE/EM approach has been implemented in the BILOG computer program. The original version of this program (Mislevy & Bock, 1984) was written for large-scale (mainframe) computers and had a limited distribution. Subsequently, the program was reprogrammed to function on the IBM PC series of microcomputers and named PC-BILOG, (Mislevy & Bock, 1986). The microcomputer version makes the program accessible to a much larger audience. While the PC-BILOG computer program has a wide range of features and analysis options, only those of interest from the MMLE point of view will be explored here. For a fuller account, the reader is referred to the PC-BILOG manual. In the discussion below, it will be assumed that the PC-BILOG version of the program is used.

Under the JMLE approach described in Chapter 4, the parameters of the items and those of the examinees were estimated via Birnbaum's two stage "back and forth" paradigm. The BILOG program also uses a two-stage procedure, but it does not use a back and forth paradigm between items and examinees. Under the MMLE/EM, the item parameters are estimated completely in one stage, and then ability parameters are estimated in subsequent stage. In the first stage, the parameters of the n items in the test are estimated using the MMLE/EM procedure described above. Since item parameter estimates cannot be obtained for items answered by all or none of the examinees, such items are eliminated at the beginning of the first stage. The expressions for $\bar{f}_{ik}$ and $\bar{r}_{ik}$ (Equations 6.23 and 6.24) involve a summation over the N examinees. In a given data set, many examinees can have the same vector of correct and incorrect responses to the n items, and the same computations are often replicated. In order to reduce the computational demands, the N examinee item response vectors are grouped into 2^n possible item response patterns. These item response patterns are indexed by l ($l = 1, 2, 3, \ldots, s$). The number of examinees possessing a given item response pattern is then given by f_l. Because of the computational efficiency, Bock and Aitkin (1981) formulated $\bar{f}_{ik}$ and $\bar{r}_{ik}$ in terms of the item response patterns rather than examinee item response vectors. This was not done in the present chapter because it makes the basic mathematics much more difficult to understand and does not coordinate well with the presentations of the previous chapters. Typically, the estimation process is initiated by assuming a unit normal prior distribution of ability. Rather than using the quadrature nodes and weights due to Stroud and Secrest (1966), a histogram approximation to these terms, due to Mislevy, is used (Mislevy & Stocking, 1989; Seong, 1990). Ten nodes on the ability scale are equally spaced over the range -4 to $+4$. These node values, X_k, remain constant throughout both the E-step and the M-step of the parameter estimation process. At each node, the ordinate of the normal density is computed and multiplied by the width (.89) of the spacing between the nodes. The areas of these rectangles are then normalized to a total area of unity by dividing each of them by the sum of the rectangles. The resulting values are then used as the quadrature weights, that is, the $A(X_k)$. Step 1 of the MMLE/EM procedure is performed for all of the items. Upon completion of the E-step, the item parameters are estimated via step b, which is the M-step. At this point, the overall likelihood is available. If the change in this likelihood between two iterations is sufficiently small or if the maximum number of EM cycles has been reached, the item parameter estimation process is terminated, and the ability estimation stage is entered. If not, the E- and M-steps are repeated. Upon reaching overall convergence, the large sample standard errors of estimation of the item parameters are available in the inverses of the information matrices of the individual items.

In the original Bock and Aitkin (1981) article, the initial set of quadrature weights $A(X_k)$ were used throughout the EM algorithm. However, in PC-BILOG, the quadrature weights are recomputed, for the fixed set of nodes, before each E-step. The posterior ability distribution, in the form of the q values of $\bar{f}_{ik}$, is used to obtain the new quadrature weights. The $\bar{f}_{ik}$ are normalized by dividing each by the sample size, a process represented in the following equation due to Mislevy and Bock (1985, p. 195):

$$
A^{(t+1)} = \frac{1}{N} \sum_{j=1}^{N} \left[\frac{L(X_k)A(X_k)^{(t)}}{\sum_{k=1}^{q} L(X_k)A(X_k)^{(t)}} \right]. \tag{6.34}
$$

These adjusted weights are rescaled to have mean zero and unit variance. It should be noted that these adjusted quadrature weights reflect the initial prior distribution of ability as well as the underlying ability distribution since the latter is embodied in the examinee's item responses. These adjusted quadrature weights, along with the current item parameter estimates, are then used in the next iteration of the EM algorithm.

In the second stage, it is assumed that the obtained item parameter estimates are the true parameter values, and the maximum likelihood estimation procedures of Chapter 3 are used to obtain an ability estimate for each examinee. The PC-BILOG handles the problem of examinees who answer all or none of the items correctly in a unique fashion. An examinee gets a raw score of zero is given a response value of $1/2$ for the easiest item and 0's for the remaining $n-1$ items. A perfect score results in a response score of $1/2$ for the most difficult item and 1's for the remaining $n-1$ items. These modified response patterns are then used to estimate the examinee's ability. Upon completion of this stage, the PC-BILOG program prints the final adjusted quadrature weights, the test characteristic curve, and the test information function.

Harwell, Baker, and Zwarts (1988) have provided a MICROSOFT QUICK-BASIC computer program for the MMLE/EM estimation of item parameters, which is reproduced in Appendix D. A computational example in which the item response patterns are used in the estimation process for a two-parameter logistic ICC model is shown. Again, sufficient detail is presented so that the reader can follow the essentials of the estimation process. Appendix H contains the MMLE/EM for the Rasch model of Thissen (1982).

The Metric Issue. One of the salient features of the MMLE/EM approach is that it does not use the Birnbaum paradigm. The item parameters are estimated in the first stage and the ability parameters in a subsequent stage. The question of interest then is: How does the MMLE/EM procedure handle the "identification problem"? The approach taken is initially to use the prior ability distribution and its representation in the form of quadrature nodes and weights to define the metric. Since the initial quadrature nodes and weights are based on an approximation to a unit normal distribution, the metric has midpoint zero and a scale of one. At the end of the first cycle of the EM algorithm, the item parameter estimates will be in a metric having a location of zero and a unit of scale of one, which corresponds to the parameters of the prior ability distribution. In each successive cycle of the EM algorithm, the adjusted quadrature weights are based on the normalized and rescaled posterior ability distribution, which also has location and scale parameters of zero and unity. Thus, the identification problem is solved via the scaling of the posterior ability distribution. In some sense, it is like the JMLE approach to the identification problem implemented in LOGIST except that, under MMLE/EM, it is the posterior ability distribution that defines the metric rather than the distribution of the examinee's individual ability estimates. It is important to note that Mislevy (1984) has shown the estimated distribution of ability and the distribution of the examinee's ability estimates are not the same. The end result is that PC-BILOG and LOGIST solve the identification problem in different ways.

In the ability estimation stage, the item parameter estimates are considered to be the "true" values of the item parameters. Thus, the metric of the ability estimates is defined by the metric of the item parameters, which is that of the estimated posterior ability distribution. What one has here is a situation in which both the item and ability parameter estimates are related to the estimated posterior ability distribution. As implemented in the PC-BILOG program, this leads to an interesting set of results first noted by Baker (1990). When the variance of the underlying ability distribution is larger than that of the prior distribution of ability, PC-BILOG compresses the distribution of the examinee's ability estimates to one having mean zero and unit variance. When the variance of the underlying ability is smaller than that of the prior distribution, quite a different phenomenon occurs. PC-BILOG sets the mean of the examinee's abilities to zero but, for some data sets, it retains the variance of the underlying distribution. This is in sharp contrast to LOGIST, that is, JMLE, where distributions with variances larger and smaller than unity are both standardized to unity. Nonetheless, the examinee ability estimates are in the same metric as that of the item parameter estimates.

6.5 Summary

Since its introduction, the JMLE paradigm for estimating item and ability parameters has been the standard estimation method in IRT. A fundamental problem with this paradigm was that estimates of the structural (item) parameters are not consistent in the presence of the incidental (ability) parameters (i.e., the Neyman-Scott problem). The marginal maximum likelihood approach of Bock and Lieberman (1970) alleviated this problem by integrating over the ability distribution and estimating the item parameters in the marginal distribution of ability. However, the simultaneous estimation of the parameters of n items was computationally burdensome and practical only for tests with a small number of items. The subsequent approach due to Bock and Aitkin (1981) is basically a reformulation of the Bock and Lieberman MMLE approach within the framework of the EM algorithm. It also produces consistent item parameter estimates and is far more manageable computationally than the earlier approach because the item parameters are estimated on a one item at a time basis.

In IRT, the basic idea of an MMLE/EM approach is this: Using maximum likelihood methods, we wish to find the values of the item parameters that maximize the log-likelihood of the complete data $(\theta, \mathbf{U})$. Since the complete data are unobservable, sufficient statistics for θ are unavailable. As a substitute, the posterior expectations of θ given the observed data, the prior ability distribution, and the item parameter values are used. Under the Bock-Aitkin model, these posterior expectations consist of the expected number of attempts and correct responses to each item (i.e., the $\bar{r}_{ik}$ and $\bar{f}_{ik}$ calculated in the E-step). These "artificial data" values are the finite-dimensional sufficient statistics for the $\bar{r}_{ik}$ and $\bar{f}_{ik}$ that are known quantities in the bioassay solution of Chapter 2. In the M-step, standard maximum likelihood estimation is performed using these sufficient statistics as the data. This leads to an iterative process in which the E-step is followed by the M-step until some convergence criterion is satisfied. Two other comments are also in order. First, the metric of the item parameter estimates is determined by the posterior distribution of ability approximated by the quadrature procedures. For the examples discussed earlier, the unit normal distribution was used. Secondly, it should be emphasized that the MMLE/EM procedure yields only item parameter estimates and, in general, must be paired with an examinee ability estimation procedure. In the PC-BILOG program, the ability estimation procedure assumes that the item parameter estimates yielded by the MMLE/EM process are the "true" values of the parameters. Then the maximum likelihood procedure of Chapter 3 or a Bayesian estimation procedure is used to obtain an ability estimate for each examinee, one examinee at a time. In contrast to the JMLE approach, there is no alternating between item and ability estimation.

While the MMLE/EM approach has many desirable features, certain limitations remain. Since the item parameters are estimated by maximum likelihood, the data must be screened to eliminate items which are answered correctly or incorrectly by all of the examinees. Despite this screening, certain data sets can yield discrimination indices that become very large. In addition, discrimination estimates near zero can result in very large absolute values of item difficulty. Such deviant values of the item parameter estimates can be produced by the MMLE/EM procedures of PC-BILOG without disrupting the overall process. However, when they are used in the second stage to estimate an examinee's ability via maximum likelihood, the estimation process fails, and no ability estimates can be obtained. Unfortunately, it is difficult to know when this will occur because it is data set dependent. An additional problem is that, in general, the EM algorithm converges slowly, and the closer it gets to the solution, the slower the rate of convergence. As a result, the run time on a personal computer can become quite long. Convergence within the M-step can be facilitated by using an acceleration technique such as that due to Ramsey(1975), which is used in PC-BILOG.

The MMLE/EM approach to item parameter estimation has a number of characteristics that are of interest. These are:

1. The marginalization over θ produces estimates of item parameters that are consistent for tests of finite length (assuming that the IRT model and the ability population model are correct).

2. Conceptualizing subjects as a random sample from a population with ability distributed in accordance with a density function $g(\theta|\tau)$ allows one to assume an arbitrary distribution of θ in the population sampled. Consequently, the metric of the item parameter estimates is defined by the location and scale parameters in τ.

3. It permits the imposition of a "Bayesian-like" structure on the estimation process such that inferences about θ are improved. As noted earlier, the EM algorithm illustrated here does not involve a classical Bayes solution since the item parameters are treated as constants, meaning the parameter estimates are generated using maximum likelihood methods.

4. The computer programming needed to implement an MMLE/EM approach is actually quite simple although the underlying mathematics are rather sophisticated. The E-step involves a very large number of computations even for a small number of items, but the computations are well structured and can be done in an orderly fashion. In fact, they were implemented in only 30 lines of BASIC code. The same was true of the M-step, in which only 21 lines of BASIC code were needed (see Appendix D for the actual BASIC program). In a production program such as PC-BILOG, a wide range of other considerations must be taken into account and, consequently, such a program would be considerably more extensive.

It should be clear that the MMLE/EM approach is a major advance in item parameter estimation techniques. While it is important in its own right, it is also important as the stepping stone to the marginalized Bayesian item parameter estimation techniques of the next chapter.

7. Bayesian Parameter Estimation Procedures

7.1 Introduction

[1]Experience with the JMLE procedure has shown that it has a number of inherent problems. Among these are item and ability parameter estimates that may assume unreasonable values and an inability to compute ability estimates for unusual examinee response patterns, for example, when all items are answered correctly or incorrectly. In the LOGIST computer program, the data are screened for such patterns, and a complex set of restrictions is imposed to keep parameter estimates within reasonable bounds. From a theoretical point of view, the primary problem is that the use of JMLE for tests of finite length may result in estimates that are not statistically consistent. In Chapter 5, it was shown that the Rasch model addressed the problem of incidental parameters by replacing an examinee's ability parameter with its sufficient statistic, which is the raw test score. However, this conditional maximum likelihood solution is only possible under the Rasch model.

A more generally applicable approach is to use the MMLE/EM procedures of Chapter 6 to resolve the problem of inconsistent item parameter estimates. Marginal maximum likelihood estimation was distinguished from JMLE by the assumption that examinee abilities have a distribution in a population. When this distribution is known or can be estimated, it permits ability to be integrated out of the likelihood function. This frees item parameters from their dependence on the ability parameters of individual examinees, resulting in maximum likelihood estimates (MLEs) of the item parameters that are consistent for tests of finite length (Bock & Aitkin, 1981). Although MMLE/EM resolved the problem of inconsistent item parameter estimates, the problems of deviant values of the item parameter estimates in some data sets and the lack of a means of estimating an examinee's ability for unusual item response patterns remain. A vehicle for attempting to prevent deviant parameter estimates from occurring is the use of estimation procedures based on a Bayesian approach. One approach to applying these methods to parameter estimation under IRT was due to Swaminathan and Gifford (1982,

---

[1] This chapter is based on the paper by M. R. Harwell and F. B. Baker (1991). The use of prior distributions in marginalized Bayesian item parameter estimation: A didactic. *Applied Psychological Measurement*, *15*, 375–389 (used with permission).

1985, 1986) who, in this set of three papers, derived estimation procedures for the one-, two-, and three-parameter logistic ICC models. The Birnbaum paradigm was used to estimate jointly the item parameters of a test and the abilities of the group of examinees. In contrast to the JMLE approach, Swaminathan and Gifford used Bayesian estimation procedures in each of the two stages. While this work employed many of the important concepts of Bayesian estimation within the context of IRT, it will not be included in the present chapter primarily because of the lack of a readily available computer program to analyze test data under the approach.

For many years, there has been a significant argument between the "frequentists" and the Bayesians in the statistical literature. In addition, there are "schools" within the Bayesian camp. While this controversy has important implications for the field of statistics, it has not yet played a major role in IRT. The researchers in IRT have adopted a more pragmatic approach in which Bayesian methods are viewed as a means of improving parameter estimation. Thus, rather than getting involved in contrasting the different statistical philosophies, the present chapter adopts the pragmatic approach.

Bayesian approaches in IRT can be distinguished by whether item parameter estimation takes place after marginalization (i.e., integration) over incidental parameters (Mislevy, 1986b; Tsutakawa & Lin, 1986) or without any marginalization (Swaminathan & Gifford, 1982, 1985, 1986). Because it is a direct extension of the MMLE/EM approach of Bock and Aitkin (1981), the item parameter estimation aspects of the present chapter are based on Mislevy's (1986b) marginalized Bayesian estimation procedure. His approach inherits the properties of MMLE but tries to constrain the parameter estimates from taking on deviant values. In addition, there is numerical evidence that marginalized solutions are superior to unmarginalized solutions (O'Hagan, 1976). Since, the estimation of an examinee's ability does not involve the incidental parameters problem, a nonmarginalized Bayesian procedure can be used to accomplish it. The availability of these Bayesian methods in the PC-BILOG computer program (Mislevy & Bock, 1986, 1989) provides a powerful yet flexible set of procedures to be used to estimate parameters in IRT. However, to take full advantage of the program's potential requires an understanding of the basis of these procedures. From a pragmatic point of view, the Bayesian estimation procedure of this chapter can be considered an extension of the MMLE procedures implemented via the EM algorithm presented in Chapter 6. Because of this, it will not be necessary to repeat all of the details of the implementation that are common to both approaches. Rather, the emphasis will be on the Bayesian aspects and how they extend the previous techniques.

7.2 The Bayesian Approach to Parameter Estimation

General introductions to Bayesian statistical methodology include Cornfield (1969), de Finetti (1974), Edwards, Lindman, and Savage (1963), Lindley (1970a, 1970b, 1971), and Novick and Jackson (1974). Examples of the use of these methods in educational settings can be found in Novick and Jackson (1974), Novick, Jackson, Thayer, and Cole (1972), and Rubin (1980). Lord (1986) compared maximum likelihood and Bayesian estimation methods in IRT.

The approach rests on Bayes' theorem, which provides a way of expressing conditional probability. The goal is to combine probabilities obtained from a likelihood function that uses sample data with probabilities obtained using one's prior information about the distribution of the set of unknown parameters. An application of Bayes' theorem produces a posterior probability distribution that is proportional to the product of the likelihood function and the prior probability distribution. The posterior probability distribution is used to make inferences about the unknown parameters (Lindley, 1971, p. 36).

To illustrate the use of Bayes' theorem in an IRT setting, suppose one wishes to estimate item parameters when ability parameters are known. Let $g(\xi|\eta)$ represent the probability distribution reflecting one's prior belief as to the distribution of the possible values of the ith item's parameters, conditional on the parameters (i.e., hyperparameter) in η. Assume that each of the n items have prior distributions of the same form. Let U be an $N \times n$ matrix of item responses and $L(U|\xi, \theta)$ the likelihood function of the sample item responses conditional on the parameters in ξ and θ. The posterior probability distribution across items and examinees can be expressed as

$$g(\xi|U, \theta, \eta) \propto L(U|\xi, \theta)g(\xi|\eta). \tag{7.1}$$

It is of interest to contrast this equation with the Bayesian Equation 6.1 employed in the previous chapter. In the earlier equation, the right and left sides were equated but, in the present case, they are defined as proportional. This is because the present equation does not employ the denominator term, which behaves essentially as if it were a proportionality constant. In addition, the present equation defines the posterior distribution for the item parameters rather than that of ability as was the case in the Equation 6.1. The likelihood function $L(U|\xi, \theta)$ of Equation 7.1 is defined as

$$L(U|\theta, \xi) = \prod_{j=1}^{N} \prod_{i=1}^{n} P_i(\theta_j)^{u_{ij}} Q_i(\theta_j)^{1-u_{ij}} = \prod_{j=1}^{N} P(U_j|\theta_j, \xi). \tag{7.2}$$

$P(U_j|\theta_j, \xi)$ is the probability of an examinee's response vector U_j, conditional on a known value of θ_j and the item parameters in ξ. This is the same likelihood that served as the basis of the JMLE and MMLE procedures of Chapters 4 and 6, respectively. Inferences about the unknown item parameters in ξ typically take the form of point estimates that maximize the posterior probability $g(\xi|U, \theta, \eta)$ with respect to the unknown parameters. In Bayesian estimation, this is often the estimated mode of the posterior probability distribution, and the resulting estimates are typically known as Bayes modal estimate or BMEs (Mislevy & Stocking, 1989).

It is typical (though not necessary) in a Bayesian approach to assume that the parameters of a given type, for example, b_i, in the ICC model are independent and identically distributed. It is also assumed that such parameters are exchangeable, meaning that the prior probability distribution for a particular parameter (e.g., b_i for item i) is no different from that of any other parameter of the same type (Swaminathan & Gifford, 1985; see Mislevy, 1988, for an alternative conceptualization of exchangeability). With these assumptions, an appropriate distributional form for the prior distribution is specified. Fortunately, educational tests designed to assess specific traits, constructs, abilities, etc., and the ICC models typically employed often provide the necessary theoretical and empirical basis for selecting appropriate prior distributions. For example, difficulty parameters are often between -4 and $+4$ standard deviations in value, suggesting that a unimodal symmetric distribution like the normal can serve as a prior distribution for difficulty. If the variance of the prior distribution of b_i is small, the prior distribution is said to be informative, and its contribution to parameter estimation is likely to be substantial. A small variance means that the values of b_i will be tightly clustered about the mean of the prior distribution, with some values of b_i more likely than others. A large variance, on the other hand, would result in an uninformative prior distribution since the values of b_i are more variable and not clustered about the mean of the prior distribution (Novick & Jackson, 1974, p. 156). Other things being equal, a prior distribution with a large variance would have less impact on parameter estimation than a prior distribution with a small variance. Hence, the variance of a prior distribution plays a key role in estimating item parameters.

In the Bayesian statistical literature, the amount of information associated with the prior distribution of a parameter is often expressed in terms of a number of observations. In the present context, this is the number of examinees. The primary effect of an informative prior distribution on parameter estimation is to shrink the estimate towards the mean of the parameter's prior distribution by an amount proportional to the information contained in the prior distribution of that parameter (Mislevy & Stocking, 1989). The more informative the prior distribution, the more the parameter estimate tends to be pulled toward the mean of its prior distribution. Pulling estimates toward the prior mean helps prevent the estimates from cycling off to unreasonable values during the estimation process (this topic is expanded on in a later section of the chapter). If the mean of the prior distribution is close to the true value of the parameter, an informative prior may also lead to BMEs that are closer than are their maximum likelihood counterparts to the true parameter value (Swaminathan & Gifford, 1985).

The relative contribution of the prior distribution $g(\xi|\eta)$ and the likelihood $L(U|\xi,\theta)$ to $g(\xi|U,\theta,\eta)$, and thus to the estimates, is an important issue. If the number of examinees is large and the contribution of the prior probability distribution is small, the likelihood will dominate the posterior probability distribution. That is, the item parameter estimates obtained from $g(\xi|U,\theta,\eta)$ in Equation 7.1 will depend almost wholly on the observed item response data represented through $L(U|\xi,\theta)$, and $g(\xi|\eta)$ will have little effect on the estimates. In this case the BMEs are likely to be almost identical to the estimates obtained via maximum likelihood estimation. If the contribution of the prior probability distribution is substantial, the BMEs will differ from the MLEs.

7.3 The Marginalized Bayesian Estimation Procedure

Recall that a Bayesian approach produces a posterior distribution that depends on the contribution of both prior information about parameters and information obtained from the sample item response data. Inferences about unknown parameters are then based on the posterior distribution. Mislevy (1986b) presented a Bayesian procedure for estimating item parameters in IRT that is a generalized form of Equation 7.1 and represents an extension of the marginalized solution of Bock and Aitkin (1981). Mislevy employed the two-stage, classical Bayesian estimation procedure attributed to Lindley and Smith (1972) in which prior information is specified in a hierarchial fashion. One begins with the joint density of all of the parameters prior to data collection. These parameters are assumed to be independent and continuous random variables with specified probability distributions:

$$g(\theta,\tau,\xi,\eta) = \prod_{j=1}^{N} g(\theta_j|\tau)g(\tau) \prod_{i=1}^{n} g(\xi_i|\eta)g(\eta). \tag{7.3}$$

First, the terms on the right-hand side of Equation 7.3 need to be examined. The $g(\theta_j|\tau)$ term is the probability distribution of an examinee's ability parameter θ_j and is conditional on the population parameters of the ability distribution contained in the vector τ. Typically, it is posited that the prior distribution of ability is a normal distribution. Since abilities are assumed to be independent and identically distributed, τ contains the common mean (μ_θ) and variance (σ_θ^2) of these prior ability distributions. In the Lindley and Smith (1972) hierarchial model, the θ_j are parameters and the population parameters μ_θ and σ_θ^2 are known as hyperparameters. The hyperparameters can also be treated as random variables having a probability distribution that, in this case, is denoted by $g(\tau)$. The third term $g(\xi_i|\eta)$ is the probability distribution for the parameters of item i contained in the vector ξ_i, conditional on the population parameters in the vector η. As was the case with ability, the a_i, b_i, and c_i, contained in ξ_i will be referred to as parameters, and η as the vector containing population parameters (i.e., hyperparameters) for the item.

Suppose one wished to make inferences about all of the unknown parameters, in this case, θ, τ, ξ, and η. After data collection, the posterior distribution across all items and examinees obtained via an application of Bayes' theorem is

$$g(\theta, \tau, \xi, \eta | U) \propto L(U | \theta, \xi) g(\theta | \tau) g(\tau) g(\xi | \eta) g(\eta). \tag{7.4}$$

In this context, specification of the prior information is accomplished through the parameters (hyperparameters) of the prior probability distributions $g(\theta | \tau)$ and $g(\xi | \eta)$ that constitute the first stage in the Lindley and Smith (1972) model. Specifying the parameters of the probability distributions of the hyperparameters, in this case, the parameters characterizing $g(\tau)$ and $g(\eta)$, constitutes the second stage.

Mislevy (1986b) pointed out that Equation 7.4 contains all of the information available about the unknown parameters. To estimate the item parameters in Equation 7.4, the marginalization approach of Bock and Lieberman (1970) can be used. Under this approach, the likelihood function is marginalized with respect to ability. If the ICC model and the prior distribution of ability are correct, the resulting item parameter estimates (for tests of finite length) are known to approach their true values as the number of examinees increases. In general, the choice of variables to marginalize over is dictated by the distinction between the parameters one wishes to estimate and those of no interest (i.e., incidental or nuisance parameters). Mislevy (1986b) suggested that, in many educational settings, the distributions of η are not especially interesting and that treating the parameters in η as nuisance parameters is appropriate. The distribution of the nuisance parameters can be removed from Equation 7.4 by integrating over their probability distributions. In the presence of moderate numbers of items and examinees, this will have little effect on the item parameter estimates. In contrast, it may be of interest to estimate the hyperparameters contained in τ, and their probability distribution is retained here temporarily.

Integrating over the probability distributions of ability $g(\theta | \tau)$ with respect to θ and item population parameters $g(\eta)$ with respect to η leads to a marginalized posterior distribution of the form

$$g(\xi, \tau | U) \propto \int \int L(U | \theta, \xi) g(\theta | \tau) g(\tau) g(\xi | \eta) g(\eta) d\theta d\eta$$
$$\propto L(U | \xi, \tau) g(\xi) g(\tau). \tag{7.5}$$

$L(U | \xi, \tau)$ is the marginal likelihood resulting from $L(U | \xi, \theta)$ having been integrated with respect to ability. Equation 7.5 is the posterior probability distribution that is appropriate for making inferences about item parameters and the parameters of the distribution of ability in the population of examinees. Readers should note that integrating over the population distribution of ability has eliminated the dependence of item parameter estimates on ability estimates of individual examinees. However, the marginal likelihood is still conditional on the hyperparameters μ_θ and σ_θ^2 of the population ability distribution and, thus, $g(\tau)$ and these hyperparameter values must be specified. Similarly, integrating over the population distribution of item parameters in Equation 7.5 has not eliminated the need to specify values for the hyperparameters η in $g(\xi)$.

7.4 Marginalized Bayesian Item Parameter Estimation in PC-BILOG

[2]As was the case in JMLE and MMLE/EM, an assumption of independence of the items is invoked. This greatly simplifies the overall approach as it allows the item parameter estimation process to proceed on an item-by-item basis. To estimate the unknown parameters of an item, the partial derivatives of Equation 7.5 are taken with respect to the item's parameters and set equal to zero. For convenience, one can work with the logarithm of the quantities in Equation 7.5. The system of Bayesian estimation equations is given by

$$\frac{\partial}{\partial v_i}\left[\log L(U|\xi,\tau)\right] + \frac{\partial}{\partial v_i}\left[\log g(\xi)\right] + \frac{\partial}{\partial v_i}\left[\log g(\tau)\right] = 0, \tag{7.6}$$

where v_i represents a parameter associated with the ith item, for example, a_i. The noteworthy characteristic of Equation 7.6 is that the effect of specifying prior distributions for the item parameters and the ability hyperparameters is to append terms for the prior distributions to the expression for the derivative of the marginal likelihood. The system of estimation equations represented in Equation 7.6 constitutes the marginalized Bayes modal equations in which the BMEs represent the joint mode of the posterior distribution. Since the distribution $g(\tau)$ does not contain item parameters, its derivative with respect to v_i will be zero. Thus, $g(\tau)$ may be dropped from Equation 7.6, resulting in the following system of equations for obtaining BMEs of the item parameters:

$$\frac{\partial}{\partial v_i}\left[\log L(U|\xi,\tau)\right] + \frac{\partial}{\partial v_i}\left[\log g(\xi)\right] = 0. \tag{7.7}$$

Since $g(\theta|\tau)$ and $g(\eta)$ were integrated out in Equation 7.5 and $g(\tau)$ was dropped here, the Bayes modal estimation equations will involve only the first stage of the Lindley and Smith (1972) two-stage hierarchial model.

---

[2] The internal workings of the PC-BILOG have not been documented in great detail. Consequently, this section is based on Mislevy (1986b) and the program manual (Mislevy & Bock, 1989) under the assumption that the two are in concert.

7.4.1 The Likelihood Component

The general equations to be solved for the item parameter estimates are composed of two components: the first term is the likelihood component and the second the prior component. Since the two components involve somewhat different concepts, they will be derived separately below and then merged to obtain the final set of solution equations. To keep the presentation as general as possible, the three-parameter ICC model will be used as the one- and two-parameter results can be obtained easily from those for this model. To provide a link with the existing literature, the a_i, b_i, c_i parameterization of the ICC will be employed. Let

$$L = L(U|\xi, \tau) = \prod_{j=1}^{N} P(U_j|\xi, \tau)$$

and $P(U_j|\xi, \tau) = \int P(U_j|\theta, \xi)g(\theta|\tau)d\theta$ is the marginalized probability of a vector of item responses U_j with respect to the item parameters ξ. After the logarithm of L is taken, the first component in Equation 7.7 can be written as

$$\frac{\partial}{\partial v_i} \left[\sum_{j=1}^{N} \log P(U_j|\xi, \tau) \right]. \tag{7.8}$$

Since the computation involved in all three solution equations is similar, the details associated with the discrimination parameter will be presented first. PC-BILOG employs the transformation $\alpha_i = \log a_i$, from which it follows that $a_i = e^{\alpha_i}$. The justification for employing this transformation will be deferred to a later section. For the three-parameter logistic ICC model,

$$P_i(\theta_j) = c_i + (1 - c_i) \left\{ \frac{\exp[e^{\alpha_i}(\theta_j - b_i)]}{1 + \exp[e^{\alpha_i}(\theta_j - b_i)]} \right\} = c_i + (1 - c_i)P_i^*(\theta_j). \tag{7.9}$$

The partial derivative of Equation 7.8 with respect to the transformed discrimination parameter is obtained first:

$$\frac{\partial}{\partial \alpha_i} [\log L] = e^{\alpha_i}(1 - c_i) \sum_{j=1}^{N} \int \frac{u_{ij} - P_i(\theta_j)}{P_i(\theta_j)Q_i(\theta_i)} \times$$
$$(\theta_j - b_i)P_i^*(\theta_j)Q_i^*(\theta_i) \left[P(\theta_j|U_j, \xi, \tau) \right] d\theta. \tag{7.10}$$

Let $W_{ij} = P_i^*(\theta_j)Q_i^*(\theta_j)/P_i(\theta_j)Q_i(\theta_j)$. The marginalized likelihood component in Equation 7.7 corresponding to α_i is

$$\alpha_i : \quad e^{\alpha_i}(1 - c_i) \sum_{j=1}^{N} \int [u_{ij} - P_i(\theta_j)] W_{ij}(\theta_j - b_i) [P(\theta_j|U_j, \xi, \tau)] d\theta. \tag{7.11}$$

The marginalized likelihood component in Equation 7.8 for the difficulty and guessing parameters are found in the same fashion and are

$$b_i : \quad -e^{\alpha_i}(1 - c_i) \sum_{j=1}^{N} \int [u_{ij} - P_i(\theta_j)]\, W_{ij}\, [P(\theta_j|U_j, \xi, \tau)]\, d\theta \qquad (7.12)$$

and

$$c_i : \quad (1 - c_i)^{-1} \sum_{j=1}^{N} \int \left[\frac{u_{ij} - P_i(\theta_j)}{P_i(\theta_j)} \right] [P(\theta_j|U_j, \xi, \tau)]\, d\theta. \qquad (7.13)$$

Except for the e^{α_i} multipliers, these equations are the same as those given in Chapter 6 for MMLE.

Because of the marginalization, all three of these equations involves an integration with respect to ability. As was the case for MMLE/EM, Hermite-Gauss quadrature will be employed to implement the numerical integration. Since the details of this substitution are the same as those presented in Chapter 6, they will not be repeated here. The use of numerical quadrature involves a change from working with individual examinee data to using "artificial" data at each of the q quadrature points. The artificial data consists of the expected number of examinees $\bar{f}_{ik}$ and the expected number of correct responses $\bar{r}_{ik}$ at each node X_k. The $\bar{f}_{ik}$ are defined as

$$\bar{f}_{ik} = \sum_{j=1}^{N} \left[\frac{L(X_k)A(X_k)}{\sum_{k=1}^{q} L(X_k)A(X_k)} \right] = \sum_{j=1}^{N} P(X_k|U_j, \xi, \tau), \qquad (7.14)$$

where $L(X_k) = \prod_{i=1}^{n} P_i(X_k)^{u_{ij}} Q_i(X_k)^{1-u_{ij}}$ is the quadrature form of the likelihood of U_j conditional on $\theta_j = X_k$ and the item parameters. In simple terms, Equation 7.14 is based on computing the expectation (probability) of each examinee having the ability X_k for all values of X_k. Then the $\bar{f}_{ik}$ are obtained by aggregating these probabilities separately at each X_k. The $\bar{r}_{ik}$ is the expected number of correct responses to item i made by the $\bar{f}_{ik}$ "artificial examinees" at ability level X_k and is defined as

$$\bar{r}_{ik} = \sum_{j=1}^{N} \left[\frac{u_{ij}L(X_k)A(X_k)}{\sum_{k=1}^{q} L(X_k)A(X_k)} \right] = \sum_{j=1}^{N} u_{ij} P(X_k|U_j, \xi, \tau). \qquad (7.15)$$

Using numerical quadrature results in replacing the $[P(\theta_j|U_j, \xi, \tau)]$ term in Equations 7.11–7.13 with $[P(X_k|U_j, \xi, \tau)]$ and the integral over ability by a summation over the X_k. It can be seen in Equations 7.14 and 7.15 that the $[P(X_k|U_j, \xi, \tau)]$ terms are then absorbed in the $\bar{r}_{ik}$ and $\bar{f}_{ik}$.

Rewriting Equations 7.11–7.13 in their numerical quadrature form using the artificial data results in the likelihood components of the solution equations for item parameters:

$$\alpha_i : \quad e^{\alpha_i}(1 - c_i) \sum_{k=1}^{q} \left[\bar{r}_{ik} - \bar{f}_{ik} P_i(X_k) \right] W_{ik}(X_k - b_i) \tag{7.16}$$

$$b_i : \quad -e^{\alpha_i}(1 - c_i) \sum_{k=1}^{q} \left[\bar{r}_{ik} - \bar{f}_{ik} P_i(X_k) \right] W_{ik} \tag{7.17}$$

$$c_i : \quad (1 - c_i)^{-1} \sum_{k=1}^{q} \frac{\left[\bar{r}_{ik} - \bar{f}_{ik} P_i(X_k) \right]}{P_i(X_k)}. \tag{7.18}$$

However, in a Bayesian solution, the components corresponding to the prior probability distributions (the second term in Equation 7.7) need to be appended to the marginal likelihood components. These terms are derived in the next section.

7.4.2 The Prior Distribution Component

One of the unique features of Bayesian statistics is that it is based on the idea that parameters have distributions, which is in sharp contrast to the usual concept that parameters are fixed constants. The role of the prior distribution in Bayesian statistics is to specify one's subjective a priori belief about the distribution of these parameters. When estimating item parameters, this requires that one specify the form of the distributions of the parameters of the ICC model of interest. Under the three-parameter logistic ICC model, this would entail specifying the form of the distributions of a_i, b_i, and c_i, respectively. In the typical cases, this form is assumed to be one of the well-known probability distributions. The choice of which distribution to assign to a particular parameter is somewhat arbitrary, however, only a limited number are used in practice. In addition, the test analyzer must specify numerical values for the hyperparameters of these distributions on subjective grounds. Since the marginalized Bayesian item parameter estimation procedures have been implemented in the PC-BILOG computer program, the only prior distributions discussed are those used in this program. The prior distributions for the item discrimination and guessing parameters will be presented below in some detail in the next two sections. Expressions for the difficulty parameter follow directly from those for discrimination.

The Discrimination Prior Distribution. PC-BILOG assumes that each a_i has a lognormal prior distribution over the range $0 < a_i < \infty$. Theoretical justification for the use of a lognormal prior distribution rests on the fact that in most testing settings the a_i are typically greater than zero, suggesting that the distribution of the a_i can be modeled by a unimodal and positively skewed distribution like the lognormal (Mislevy, 1986b). The transformation $\alpha_i = \log a_i$ results in each α_i having a normal prior distribution with a density that is proportional to $\exp[-.5(\alpha_i - \mu_\alpha)^2/\sigma_\alpha^2]$. The normal prior distribution of each α_i is defined by its hyperparameters, μ_α and σ_α. Since the assumption

of exchangeability is employed, these prior distributions are all assumed to
have the same form and the same hyperparameter values.

Since the antilog of the mean of α_i is not the mean of a_i, readers may
find it helpful to see the mean and variance of the lognormal distribution of
a_i expressed in terms of the transformed discrimination indices α_i. Aitchison
and Brown (1957, p. 8) provide the following:

$$\mu_a = \exp\left(\mu_\alpha + 0.5\sigma_\alpha\right) \quad \text{and} \quad \sigma_\alpha^2 = \exp\left(2\mu_\alpha + \sigma_\alpha\right)\left[\exp(\sigma_\alpha^2) - 1\right]. \quad (7.19)$$

In PC-BILOG, the default values of the hyperparameters are $\mu_\alpha = 0$ and
$\sigma_\alpha = 0.5$, which results in $\mu_a = 1.13$ and $\sigma_a = 0.6$.

To examine the effect of a normal prior distribution on estimating α_i,
recall that, under a Bayesian approach, Equation 7.7 results in appending
the likelihood component for α_i (see Equation 7.11) with a partial derivative term that is associated with the joint prior distribution (across items)
of α_i. This means that the prior information associated with all n items is
used in estimating α_i. Under the assumption of independent and identically
distributed parameters, the joint prior distribution of each type of item parameter can be examined (see Swaminathan & Gifford, 1986). Let α represent
an $n \times 1$ vector of transformed discrimination parameters. Then, the portion
of the expression $\log g(\xi)$ in Equation 7.7 that pertains to discrimination can
be written as $\log g(\alpha)$, which is the logarithm of the joint prior probability distribution of n transformed discrimination parameters. The appended
term in Equation 7.7 is the partial derivative of this expression taken with
respect to the discrimination parameter for the ith item. Since α_i is normally
distributed, the derivative is obtained as follows:

$$g(\alpha_i) = \frac{1}{\sqrt{2\pi\sigma_\alpha^2}} \exp\left[-\frac{1}{2}\left(\frac{\alpha_i - \mu_\alpha}{\sigma_\alpha}\right)^2\right],$$

then,

$$\frac{\partial}{\partial \alpha_i}\left[\log g(\alpha)\right] = \frac{\partial}{\partial \alpha_i}\left[-\frac{1}{2}\left(\frac{\alpha_i - \mu_\alpha}{\sigma_\alpha}\right)^2\right] = -\frac{\alpha_i - \mu_\alpha}{\sigma_\alpha^2}. \quad (7.20)$$

Appending the likelihood component in Equation 7.16 with the right-hand
side of Equation 7.20 results in the Bayes modal estimation equation for α_i
used in PC-BILOG:

$$L_1 = \frac{\partial}{\partial \alpha_i}\left\{\log\left[L(U|\xi,\tau)g(\xi)\right]\right\}$$

$$= e^{\alpha_i}(1 - c_i)\sum_{k=1}^{q}\left[\bar{r}_{ik} - \bar{f}_{ik}P_i(X_k)\right]W_{ik}(X_k - b_i) - \frac{\alpha_i - \mu_\alpha}{\sigma_\alpha^2} = 0. \quad (7.21)$$

In the likelihood component section above, the transformation $\alpha_i = \log a_i$
was used but not justified. It can now be seen that the transformation was
convenient because it keeps the metric of the discrimination parameter the
same in both components of the Bayes modal estimation equations.

The Difficulty Prior Distribution. Since a normal prior is used for the item difficulty parameter b_i, the Bayes modal estimation equation for b_i can be written by inspection from Equation 7.21:

$$L_2 = -e^{\alpha_i}(1 - c_i) \sum_{k=1}^{q} \left[\bar{r}_{ik} - \bar{f}_{ik} P_i(X_k) \right] W_{ik} - \frac{b_i - \mu_b}{\sigma_b^2} = 0. \qquad (7.22)$$

Guessing Parameter Prior Distribution. Next, consider the role of a prior distribution when estimating "guessing" parameters. Since c_i is bounded by 0 and 1, a beta prior distribution was proposed by Swaminathan and Gifford (1986) and is incorporated into PC-BILOG. The beta distribution is given by

$$f(c_i) = \frac{(s + t + 1)!}{s!t!} c_i^s (1 - c_i)^t,$$

where the random variable is c_i, while s and t are the parameters of the distribution. The mean of the beta distribution is defined as

$$p = \frac{s + 1}{s + t + 2}.$$

Depending on the values of the parameters, a beta distribution can assume a variety of shapes (see Novick & Jackson, 1974, for an example). In PC-BILOG, the two parameters of a beta distribution have been redefined as ALP (i.e., ALPHA) and BET (i.e., BETA). These parameters are given by ALP $= mp+1$ and BET $= m(1-p)+1$, where m is an a priori weight assigned to the prior information and p is the mean of the beta prior distribution. In order to connect these two parameterizations, let $m = s+t+2$, $p = (s+1)/m$; then $s = pm - 1$ and $t = m - s - 2 = m(1 - p) - 1$. Substituting terms yields $s =$ ALP $- 2$ and $t =$ BET $- 2$. Then,

$$f(c_i) = \frac{[(\text{ALP} - 2) + (\text{BET} - 2) + 1]!}{(\text{ALP} - 2)!(\text{BET} - 2)!} c_i^{\text{ALP}-2}(1 - c_i)^{\text{BET}-2}.$$

The use of a beta prior for guessing parameters revolves around interpreting the mean p as the probability that a low-ability examinee will respond correctly to an item; that is, it is the prior value of c_i. The idea is to specify ALP and BET values that result in the desired p value. This is achieved through the relationship $p = (\text{ALP} - 1)/(\text{ALP} + \text{BET} - 2)$. By default PC-BILOG assumes that the number of alternatives (i.e., NALT) for each dichotomously scored item is five and that $p = 1/\text{NALT} = 1/5 = .2$. Note that, for the other numbers of alternatives, the values of ALP and BET will be different. Swaminathan and Gifford (1986) indicated that weights of $m = 15$ to 20 are preferred; by default PC-BILOG assigns $m = 20$. The default values then are ALP $= 20(.2) + 1 = 5$ and BET $= 20(1 - .2) + 1 = 17$, from which $p = .2$ is obtained [i.e., $.2 = (5 - 1)/(5 + 17 - 2)$]. If one specified $p = .15$ with an a priori weight of $m = 26$, then ALP $= (.26)(.15) + 1 = 4.9 \doteq 5$ and BET $= (26)(1 - .15) + 1 = 23.1 \doteq 23$. Thus $p = (5 - 1)/(5 + 23 - 2) = .15$.

The assigned values of m also influence the credibility intervals (akin to confidence intervals) constructed about c_i (see Swaminathan & Gifford, 1986; Novick & Jackson, 1974, p. 119). These intervals take into account prior information and provide a measure of the strength of the belief that c_i lies within a range of values. Other things being equal, larger a priori weights lead to narrower intervals. This suggests that credibility intervals can be used in lieu of specifying the variance of a prior distribution for guessing parameters. These intervals are tabled in Novick and Jackson (1974, pp. 402–409), and interested readers should consult this reference for additional information.

Generation of the Bayes modal estimation equation for c_i requires that the likelihood component in Equation 7.18 be appended by a term based on the second term in Equation 7.7. Let c represent a $n \times 1$ vector of guessing parameters, and write the expression $\log g(\xi)$ in terms of c as $\log g(c)$. Assuming that the guessing parameters are independent and identically distributed, the partial derivative with respect to c_i is

$$
\frac{\partial}{\partial c_i} [\log g(c)] = \frac{\partial}{\partial c_i} [(\mathrm{ALP} - 2) \log c_i + (\mathrm{BET} - 2) \log(1 - c_i)]
$$

$$
= \frac{\mathrm{ALP} - 2}{c_i} - \frac{\mathrm{BET} - 2}{1 - c_i}. \tag{7.23}
$$

Appending Equation 7.18 with the right-hand side of Equation 7.23 yields the Bayes modal estimation equation for c_i:

$$
L_3 = (1 - c_i)^{-1} \sum_{k=1}^{q} \frac{\left[\bar{r}_{ik} - \bar{f}_{ik} P_i(X_k) \right]}{P_i(X_k)} +
$$

$$
\left[\frac{\mathrm{ALP} - 2}{c_i} - \frac{\mathrm{BET} - 2}{1 - c_i} \right] = 0. \tag{7.24}
$$

The role of the appending term in Equation 7.24 is similar to that in the estimation of discrimination and difficulty.

7.4.3 Bayesian Modal Estimation via EM

The set of marginalized Bayesian modal estimation equations defined by Equations 7.21, 7.22, and 7.24 must be solved simultaneously for each item to obtain the item parameter estimates. Since marginal Bayesian modal estimation can be considered an extension of MMLE, it will be accomplished within the framework of the same EM algorithm as described in Chapter 6. In the expectation (E) step of the first cycle of the algorithm, the "artificial data" are computed using a priori values for the item parameters. In the maximization (M) step, the Bayesian modal estimates of the item parameters are computed on a one item at a time basis using the artificial data. Since Equations 7.21, 7.22, and 7.24 are nonlinear in the parameters, a procedure such as Newton-Raphson or Fisher scoring for parameters is used within the M step of the EM algorithm to obtain the item parameter estimates. These item parameter estimates then are used in the expectation phase of the second EM cycle. This process is repeated until a suitable convergence criterion is met. In the case of PC-BILOG, the convergence criterion is typically specified as a number of EM cycles.

The M step Fisher Scoring Equations. Since the first derivatives of the logarithm of the posterior item parameter distribution, L_1, L_2, and L_3, are given by Equations 7.21, 7.22, and 7.24, respectively, only the second, and cross derivatives with respect to the item parameters are needed to complete the elements of the Fisher scoring equations solved iteratively in the M step. The mathematical details for obtaining these derivative terms are the same as those presented in Chapter 2 except for the use of log transform of the a_i parameter. In addition, the quadrature form of these derivatives using the artificial data is of interest.

From Equation 2.27, the second derivative with respect to α_i is

$$L_{11} = (e^{\alpha_i})^2 \sum_{k=1}^{q} (X_k - b_i)^2 \frac{[P_i(X_k) - c_i]^2}{(1 - c_i)^2} \frac{Q_i(X_k)}{P_i(X_k)} \times$$

$$[-\bar{f}_{ik} P_i(X_k)^2 + \bar{r}_{ik} c_i] - \frac{1}{\sigma_\alpha^2}. \tag{7.25}$$

Let $\bar{r}_{ik} = \bar{f}_{ik} p_i(X_k)$, where $p_i(X_k)$ is the "observed" proportion of correct response at X_k, and taking the expectation of the observed value yields

$$E[\bar{f}_{ik} p_i(X_k)] = \bar{f}_{ik} P_i(X_k).$$

Then the expectation of L_{11} is

$$\Lambda_{11} = -(e^{\alpha_i})^2 \sum_{k=1}^{q} \bar{f}_{ik} (X_k - b_i)^2 \left[\frac{P_i(X_k) - c_i}{1 - c_i} \right]^2 \frac{Q_i(X_k)}{P_i(X_k)} - \frac{1}{\sigma_\alpha^2}. \tag{7.26}$$

From Equation 2.28, the second derivative with respect to b_i is

$$L_{22} = (e^{\alpha_i})^2 \sum_{k=1}^{q} \frac{P_i(X_k) - c_i}{(1 - c_i)^2} \frac{Q_i(X_k)}{P_i(X_k)} \left[-\bar{f}_{ik} P_i(X_k)^2 + \bar{r}_{ik} c_i \right] - \frac{1}{\sigma_\beta^2}. \tag{7.27}$$

Again, taking the expectation, yields

$$\Lambda_{22} = -(e^{\alpha_i})^2 \sum_{k=1}^{q} \bar{f}_{ik} \left[\frac{P_i(X_k) - c_i}{1 - c_i} \right]^2 \frac{Q_i(X_k)}{P_i(X_k)} - \frac{1}{\sigma_\beta^2}. \tag{7.28}$$

Using Equation 2.29 as the starting point, the second derivative with respect to c_i is

$$L_{33} = -\sum_{k=1}^{q} \frac{\bar{r}_{ik}}{(1 - c_i)^2} \frac{Q_i(X_k)}{P_i(X_k)^2} - \frac{ALP - 2}{c_i^2} - \frac{BET - 2}{(1 - c_i)^2}. \tag{7.29}$$

Taking the expectation, yields

$$\Lambda_{33} = -\sum_{k=1}^{q} \frac{\bar{f}_{ik}}{(1-c_i)^2} \frac{Q_i(X_k)}{P_i(X_k)} - \frac{\text{ALP}-2}{c_i^2} - \frac{\text{BET}-2}{(1-c_i)^2}. \tag{7.30}$$

The expressions for the cross-derivative terms fall directly from those given in Chapter 2 and, in the present context, are:

$$\Lambda_{12} = \Lambda_{21} = (e^{\alpha_i})^2 \sum_{k=1}^{q} \bar{f}_{ik}(X_k - b_i) \left[\frac{P_i(X_k) - c_i}{1-c_i} \right]^2 \frac{Q_i(X_k)}{P_i(X_k)} \tag{7.31}$$

$$\Lambda_{13} = \Lambda_{31} = -(e^{\alpha_i})^2 \sum_{k=1}^{q} \bar{f}_{ik}(X_k - b_i) \frac{P_i(X_k) - c_i}{(1-c_i)^2} \frac{Q_i(X_k)}{P_i(X_k)} \tag{7.32}$$

$$\Lambda_{23} = \Lambda_{32} = e^{\alpha_i} \sum_{k=1}^{q} \bar{f}_{ik} \frac{P_i(X_k) - c_i}{(1-c_i)^2} \frac{Q_i(X_k)}{P_i(X_k)}. \tag{7.33}$$

An interesting aspect of employing prior distributions on the item parameters becomes apparent when taking the cross derivatives. In the first derivatives, a prior term is appended to the likelihood component. But each of these appended terms only involves a distribution for the parameter of interest. There are no joint distributions with the remaining parameters. Thus, in the prior terms it has been assumed that, within an item, the item parameters are independent. Given these terms, the Fisher scoring equations to be solved iteratively are

$$\begin{bmatrix} \hat{a}_i \\ \hat{b}_i \\ \hat{c}_i \end{bmatrix}_{(t+1)} = \begin{bmatrix} \hat{a}_i \\ \hat{b}_i \\ \hat{c}_i \end{bmatrix}_{(t)} - \begin{bmatrix} \Lambda_{11} & \Lambda_{12} & \Lambda_{13} \\ \Lambda_{21} & \Lambda_{22} & \Lambda_{23} \\ \Lambda_{31} & \Lambda_{32} & \Lambda_{33} \end{bmatrix}_{(t)}^{-1} \times \begin{bmatrix} L_1 \\ L_2 \\ L_3 \end{bmatrix}_{(t)}. \tag{7.34}$$

Again, the large sample standard errors of estimation of the item parameters are provided by the inverse of the information matrix.

Appendix E contains a computer program written in MICROSOFT QUICKBASIC that implements the BME/EM process for item parameter estimation. The LSAT-6 data set is used again as the data set to be analyzed. Sufficient detail of the estimation computations will be presented so that the reader can follow the process.

7.5 Estimation of Ability

As was the case with MMLE, the Bayesian item parameter estimates were obtained in a separate stage from the examinee ability estimates. As a result, the obtained item parameters can be considered to be the true values in the ability estimation procedures. The PC-BILOG program implements the two Bayesian ability estimation procedures due to Bock and Aitkin (1981) that are presented below.

7.5.1 Bayesian Modal Estimation

The Bayesian modal or maximum a posteriori (MAP) estimate of an examinee's ability is based on the following form of Bayes' theorem:

$$g(\theta_j|U_j,\xi) \propto L(U_j|\theta_j,\xi)g(\theta).$$

It will be assumed that the prior distribution $g(\theta)$ is normal with hyperparameters μ_θ and σ_θ^2, which are specified in advance. As was the case in item parameter estimation, the logarithm will be taken, yielding

$$\log g(\theta_j|U_j,\xi) \propto \log L(U_j|\theta_j,\xi) + \log g(\theta). \tag{7.35}$$

The likelihood component is given by

$$L(U_j|\theta_j,\xi) = \prod_{i=1}^{n} P_i(\theta)^{u_{ij}} Q_i(\theta)^{1-u_{ij}}. \tag{7.36}$$

To obtain an examinee's MAP estimator, the Fisher scoring procedure will be used. Thus, the first and second derivatives of Equation 7.35 with respect to θ_j are needed. The mathematical details of these derivatives of the likelihood component have been presented in Chapter 3. Since the item parameters are known, there is no need for transformation of the discrimination parameter; hence the a_i, b_i, c_i parameterization can be employed. The first derivative of Equation 7.35 is given by

$$L_\theta = \sum_{i=1}^{n} a_i \left[\frac{P_i(\theta_j) - c_i}{P_i(\theta_j)(1 - c_i)} \right] [u_{ij} - P_i(\theta_j)] - \left(\frac{\theta_j - \mu_\theta}{\sigma_j^2} \right). \tag{7.37}$$

The second derivative is

$$L_{\theta\theta} = \sum_{i=1}^{n} a_i^2 \frac{(P_i(\theta_j) - c_i)}{(1 - c_i)^2} \frac{Q_i(\theta_j)}{P_i(\theta_j)} \left[\frac{u_{ij}c_i - P_i(\theta_j)^2}{P_i(\theta_j)} \right] - \frac{1}{\sigma_\theta^2}. \tag{7.38}$$

For Fisher's method of scoring, replacing the u_{ij} by its expectation yields

$$\Lambda_{\theta\theta} = -\sum_{i=1}^{n} a_i^2 \left[\frac{P_i(\theta_j) - c_i}{1 - c_i} \right]^2 \frac{Q_i(\theta_j)}{P_i(\theta_j)} - \frac{1}{\sigma_\theta^2}. \tag{7.39}$$

The estimate of θ_j is obtained by solving the following equation iteratively, beginning with an initial value of $\theta_j = \theta_j^{(0)}$:

$$[\hat{\theta}_j]_{(t+1)} = [\hat{\theta}_j]_{(t)} - [\Lambda_{\theta\theta}]_{(t)}^{-1} \times [L_\theta]_{(t)}, \tag{7.40}$$

where t indexes the iteration. Given that the examinees are independent, there will be one such equation for each examinee. The Bayes modal estimation procedure will always converge for all possible item response patterns. Thus, ability estimates can be obtained even when an examinee answers all items correctly or incorrectly (Mislevy & Bock, 1989).

7.5.2 Bayes EAP Estimation

The Bayes expected a posteriori (EAP) estimation procedure is based on the following form of Bayes' theorem:

$$g(\theta_j|U_j,\xi) = \frac{P(U_j|\theta_j,\xi)g(\theta)}{P(U_j)}.$$

(7.41)

Under the local independence assumption, the probability of examinee j responding with the response vector $U_j = [u_{1j}, u_{2j}, \ldots, u_{nj}]'$ is

$$P(U_j|\theta_j,\xi) = \prod_{i=1}^{n} P_i(\theta_j)^{u_{ij}} Q_i(\theta_j)^{1-u_{ij}}$$

which is the likelihood conditional on the value of θ_j. For a subject randomly sampled from a population with ability distribution $g(\theta)$, the denominator of Equation 7.41 is given by

$$P(U_j) = \int_{\Theta} P(U_j|\theta)g(\theta)d\theta$$

which is the unconditioned probability. From Equation 7.41, the unconditional expectation of θ_j, given U_j, is

$$E(\theta_j|U_j,\xi) = \frac{\displaystyle\int \theta_j g(\theta) \prod_{i=1}^{n} P_i(\theta_j)^{u_{ij}} Q_i(\theta_j)^{1-u_{ij}} d\theta}{\displaystyle\int g(\theta) \prod_{i=1}^{n} P_i(\theta_j)^{u_{ij}} Q_i(\theta_j)^{1-u_{ij}} d\theta}.$$

(7.42)

Since this equation involves integrals, the Hermite-Gauss quadrature approximation will be used to approximate the normal distribution $g(\theta)$. Mislevy and Bock (1982b) have given the quadrature form as

$$E(\theta_j|U_j,\xi) = \bar{\theta}_j = \frac{\displaystyle\sum_{k=1}^{q} X_k L(X_k) A(X_k)}{\displaystyle\sum_{k=1}^{q} L(X_k) A(X_k)}.$$

(7.43)

This equation has a number of interesting characteristics. First, it is non-iterative and yields the EAP estimate of ability directly. Second, the values of $A(X_k)$ employed are the final adjusted quadrature weights for the fixed X_k as produced by the final EM cycle of the Bayesian modal item parameter estimation stage. Hence, the prior distribution of the ability employed here is the final adjusted quadrature distribution. Third, the values of $L(X_k)$ are also readily available from the previous stage. The end result is that the EAP ability estimates are easily obtained. Mislevy and Stocking (1989) recommend using EAP ability estimators as the method of choice.

Appendix E contains a computer program written in MICROSOFT QUICKBASIC that implements the estimation of an examinee's ability. The estimation process is performed for one of the vectors in the LSAT-6 data set.

7.6 Role of the Prior Distributions: Shrinkage

Recall that the role of prior distributions is to supplement the information contained in the sample data. If the prior distribution provides a good deal of information, we wish the appending term to have an impact on item parameter estimation. As noted earlier, this is accomplished through the Bayesian concept of "shrinkage." The contribution of a prior distribution to the estimation of an item parameter, say α_i, depends on the amount of shrinkage of the item parameter estimate toward the mean, say μ_α, of its prior distribution. The amount of shrinkage is associated with the loss function $(\alpha_i - \mu_\alpha)$ and with the size of σ_α (see Novick & Jackson, 1974, pp. 3–15). Other things being equal, the more similar α_i and μ_α are, the smaller the loss and the less the shrinkage. Greater shrinkage occurs with estimates of α_i that differ substantially from μ_α. This tends to restrain estimates from assuming unreasonable values. Suppose, for example, that $\sigma_\alpha = 0.5$ and $\mu_\alpha = 0$. If the estimated $\alpha_i = 5$, the contribution of the appending term in Equation 7.22 (i.e., the weight of the prior) is -20; if $\alpha_i = 2$, the contribution is -8; for $\alpha_i = \mu_\alpha = 0$, the appending term makes no contribution and there is no shrinkage. These examples illustrate that, other things being equal, the closer the estimated α_i is to the mean of its prior distribution, the less impact the prior has on the estimation process. These examples also help illustrate the key role that σ_α plays. A noninformative prior distribution would tend to have a large variance and would reduce the value of $(\alpha_i - \mu_\alpha)/\sigma_\alpha^2$ to a small contribution, whereas an informative prior would tend to have a small variance and, other things being equal, the $(\alpha_i - \mu_\alpha)/\sigma_\alpha^2$ term would make a larger contribution to estimating the item parameter. In addition, the $1/\sigma^2$ term appears in Λ_{11} in the information matrix and, hence, it has a role in the standard error of estimation of α_i.

It is important to emphasize that an informative prior is not necessarily an appropriate prior. For example, a user-specified or a default value of μ_α could differ considerably from the true underlying mean discrimination. When combined with a small σ_α, the result would be to pull the α_i estimate toward an inappropriate mean of the prior distribution of discrimination, clearly an undesirable effect. Mislevy (1986b) pointed out that incorrectly specifying μ_α is likely to result in an "ensemble bias." This means that all the estimated α_i will be biased in some fashion and statistical properties like consistency are unlikely to hold. The same holds true for other item parameter estimates. Mislevy and Stocking (1989) urge PC-BILOG users to avail themselves of the diagnostic features of PC-BILOG to check on the correctness of the ICC model and the prior distributions to minimize the possibility of ensemble bias.

The possibility of incorrectly specifying the mean of a prior distribution can sometimes be lessened by using the FLOAT option available in PC-BILOG. Under this option, the hyperparameter μ_α is estimated from the sample's item response data. The formula for estimating μ_α is obtained by finding the partial derivative of the densities $g(\alpha|\mu_\alpha, \sigma_\alpha)$ and $g(\mu_\alpha|\nu_\alpha, \zeta_\alpha)$ with respect to μ_α, setting the resulting equation equal to zero, and solving for μ_α. The latter density arises when μ_α is treated as a continuous (normally distributed) random variable with mean ν_α and variance ζ_α (see Anderson, 1984, p. 272). The expression for μ_α is a weighted average of the mean of the n estimated (transformed) discrimination parameters ($\bar{\alpha}$) and ν_α:

$$\mu_\alpha = \frac{n\bar{\alpha} + d\nu_\alpha}{n + d}. \tag{7.44}$$

The scalar d is the weight or believability of the prior information, expressed in terms of the number of items that the prior information is considered to be worth (see Equation 26 in Mislevy, 1986b). When the number of items in a test is large relative to the weight allocated to the mean of a prior distribution, the estimate of μ_α will be close to the mean of the estimates of the α_i obtained in the previous cycle of the parameter estimation algorithm. When the number of items is not large or the prior information is weighted more heavily through d, the effect of ν_α in Equation 7.44 will be greater. In PC-BILOG, the value of d is set to zero and cannot be changed by the user. Thus, the FLOAT option estimates μ_α as the average of the n sample item discriminations in a test. When the FLOAT option is selected, the estimate of μ_α obtained via Equation 7.44 is employed in the appended term of the Bayes modal estimation Equation 7.21. The estimated value of μ_α is then used as the mean of the prior distribution of each item in the test. If the FLOAT option is not selected, the μ_α used in Equation 7.21 is either specified by a user or assigned the program default value.

An additional comment concerning the estimation of μ_α from the sample item response data is in order. Mislevy and Stocking (1989) recommended the use of the FLOAT option in PC-BILOG unless one is confident about the appropriateness of the values of the prior distribution's hyperparameters. However, the obtained value of μ_α may still be inappropriate since there is no guarantee that the sample data will accurately estimate μ_α. Also, the PC-BILOG manual recommends that the FLOAT option not be used for small data sets because it can induce "item parameter drift" and the EM algorithm will not converge. Unfortunately, exhaustive comparisons of the effects of the FLOAT option on parameter estimation across varying number of items and examinees are not yet available. For some initial results comparing the use of the FLOAT option and the default value of μ_α see Baker (1990). It seems safe to conclude that a good deal of computer simulation work needs to be done before questions of this nature can be adequately addressed.

Although only the role of the discrimination prior has been discussed, the priors on b_i and on c_i behave in a similar manner. It is not necessary to impose a prior on all of the parameters of an ICC model. In the case of the three-parameter ICC model, PC-BILOG imposes priors only on a_i and c_i. In Chapter 2, it was seen that b_i is generally easy to estimate and a prior distribution would not be needed to prevent deviant values, especially when a_i was shrunk away from near-zero values by it's prior. It should be noted that effective use of priors on the item parameters requires a good understanding of the characteristics of the test and the group of examinees. One cannot blindly employ either the PC-BILOG default values of the hyperparameters of the priors or user-specified values; either case, one can do unsuspected violence to the item parameter estimates.

7.7 Research on Marginal Bayesian Parameter Estimation

Because the PC-BILOG computer program has so recently become available, research into the properties of the obtained parameter estimates is not as extensive as that associated with the LOGIST program. Yen (1987) compared the results yielded by the mainframe versions of BILOG and LOGIST. Simulated data were generated under a three-parameter model for 1000 examinees and four 20-item and four 40-item tests. In three of the tests, in each set, the values of a_i and c_i were fixed, and the values of b_i varied. The fourth test was modeled after an existing reading vocabulary test. The item parameters were estimated using the default values of the hyperparameters for a_i and c_i. When BILOG employed maximum likelihood estimation of ability, LOGIST was about 25% faster,in terms of CPU time than BILOG for the 20-item tests and about the same for the 40-item tests. When options in BILOG for allowing omitted items to be given partial credit were used, it took 62–94% more time than when omitted items were ignored. Hence, BILOG was considerably slower that LOGIST in this situation. The item parameter estimates yielded by BILOG were generally more accurate estimates of the underlying parameter values than those yielded by LOGIST. Yen attributed this to the procedures used in LOGIST to estimate the c_i parameter, which results in a positive correlation between the true b_i and the amount of error in the estimates of b_i and c_i. The ICCs defined by the BILOG item parameter estimates were nearly identical to those yielded by LOGIST. Other than execution time when omitted items are analyzed, BILOG and LOGIST generally appeared to behave comparably.

Mislevy and Stocking (1989) compared BILOG and LOGIST with respect to the recovery of the item parameters for a 15-item test and 45-item test, both with a sample size of 1500. They found that BILOG estimated the parameters of the 15-item test more accurately than did LOGIST but, with the 45-item test, the two programs were comparable. From the limited results, it appears that, for tests having around 40 items and large sample sizes ($N \geq 1000$), the BILOG and LOGIST programs have similar item parameter recovery characteristics.

Baker[3] (1990) performed a simulation study to investigate the effects of the discrimination prior and the FLOAT option on PC-BILOG results. Three sets of item response data, based on 45 item tests and groups of 500 examinees, were generated under a two-parameter logistic ICC model in a $N(0,1)$ metric via a computer program. Data set 1 represented a discriminating ($\bar{a} = 1.537$) test whose difficulty was matched to the mean ability of a group of examinees whose ability distribution had a unit variance. Data set 2 represented a difficult test with moderate discrimination ($\bar{a} = 0.959$) administered to a group of examinees having a variance of 1.5. Data set 3 represented an easy test with low discrimination ($\bar{a} = 0.469$) given to a group having a variance of 0.75. In all cases, the underlying item difficulty and ability parameters were normally distributed while the item discrimination parameters were uniformly distributed.

In the PC-BILOG analysis of the data sets, the following specifications for the prior distribution of item discrimination were employed for each item in a test:

Run A	No prior distribution
Run B	Default prior $\mu_\alpha = 0$, $\sigma_\alpha = 0.5$; no FLOAT option
Run C	Default prior $\mu_\alpha = 0$, $\sigma_\alpha = 0.5$; with FLOAT option
Run D	Prior $\mu_\alpha = 0$, $\sigma_\alpha = 0.75$; no FLOAT option
Run E	Prior $\mu_\alpha = 0$, $\sigma_\alpha = 0.75$; with FLOAT option
Run F	Prior $\mu_\alpha = 0$, $\sigma_\alpha = 0.25$; no FLOAT option
Run G	Prior $\mu_\alpha = 0$, $\sigma_\alpha = 0.25$; with FLOAT option

The no prior distribution analysis was used to obtain a baseline MMLE/EM solution. The 0.5, 0.75 and 0.25 values of the standard deviation were used to provide several levels of "strength" of the prior distribution for discrimination. In particular, the 0.25 value is "strong," that is, defines an informative prior and should pull the item discrimination estimates towards the mean value of the prior. When the FLOAT option was not used, the user-specified value of zero (in a logarithmic metric) for the mean of the prior was employed. In data set 1, this value is well below the underlying value of $\bar{a}$. In data set 2, it is matched to $\bar{a}$. In data set 3, it is well above the underlying value of $\bar{a}$. The interest was in the degree to which the prior pulls the means of the estimated item discriminations towards the user-specified mean of the prior. When the FLOAT option is used, the mean of the prior distribution is estimated from the item response data, it will be matched to that of the estimated discriminations, and there should be no pulling effect on their mean.

---

[3] This section is based on the results section of the paper by F. B. Baker (1990). Some observations on the metric of PC-BILOG results. *Applied Psychological Measurement*, *14*, 139–150 (used with permission).

Since the ratio of the mean of the item discrimination estimates to the mean of the underlying item discrimination parameters contains the information about the ability scale's unit of measurement, the FLOAT option is an important factor in the test equating process. Each of the three data sets were analyzed via the PC-BILOG program under all seven of the above specifications on the discrimination priors. The ability estimates were obtained via the Bayesian EAP procedure.

When the summary statistics for the PC-BILOG results were equated to the underlying metric, they were generally close to those of the underlying parameters. In data set 1, the unit variance of the underlying ability distribution matched that of the initial quadrature distribution, and the average item discrimination was high. In this situation, the effect of the three different values of the standard deviation of the item discrimination priors on the equated means of item difficulty and discrimination was slight. Not using the FLOAT option set the mean of the discrimination prior at a value of $\bar{a} = 1.13$, and had the effect of pulling the mean equated item discriminations below the underlying value of 1.537 by about 0.03 to 0.07. When the FLOAT option was used, the mean item discrimination was underestimated by about 0.03. Thus, this option had little effect here on the equated mean discrimination estimates. The mean equated item difficulties were only slightly affected by the characteristics of the prior discrimination distributions since the obtained values were little different from those observed when no prior was employed. The equated means of the ability estimates were generally slightly larger than the underlying mean of 0.068, both when priors were and were not imposed on the item discriminations. The equated standard deviations of the ability estimates generally were a slight underestimate of the underlying value of unity; a result that is consistent with using Bayesian EAP to estimate the individual examinee abilities.

In data set 2, the standard deviation of the underlying ability distribution was larger than that of the initial quadrature distribution, and the item discrimination was moderate. In this data set, the effect of the three different values of the standard deviations of the item discrimination priors on the equated mean item difficulty and discrimination also was slight. Not using the FLOAT option set the mean of the discrimination prior at a value of $\bar{a} = 1.13$, which roughly matches the underlying mean discrimination of one. Using the FLOAT option in this data set should also have matched the mean item discrimination. However, using the FLOAT option had the effect of slightly lowering the equated mean discriminations. In this data set, the means of the equated ability estimates were not affected by the characteristics of the discrimination prior distributions. The obtained values were little different from those obtained when no prior was employed and were close to the underlying value of -0.570. As anticipated, the equated standard deviations of the ability estimates were consistently a slight underestimate of the underlying value.

In data set 3, the standard deviation of the underlying ability distribution was smaller than that of the initial quadrature distribution, and the mean item discrimination was low. In this situation, the effect of the three different values of the standard deviations of the item discrimination priors on the equated mean item difficulty and discrimination was inconsistent. Four of the equated mean item difficulties were more negative than the underlying value. The more diffuse the discrimination prior, the more negative the mean equated item difficulty. All of the equated mean discriminations were overestimates of the underlying value. This reflects the fact that the mean of the discrimination prior had a value of $\bar{a} = 1.13$, which was much larger than the underlying value of 0.469. Thus, it would pull the discrimination estimates in an upward direction. When the standard deviation of the discrimination prior distribution was 0.5 or 0.75, using the FLOAT option appeared to reduce the mean equated discrimination by about 0.10 from the values obtained when it was not used. In run 3F (data set 3 with $\sigma_\alpha = .25$ and no FLOAT option), the equated mean discrimination had a value of 0.725, while run 3G ($\sigma_\alpha = .25$ and FLOAT option) yielded a mean of 0.515. Here, the use of the FLOAT option in conjunction with a strong prior resulted in a mean discrimination that was 0.21 closer to the underlying value. When no prior distributions were imposed on the item discriminations, the mean value of the estimated examinees abilities (0.684) over estimated the underlying value of 0.510. When item discriminations priors were used, the equated means of the ability estimates were close to the underlying values. The values of the equated standard deviations of examinee ability estimates were smaller than the underlying values in five of the seven cases. It is of interest to note that, when the FLOAT option was used, the standard deviation of the final adjusted quadrature distribution was essentially unity. When the FLOAT option was not used, the standard deviation of this distribution was smaller and, in the case of run 3F, it was only 0.718. A similar but not quite so dramatic result can be seen in run 3B, where the standard deviation was 0.866.

From these results, it is clear that specifying a prior on the item discriminations that is both wrong and strong is to be avoided. The plotted data for all seven analyses of data set 3 also indicated that, with low underlying item discriminations, the wrong prior mean also has the effect of scattering the discrimination and ability estimates, regardless of the strength of the prior. There also appears to be some interaction between the mean difficulty of the test relative to the mean ability of the examinees and the effect of the discrimination prior distributions. In data set 1, the test difficulty was matched to the mean ability of the group. The mean item discrimination was 0.4 above the mean of the prior discrimination distributions. Yet, there was little "pulling" effect due to the mean of the prior in any of the six cases. In data set 3, the test was very easy for the group $(\bar{\beta} - \bar{\theta}) = -0.977$. The mean item discrimination was 0.66 below the mean of the discrimination prior distributions. Here, the pulling effect was pronounced when the FLOAT option was not used.

A simulation study due to Harwell and Janosky (1991) examined the effect of three factors on the recovery of underlying item and ability parameters. The three factors were: the number of examinees (75, 100, 150, 250), number of items (15, 25), and the variance of the prior distribution for the discrimination parameter (no prior, 0.75^2, 0.5^2, 0.25^2, 0.1^2, all in a logarithmic metric). The two-parameter logistic ICC model was used, and no prior was imposed on the item difficulties. The ability and difficulty parameters were sampled from a normal distribution over the range of -3 to $+3$. The discrimination parameters were sampled from a uniform distribution over a range of 0.6 to 0.9 (in a logistic metric). These specifications resulted in 50 data sets that were analyzed via PC-BILOG using the Bayesian modal estimation procedure for both items and examinees. The obtained parameter estimates were equated to the underlying metric. Then the correlations of the estimates and the underlying parameters, as well as the root mean square differences (RMSD), were computed.

For a 15-item test, smaller prior variances led to smaller discrimination RMSDs for $N = 75$, 100, 150. However, $N = 250$ the smaller prior variances did not further reduce the value of the discrimination RMSDs. There was little difference in the accuracy of item parameter estimation for $\sigma_\alpha^2 = .25^2$ and 0.1^2 over the levels of the number of examinees. The results for the 25-item case indicate smaller RMSDs for $N = 75$ and $N = 100$ than for the 15-item tests. In addition, the several prior variances did not noticeable affect the magnitude of the RMSDs once the sample size was greater than 100. An unusual result was that the smallest prior variance showed the largest RMSD values for the larger number of examinees. This was probably related to the shrinkage due to a highly informative prior that tightly clustered the discrimination estimates about the mean of the prior distribution rather than about the underlying mean value, in effect, producing a one-parameter model. Harwell and Janosky (1991) concluded that, when $N = 250$, the size of the variance of the discrimination prior distribution had little effect on the magnitude of the RMSDs. For samples of $N \geq 250$, the sample size neutralizes the effect of the prior variances on the discrimination estimates. This result held for both the 15- and 25-item tests and, when $N > 100$, the RMSDs were acceptably small. Harwell and Janosky felt that, other things being equal, increasing the number of examinees increases the contribution of the likelihood component to the Bayes modal estimate of the item parameters relative to that of the prior component. However, for smaller number of examinees ($N < 150$) and the 15-item test, the smaller discrimination prior variances noticeably improved the accuracy of estimation of this parameter.

While limited in scope, these two simulation studies suggest that the users of the Bayesian modal estimation procedures, as implemented in PC-BILOG, need to attend to the characteristics of the data being analyzed. The Harwell and Janosky (1991) results indicate that the variances of the discrimination prior had little effect once the sample size was 250 or greater. However, they did not directly examine the pulling effect of the prior on the mean value of the estimates. The Baker (1990) results indicate that, even for larger data sets, the priors can introduce an ensemble bias into the results. In some cases, the FLOAT option was effective in preventing this bias. In addition, Baker showed that, for some data sets, there was little difference between the MMLE/EM estimates and those obtained via the Bayesian procedures.

Since it is quite difficult to specify the mean and variance of a prior distribution a priori, the test analyzer needs to develop techniques to ensure that the prior specifications are appropriate. Mislevy and Stocking (1989) suggest that the users of PC-BILOG avail themselves of the diagnostic features of the program to accomplish this goal. Recently, Swaminathan, Hambleton, Sireci, Xing, and Rizavi (2003) investigated the effect of priors on the accuracy of item parameter estimation in small samples. Three dichotomous item response models (i.e., one-, two-, and three-parameter models) were used to generate data for a test with 21 items by six small sample sizes (100, 150, 200, 300, 400, and 500 examinees). Ten prior specifications were used, that included specifications based on judgmental ratings by subject matter specialists and test developers. Item parameters were estimated using BILOG (Mislevy & Bock, 1989), and the accuracy was evaluated using mean squared errors, bias, and the variance of estimates. The results indicated that parameter estimation was improved considerably by the use of priors employed, especially for the two- and three-parameter models.

7.8 Summary

The employment of Bayesian statistical procedures within IRT results in a powerful, yet flexible methodology for estimating both item and ability parameters. In the case of item parameter estimation, the marginal Bayesian modal estimation procedure can be considered an extension of the MMLE procedure of Chapter 6. The elements in the Fisher scoring equations simply have terms appended to them that implement the effect of the prior distributions on the item parameter estimation process. With respect to the priors, the parameters of a given item are considered to be independent. As a result, the prior terms are appended only to the first derivatives and to the diagonal elements in the matrix of second derivatives in these equations. The item parameters are estimated using the Bock and Aitkin (1981) version of an EM algorithm, and the ability parameters are estimated in a separate subsequent stage. The marginal Bayesian approach does an effective job of "shrinking" the item parameter estimates towards the mean of the imposed prior distributions. In particular, it appears to do a good job of coping with the estimation of the c_i (guessing) parameter under the three-parameter ICC model. In Chapter 4, it was seen that the LOGIST program had to impose severe constraints on the increments to the estimates of c_i and a complex sequence of item parameter estimation in order to obtain estimates of c_i. Despite this, for many items, c_i could not be estimated, and the average value from those items were it could be estimated was substituted for the inestimable cases. In some sense, this scheme was imposing "heuristic" priors on the item parameters. Empirical results have shown that, when the sample size is large, the item parameter estimates yielded by LOGIST and BILOG are remarkably similar. Thus, the Bayesian approach is the simpler of the two with respect to the overall scheme of things, especially in its handling of the estimation of the guessing parameter. For the unwary, however, there are traps within the Bayesian approach. The specification of values of the hyperparameters of the prior distributions requires an intimate understanding of both the test instrument and the group of examinees. Values of the hyperparameters that are poorly specified by the user or improper use of the default values can result in inappropriate item parameter estimates. One useful approach is to analyze the data first using MMLE/EM to obtain a baseline set of item parameter estimates. Upon inspection of the item parameter estimates, an appropriate set of hyperparameter values can be specified that can make appropriate adjustments to the item parameter estimates. Essentially, what one does is to use the priors to fine-tune the estimation process. Little is known about the quality of item parameter estimates produced by PC-BILOG for small data sets or the impact of prior distributions other than those employed in PC-BILOG on the item parameter estimates.

Since the examinee abilities are estimated in a separate stage, the user of PC-BILOG can select from among maximum likelihood, Bayesian modal, and Bayesian EAP estimation. The two Bayesian methods differ in the manner in which the prior distribution is incorporated into the estimation equations. Unfortunately, there is little empirical evidence to serve as the basis for selecting between these two Bayesian ability estimation procedures or maximum likelihood. Mislevy and Stocking (1989) suggest that EAP estimation be the method of choice. Additional computer simulation studies are needed to provide a set of guidelines for selecting an ability estimation procedure.

A number of other Bayesian approaches to parameter estimation have appeared in the literature (see Tsutakawa, 1984; Tsutakawa & Lin, 1986; Swaminathan & Gifford, 1982, 1985, 1986). These may have certain advantages and disadvantages relative to the marginalized procedures presented above. However, these alternative approaches have not been implemented in well-supported computer programs, and access to them is limited. In contrast, PC-BILOG is well supported and readily available, which makes it the program of choice. To explore the full range of analysis options of this program is beyond the scope of the present book. Only the essential features of the marginalized Bayesian approach have been presented here. Hopefully, the discussion will provide the reader a basis for fully exploiting the range of capabilities provided.

8. The Graded Item Response

8.1 Introduction

Up to this point, the presentation has been limited to free response items that have been dichotomously scored. This scoring procedure was based on the assumption that a continuous hypothetical item variable for item i, Γ_i, underlies the examinee's response, but the only manifest data available is the correctness of the response. Such binary items are the mainstay of educational and psychological testing; however, other item scoring procedures, such as the graded and nominal response cases, can also be employed. Under the graded scoring procedure, the hypothetical item variable scale is divided into ordered categories. Thus, the lowest category would contribute the least to a person's test score while the highest category would contribute the most. The Likert-type item traditionally used in questionnaires, attitude inventories, and surveys is a classic example of an item whose responses are scored in a graded fashion. In such items, a scale is used, and the labeling imparts the order, such as strongly disagree, disagree, indifferent, agree, and strongly agree. A weight of 1 would be assigned to strongly disagree, 2 to disagree, and so on until strongly agree is assigned a weight of 5. When nominal scoring is used, it is assumed that item responses can be categorized, but the categories are not ordered. From an intuitive point of view, there is a progression from dichotomous to graded to nominal scoring of item responses. The present chapter will deal only with the graded response case, while the nominal response will be presented in the next chapter.

8.2 Some Fundamentals of the Graded Response Case

The underlying logic of modeling the graded response case will be introduced as an extension of Lord's justification of the normal ogive for dichotomously scored items as depicted in Figure 1.3. When an examinee's free response to an item is scored on a graded basis, it is assumed that a hypothetical continuous item variable underlies the response. This item continuum ranges from $-\infty$ to $+\infty$ and has been divided into m response categories for a given item. Because of the underlying continuity, the resultant response categories are ordered, with k denoting an arbitrary category, $k = 1, 2, \ldots, m_i$, where m_i is the number of response categories for item i. Note that, without loss of generality, m will be used for m_i. From a theoretical point of view, the examinee gives a free response to the item. The response is then inspected and determined to belong to the range of responses encompassed by a particular response category. Thus, the jth examinee's response can be assigned only to one of the m item response categories, and $u_{ijk} = 1$ if the response is assigned to category k of item i, and zero otherwise. If one were to obtain a raw test score from a test composed of such items, the category $k = m$ would have the largest weight associated with it and, hence, is the "highest category." The "lowest category" denoted by $k = 1$ would have the lowest weight. If one were interested in obtaining a raw score, it would be the sum of the weights associated with the item response categories chosen by the examinees for each of the n items in the instrument. The graded response case can be modeled using an approach very similar to that used earlier for the dichotomous response case. Figure 8.1 shows the graded response analogy to Figure 1.3 for the dichotomous response case. The values of the item variable demarcating the response categories are denoted by γ_k, where $\gamma_0 = -\infty$ and $\gamma_m = +\infty$. The regression of the item variable on the ability scale is denoted by $\gamma'|\theta$. At each ability level, there exists a conditional normal distribution of the responses of the examinees. These conditional distributions are intersected by the response category boundaries. Thus, the probability of an examinee's response falling in category k is denoted by the area of the conditional distributions falling between the limits γ_{k-1} and γ_k and is denoted by $P_{ik}(\theta)$. It is important to observe that, at each ability level,

$$\sum_{k=1}^{m} P_{ik}(\theta) = 1. \tag{8.1}$$

Now, if one plots the $P_{ik}(\theta)$ for a given response category as a function of ability, the operating characteristic of an item response category (Samejima, 1969) will be obtained. This particular terminology is not in keeping with previous usage, and the term "item response category characteristic curve" (IRCCC) will be used.[1] As in the dichotomous response case, it will be assumed that both the distribution of the item variable Γ_i and the conditional distribution of Γ_i at each θ are normal densities, but other distributions could have been used. Again, no assumption is needed as to the frequency distribution of the examinees over the ability scale. Given these assumptions and specific values for the category boundaries γ_k, the IRCCC for the several item response categories can be obtained. Figure 8.1 depicts a graded item having four categories with $\gamma_1 = -1, \gamma_2 = 0, \gamma_3 = +1$, and a regression of the item variable on ability given by $\gamma' = \theta/2$. Then, the probabilities associated with each category can be obtained over the range of θ. Table 8.1 contains $P_{ik}(\theta)$

---

[1] A very recent terminology is item response function (IRF), which encompasses the characteristic curves for all the item scoring procedures used in IRT.

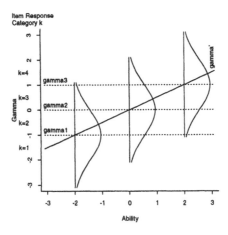

Fig. 8.1. Partitioning the conditional response distributions under a graded response.

for this item over the range $-3 \leq \theta \leq 3$ in ability increments of 0.5. These response category probabilities are plotted as a function of θ in Figure 8.2.

Table 8.1. Response Category Probabilities for the Item of Figure 8.1

θ	$P_{i1}(\theta)$	$P_{i2}(\theta)$	$P_{i3}(\theta)$	$P_{i4}(\theta)$
-3.0	0.6915	0.2417	0.0606	0.0062
-2.5	0.5987	0.2956	0.0934	0.0122
-2.0	0.5000	0.3413	0.1359	0.0228
-1.5	0.4013	0.3721	0.1866	0.0401
-1.0	0.3085	0.3829	0.2417	0.0668
-0.5	0.2266	0.2721	0.2956	0.1056
0.0	0.1587	0.3413	0.3413	0.1587
0.5	0.1056	0.2956	0.3721	0.2266
1.0	0.0668	0.2417	0.3829	0.3085
1.5	0.0401	0.1866	0.3721	0.4013
2.0	0.0228	0.1359	0.3413	0.5000
2.5	0.0122	0.0934	0.2956	0.5987
3.0	0.0062	0.0606	0.2417	0.6915

A number of interesting features of the IRCCC are apparent in Figure 8.2. Foremost, the curves do not share a common form. The IRCCC for category 4 is a monotone-increasing function of θ similar in appearance to the usual ICC. The curves for the center two response categories, $k = 2$ and $k = 3$, are unimodal functions of θ. The final category $k = 1$ yields a monotone-decreasing function of θ similar in appearance to the ICC for the incorrect response to a dichotomously scored item. In the graded response case, this

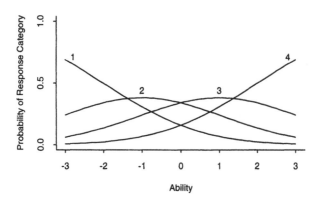

Fig. 8.2. Item response characteristic curves for an item with four graded response categories.

pattern will always appear when more than two response categories are used. The nonextreme categories behave in a nonmonotone fashion, that is, the probability of choosing an intermediate category increases with increasing ability up to a point and then decreases, and the IRCCC is asymptotically zero at both $+$ and $-$ infinity.

When plotted simultaneously on the same graph, as in Figure 8.2, the interrelationship among the response categories is apparent. At low ability levels, the lowest category, $k = 1$, has a high probability. At middle levels of θ, the four categories all have moderate probabilities. At high ability levels, the highest category has a large probability while the probability associated with each of the $m - 1$ remaining categories is asymptotically zero. The source of this pattern is evident in Figure 8.1 where the conditional distributions take on different positions relative to the fixed category boundaries as a function of the item-ability regression.

The lack of a consistent form for the IRCCC of all response categories poses a problem as it would greatly complicate the modeling of the IRCCC and parameter estimation. A rather simple mechanism is used to resolve this difficulty. In Figure 8.1, if only the category limit γ_3 is used, the net effect is to treat the item as dichotomously scored. Category 4 would be the correct response, and the remaining $m - 1$ categories would be pooled as the incorrect response. Thus, under the conditions specified here, the characteristic curve of category 4 would be a normal ogive identical to the ICC for the dichotomous case. If only γ_2 were used as the delimiter for dichotomous scoring, response categories 3 and 4 would be pooled to form the correct response, and categories 1 and 2 pooled to act as the incorrect response. It is important to note here that the probability of a "correct" response is actually the probability of choosing a category greater than category 2. Since we now have the equivalent of dichotomous scoring, the characteristic curve of the pooled category will be a normal ogive. When γ_1 is used as the limit, categories 2, 3, and 4 are pooled as the correct response, and the characteristic curve again is a normal ogive. What has been shown here is that, when one considers the cumulative probability of the response categories (from highest-weighted category towards lowest-weighted category) as a function of ability, the form of these probabilities will be that of a normal ogive. The cumulative probabilities are reported in Table 8.2, and the resulting curves are plotted in Figure 8.3. The normal ogives of Figure 8.3 represent the boundaries on the cumulative probabilities of the response categories. While there are four response categories in the example, there are only three boundaries between categories, as well as an upper bound of one and a lower bound of zero. Thus, the probability of a response belonging to a given category is given by the difference between any two adjacent boundaries. If one lets $P_{ik}^*(\theta)$ denote a boundary and $P_{ik}(\theta)$ the probability of a category, then

$$P_{i4}^*(\theta) = 0$$
$$P_{i3}^*(\theta) = P_{i4}(\theta)$$
$$P_{i2}^*(\theta) = P_{i3}(\theta) + P_{i4}(\theta)$$
$$P_{i1}^*(\theta) = P_{i2}(\theta) + P_{i3}(\theta) + P_{i4}(\theta)$$
$$P_{i0}^*(\theta) = P_{i1}(\theta) + P_{i2}(\theta) + P_{i3}(\theta) + P_{i4}(\theta) = 1, \tag{8.2}$$

and

$$P_{ik}^*(\theta) = \sum_{k'=k+1}^{m} P_{ik'}(\theta).$$

In general,

$$P_{ik}(\theta) = P_{i,k-1}^*(\theta) - P_{ik}^*(\theta). \tag{8.3}$$

The normal ogives representing the boundaries of the cumulative probabilities can be characterized by the parameters α_i, β_{ik}. Since these ogives represent the sum of the response category probabilities, negative differences between curves are not possible. Thus, these boundary curves cannot cross, and they must share a common value of the discrimination parameter α_i. Also, in order to maintain the underlying order of the response categories, the location parameters of the boundary ogives must be ordered:

$$\beta_{i1} < \cdots < \beta_{ik} < \cdots < \beta_{i,m-1},$$

where $-\infty < \beta_{ik} < +\infty$. There is no requirement that the location parameters be equally spaced, only that they be monotonically increasing. Equal values of successive βs indicates that no responses were allocated to the category. In the data underlying Tables 8.1 and 8.2, the parameters of the boundary curves were $\beta_{i1} = -2.0$, $\beta_{i2} = 0.0$, $\beta_{i3} = +2.0$, and the common value

Table 8.2. Boundary Characteristic Curve Values for a Graded Response Item Having Four Categories ($\alpha_i = 0.5$, $\beta_{i1} = -2.0$, $\beta_{i2} = 0.0$, $\beta_{i3} = 2.0$)

θ	$P_{i1}^*(\theta)$	$P_{i2}^*(\theta)$	$P_{i3}^*(\theta)$
-3.0	0.3085	0.0668	0.0062
-2.5	0.4013	0.1056	0.0122
-2.0	0.5000	0.1587	0.0228
-1.5	0.5987	0.2266	0.0401
-1.0	0.6915	0.3085	0.0668
-0.5	0.7734	0.4013	0.1056
0.0	0.8413	0.5000	0.1587
0.5	0.8944	0.5987	0.2266
1.0	0.9332	0.6915	0.3085
1.5	0.9599	0.7734	0.4013
2.0	0.9772	0.8413	0.5000
2.5	0.9878	0.8944	0.5987
3.0	0.9938	0.9332	0.6915

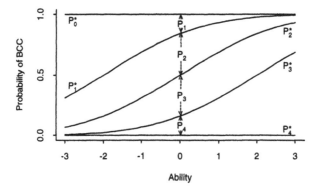

Fig. 8.3. Boundary characteristic curves for a graded response item having four categories.

of the discrimination parameter was $\alpha_i = 0.5$. The interpretation of these parameters is not direct since the use of the boundary curves is actually a mathematical device for parameter estimation purposes. Unfortunately, there will always be one less location parameter than there are item response categories due to the restriction,

$$\sum_{k=1}^{m} P_{ik}(\theta) = 1.$$

The problem then is how to use the $m - 1$ values of β_k to describe the m item response categories. Samejima (1969, p. 34) shows that the modal point of an item response category characteristic curve is given by $\beta'_{ik} = (\beta_{ik} + \beta_{i,k-1})/2$ except for response categories $k = 1$ and $k = m$. Thus, to specify the locations of the response categories, one would use:

k location index

1    $\beta'_{i1} = \beta_{i1}$

2    $\beta'_{i2} = (\beta_{i2} + \beta_{i1})/2$

3    $\beta'_{i3} = (\beta_{i3} + \beta_{i2})/2$

$\vdots$

k $\beta'_{ik} = (\beta_{ik} + \beta_{i,k-1})/2$

$\vdots$

m $\beta'_{im} = \beta_{i,m-1}.$

Only in the case of β'_{i1} and β'_{im}, do these parameters retain their interpretation as the point on the ability scale at which the probability is that the response will be allocated to that category is .5. When β'_{ik} is between β'_{i1} and β'_{im}, the parameter simply locates the modal point of the nonmonotone IRCCC. The maximum probability at that point will be a function of the number of response categories of the item and will always be less than .5. The discrimination parameter poses no interpretive difficulties as it is the same for all item response categories. Samejima (1969) refers to α_i as the item discrimination index rather than relating it to the separate response categories. The overall result is that the parameters β'_{ik} can be used to describe the location of an item response category on the ability scale in a manner analogous to that used with dichotomously scored items.

Given this background, the item depicted in Tables 8.1 and 8.2 and in Figures 8.2 and 8.3 can be interpreted using a normal ogive model for the boundary characteristic curves. The value of α_i is such that the item response categories discriminate quite well between examinees who choose adjacent response categories. The location parameters are centered on $\beta_{i2} = 0.0$ indicating the item functions at the middle of the ability scale. The location indexes of the item response categories would be $\beta'_{i1} = -2.0$, $\beta'_{i2} = -1.0$, $\beta'_{i3} = +1.0$, and $\beta'_{i4} = +2.0$. These indicate that the ordered item response categories are spread out along the ability continuum. Thus, this item would be a highly desirable item from the point of view of its item parameters. There is no requirement that the item be centered at the midpoint of the ability scale, and the item response category location parameters could be centered on some other value. Because of the restriction that α_i is the same for all response categories, the category order will always be the same. However, the distance between the β'_{ik} of adjacent categories can vary. The ideal would be equal spacing of the response category location parameters.

While the graded response model was introduced above via an extension of Lord's justification of the normal ogive ICC model, this approach does not match actual testing practice very well. The usual situation is one in which the item stem is followed by a number of response alternatives that are obviously ordered. Sometimes, a scale is used and the labeling imparts the order, such as strongly disagree, disagree, indifferent, agree, and strongly agree. In other cases, the instructions indicate the ordering of the response alternatives. As a result, rather than making a free response, the examinee employs Thurstone's (1927) threshold concept: examining and rejecting lower-weighted response categories until deciding on a category. Thus, when the examinee's response propensity exceeds a category boundary γ_{k-1}, category k is selected. In this case, $P_{ik}(\theta)$ is the probability of an examinee of ability θ choosing response category k and, again, it is defined as the difference between successive boundary curves. Samejima (1969, 1972) based her development of the graded response model on this equivalent definition.

8.3 Parameter Estimation Procedures for the Graded Response Case

In the interests of simplicity, the JMLE paradigm used in Chapter 4 will be employed in the graded response case. Because of the independence assumptions invoked, only the Newton-Raphson equations for a single item and a single examinee will be of interest. Therefore, the presentation of the maximum likelihood estimation procedures can be limited again to these two building blocks. The mathematics of the maximum likelihood approach was presented by Kolakowski and Bock (1973b) under a logistic model for the boundary curves and by Aitchison and Silvey (1957) for a normal model. Gurland, Lee, and Dahm (1960) presented the derivation of the minimum transformed χ^2 approach under a normal and logistic model in the context of bioassay. Since the logistic model for the boundary curves has computational advantages, it will be employed here also. The derivation of the maximum likelihood estimators under this model will be presented below in a notation that is consistent with the previous chapters.

8.3.1 Maximum Likelihood Estimation of the Item Parameters

Although the estimation equations could be established as a function of the ability scores of individual examinees, the computational aspects of the estimation process are simplified when examinees are grouped with respect to ability. Again, this requires an assumption of local independence in the neighborhood of the midpoint of the ability interval. Since only a single item is of interest, the item subscript i will not be incorporated in the notation. Because of the complexity of the notation needed in the graded response case, the following symbols will be used:

1. θ_g is the midpoint of ability interval g, and $g = 1, 2, \ldots, G$.
2. The number of examinees in the ability groupings are $f_1, f_2, \ldots, f_g, \ldots, f_G$, and $\sum_{g=1}^{G} f_g = N$.
3. Within a given ability group, the number of examinees choosing each of the m item response categories is given by $r_{g1}, r_{g2}, \ldots, r_{gk}, \ldots, r_{gm}$, and $\sum_{k=1}^{m} r_{gk} = f_g$.
4. The observed proportion of choice of an item response category is

$$p_{gk} = \frac{r_{gk}}{f_g},$$

where

$$\sum_{k=1}^{m} p_{gk} = 1$$

and

$$p_{gm} = 1 - \sum_{k=1}^{m-1} p_{gk}.$$

5. The probability of an examinee of ability θ_g selecting item response category k is

$$P_k(\theta_g) = E(p_{gk}),$$

where

$$\sum_{k=1}^{m} P_k(\theta_g) = 1$$

and

$$P_m(\theta_g) = 1 - \sum_{k=1}^{m-1} P_k(\theta_g).$$

6. R is the $G \times n$ matrix of r_{gk}.

Because of the restriction on the sum of the $P_k(\theta_g)$, there are more item location parameters to be estimated than there are degrees of freedom. This problem can be avoided by defining the probability of selecting an item response category in terms of the boundaries on the probabilities. Then,

$$P_m^*(\theta_g) = 0$$
$$P_{m-1}^*(\theta_g) = P_m(\theta_g) + 0$$
$$P_{m-2}^*(\theta_g) = P_{m-1}(\theta_g) + P_m(\theta_g) + 0$$

$$\vdots$$

$$P_k^*(\theta_g) = \sum_{v=k+1}^{m} P_v(\theta_g)$$

$$\vdots$$

$$P_1^*(\theta_g) = \sum_{k=2}^{m} P_k(\theta_g)$$

$$P_0^*(\theta_g) = \sum_{k=1}^{m} P_k(\theta_g) = 1$$

and

$$P_m(\theta_g) = P_{m-1}^*(\theta_g) - P_m^*(\theta_g)$$
$$P_{m-1}(\theta_g) = P_{m-2}^*(\theta_g) - P_{m-1}^*(\theta_g)$$

$$\vdots$$

$$P_k(\theta_g) = P_{k-1}^*(\theta_g) - P_k^*(\theta_g)$$

$$\vdots$$

$$P_1(\theta_g) = P_0^*(\theta_g) - P_1^*(\theta_g).$$

To simplify notation, let $P_k(\theta_g) = P_{gk}$ and $P_k^*(\theta_g) = P_{gk}^*$. As was the case with dichotomously scored items, the two-parameter logistic ICC for the boundary curves will be reparameterized in linear terms as it leads to simpler expressions. Let

$$\alpha(\theta_g - \beta_k) = \zeta_k + \lambda\theta_g$$

and, then,

$$P_{gk}^* = \frac{1}{1 + e^{-(\zeta_k + \lambda\theta_g)}}$$

and

$$P_{gk} = P_{g,k-1}^* - P_{gk}^* = \frac{1}{1 + e^{-(\zeta_{k-1} + \lambda\theta_g)}} - \frac{1}{1 + e^{-(\zeta_k + \lambda\theta_g)}}.$$

For a given item the likelihood function is

$$\text{Prob}(R|\zeta, \lambda, \theta) = \prod_{g=1}^{G} \frac{f_g!}{r_{g1}! r_{g2}! \cdots r_{gk}! \cdots r_{gm}!} P_{g1}^{r_1} P_{g2}^{r_2} \cdots P_{gk}^{r_{gk}} \cdots P_{gm}^{r_{gm}}. \quad (8.4)$$

The log-likelihood is

$$L = \log \text{Prob}(R|\zeta, \lambda, \theta) = \text{constant} + \sum_{g=1}^{G} \sum_{k=1}^{m} r_{gk} \log P_{gk}. \quad (8.5)$$

As usual, the derivatives of L with respect to the parameters are needed. In order to obtain the derivatives of the log-likelihood, it is useful to show the following relationship:

$$\sum_{k=1}^{m} r_{gk} \log P_{gk} = r_{g1} \log(1 - P_{g1}^*) + r_{g2} \log(P_{g1}^* - P_{g2}^*) + \cdots +$$

$$r_{g,k-1} \log(P_{g,k-2}^* - P_{g,k-1}^*) + r_{gk} \log(P_{g,k-1}^* - P_{gk}^*) +$$

$$r_{g,k+1} \log(P_{gk}^* - P_{g,k+1}^*) + \cdots r_{g,m-1} \log(P_{g,m-2}^* - P_{g,m-1}^*) +$$

$$r_{gm} \log(P_{g,m-1}^* - 0). \tag{8.6}$$

In this form, it is clear that derivatives with respect to the parameters of an arbitrary boundary ogive (denoted by y) will involve only two adjacent terms in this expression.

The first derivatives of the log-likelihood with respect to the parameters of the boundary ogives are

$$\frac{\partial L}{\partial \zeta_y} = \frac{\partial}{\partial \zeta_y} \sum_{g=1}^{G} \sum_{k=1}^{m} r_{gk} \log P_{gk} = \sum_{g=1}^{G} \sum_{k=1}^{m} \frac{r_{gk}}{P_{gk}} \frac{\partial P_{gk}}{\partial \zeta_y}.$$

When $y = k$, we have

$$\frac{\partial(P_{g,k-1}^* - P_{gk}^*)}{\partial \zeta_k} = -P_k^* Q_k^*$$

and

$$\frac{\partial(P_{gk}^* - P_{g,k+1}^*)}{\partial \zeta_k} = P_k^* Q_k^*.$$

Substituting yields

$$\frac{\partial L}{\partial \zeta_k} = \sum_{g=1}^{G} \left(-\frac{r_{gk} P_{gk}^* Q_{gk}^*}{P_{g,k-1}^* - P_{gk}^*} + \frac{r_{g,k+1} P_{gk}^* Q_{gk}^*}{P_{gk}^* - P_{g,k+1}^*} \right)$$

$$= \sum_{g=1}^{G} P_{gk}^* Q_{gk}^* \left(-\frac{r_{gk}}{P_{gk}^*} + \frac{r_{g,k+1}}{P_{g,k+1}^*} \right).$$

Let $w_{gk} = P_{gk}^* Q_{gk}^*$, $r_{gk} = f_g p_{gk}$, and $r_{g,k+1} = f_g p_{g,k+1}$, then

$$\frac{\partial L}{\partial \zeta_k} = \sum_{g=1}^{G} f_g w_{gk} \left(-\frac{p_{gk}}{P_{gk}^*} + \frac{p_{g,k+1}}{P_{g,k+1}^*} \right) = L_{\zeta_k}.$$

Without loss of generality, we may use L_k for L_{ζ_k}.

For example,

$$\frac{\partial L}{\partial \zeta_1} = \sum_{g=1}^{G} f_g w_{g1} \left(-\frac{p_{g1}}{P_{g1}^*} + \frac{p_{g2}}{P_{g2}^*} \right) = L_1$$

and

$$\frac{\partial L}{\partial \zeta_{m-1}} = \sum_{g=1}^{G} f_g w_{g,m-1} \left(-\frac{p_{g,m-1}}{P_{g,m-1}^*} + \frac{p_{gm}}{P_{gm}^*} \right) = L_{m-1},$$

where $w_{g,m-1} = P_{g,m-1}^* Q_{g,m-1}^*$. Also

$$\frac{\partial L}{\partial \lambda} = \sum_{g=1}^{G} \sum_{k=1}^{m} \frac{\partial}{\partial \lambda} \left(r_{gk} \log P_{gk} \right) = \sum_{g=1}^{G} \sum_{k=1}^{m} \frac{r_{gk}}{P_{gk}} \frac{\partial P_{gk}}{\partial \lambda}.$$

But

$$\frac{\partial P_{gk}}{\partial \lambda} = \frac{\partial}{\partial \lambda} \left(P_{g,k-1}^{*} - P_{gk}^{*} \right) = \theta_g P_{g,k-1}^{*} Q_{g,k-1}^{*} - \theta_g P_{gk}^{*} Q_{gk}^{*},$$

then

$$\frac{\partial L}{\partial \lambda} = \sum_{g=1}^{G} \theta_g \sum_{k=1}^{m} \frac{r_{gk}}{P_{gk}} \left(P_{g,k-1}^{*} Q_{g,k-1}^{*} - P_{gk}^{*} Q_{gk}^{*} \right)$$

$$= \sum_{g=1}^{G} f_g \theta_g \sum_{k=1}^{m} \frac{p_{gk}}{P_{gk}} \left(w_{g,k-1} - w_{gk} \right) = L_\lambda,$$

where $\partial \lambda$ denotes the derivative with respect to the common slope (discrimination index).

The second derivatives are

$$\frac{\partial^2 L}{\partial \zeta_y^2} = \frac{\partial}{\partial \zeta_y} \left[\sum_{g=1}^{G} \left(-\frac{r_{gk} P_{gk}^{*} Q_{gk}^{*}}{P_{gk}} + \frac{r_{g,k+1} P_{g,k+1}^{*} Q_{g,k+1}^{*}}{P_{g,k+1}} \right) \right].$$

For $y = k$, we have

$$\frac{\partial^2 L}{\partial \zeta_y^2} = \sum_{g=1}^{G} \left\{ -r_{gk} \left[\frac{P_{gk} P_{gk}^{*} Q_{gk}^{*} (Q_{gk}^{*} - P_{gk}^{*}) + (P_{gk}^{*} Q_{gk}^{*})^2}{P_{gk}^2} \right] + \right.$$

$$\left. r_{g,k+1} \left[\frac{P_{g,k+1} P_{gk}^{*} Q_{gk}^{*} (Q_{gk}^{*} - P_{gk}^{*}) - (P_{gk}^{*} Q_{gk}^{*})^2}{P_{g,k+1}^2} \right] \right\}$$

$$= \sum_{g=1}^{G} \left\{ \left[-\frac{r_{gk} (Q_{gk}^{*} - P_{gk}^{*}) P_{gk}^{*} Q_{gk}^{*}}{P_{gk}} - \frac{r_{gk} (P_{gk}^{*} Q_{gk}^{*})^2}{P_{gk}^2} \right] + \right.$$

$$\left. \left[\frac{r_{g,k+1} (Q_{gk}^{*} - P_{gk}^{*}) P_{gk}^{*} Q_{gk}^{*}}{P_{g,k+1}} - \frac{r_{g,k+1} (P_{gk}^{*} Q_{gk}^{*})^2}{P_{g,k+1}^2} \right] \right\}$$

$$= \sum_{g=1}^{G} \left\{ \left[\left(-\frac{r_{gk}}{P_{gk}} + \frac{r_{g,k+1}}{P_{g,k+1}} \right) (Q_{gk}^{*} - P_{gk}^{*})(P_{gk}^{*} Q_{gk}^{*}) \right] + \right.$$

$$\left. \left[\left(-\frac{r_{gk}}{P_{gk}^2} - \frac{r_{g,k+1}}{P_{g,k+1}^2} \right) (P_{gk}^{*} Q_{gk}^{*})^2 \right] \right\}.$$

Because $r_{gk} = f_{gk} P_{gk}$ and $r_{g,k+1} = f_{g,k+1} P_{g,k+1}$, we have

$$\frac{\partial^2 L}{\partial \zeta_y^2} = \sum_{g=1}^{G} \left\{ \left[\left(-\frac{f_g p_{gk}}{P_{gk}} + \frac{f_g p_{g,k+1}}{P_{g,k+1}} \right) (Q_{gk}^{*} - P_{gk}^{*})(P_{gk}^{*} Q_{gk}^{*}) \right] + \right.$$

$$\left. \left[\left(-\frac{f_g p_{gk}}{P_{gk}^2} - \frac{f_g p_{g,k+1}}{P_{g,k+1}^2} \right) (P_{gk}^{*} Q_{gk}^{*})^2 \right] \right\}.$$

There are $m - 1$ location parameters, but one does not need to obtain all possible cross derivatives. Since

$$P_{g,k-1} = P^*_{g,k-2} - P^*_{g,k-1},$$
$$P_{gk} = P^*_{g,k-1} - P^*_{gk},$$
$$P_{g,k+1} = P^*_{gk} - P^*_{g,k+1},$$
$$P_{g,k+2} = P^*_{g,k+1} - P^*_{g,k+2},$$

the cross derivative with respect to ζ_{k-1}, ζ_k or ζ_k, ζ_{k+1} will exist, but the cross derivative with respect to ζ_{k-2}, ζ_k or ζ_k, ζ_{k+2} will not exist nor will any exist where the difference in subscripts is greater than unity. As a result, in the matrix of second derivatives with respect to ζ, only the diagonal and the first term above and below the diagonal will exist. The off-diagonal terms will have the general form:

$$\frac{\partial^2 L}{\partial \zeta_y \partial \zeta_{y-1}} = \frac{\partial}{\partial \zeta_{y-1}} \left[\sum_{g=1}^{G} \left(-\frac{r_{gk} P^*_{gk} Q^*_{gk}}{P_{gk}} + \frac{r_{g,k+1} P^*_{gk} Q^*_{gk}}{P_{g,k+1}} \right) \right]$$

$$= \sum_{g=1}^{G} \frac{r_{gk}}{P^2_{gk}} (P^*_{gk} Q^*_{gk})(P^*_{g,k-1} Q^*_{g,k-1})$$

$$= \sum_{g=1}^{G} \frac{f_g p_{gk}}{P^2_{gk}} (P^*_{gk} Q^*_{gk})(P^*_{g,k-1} Q^*_{g,k-1}),$$

where $r_{gk} = f_g p_{gk}$; and

$$\frac{\partial^2 L}{\partial \zeta_y \partial \zeta_{y+1}} = \frac{\partial}{\partial \zeta_{y+1}} \left[\sum_{g=1}^{G} \left(-\frac{r_{gk} P^*_{gk} Q^*_{gk}}{P_{gk}} + \frac{r_{g,k+1} P^*_{gk} Q^*_{gk}}{P_{g,k+1}} \right) \right]$$

$$= \sum_{g=1}^{G} \frac{r_{g,k+1}}{P^2_{g,k+1}} (P^*_{gk} Q^*_{gk})(P^*_{g,k+1} Q^*_{g,k+1})$$

$$= \sum_{g=1}^{G} \frac{f_g p_{g,k+1}}{P^2_{g,k+1}} (P^*_{gk} Q^*_{gk})(P^*_{g,k+1} Q^*_{g,k+1}),$$

where $r_{g,k+1} = f_g p_{g,k+1}$. Also

$$\frac{\partial^2 L}{\partial \lambda^2} = \frac{\partial}{\partial \lambda} \left[\sum_{g=1}^{G} \sum_{k=1}^{m} \frac{r_{gk}}{P_{gk}} \frac{\partial P_{gk}}{\partial \lambda} \right]$$

$$= \sum_{g=1}^{G} \sum_{k=1}^{m} \left[\frac{r_{gk}}{P_{gk}} \frac{\partial^2 P_{gk}}{\partial \lambda^2} - \frac{r_{gk}}{P^2_{gk}} \left(\frac{\partial P_{gk}}{\partial \lambda} \right)^2 \right].$$

The first term in the bracket will be kept in its definitional form since it vanishes when the summation over the response categories is performed:

$$\frac{\partial^2 L}{\partial \lambda^2} = \sum_{g=1}^{G} \sum_{k=1}^{m} \frac{r_{gk}}{P_{gk}} \frac{\partial^2 P_{gk}}{\partial \lambda^2} -$$

$$\sum_{g=1}^{G} \sum_{k=1}^{m} \frac{r_{gk}}{P_{gk}^2} \left(-\theta_g P_{gk}^* Q_{gk}^* + \theta_g P_{g,k-1}^* Q_{g,k-1}^* \right)^2.$$

Let $r_{gk} = f_g p_{gk}$, then

$$\frac{\partial^2 L}{\partial \lambda^2} = \sum_{g=1}^{G} \sum_{k=1}^{m} \frac{f_g p_{gk}}{P_{gk}} \frac{\partial^2 P_{gk}}{\partial \lambda^2} -$$

$$\sum_{g=1}^{G} \sum_{k=1}^{m} \frac{f_g p_{gk}}{P_{gk}^2} \left(-P_{gk}^* Q_{gk}^* + P_{g,k-1}^* Q_{g,k-1}^* \right)^2 \theta_g^2.$$

And

$$\frac{\partial^2 L}{\partial \zeta_k \partial \lambda} = \frac{\partial}{\partial \lambda} \left[\sum_{g=1}^{G} \left(-\frac{r_{gk} P_{gk}^* Q_{gk}^*}{P_{gk}} + \frac{r_{g,k+1} P_{gk}^* Q_{gk}^*}{P_{g,k+1}} \right) \right]$$

$$= \frac{\partial}{\partial \lambda} \left[\sum_{g=1}^{G} \left(-\frac{r_{gk}}{P_{gk}} \frac{\partial P_{gk}}{\partial \zeta_k} + \frac{r_{g,k+1}}{P_{g,k+1}} \frac{\partial P_{gk}}{\partial \zeta_k} \right) \right]$$

$$= \sum_{g=1}^{G} \left[-\frac{\left(r_{gk} P_{gk} \frac{\partial^2 P_{gk}}{\partial \zeta_k \partial \lambda} - r_{gk} \frac{\partial P_{gk}}{\partial \zeta_k} \frac{\partial P_{gk}}{\partial \lambda} \right)}{P_{gk}^2} + \right.$$

$$\left. \frac{\left(r_{g,k+1} P_{g,k+1} \frac{\partial^2 P_{gk}}{\partial \zeta_k \partial \lambda} - r_{g,k+1} \frac{\partial P_{gk}}{\partial \zeta_k} \frac{\partial P_{gk}}{\partial \lambda} \right)}{P_{g,k+1}^2} \right]$$

$$= \sum_{g=1}^{G} \left[-\left(\frac{r_{gk}}{P_{gk}} \frac{\partial^2 P_{gk}}{\partial \zeta_k \partial \lambda} - \frac{r_{g,k+1}}{P_{g,k+1}} \frac{\partial^2 P_{gk}}{\partial \zeta_k \partial \lambda} \right) + \right.$$

$$\left. \left(\frac{r_{gk}}{P_{gk}^2} \frac{\partial P_{gk}}{\partial \zeta_k} \frac{\partial P_{gk}}{\partial \lambda} - \frac{r_{g,k+1}}{P_{g,k+1}^2} \frac{\partial P_{gk}}{\partial \zeta_k} \frac{\partial P_{g,k+1}}{\partial \lambda} \right) \right]$$

$$= \sum_{g=1}^{G} \left[-\left(\frac{r_{gk}}{P_{gk}} - \frac{r_{g,k+1}}{P_{g,k+1}} \right) \frac{\partial^2 P_{gk}}{\partial \zeta_k \partial \lambda} + \right.$$

$$\left. \left(\frac{r_{gk}}{P_{gk}^2} \frac{\partial P_{gk}}{\partial \zeta_k} - \frac{r_{g,k+1}}{P_{g,k+1}^2} \frac{\partial P_{g,k+1}}{\partial \lambda} \right) \frac{\partial P_{gk}}{\partial \zeta_k} \right],$$

where

$$\frac{\partial P_{gk}}{\partial \lambda} = -\theta_g P_{gk}^* Q_{gk}^* + \theta_g P_{g,k-1}^* Q_{g,k-1}^*$$

and

$$\frac{\partial P_{g,k+1}}{\partial \lambda} = -\theta_g P^*_{g,k+1} Q^*_{g,k+1} + \theta_g P^*_{gk} Q^*_{gk}.$$

The first term on the right will be kept in its present form since it also will vanish when its expectation is taken; then,

$$\frac{\partial^2 L}{\partial \zeta_k \partial \lambda} = -\sum_{g=1}^{G} \left(\frac{r_{gk}}{P_{gk}} - \frac{r_{g,k+1}}{P_{g,k+1}} \right) \frac{\partial^2 P_{gk}}{\partial \zeta_k \partial \lambda} +$$

$$\sum_{g=1}^{G} \left\{ \left[\frac{r_{gk}}{P^2_{gk}} (-\theta_g P^*_{gk} Q^*_{gk} + \theta_g P^*_{g,k-1} Q^*_{g,k-1}) - \right. \right.$$

$$\left. \left. \frac{r_{g,k+1}}{P^2_{g,k+1}} (-\theta_g P^*_{g,k+1} Q^*_{g,k+1} + \theta_g P^*_{gk} Q^*_{gk}) \right] (P^*_{gk} Q^*_{gk}) \right\}.$$

Let $r_{gk} = f_g p_{gk}$ and $r_{g,k+1} = f_g p_{g,k+1}$, then

$$\frac{\partial^2 L}{\partial \zeta_k \partial \lambda} = -\sum_{g=1}^{G} f_g \left(\frac{p_{gk}}{P_{gk}} - \frac{p_{g,k+1}}{P_{g,k+1}} \right) \frac{\partial^2 P_{gk}}{\partial \zeta_k \partial \lambda} +$$

$$\sum_{g=1}^{G} f_g P^*_{gk} Q^*_{gk} \left[\frac{p_{gk}}{P^2_{gk}} (-\theta_g P^*_{gk} Q^*_{gk} + \theta_g P^*_{g,k-1} Q^*_{g,k-1}) - \right.$$

$$\left. \frac{p_{g,k+1}}{P^2_{g,k+1}} (-\theta_g P^*_{g,k+1} Q^*_{g,k+1} + \theta_g P^*_{gk} Q^*_{gk}) \right].$$

As is usual maximum likelihood practice, the observed proportions for the choice of a category will be replaced by their expectations. With $w_{gk} = P^*_{gk} Q^*_{gk}$, $w_{g,k-1} = P^*_{g,k-1} Q^*_{g,k-1}$, and $w_{g,k+1} = P^*_{g,k+1} Q^*_{g,k+1}$, the elements of the information matrix needed in the Fisher-scoring procedure are:

$$E \left(\frac{\partial^2 L}{\partial \zeta_k^2} \right) = -\sum_{g=1}^{G} f_g w_{gk}^2 \left(\frac{1}{P_{gk}} + \frac{1}{P_{g,k+1}} \right) = \Lambda_{kk}$$

$$E \left(\frac{\partial^2 L}{\partial \zeta_k \partial \zeta_{k-1}} \right) = \sum_{g=1}^{G} f_g w_{gk} w_{g,k-1} \left(\frac{1}{P_{gk}} \right) = \Lambda_{k,k-1}$$

$$E \left(\frac{\partial^2 L}{\partial \zeta_k \partial \zeta_{k+1}} \right) = \sum_{g=1}^{G} f_g w_{gk} w_{g,k+1} \left(\frac{1}{P_{g,k+1}} \right) = \Lambda_{k,k+1}$$

$$E \left(\frac{\partial^2 L}{\partial \lambda^2} \right) = -\sum_{g=1}^{G} f_g \theta_g^2 \sum_{k=1}^{m} (w_{g,k-1} - w_{gk})^2 \left(\frac{1}{P_{gk}} \right) = \Lambda_{\lambda\lambda}$$

$$E \left(\frac{\partial^2 L}{\partial \zeta_k \lambda} \right) = \sum_{g=1}^{G} f_g w_{gk} \theta_g \left[\frac{(w_{g,k-1} - w_{gk})}{P_{gk}} - \frac{(w_{gk} - w_{g,k+1})}{P_{g,k+1}} \right] = \Lambda_{k\lambda}.$$

In order to obtain the information matrix, a negative sign is factored out of each expectation. Then the Fisher-scoring equation to be solved iteratively for the parameters of the boundary curves for a graded response item is

$$
\begin{bmatrix}
\hat{\zeta}_1 \\
\hat{\zeta}_2 \\
\hat{\zeta}_3 \\
\vdots \\
\hat{\zeta}_{m-1} \\
\hat{\lambda}
\end{bmatrix}_{(t+1)}
=
\begin{bmatrix}
\hat{\zeta}_1 \\
\hat{\zeta}_2 \\
\hat{\zeta}_3 \\
\vdots \\
\hat{\zeta}_{m-1} \\
\hat{\lambda}
\end{bmatrix}_{(t)}
-
$$

$$
\begin{bmatrix}
\Lambda_{11} & \Lambda_{12} & 0 & \cdots & 0 & \Lambda_{1\lambda} \\
\Lambda_{21} & \Lambda_{22} & \Lambda_{23} & \cdots & 0 & \Lambda_{2\lambda} \\
0 & \Lambda_{32} & \Lambda_{33} & \cdots & 0 & \Lambda_{3\lambda} \\
\vdots & \vdots & \vdots & \ddots & \vdots & \vdots \\
0 & 0 & 0 & \cdots & \Lambda_{m-1,m-1} & \Lambda_{m-1,\lambda} \\
\Lambda_{\lambda 1} & \Lambda_{\lambda 2} & \Lambda_{\lambda 3} & \cdots & \Lambda_{\lambda,m-1} & \Lambda_{\lambda\lambda}
\end{bmatrix}_{(t)}^{-1}
\begin{bmatrix}
L_1 \\
L_2 \\
L_3 \\
\vdots \\
L_{m-1} \\
L_\lambda
\end{bmatrix}
, \qquad (8.7)
$$

where t indexes the stage in the iterative process. Since the linear parameterization was employed above, it is necessary to convert to the α, β_k parameterization for interpretive purposes. Again, $\beta_k = -\zeta_k/\lambda$, and then $\beta_{k'} = (\beta_k + \beta_{k-1})/2$, and $\alpha = \lambda$.

8.3.2 Ability Estimation in the Graded Response Case

The likelihood function for a given examinee of ability θ_j is the likelihood of a particular item response vector $U_j = (u_{1k}, u_{2k}, \ldots, u_{nk})$, where $u_{ik} = 1$ if the examinee chose response k to item i and $u_{ik} = 0$ otherwise. To simplify the notation, the examinee subscript will not be shown on the right-hand side of the equations that follow. Then,

$$
\text{Prob}(U_j | \zeta, \lambda, \theta_j) = \prod_{i=1}^{n} \prod_{k=1}^{m_i} P_{ik}^{u_{ik}} \qquad (8.8)
$$

and the log-likelihood is

$$
L = \log \text{Prob}(U_j | \zeta, \lambda, \theta_j) = \sum_{i=1}^{n} \sum_{k=1}^{m_i} u_{ik} \log P_{ik}. \qquad (8.9)
$$

The first derivative of L with respect to θ_j is

$$
\frac{\partial L}{\partial \theta_j} = \sum_{i=1}^{n} \sum_{k=1}^{m_i} \frac{u_{ik}}{P_{ik}} \frac{\partial P_{ik}}{\partial \theta_j},
$$

but

$$\frac{\partial P_{ik}}{\partial \theta_j} = \frac{\partial}{\partial \theta_j} \left(P_{i,k-1}^* - P_{ik}^* \right) = \lambda_i P_{i,k-1}^* Q_{i,k-1}^* - \lambda_i P_{ik}^* Q_{ik}^*.$$

Substituting the above term yields

$$\frac{\partial L}{\partial \theta_j} = \sum_{i=1}^{n} \sum_{k=1}^{m_i} u_{ik} \lambda_i \left(\frac{w_{i,k-1} - w_{ik}}{P_{ik}} \right),$$

where $w_{i,k-1} = P_{i,k-1}^* Q_{i,k-1}^*$ and $w_{ik} = P_{ik}^* Q_{ik}^*$. The second derivative of L with respect to θ_j is

$$\frac{\partial^2 L}{\partial \theta_j^2} = \frac{\partial}{\partial \theta_j} \sum_{i=1}^{n} \sum_{k=1}^{m_i} u_{ik} \lambda_i \left(\frac{P_{i,k-1}^* Q_{i,k-1}^* - P_{ik}^* Q_{ik}^*}{P_{ik}} \right)$$

$$= \sum_{i=1}^{n} \sum_{k=1}^{m_i} u_{ik} \lambda_i \frac{1}{P_{ik}^2} \times$$

$$\left[P_{ik} \frac{\partial}{\partial \theta_j} \left(P_{i,k-1}^* Q_{i,k-1}^* - P_{ik}^* Q_{ik}^* \right) - \left(P_{i,k-1}^* Q_{i,k-1}^* - P_{ik}^* Q_{ik}^* \right) \frac{\partial P_{ik}}{\partial \theta_j} \right]$$

$$= \sum_{i=1}^{n} \sum_{k=1}^{m_i} u_{ik} \lambda_i \frac{1}{P_{ik}^2} \times$$

$$\left[P_{ik} \left(\lambda_i P_{ik}^{*2} Q_{ik}^* - \lambda_i P_{ik}^* Q_{ik}^{*2} - \lambda_i P_{i,k-1}^{*2} Q_{i,k-1}^* + \lambda_i P_{i,k-1}^* Q_{i,k-1}^{*2} \right) - \left(P_{i,k-1}^* Q_{i,k-1}^* - P_{ik}^* Q_{ik}^* \right) \lambda_i \left(P_{i,k-1}^* Q_{i,k-1}^* - P_{ik}^* Q_{ik}^* \right) \right]$$

$$= \sum_{i=1}^{n} \lambda_i^2 \sum_{k=1}^{m_i} u_{ik} \times$$

$$\left[\frac{-P_{ik}^* Q_{ik}^* (Q_{ik}^* - P_{ik}^*) + P_{i,k-1}^* Q_{i,k-1}^* (Q_{i,k-1}^* - P_{i,k-1}^*)}{P_{ik}} - \frac{(P_{i,k-1}^* Q_{i,k-1}^* - P_{ik}^* Q_{ik}^*)^2}{P_{ik}^2} \right]$$

$$= \sum_{i=1}^{n} \lambda_i^2 \sum_{k=1}^{m_i} u_{ik} \left[\frac{-w_{ik} (Q_{ik}^* - P_{ik}^*) + w_{i,k-1} (Q_{i,k-1}^* - P_{i,k-1}^*)}{P_{ik}} - \frac{(w_{i,k-1} - w_{ik})^2}{P_{ik}^2} \right].$$

The Fisher-scoring equation is

$$\left[\hat{\theta}_j \right]_{(t+1)} = \left[\hat{\theta}_j \right]_{(t)} - \left[\frac{\partial L / \partial \theta_j}{\partial^2 L / \partial \theta_j^2} \right]_{(t)}, \tag{8.10}$$

where t denotes the iteration. There will be one such equation for each examinee.

The JMLE procedure would be implemented in the same manner as for the dichotomous response case. The examinees could be grouped by raw test score and the standard score for the group used as the initial value of θ_j. Equation 8.7 would be solved iteratively for each item. Using the resulting estimates as item parameters, Equation 8.10 would be solved iteratively to obtain a value of $\hat{\theta}_j$ for each examinee. As in the case under dichotomous response, the identification problem can be resolved by standardizing the ability estimates to mean zero and unit variance. The whole process would be repeated until some convergence criterion, such as a specified difference between two successive values of the likelihood function, was met. Thus, the basic paradigm is unchanged, only the Newton-Raphson equations differ from those of the dichotomous response case.

The JMLE procedure for the graded response case was implemented in the LOGOG computer program due to Kolakowski and Bock (1973b). The program was written for a large-scale (mainframe) computer, and a personal computer version does not appear to have been developed. Since the program for JMLE under the graded response model would be rather large, Appendix F contains only the two basic building blocks written in MICROSOFT QUICKBASIC. Again, sufficient details of the analysis of a data set are presented so that the reader can follow the estimation process. At the present time, the most readily available program for estimating parameters under a graded response model is MULTILOG (Thissen, 1991; Thissen, Chen, & Bock, 2003), which executes on a personal computer. It is a generalized program that employs MMLE to obtain the item parameter estimates for a variety of ICC models and scoring procedures. Because MMLE is used, the examinee ability estimates are obtained in a separate stage.

8.4 The Information Function for the Graded Response Case

Although one's real interest is in the test information as an indicator of the precision with which a test estimates ability, test information is defined in terms of item information functions.

As a result, the discussion of information will focus initially on the item information function in the graded response case. When an item is scored dichotomously, the item information function was defined by Birnbaum (1968) to be

$$I_i(\theta) = -E\left(\frac{\partial^2 \log P_i(\theta)}{\partial \theta^2}\right) = \frac{[P_i'(\theta)]^2}{P_i(\theta)Q_i(\theta)}, \tag{8.11}$$

where $P_i'(\theta)$ is $\partial P_i(\theta)/\partial\theta)$, $P_i(\theta)$ is the probability of a correct response, and $Q_i(\theta)$ the probability of an incorrect response. A dichotomously scored item can be considered a special case of a graded response item having two categories with $P_{i1}(\theta) = P_i(\theta)$, $P_{i2}(\theta) = Q_i(\theta)$. Also, there will be information associated with each response category. Let $I_{ik}(\theta)$ denote the amount of information associated with a particular item response category. However, an examinee can choose only one of the two response categories with probability $P_i(\theta)$ or $Q_i(\theta)$. As a result, the amount of information contributed by a response category to the item information is given by $I_{i1}(\theta)P_i(\theta)$ for the correct response and $I_{i2}(\theta)Q_i(\theta)$ for the incorrect response. Then the amount of information yielded by the item at ability level θ is

$$I_i(\theta) = I_{i1}(\theta)P_i(\theta) + I_{i2}(\theta)Q_i(\theta).$$

Generalizing to the case of m_i response categories, we have

$$I_i(\theta) = \sum_{k=1}^{m_i} I_{ik}(\theta)P_{ik}(\theta), \tag{8.12}$$

where the quantity $I_{ik}(\theta)P_{ik}(\theta)$ is the amount of information share of category k. Samejima (1969, 1972) has defined the information function of an item response category as

$$I_{ik}(\theta) = -\frac{\partial^2 \log P_{ik}(\theta)}{\partial \theta^2} = -\frac{\partial}{\partial \theta}\left[\frac{P'_{ik}(\theta)}{P_{ik}(\theta)}\right], \tag{8.13}$$

where $P'_{ik}(\theta) = \partial P_{ik}(\theta)/\partial \theta$. Then,

$$I_{ik}(\theta) = \frac{[P'_{ik}(\theta)]^2 - P_{ik}(\theta)P''_{ik}(\theta)}{[P_{ik}(\theta)]^2}, \tag{8.14}$$

where $P''_{ik}(\theta) = \partial P'_{ik}(\theta)/\partial \theta$. The item response category information share is

$$I_{ik}(\theta)P_{ik}(\theta) = \frac{[P'_{ik}(\theta)]^2}{P_{ik}(\theta)} - P''_{ik}(\theta). \tag{8.15}$$

The item information function then becomes

$$I_i(\theta) = \sum_{k=1}^{m_i} \frac{[P'_{ik}(\theta)]^2 - P_{ik}(\theta)P''_{ik}(\theta)}{[P_{ik}(\theta)]^2} P_{ik}(\theta)$$

$$= \sum_{k=1}^{m_i} \left(\frac{[P'_{ik}(\theta)]^2}{P_{ik}(\theta)} - P''_{ik}(\theta)\right). \tag{8.16}$$

Both the item response category information function and the item information function can be expressed in terms of the boundary probabilities $P^*_{ik}(\theta)$. From Equation 8.3, we have

$$P_{ik}(\theta) = P^*_{i,k-1}(\theta) - P^*_{ik}(\theta)$$
$$P'_{ik}(\theta) = P^{*'}_{i,k-1}(\theta) - P^{*'}_{ik}(\theta)$$
$$P''_{ik}(\theta) = P^{*''}_{i,k-1}(\theta) - P^{*''}_{ik}(\theta).$$

Substituting into Equation 8.14

$$I_{ik}(\theta) =$$

$$\frac{[P_{i,k-1}^{*\prime}(\theta) - P_{ik}^{*\prime}(\theta)]^2 - [P_{i,k-1}^{*}(\theta) - P_{ik}^{*}(\theta)][P_{i,k-1}^{*\prime\prime}(\theta) - P_{ik}^{*\prime\prime}(\theta)]}{[P_{i,k-1}^{*}(\theta) - P_{ik}^{*}(\theta)]^2}, \quad (8.17)$$

and the item response information share is

$$I_{ik}(\theta)P_{ik}(\theta) = \frac{[P_{i,k-1}^{*\prime}(\theta) - P_{ik}^{*\prime}(\theta)]^2}{P_{i,k-1}^{*}(\theta) - P_{ik}^{*}(\theta)} - [P_{i,k-1}^{*\prime\prime}(\theta) - P_{ik}^{*\prime\prime}(\theta)]. \quad (8.18)$$

Then the item information function is

$$
\begin{aligned}
I_i(\theta) &= \sum_{k=1}^{m_i} I_{ik}(\theta)P_{ik}(\theta) \\
&= \sum_{k=1}^{m_i} I_{ik}(\theta)[P_{i,k-1}^{*}(\theta) - P_{ik}^{*}(\theta)] \\
&= \sum_{k=1}^{m_i} \frac{[P_{i,k-1}^{*\prime}(\theta) - P_{ik}^{*\prime}(\theta)]^2 - [P_{i,k-1}^{*}(\theta) - P_{ik}^{*}(\theta)][P_{i,k-1}^{*\prime\prime}(\theta) - P_{ik}^{*\prime\prime}(\theta)]}{P_{i,k-1}^{*}(\theta) - P_{ik}^{*}(\theta)} \\
&= \sum_{k=1}^{m_i} \frac{[P_{i,k-1}^{*\prime}(\theta) - P_{ik}^{*\prime}(\theta)]^2}{P_{i,k-1}^{*}(\theta) - P_{ik}^{*}(\theta)} - \sum_{k=1}^{m_i}[P_{i,k-1}^{*\prime\prime}(\theta) - P_{ik}^{*\prime\prime}(\theta)].
\end{aligned}
$$

Samejima (1969, p. 39) shows that the second term on the right vanishes and then the item information function is

$$I_i(\theta) = \sum_{k=1}^{m_i} \frac{[P_{i,k-1}^{*\prime}(\theta) - P_{ik}^{*\prime}(\theta)]^2}{P_{i,k-1}^{*}(\theta) - P_{ik}^{*}(\theta)}. \quad (8.19)$$

Although Equations 8.15 and 8.18 are equivalent as are Equations 8.16 and 8.19, Equations 8.15 and 8.16 are definitional in form. This is due to the fact that direct mathematical expressions for the $P_{ik}(\theta)$ do not exist and the $P_{ik}(\theta)$ can be expressed only in terms of the functions for the boundary curves. As a result, Equation 8.18 is used to compute the item response information share and Equation 8.19 the item information function.

An interesting exercise is to specialize Equation 8.12 to two response categories. After algebraic simplification, it should yield Equation 8.11. Substituting Equation 8.14 into Equation 8.12, we have

$$
\begin{aligned}
I_i(\theta) &= \frac{[P_i^{\prime}(\theta)]^2 - P_i(\theta)P_i^{\prime\prime}(\theta)}{[P_i(\theta)]^2}P_i(\theta) + \frac{[-P_i^{\prime}(\theta)]^2 - Q_i(\theta)P_i^{\prime\prime}(\theta)}{[Q_i(\theta)]^2}Q_i(\theta) \\
&= \frac{\{[P_i^{\prime}(\theta)]^2 - P_i(\theta)P_i^{\prime\prime}(\theta)\}Q_i(\theta) + \{[-P_i^{\prime}(\theta)]^2 + Q_i(\theta)P_i^{\prime\prime}(\theta)\}P_i(\theta)}{P_i(\theta)Q_i(\theta)} \\
&= \frac{[P_i^{\prime}(\theta)]^2[P_i(\theta) + Q_i(\theta)]}{P_i(\theta)Q_i(\theta)} \\
&= \frac{[P_i^{\prime}(\theta)]^2}{P_i(\theta)Q_i(\theta)}.
\end{aligned}
$$

This result is identical to the expression due to Birnbaum (1968) for the item information function when an item is scored dichotomously. However, he approached the derivation from a very different frame of reference. Thus, the item information function presented in earlier sections is a particular case of the general expression (Equation 8.12) provided by Samejima (1969). In addition, the usual item information function expression for a dichotomously scored item obscures the contribution of the two response categories. This present result also highlights the distinction between the amount of information due to an item response category and the information share of an item response category. The amount of information due to an item response category $I_{ik}(\theta)$ is a measure of how well responses in that category estimate the examinee's ability. The information share of a response category $I_{ik}(\theta)P_{ik}(\theta)$ is the amount of information contributed by the category to the item information. In the former case, $I_i(\theta) \neq \sum_{k=1}^{m_i} I_{ik}(\theta)$, and interpretation of graphs of $I_i(\theta)$ and $I_{ik}(\theta)$ would be difficult. In the latter case, $I_i(\theta) = \sum_{k=1}^{m_i} I_{ik}(\theta)P_{ik}(\theta)$, and the relative contribution of each response category is clear. Thus, from an interpretive point of view, the amount of information share of each response category is the most informative. This also provides continuity of meaning from dichotomous to graded response and, as we shall see later, to nominal response.

8.4.1 Normal ICC Model and Graded Response Information

Although the normal ogive model is primarily of historical interest, information for the graded response model will be derived initially under this model to link the discussion to the presentation of Chapter 3. Under a normal ogive model for the boundary curves, working expressions for the information share of an item response category and for the item information can be obtained. Recall that, under the α_i, β_{ik} parameterization,

$$\frac{\partial P_{ik}^*(\theta)}{\partial \theta} = P_{ik}^{*'}(\theta) = \alpha_i h_{ik}^*(\theta)$$

and

$$\frac{\partial^2 P_{ik}^*(\theta)}{\partial \theta^2} = \alpha_i^2 h_{ik}^*(\theta).$$

Then,

$$P_{i,k-1}^{*'}(\theta) - P_{ik}^{*'}(\theta) = \alpha_i h_{i,k-1}^*(\theta) - \alpha_i h_{ik}^*(\theta)$$

$$P_{i,k-1}^{*''}(\theta) - P_{ik}^{*''}(\theta) = \alpha_i^2 h_{i,k-1}^*(\theta) - \alpha_i^2 h_{ik}^*(\theta).$$

Substituting in Equation 8.11, the item response information share is

$$I_{ik}(\theta)P_{ik}(\theta) = \frac{[\alpha_i h_{i,k-1}^*(\theta) - \alpha_i h_{ik}^*(\theta)]^2}{P_{i,k-1}^*(\theta) - P_{ik}^*(\theta)} - \alpha_i^2 [h_{i,k-1}^*(\theta) - h_{ik}^*(\theta)]. \quad (8.20)$$

The item information function is

$$I_i(\theta) = \sum_{k=1}^{m_i} \frac{[\alpha_i h_{i,k-1}^*(\theta) - \alpha_i h_{ik}^*(\theta)]^2}{P_{i,k-1}^*(\theta) - P_{ik}^*(\theta)}, \tag{8.21}$$

and the information shares of the several response categories sum to the amount of item information at a given ability level.

Figure 8.4 contains a plot of the item response information shares and the item information for the graded response item used in the examples above. The item information function was unimodal, with a maximum at $\theta = -0.5$ and, because of the values of β_k, was symmetric about this point. The information shares of the individual response categories revealed some interesting characteristics. The curves for the two extreme categories, $k = 1$ and $k = 4$, were unimodal, with maximums at $\theta = -0.5$ and $\theta = +1.0$, respectively. The curves for the central two response categories were both bimodal, with a pronounced trough between the two modes. From an interpretive point of view, these curves indicate that there are regions on the ability scale where the intermediate response categories yield poor estimates of ability. However, these same categories yield more information in regions both above and below the trough. At the same time, the curves for the extreme categories are much better behaved. Samejima (1969, p. 39) indicates that the existence of the trough is a function of the width of the interval $(\beta_{ik} + \beta_{i,k-1})/2$. The wider this interval, the wider the trough; the narrower the interval, the less pronounced the trough. With reasonably small intervals (say, < 1), the trough disappears and the curve becomes unimodal. Thus, with closely spaced values of β_k, all the intermediate information share curves could be unimodal. With widely spaced values, the nonextreme response categories could all yield information share curves possessing a trough. From a precision of estimation framework, it would appear that the graded response case requires some attention to the spacing of the response category boundaries.

An interesting issue examined by Samejima (1969) was the effect on the item information function of increasing the number of response categories. Let us use the data of Table 8.1 to illustrate the issue and the results. If this item were dichotomously scored, with response category 2 or greater as the correct response, the item information function would be as shown in Figure 8.5. The item information function based on four response categories is also plotted in this figure. The amount of item information yielded by the four ordered response categories is uniformly larger than that for the same item scored dichotomously. Samejima (1969, p. 40) indicates that increasing the number of response categories will result in an increase in the amount of item information. This holds if an existing category is divided into more categories or if the item is completely recategorized (Samejima 1969, p. 42). Thus, the graded response case will yield a smaller standard error for the estimate of an examinees ability than the dichotomous case (see Figure 8.5).

The role of the item discrimination parameter α_i common to the $m_i - 1$ boundary ogives has not been included in the discussion of the information function. Under a normal ogive model, α_i appears as a multiplier in the

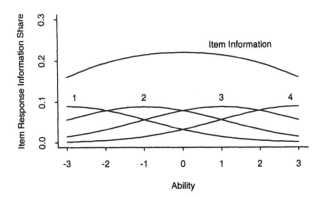

Fig. 8.4. Information functions for an item having four graded response categories using data from Table 8.1.

expression for the item response information function (see Equation 8.11). Hence, increasing the value of α_i will increase the amount of item information over the range of ability of interest.

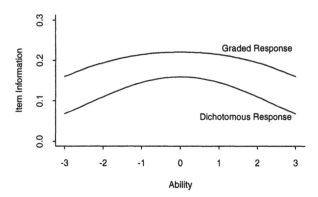

Fig. 8.5. Item information function for graded response and dichotomous response (category 2 or greater scored as correct).

8.4.2 Logistic ICC Model and Graded Response Information

In the preceding section, the boundary curves were normal ogives. As was the case when items were dichotomously scored, the logistic ogive can be substituted for the normal ogive when a graded response is used. Under a logistic model and the α_i, β_{ik} parameterization,

$$P_{ik}(\theta) = P^*_{i,k-1}(\theta) - P^*_{ik}(\theta)$$

and

$$P'_{ik}(\theta) = P^{*'}_{i,k-1}(\theta) - P^{*'}_{ik}(\theta) = \alpha_i P^*_{i,k-1}(\theta)Q^*_{i,k-1}(\theta) - \alpha_i P^*_{ik}(\theta)Q^*_{ik}(\theta)$$

$$P''_{ik}(\theta) = P^{*''}_{i,k-1}(\theta) - P^{*''}_{ik}(\theta)$$
$$= \alpha_i^2 P^*_{i,k-1}(\theta)Q^*_{i,k-1}(\theta)[Q^*_{i,k-1}(\theta) - P^*_{i,k-1}(\theta)] -$$
$$\alpha_i^2 P^*_{ik}(\theta)Q^*_{ik}(\theta)[Q^*_{ik}(\theta) - P^*_{ik}(\theta)].$$

Substituting in Equation 8.15, the item response information share is

$$I_{ik}(\theta)P_{ik}(\theta) = \frac{[\alpha_i P^*_{i,k-1}(\theta)Q^*_{i,k-1}(\theta) - \alpha_i P^*_{ik}(\theta)Q^*_{ik}(\theta)]^2}{P^*_{i,k-1}(\theta) - P^*_{ik}(\theta)} -$$
$$\{\alpha_i^2 P^*_{i,k-1}(\theta)Q^*_{i,k-1}(\theta)[Q^*_{i,k-1}(\theta) - P^*_{i,k-1}(\theta)] -$$
$$\alpha_i^2 P^*_{ik}(\theta)Q^*_{ik}(\theta)[Q^*_{ik}(\theta) - P^*_{ik}(\theta)]\}. \qquad (8.22)$$

Although the expression has a large number of elements, all of them are expressed in terms of the boundary probabilities, which are easily evaluated. The item information function is given by

$$I_i(\theta) = \sum_{k=1}^{m_i} I_{ik}(\theta)P_{ik}(\theta)$$
$$= \sum_{k=1}^{m_i} \frac{[\alpha_i P^*_{i,k-1}(\theta)Q^*_{i,k-1}(\theta) - \alpha_i P^*_{ik}(\theta)Q^*_{ik}(\theta)]^2}{P^*_{i,k-1}(\theta) - P^*_{ik}(\theta)}$$
$$= \sum_{k=1}^{m_i} \frac{\alpha_i^2 (w_{i,k-1} - w_{ik})^2}{P^*_{i,k-1}(\theta) - P^*_{ik}(\theta)} \qquad (8.23)$$

and the appended terms in Equation 8.22 summed to zero.

8.5 Other Models for Polytomous Items

In addition to the graded response model of this chapter and the nominal response model that is presented in Chapter 9, there are several alternative models for polytomously scored items. In this section, several alternative models are briefly presented as well as estimation methods and information functions; the models presented include Muraki's rating scale model, the generalized partial credit model, the partial credit model, and the rating scale model. For detailed discussions about the polytomous models and additional information, interested readers should refer to De Ayala (1993) and van der Linden and Hambleton (1997) and references therein.

Let us use a trivially different notation in this section. Let the item category response function be

$$\text{Prob}\{u_i = k|\theta, \xi_i\} \equiv P_{ik}(\theta), \tag{8.24}$$

where u_i is the response random variable, k is the realization of u_i, θ is the ability parameter, and ξ_i is the vector of item parameters. Without the loss of generality, we assume that $u_i = 1, \ldots, K_i$, where K_i is the total number of categories of the item.

In the Muraki's (1990) rating scale model, the item category response function is defined as

$$P_{ik}(\theta) = P_{ik}^+(\theta) - P_{i(k-1)}^+(\theta), \tag{8.25}$$

where the boundary response function is defined as

$$P_{ik}^+(\theta) = \text{Prob}\{u_i \leq k|\theta, \xi_i\}, \tag{8.26}$$

where $P_{i0}^+(\theta) = 0$ and $P_{iK_i}^*(\theta) = 1$. The logistic model of the boundary response function is

$$P_{ik}^+(\theta) = \frac{\exp[Da_i(\theta - b_i + c_k)]}{1 + \exp[Da_i(\theta - b_i + c_k)]}, \tag{8.27}$$

where c_k is the category threshold parameter. Note that the total number of item categories is fixed (e.g., $K_i = K$) and that the total number of c_k is $K - 1$.

In the generalized partial credit model (Muraki, 1992, 1993, 1997; see also Park & Muraki, 2003), the item category response function is defined as

$$P_{ik}(\theta) = \frac{\exp\left[\sum_{v=1}^{k} Da_i(\theta - b_{iv})\right]}{\sum_{h=1}^{K_i} \exp\left[\sum_{v=1}^{h} Da_i(\theta - b_{iv})\right]}, \tag{8.28}$$

where $b_{i1} = 0$. There will be $K_i - 1$ number of step parameters, $b_{i2}, \ldots, b_{iK_i}$. The item step parameter b_{ik} is the point on the θ scale where two item response functions cross. We may write

$$Z_{ik}(\theta) = Da_i(\theta - b_{ik}) = Da_i(\theta - b_i + d_k), \tag{8.29}$$

where b_i is the location parameter and d_k is the category parameter. The item parameters are a_i, b_i, and d_k (the total number of d_k is $K_i - 1$) with a constraint,

$$\sum_{k=2}^{K_i} d_k = 0. \tag{8.30}$$

For the partial credit model, we may use the notations by Masters (1982) and Masters and Wright (1997) in this section. Let the polytomous item scores are $0, 1, \ldots, m_i$, where i designates item. Let the item score be x. The item category response function is defined as

$$P_{ix}(\theta) = \frac{\exp\left[\sum_{k=0}^{x}(\theta - b_{ik})\right]}{\sum_{h=0}^{m_i} \exp\left[\sum_{k=0}^{h}(\theta - b_{ik})\right]}, \tag{8.31}$$

where b_{ik}'s are item step parameters. If we define

$$\sum_{k=0}^{0}(\theta - b_{ik}) = 0, \tag{8.32}$$

then

$$\sum_{k=0}^{h}(\theta - b_{ik}) = \sum_{k=1}^{h}(\theta - b_{ik}). \tag{8.33}$$

When the total number of item categories is $m_i + 1$, there are m_i number of item step parameters, that is, $b_{i1}, \ldots, b_{im_i}$. As in the generalized partial credit model, b_{ik} is the point on the θ scale where two item category response function cross.

In the rating scale model (Andersen, 1997; Andrich, 1978), the rating scale model is applicable to a test in which all items have the same number of categories. In this model

$$b_{ik} = b_i + t_k, \tag{8.34}$$

where b_i is the location parameter and t_k is the intercept parameter. Let

$$k_x = \sum_{k=1}^{x} t_k \tag{8.35}$$

and $k_0 = 0$, then the item response category function is

$$P_{ix}(\theta) = \frac{\exp[k_x + x(\theta - b_i)]}{\sum_{h=0}^{m_i} \exp[k_h + h(\theta - b_i)]}, \tag{8.36}$$

where each item has one location parameter b_i. The number of category intercept parameters is m_i.

Let us also briefly look at the parameter estimation for polytomous item response models. Let the index variable be

$$y_{ijk} = \begin{cases} 1 \text{ if } u_{ij} = k \\ 0 \text{ otherwise,} \end{cases} \tag{8.37}$$

then

$$P_{ijk} = \text{Prob}\{u_{ij} = k|\theta_j, \xi_i\} = \prod_{k=1}^{K_i} P_{ijk}^{y_{ijk}}. \tag{8.38}$$

Let's define the following vectors and matrices that contain all cases with subscripts: ξ is the vector of item parameters, θ is the vector of ability parameters, and U is the matrix of item responses that is equivalent to Y which is the matrix of all index variables. The probability of obtaining U given θ and ξ is

$$\text{Prob}\{U|\theta, \xi\} = \prod_{i=1}^{n}\prod_{j=1}^{N}\prod_{k=1}^{K_i} P_{ijk}^{y_{ijk}} = p(U|\theta, \xi). \tag{8.39}$$

The three estimation methods that can be applied, based on respective models, are joint maximum likelihood estimation, conditional maximum likelihood estimation, and marginal maximum likelihood estimation. Marginal maximum likelihood estimation can be seen as a standard estimation method. For marginal maximum likelihood estimation, the ability parameter will be integrated out from the likelihood function using

$$p(u_j|\xi) = \int p(u_j|\theta_j, \xi)p(\theta_j)d\theta_j, \tag{8.40}$$

where u_j is the response vector of examinee j for n items. The marginal probability (i.e., marginal likelihood) is defined as

$$p(U|\xi) = \prod_{j=1}^{N} p(u_j|\xi) = m(\xi). \tag{8.41}$$

Item parameter estimates are the values that maximize the marginal likelihood function. Ability parameters are estimated using the methods of expected a posteriori (EAP), maximum a posteriori (MAP), or maximum likelihood (ML) after obtaining item parameter estimates.

In polytomous item response theory models, item category information function can be defined as (see Samejima, 1969; Bock, 1972), in a general sense,

$$I_{ik}(\theta) = \frac{[P_{ik}'(\theta)]^2}{[P_{ik}(\theta)]^2} - \frac{P_{ik}''(\theta)}{P_{ik}(\theta)}, \tag{8.42}$$

where

$$P_{ik}'(\theta) = \frac{\partial P_{ik}(\theta)}{\partial \theta}, \quad \text{and} \quad P_{ik}''(\theta) = \frac{\partial^2 P_{ik}(\theta)}{\partial \theta^2}. \tag{8.43}$$

The item information function is

$$I_i(\theta) = \sum_{k=1}^{K_i} I_{ik}(\theta)P_{ik}(\theta). \tag{8.44}$$

Note that there are still other models for polytomously scored items. These include, for example, the steps model (Verhelst, Glas, & de Vries, 1997), the sequential model (Tutz, 1997), and the continuation ratio model (Kim, 2002).

8.6 Research on the Graded Response Model

Compared to the dichotomous response case, there is very little research dealing with the properties of parameter estimates under a graded response model. In their study of computerized adaptive testing using graded response items, Dodd, Koch and De Ayala (1989) incidentally examined the parameter estimates yielded by MULTILOG (Thissen, 1986). They defined a pool of 30 items, each having five response categories. The difficulty parameters of the boundary curves were varied with respect to range within an item, and the items were uniformly located over the ability scale. The item discrimination parameters ranged from 0.90 to 2.15 in 0.05 increments. A 60-item pool was created by replicating the characteristics of the existing 30 items. The simulated examinees' abilities were selected from a unit normal distribution. Item response vectors of length 60 were generated for each of the 1000 examinees under the item parameters specified. The item response vectors for a 30-item pool and a 60-item pool were then analyzed via MULTILOG under a graded response model. Dodd and co-workers found that the negatively valued boundary curve difficulties tended to be underestimated while the positively valued ones were overestimated. The obtained boundary curves were reported as similar to those yielded by the underlying item parameters. They did not report the characteristics of the obtained item discrimination indices. However, the maxima of the obtained test information functions for the 30- and 60-item pools were lower than those yielded by the underlying parameters. This suggests that the obtained item discriminations may have been smaller than the underlying values. Dodd et al. did not equate the metric of the obtained item parameters into the underlying metric. Thus, the lower values may simply be reflecting a slight change in ability scale metric. The correlation of obtained and underlying ability for the 1000 examinees was reported as $r = .99$. From this very limited evidence, it would appear that MULTILOG was able to recover the underlying parameters reasonably well for graded response items.

Koch (1983) employed the graded response model to analyze an attitude inventory that employed Likert-scale items. The instrument consisted of 40 five-point scales that rated a teacher's attitude towards his/her administrator's communication skills. The responses of 491 teachers to the instrument were analyzed using the LOGOG (Kolakowski & Bock, 1973b) program. For purposes of item parameter estimation, the ability distribution was divided into 10 intervals, and the examinees were grouped accordingly. One purpose of the study was to compare the item parameter estimates yielded by LOGOG with a factor analytic technique (Samejima, 1976) and the traditional CTT approach. A second purpose was to compare three ability estimation procedures: (1) the sum of the item response category weights for an examinee, (2) the LOGOG JMLE of ability, and (3) an empirical MLE approach based on Reckase's (1974) approach for dichotomously scored items. The obtained χ^2 goodness-of-fit indices were used to edit 10 items out of the instrument. Thirteen examinees were eliminated because of nonconvergence of the MLE ability estimation process. As would be anticipated, the agreement among the three item parameter estimation procedures was quite high. The lowest correlation ($r = .54$) was between the traditional and LOGOG estimates of item discrimination. The correlations for the boundary curve difficulty parameters yielded by the three methods all were $r = .99$. The results yielded by LOGOG and Samejima's factor analytic approach were quite similar in the case of item discriminations and nearly identical in the case of the item boundary curve difficulty parameters. In the case of ability estimates, all three estimation procedures yielded estimates that correlated very highly ($r \geq .94$) with each other. Unfortunately, this study does not provide any indication of the quality of parameter recovery under the graded response model. It does, however, show that the JMLE parameter estimates are consistent with expectations.

Reise and Yu (1990) conducted a Monte Carlo simulation study using MULTILOG (Thissen, 1986) to answer how many subjects are required to estimate parameters in the graded response model. They showed that the graded response model item parameters can be estimated with as few as 250 examinees but recommended 500 examinees to achieve an adequate calibration.

8.7 Summary

The graded response case makes an important contribution to IRT in a number of ways: First, it extends the scoring of an item from the dichotomous to graded response, thus opening the way to the analysis of rating scales of the Likert type and items that are scored under partial credit models. Doing so should aid in the interpretation of such instruments. Each item will be positioned along the ability scale by its set of response category location parameters, which can be used to evaluate the spacing of the categories both within and across items. Perhaps the ideal item would be one with equally spaced location parameters centered on some point of interest on the ability continuum. Second, Samejima showed that graded response items yield higher item and test information functions than do dichotomously scored items. Thus, ability can be estimated with greater precision, or the same level of precision can be obtained with fewer items. Third, Samejima's derivation of the information function for the graded response case provided a better understanding of the information function for the dichotomous case. Although the mathematics of the resulting item information function was the same as that due to Birnbaum (1968), it showed that both the correct and incorrect response categories contribute to the amount of information.

At the present time, the graded response model is not supported by the wealth of studies that are associated with dichotomously scored items. There are only a few reported applications of the model in applied settings. In addition, examinations of the parameter estimation properties of the various computer programs for the graded response and partial credit models are lacking. However, the ready availability of computer programs, such as MULTILOG and PARSCALE (Muraki & Bock, 2003), for these models that execute on personal computers should make such studies feasible and exploration of the domain possible.

9. Nominally Scored Items

9.1 Introduction

In previous chapters, the examinee's free response to an item was scored on either a right-wrong or a graded basis. The repertoire of scoring procedures will now be extended to include nominal scoring of the responses to an item. Under nominal scoring, the possible responses to an item are allocated to m mutually exclusive, exhaustive, and nonordered categories. Thus, an examinee's free response to an item can be designated according to the category in which it belongs, but the category is represented by a nominal measure. For example, an item may consist of a complex mathematics problem to be solved and the examinee's solution scored as (A) heuristic, (B) algorithmic, (C) eclectic, or (D) trial and error, and no intrinsic ordering of these categories is assumed. As was the case with previous scoring procedures, the idea of scoring the free responses of an examinee does not match actual practice. Thus, an item will be considered to have m response categories, and the examinee is free to choose any one of them. In the nominal response case, the examinee does not accumulate a total score since nominal measures cannot be summed. However, as shall be shown later, it is possible to estimate an examinee's ability using nominally scored items, even though a total score cannot be obtained. The item parameter estimation procedures associated with the nominal response case provide the solution to a problem that has long vexed educational measurement specialists, namely, how to estimate simultaneously the parameters of all the response alternatives to a multiple choice item. Thus, although items that are actually scored on a nominal basis are rare, this latter application has immediate and potentially widespread utility.

When item responses are nominally scored, a number of interesting issues arise with respect to developing a model for the item characteristic curves. One can assume again that an underlying ability scale exists that can be measured in the usual θ metric and that an examinee of ability, say θ_j, will select only one of the m item response categories. There will be a probability $P_{ik}(\theta_j)$ that the response selected will be in category k of item i. However, the numerical score received by the examinee for the response has no meaning other than to designate the response category. Figure 9.1 depicts a plot of the proportion of responses assigned to each of three nominally scored response categories as a function of ability. As was the case for graded response, an item response category characteristic curve (IRCCC) will be obtained for each item response category.

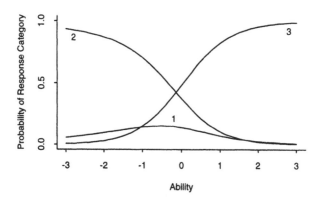

Fig. 9.1. Item response category characteristic curves for the data of Table 9.1.

The immediate problem then is to find a mathematical model that can simultaneously represent this melange of curves. In the case of graded response, it can be assumed that the underlying psychological process is the threshold concept and that the response categories are ordered. As a result, the probability of interest can be redefined in terms of boundary curves. Then, either the normal or logistic ogive can be used to model the cumulative probabilities up to the upper bound of a response category as was done in Chapter 8. Under nominal scoring, one cannot rely on such a tactic as there is no intrinsic ordering to the response categories. The elegant solution to modeling the IRCCCs of a nominally scored item was presented by Bock (1972, 1997) and employed the multivariate logistic function, which was a generalization of the bivariate logistic function derived by Gumbel (1961). The mathematical definition of the multivariate logistic function is

$$P_k(\theta_j) = \frac{e^{Z_k(\theta_j)}}{\sum_{v=1}^{m} e^{Z_v(\theta_j)}}, \tag{9.1}$$

where j denotes the examinee $(j = 1, 2, \ldots, N)$, v denotes the item response category $(v = 1, 2, \ldots, k, \ldots, m)$, and $Z_v(\theta_j) = \zeta_v + \lambda_v \theta_j$ is called the multivariate logit and there will be a vector of m such logits for each item.

$P_k(\theta_j)$ is the probability of an examinee of ability θ_j choosing item response category k. Since only one item at a time is of interest, a subscript designating the item will not be employed although it will be understood. Equation 9.1 can be further simplified by letting

$$D = \sum_{v=1}^{m} e^{Z_v(\theta_j)}$$

and then

$$P_k(\theta_j) = \frac{e^{Z_k(\theta_j)}}{D}. \tag{9.2}$$

However, at any given ability level θ_j,

$$\sum_{k=1}^{m} P_k(\theta_j) = \sum_{k=1}^{m} \frac{e^{Z_k(\theta_j)}}{D} = \frac{1}{D} \sum_{k=1}^{m} e^{Z_k(\theta_j)} = \frac{D}{D} = 1. \tag{9.3}$$

Since the term $Z_k(\theta_j)$ appears in both the numerator and denominator of Equation 9.2, this function is invariant with respect to translation of the logit. As a result, it is necessary to "anchor" the multivariate logits in some manner. Bock (1972) imposed the arbitrary linear restriction

$$\sum_{v=1}^{m} Z_v(\theta_j) = 0 \tag{9.4}$$

which also implies that

$$\sum_{v=1}^{m} \zeta_v = 0 \quad \text{and} \quad \sum_{v=1}^{m} \lambda_v = 0.$$

Given this introduction, let us examine the behavior of the multivariate logistic model for the IRCCC of a nominally scored item. Table 9.1 contains the values of $P_k(\theta_j)$ for a three-category item. Since all the elements in Equation 9.1 are in terms of e^x, the multivariate logistic function can be evaluated easily with an inexpensive slide-rule type of calculator having one memory register to hold D. The terms corresponding to the fourth row of Table 9.1, where $\theta_j = -2.0$ and $\zeta_1 = -0.75$, $\lambda_1 = -0.20$, $\zeta_2 = 0.25$, $\lambda_2 = -0.80$, $\zeta_3 = 0.50$, and $\lambda_3 = 1.00$ will be calculated. Then

$$Z_1(\theta_j) = -0.75 + (-0.20)(-2.0) = -0.35$$
$$Z_2(\theta_j) = 0.25 + (-0.80)(-2.0) = 1.85$$
$$Z_3(\theta_j) = 0.50 + (1.00)(-2.0) = -1.50$$
$$e^{Z_1(\theta_j)} = 0.704688$$
$$e^{Z_2(\theta_j)} = 6.359819$$
$$e^{Z_3(\theta_j)} = 0.223130$$
$$D = 7.287637.$$

And

$$P_1(\theta_j) = \frac{e^{Z_1(\theta_j)}}{D} = .0967$$

$$P_2(\theta_j) = \frac{e^{Z_2(\theta_j)}}{D} = .8727$$

$$P_3(\theta_j) = \frac{e^{Z_3(\theta_j)}}{D} = .0306.$$

Similar calculations can easily be performed for the other rows, and the value of D has been reported to provide an accuracy check. Figure 9.1 depicts the item response category characteristic curves for the item specified in Table 9.1.

Table 9.1. Evaluation of the Multivariate Logistic Function with Three Response Categories

θ	Category 1 $\zeta_1 = -0.75$ $\lambda_1 = -0.20$		Category 2 $\zeta_2 = 0.25$ $\lambda_2 = -0.80$		Category 3 $\zeta_3 = 0.50$ $\lambda_3 = 1.00$		D
	Z_1	$P_1(\theta)$	Z_2	$P_2(\theta)$	Z_3	$P_3(\theta)$	
−3.0	−0.15	.0570	2.65	.9376	−2.50	.0054	15.0968
−2.5	−0.25	.0749	2.25	.9121	−2.00	.0130	10.4019
−2.0	−0.35	.0967	1.85	.8727	−1.50	.0306	7.2876
−1.5	−0.45	.1210	1.45	.8092	−1.00	.0698	5.2686
−1.0	−0.55	.1428	1.05	.7071	−0.50	.1501	4.0411
−0.5	−0.65	.1519	0.65	.5572	0.00	.2909	3.4376
0.0	−0.75	.1387	0.25	.3771	0.50	.4842	3.4051
0.5	−0.85	.1067	−0.15	.2148	1.00	.6785	4.0064
1.0	−0.95	.0710	−0.55	.1060	1.50	.8230	5.4454
1.5	−1.05	.0431	−0.95	.0476	2.00	.9093	8.1257
2.0	−1.15	.0248	−1.35	.0203	2.50	.9549	12.7584
2.5	−1.25	.0139	−1.75	.0085	3.00	.9776	20.5458
3.0	−1.35	.0077	−2.15	.0035	3.50	.9888	33.4912

For mathematical simplicity, the multivariate logits have been expressed in their linear form. However, the item parameters α_k, β_k are used. The transformation is the same as that for the dichotomous case and yields

$$\beta_1 = -\frac{\zeta_1}{\lambda_1}, \qquad \beta_2 = -\frac{\zeta_2}{\lambda_2}, \qquad \ldots, \qquad \beta_k = -\frac{\zeta_k}{\lambda_k}, \qquad \ldots, \qquad \beta_m = -\frac{\zeta_m}{\lambda_m}$$

and

$$\alpha_1 = \lambda_1, \qquad \alpha_2 = \lambda_2, \qquad \ldots, \qquad \alpha_k = \lambda_k, \qquad \ldots, \qquad \alpha_m = \lambda_m.$$

These parameter values underlying the item of Table 9.1 are

$$\beta_1 = -3.7500 \qquad \alpha_1 = -0.20$$
$$\beta_2 = 0.3125 \qquad \alpha_2 = -0.80$$
$$\beta_3 = -0.5000 \qquad \alpha_3 = 1.00.$$

The interpretation of the values of these parameters is not as simple as it was in the case of the dichotomous or graded response models. The location of the IRCCCs and their forms depends upon the combinations of parameter values for the item's full set of response categories. Perhaps the simplest way to interpret the item response category parameters is via the graph of the IRCCCs such as given in Figure 9.1. In the present example, the probability of choosing response category one is a unimodal curve with a maximum of .15 at $\theta_j = -0.5$. Response category two yielded a monotonically decreasing IRCCC with $P_k(\theta_j) = .5$ at $\theta_j = 0.34$. Response category three yielded a monotonically increasing IRCCC with $P_k(\theta_j) = .5$ at $\theta_j = 0.039$. These two curves are similar to those for the incorrect and correct responses to a dichotomously scored item. However, it should be noted that the β_k do not have a probabilistic interpretation under the nominal response model since their values do not correspond to the ability level were $P_k(\theta_j) = .5$. In addition, $\alpha_2 = -0.8$ and $\alpha_3 = 1.0$ which are not complementary.

Given m response categories for an item, there would be one monotonically increasing IRCCC, one monotonically decreasing IRCCC, and the remaining $m - 2$ curves should be unimodal curves with their upper and lower tails asymptotically equal to zero. However, there is one anomaly of the multivariate logistic function that violates this pattern. When two like signed values of λ_k or α_k are equal, the IRCCC of each category will become asymptotic to a probability that falls between zero and one. The values of these asymptotes depend upon the location parameters. For example, in a three response category item with $\beta_1 = 1.5$, $\alpha_1 = 0.5$, $\beta_2 = 0.25$, $\alpha_2 = -1.0$, $\beta_3 = -1.0$, $\alpha_3 = 0.5$, the asymptotes for categories 1 and 3 were .22 and .78, respectively. In some sense, the examinees responded to these two categories as though they dealt with the same thing. Thissen and Steinberg (1988) provide an example and interpretation of this phenomenon in an actual data set.

The multivariate logistic function can be specialized to the case of two response categories as follows:

$$P_1(\theta_j) = \frac{e^{Z_1}}{e^{Z_1} + e^{Z_1}} \quad \text{and} \quad P_2(\theta_j) = \frac{e^{Z_2}}{e^{Z_1} + e^{Z_1}}, \tag{9.5}$$

where $Z_1 = Z_1(\theta_j)$ and $Z_2 = Z_2(\theta_j)$. In the case of a dichotomously scored item, let category 2 denote the correct response and category 1 denote the incorrect response. Then $P_2(\theta_j) = P(\theta_j)$ $P_1(\theta_j) = Q(\theta_j)$, where $P_2(\theta_j) + P_1(\theta_j) = P(\theta_j) + Q(\theta_j) = 1$. Now due to the restriction given in Equation 9.4, $Z_1 + Z_2 = 0$ and then $Z_1 = -Z_2$, which was the case for dichotomously scored items (see Figure 1.3). Substituting in Equation 9.5 yields

$$P_2(\theta_j) = \frac{e^{Z_2}}{e^{Z_1} + e^{Z_2}} = \frac{1}{1 + e^{(Z_1 - Z_2)}} = \frac{1}{1 + e^{-2Z_2}}. \tag{9.6}$$

However, in the case of dichotomously scored items, the logistic ogive model was defined as

$$P(\theta_j) = \frac{1}{1 + e^{-Z}}, \tag{9.7}$$

where

$$Z = \log\left[\frac{P(\theta_j)}{Q(\theta_j)}\right] = \log P(\theta_j) - \log Q(\theta_j) = \zeta + \lambda\theta_j.$$

Thus, multivariate logistic function does not reduce to the univariate logistic function. It can be seen that they differ by a factor of 2 in the exponent. As a result,

$$Z = \zeta + \lambda\theta_j = 2Z_2 = 2(\zeta_2 + \lambda_2\theta_j)$$

and

$$\zeta = 2\zeta_2 \quad \text{and} \quad \lambda = 2\lambda_2. \tag{9.8}$$

In the dichotomous response case, the location parameter is given by

$$\beta = -\frac{\zeta}{\lambda}$$

and, from the multivariate logistic function,

$$\beta = -\frac{2\zeta_2}{2\lambda_2} = -\frac{\zeta_2}{\lambda_2}.$$

Thus, the value of the location parameter is unaffected. The discrimination parameter α_2 will be half that of the univariate logistic. Despite these differences, the univariate and multivariate logistic functions can both be used as a model for dichotomously scored items. This relationship will appear again in Section 9.5, when extensions of the nominal response model are considered.

9.2 Maximum Likelihood Estimation of Item Parameters

As was the case with dichotomously scored and graded items, one way to estimate the item and ability parameters simultaneously is via the Birnbaum joint maximum likelihood estimation (JMLE) paradigm as was done in the LOGOG computer program (Kolakowski & Bock, 1973b). Thus, when estimating the item parameters via JMLE, one assumes that the items are independent and that the ability levels of the examinees are known. Again, each item will be handled separately, and only $2m$ item parameters need to be estimated simultaneously. When estimating ability, the item parameters for all n items are assumed known, and the ability level of each examinee can be estimated separately. The general framework for the JMLE of the item and ability parameters is the same as that in Chapter 4. Because of the nominal scoring and multiple response categories, the mathematical expressions are more complex, and matrix algebra will be used to simplify the overall results. In addition, simple examples employing an item with three response categories will be used to illustrate the underlying structure of the formulas.

Following Bock (1972) and Kolakowski and Bock (1973b), the ability scores will be grouped into G groups to facilitate the computational procedures. This requires the assumption that local independence holds in the neighborhood of the value θ_g used to represent the ability of the group. Thus, along the ability scale, there will be G groups with $f_1, f_2, \ldots, f_g, \ldots, f_G$ examinees in each group. No distributional assumptions are made with respect to these frequencies. Then, for a given item and a specific ability group g, the probability that $r_{g1}, r_{g2}, \ldots, r_{gk}, \ldots, r_{gm}$ out of the f_g subjects in the group will select response categories $1, 2, \ldots, k, \ldots, m$, respectively is

$$\text{Prob}(r_v | \theta_g) = \frac{f_g!}{\prod\limits_{v=1}^{m} r_{gv}!} \prod_{v=1}^{m} [P_v(\theta_g)]^{r_{gv}}.$$

Then, the likelihood of the $m \times G$ response matrix R is

$$\text{Prob}(R | \zeta_v, \lambda_v, \theta_g) = \prod_{g=1}^{G} \frac{f_g!}{\prod\limits_{v=1}^{m} r_{gv}!} \prod_{v=1}^{m} [P_v(\theta_g)]^{r_{gv}} \tag{9.9}$$

and the item log-likelihood is

$$L = \log[\text{Prob}(R | \zeta_v, \lambda_v, \theta_g)] = \text{constant} + \sum_{g=1}^{G} \sum_{v=1}^{m} r_{gv} \log P_v(\theta_g). \tag{9.10}$$

To simplify the notation, let $P_{gv} = P_v(\theta_g)$. At this point, there is a mathematical problem in that the P_{gv} are invariant with respect to a translation in the multivariate logits. This identification problem is handled by letting $\sum_{v=1}^{m} Z_v = 0$. This restriction must now be embedded in the definitions of the multivariate logits in order to estimate the item parameters. Bock (1972) derived the necessary procedures based upon the generalized procedures for nominal response (Bock, 1970). In addition, our interest will be restricted to a single item. The logits for a single item at a given ability level θ_g can be expressed as the elements of a row vector

$$\underset{1 \times m}{Z'} = \underset{1 \times 2}{K} \cdot \underset{2 \times m}{B} \cdot \underset{m \times m}{A}, \tag{9.11}$$

where

$$\underset{1 \times 2}{K} = \begin{bmatrix} 1 & \theta_g \end{bmatrix},$$

$$\underset{2 \times m}{B} = \begin{bmatrix} \zeta_1 & \zeta_2 & \cdots & \zeta_k & \cdots & \zeta_m \\ \lambda_1 & \lambda_2 & \cdots & \lambda_k & \cdots & \lambda_m \end{bmatrix},$$

and

$$
\underset{m\times m}{A} = \begin{bmatrix}
1-\frac{1}{m} & -\frac{1}{m} & \cdots & -\frac{1}{m} & \cdots & -\frac{1}{m} \\
-\frac{1}{m} & 1-\frac{1}{m} & \cdots & -\frac{1}{m} & \cdots & -\frac{1}{m} \\
\vdots & \vdots & \ddots & \vdots & \ddots & \vdots \\
-\frac{1}{m} & -\frac{1}{m} & \cdots & 1-\frac{1}{m} & \cdots & -\frac{1}{m} \\
\vdots & \vdots & & \ddots & \vdots & \ddots & \vdots \\
-\frac{1}{m} & \frac{1}{m} & \cdots & -\frac{1}{m} & \cdots & 1-\frac{1}{m}
\end{bmatrix},
$$

where A is the projection operator implementing the restriction of Equation 9.4. Unfortunately, this matrix is of less than full rank and must be reparameterized. This can be accomplished by defining a set of simple contrasts against an arbitrary response category, the first here. Let

$$
\underset{m\times(m-1)}{S} = \begin{bmatrix}
1 & 1 & 1 & \cdots & 1 \\
-1 & 0 & 0 & \cdots & 0 \\
0 & -1 & 0 & \cdots & 0 \\
0 & 0 & -1 & \cdots & 0 \\
\vdots & \vdots & \vdots & \ddots & \vdots \\
0 & 0 & 0 & \cdots & -1
\end{bmatrix}.
$$

Then, let

$$ A = ST $$

and reparameterize Equation 9.4 as follows:

$$ Z' = K(BS)T, $$

where

$$
\underset{(m-1)\times m}{T} = (S'S)^{-1}S'A = \begin{bmatrix}
\frac{1}{m} & \frac{1}{m}-1 & \frac{1}{m} & \cdots & \frac{1}{m} \\
\frac{1}{m} & \frac{1}{m} & \frac{1}{m}-1 & \cdots & \frac{1}{m} \\
\vdots & \vdots & \vdots & \ddots & \vdots \\
\frac{1}{m} & \frac{1}{m} & \frac{1}{m} & \cdots & \frac{1}{m}-1
\end{bmatrix}.
$$

Let

$$ \underset{1\times m}{Z'} = \underset{1\times m}{K} \cdot \underset{1\times 2}{} \underset{2\times(m-1)}{\Gamma} \cdot \underset{(m-1)\times m}{T} \tag{9.12} $$

and

$$
\Gamma = BS = \begin{bmatrix}
\gamma_1 & \gamma_2 & \cdots & \gamma_k & \cdots & \gamma_{m-1} \\
\xi_1 & \xi_2 & \cdots & \xi_k & \cdots & \xi_{m-1}
\end{bmatrix}.
$$

This reparameterization of the vector of multinomial logits, at a given ability level θ, has a number of interesting features. The new parameters γ_k and ξ_k[1] are contrasts in each of the original parameters ζ_k, λ_k. To illustrate this, let $m = 3$; then,

---

[1] The γ_k and ξ_k used here are not the elementary symmetric functions and situational parameters of the Rasch model in Chapter 5.

$$B = \begin{bmatrix} \zeta_1 & \zeta_2 & \zeta_3 \\ \lambda_1 & \lambda_2 & \lambda_3 \end{bmatrix} \quad \text{and} \quad S = \begin{bmatrix} 1 & 1 \\ -1 & 0 \\ 0 & -1 \end{bmatrix}.$$

Then,

$$\Gamma = BS = \begin{bmatrix} \zeta_1 - \zeta_2 & \zeta_1 - \zeta_3 \\ \lambda_1 - \lambda_2 & \lambda_1 - \lambda_3 \end{bmatrix} = \begin{bmatrix} \gamma_1 & \gamma_2 \\ \xi_1 & \xi_2 \end{bmatrix}.$$

Since the Newton-Raphson procedure will be used to implement the maximum likelihood estimation process, derivatives of the log-likelihood with respect to these new parameters, γ_v and ξ_v, are needed. The likelihood equations then are:

$$\frac{\partial L}{\partial \gamma_k} = \frac{\partial}{\partial \gamma_k} \sum_{g=1}^{G} (r_{g1} \log P_{g1} + r_{g2} \log P_{g2}$$

$$+ \cdots + r_{gk} \log P_{gk} + \cdots + r_{gm} \log P_{gm})$$

$$= \sum_{g=1}^{G} \left(\frac{r_{g1}}{P_{g1}} \frac{\partial P_{g1}}{\partial \gamma_k} + \frac{r_{g2}}{P_{g2}} \frac{\partial P_{g2}}{\partial \gamma_k} \right.$$

$$\left. + \cdots + \frac{r_{gk}}{P_{gk}} \frac{\partial P_{gk}}{\partial \gamma_k} + \cdots + \frac{r_{gm}}{P_{gm}} \frac{\partial P_{gm}}{\partial \gamma_k} \right)$$

$$\frac{\partial L}{\partial \xi_k} = \frac{\partial}{\partial \xi_k} \sum_{g=1}^{G} (r_{g1} \log P_{g1} + r_{g2} \log P_{g2}$$

$$+ \cdots + r_{gk} \log P_{gk} + \cdots + r_{gm} \log P_{gm})$$

$$= \sum_{g=1}^{G} \left(\frac{r_{g1}}{P_{g1}} \frac{\partial P_{g1}}{\partial \xi_k} + \frac{r_{g2}}{P_{g2}} \frac{\partial P_{g2}}{\partial \xi_k} \right.$$

$$\left. + \cdots + \frac{r_{gk}}{P_{gk}} \frac{\partial P_{gk}}{\partial \xi_k} + \cdots + \frac{r_{gm}}{P_{gm}} \frac{\partial P_{gm}}{\partial \xi_k} \right). \tag{9.13}$$

The derivatives of P_{gk} with respect to γ_k and ξ_k are needed. Recall that

$$P_{gk} = \frac{e^{Z_{gk}}}{e^{Z_{g1}} + e^{Z_{g2}} + \cdots + e^{Z_{gk}} + \cdots + e^{Z_{gm}}} = \frac{e^{Z_{gk}}}{\sum_{v=1}^{m} e^{Z_{gv}}} = \frac{e^{Z_{gk}}}{D}.$$

In general,

$$\frac{\partial P_{gv}}{\partial \gamma_k} = \frac{D \left(\partial e^{Z_{gv}} / \partial \gamma_k \right) - e^{Z_{gv}} \left(\partial D / \partial \gamma_k \right)}{D^2}.$$

Now, when $v = k$,

$$\frac{\partial P_{gk}}{\partial \gamma_k} = \frac{D \left(\partial e^{Z_{gk}} / \partial \gamma_k \right) - e^{Z_{gk}} \left(\partial D / \partial \gamma_k \right)}{D^2}$$

and, when $v \neq k$,

$$\frac{\partial P_{gv}}{\partial \gamma_k} = \frac{D \left(\partial e^{Z_{gv}} / \partial \gamma_k\right) - e^{Z_{gv}} \left(\partial D / \partial \gamma_k\right)}{D^2}.$$

Consequently, the derivatives of D and of the multivariate logit with respect to the parameters, γ_k and ξ_k, are needed. To do this, Equation 9.12 is expanded as follows:

$$\underset{1 \times m}{Z'} = \underset{1 \times 2}{\left[1 \ \theta_g\right]} \times \left[\begin{matrix} \gamma_1 \ \gamma_2 \cdots \gamma_k \cdots \gamma_{m-1} \\ \xi_1 \ \xi_2 \cdots \xi_k \cdots \xi_{m-1} \end{matrix}\right]_{2 \times (m-1)}$$

$$\times \left[\begin{matrix} \frac{1}{m} & \frac{1}{m} - 1 & \frac{1}{m} & \cdots & \frac{1}{m} \\ \frac{1}{m} & \frac{1}{m} & \frac{1}{m} - 1 & \cdots & \frac{1}{m} \\ \vdots & \vdots & \vdots & \ddots & \vdots \\ \frac{1}{m} & \frac{1}{m} & \frac{1}{m} & \cdots & \frac{1}{m} - 1 \end{matrix}\right]_{(m-1) \times m}$$

$$= \left[1 \ \theta_g\right]$$

$$\times \left[\begin{matrix} \frac{\sum \gamma_v}{m} & \frac{\sum \gamma_v}{m} - \gamma_1 & \cdots & \frac{\sum \gamma_v}{m} - \gamma_k & \cdots & \frac{\sum \gamma_{m-1}}{m} - \gamma_{m-1} \\ \frac{\sum \xi_v}{m} & \frac{\sum \xi_v}{m} - \xi_1 & \cdots & \frac{\sum \xi_v}{m} - \xi_k & \cdots & \frac{\sum \xi_{m-1}}{m} - \xi_{m-1} \end{matrix}\right],$$

where the sums are over the range $v = 1$ to $m - 1$. Then, for $k = 1$,

$$Z_{g1} = \frac{\sum \gamma_v}{m} + \frac{\sum \xi_v}{m} \theta_g$$

and, for $k = 2, 3, \ldots, m$,

$$Z_{gk} = \left(\frac{\sum \gamma_v}{m} - \gamma_{k-1}\right) + \left(\frac{\sum \xi_v}{m} - \xi_{k-1}\right) \theta_g.$$

The derivatives of the multivariate logits with respect to γ_k will be

$$\frac{\partial Z_{g1}}{\partial \gamma_1} = \frac{1}{m}, \quad \frac{\partial Z_{g2}}{\partial \gamma_1} = \frac{1}{m} - 1, \quad \frac{\partial Z_{g3}}{\partial \gamma_1} = \frac{1}{m}, \quad \cdots, \quad \frac{\partial Z_{gm}}{\partial \gamma_1} = \frac{1}{m},$$

$$\frac{\partial Z_{g1}}{\partial \gamma_2} = \frac{1}{m}, \quad \frac{\partial Z_{g2}}{\partial \gamma_2} = \frac{1}{m}, \quad \frac{\partial Z_{g3}}{\partial \gamma_2} = \frac{1}{m} - 1, \quad \cdots, \quad \frac{\partial Z_{gm}}{\partial \gamma_2} = \frac{1}{m},$$

and, in general, when $v \neq k + 1$,

$$\frac{\partial Z_{gv}}{\partial \gamma_k} = \frac{1}{m}, \tag{9.14}$$

but, when $v = k + 1$,

$$\frac{\partial Z_{gv}}{\partial \gamma_k} = \frac{1}{m} - 1.$$

Since $D = e^{Z_{g1}} + e^{Z_{g2}} + \cdots + e^{Z_{gk}} + \cdots + e^{Z_{gm}}$,

$$\frac{\partial D}{\partial \gamma_k} = e^{Z_{g1}}\frac{\partial Z_{g1}}{\partial \gamma_k} + e^{Z_{g2}}\frac{\partial Z_{g2}}{\partial \gamma_k} + \cdots + e^{Z_{gk}}\frac{\partial Z_{gk}}{\partial \gamma_k} + \cdots + e^{Z_{gm}}\frac{\partial Z_{gm}}{\partial \gamma_k}.$$

Substituting from Equation 9.14,

$$\frac{\partial D}{\partial \gamma_k} = \frac{e^{Z_{g1}}}{m} + \frac{e^{Z_{g2}}}{m} + \cdots + \frac{e^{Z_{gk}}}{m}$$
$$+ e^{Z_{g,k+1}}\left(\frac{1}{m} - 1\right) + \frac{e^{Z_{g,k+2}}}{m} + \cdots + \frac{e^{Z_{gm}}}{m} \tag{9.15}$$

$$= \frac{\sum_{v=1}^{m} e^{Z_{gv}}}{m} - e^{Z_{g,k+1}}$$

$$= \frac{D}{m} - e^{Z_{g,k+1}}.$$

Now, substituting Equations 9.14 and 9.15 in the derivatives of P_{gv}, we obtain, when $v \neq k+1$,

$$\frac{\partial P_{gv}}{\partial \gamma_k} = \frac{De^{Z_{gk}}(1/m) - e^{Z_{gk}}[(D/m) - e^{Z_{g,k+1}}]}{D^2} = P_{gk}P_{g,k+1} \tag{9.16}$$

and, when $v = k+1$,

$$\frac{\partial P_{g,k+1}}{\partial \gamma_k} = \frac{D(\partial e^{Z_{g,k+1}}/\partial \gamma_k) - e^{Z_{g,k+1}}(\partial D/\partial \gamma_k)}{D^2}$$

$$= \frac{De^{Z_{g,k+1}}[(1/m) - 1] - e^{Z_{g,k+1}}[(D/m) - e^{Z_{g,k+1}}]}{D^2}$$

$$= \frac{P_{g,k+1}}{m} - P_{g,k+1} - \frac{P_{g,k+1}}{m} + P_{g,k+1}P_{g,k+1}$$

$$= P_{g,k+1}(P_{g,k+1} - 1). \tag{9.17}$$

Finally, substituting in Equation 9.13 yields

$$\frac{\partial L}{\partial \gamma_k} = \sum_{g=1}^{G}\left\{\frac{r_{g1}}{P_{g1}}\left(P_{g1}P_{g,k+1}\right) + \frac{r_{g2}}{P_{g2}}\left(P_{g2}P_{g,k+1}\right) + \cdots + \frac{r_{gk}}{P_{gk}}\left(P_{gk}P_{g,k+1}\right)\right.$$

$$\left. + \frac{r_{g,k+1}}{P_{g,k+1}}\left[P_{g,k+1}\left(P_{g,k+1} - 1\right)\right] + \cdots + \frac{r_{gm}}{P_{gm}}\left(P_{gm}P_{g,k+1}\right)\right\}$$

$$= \sum_{g=1}^{G}\left[r_{g1}P_{g,k+1} + r_{g2}P_{g,k+1} + \cdots + r_{gk}P_{g,k+1}\right.$$

$$\left. + r_{g,k+1}\left(P_{g,k+1} - 1\right) + \cdots + r_{gm}P_{g,k+1}\right]$$

$$= \sum_{g=1}^{G}\left[\left(\sum_{v=1}^{m}r_{gv}P_{g,k+1}\right) - r_{g,k+1}\right]$$

$$= \sum_{g=1}^{G}\left(-r_{g,k+1} + f_g P_{g,k+1}\right). \tag{9.18}$$

Examining some particular cases,

$$\frac{\partial L}{\partial \gamma_1} = \sum_{g=1}^{G} (-r_{g2} + f_g P_{g2})$$

$$\frac{\partial L}{\partial \gamma_2} = \sum_{g=1}^{G} (-r_{g3} + f_g P_{g3})$$

$$\vdots$$

$$\frac{\partial L}{\partial \gamma_{m-1}} = \sum_{g=1}^{G} (-r_{gm} + f_g P_{gm}).$$

The fact that r_{g1} and P_{g1} do not appear in this vector is because of the simple contrasts in the S matrix used to reparameterize the item parameters in order to insure that $\sum P_k = 1$. Bock (1972) has presented these results in matrix algebra form, and these are shown below. First, however, let us define the vector

$$[R_{gv} - f_g \Omega_{gv}],$$

where

$$R'_{gv} = (\theta_{g1}, \theta_{g2}, \ldots, \theta_{gm}),$$
$$\Omega'_{gv} = (P_{g1}, P_{g2}, \ldots, P_{gm}),$$

and the vector

$$\gamma' = (\gamma_1, \gamma_2, \ldots, \gamma_{m-1}).$$

Then the likelihood equation for the vector of derivatives can be written as

$$\left[\frac{\partial L}{\partial \gamma_k}\right] = \sum_{g=1}^{G} T [R_{gv} - f_g \Omega_{gv}]. \tag{9.19}$$

The equivalence of this result with those given above is easily shown. Let $m = 3$; then,

$$T[R_{gv} - f_g \Omega_{gv}] = \begin{bmatrix} \frac{1}{m} & \frac{1}{m} - 1 & \frac{1}{m} \\ \frac{1}{m} & \frac{1}{m} & \frac{1}{m} - 1 \end{bmatrix} \cdot \begin{bmatrix} r_{g1} - f_g P_{g1} \\ r_{g2} - f_g P_{g2} \\ r_{g3} - f_g P_{g3} \end{bmatrix}$$

and

$$\frac{\partial L}{\partial \gamma_1} = \sum_{g=1}^{G} \left[\frac{1}{m}(r_{g1} - f_g P_{g1}) + \left(\frac{1}{m} - 1\right)(r_{g2} - f_g P_{g2}) + \frac{1}{m}(r_{g3} - f_g P_{g3})\right]$$

$$= \sum_{g=1}^{G} \left(\frac{r_{g1}}{m} - \frac{f_g P_{g1}}{m} + \frac{r_{g2}}{m} - \frac{f_g P_{g2}}{m} - r_{g2} + f_g P_{g2} + \frac{r_{g3}}{m} - \frac{f_g P_{g3}}{m}\right)$$

$$= \sum_{g=1}^{G} \left(\sum_{v=1}^{m} \frac{r_{gv}}{m} - \sum_{v=1}^{m} \frac{f_g P_{gv}}{m} - r_{g2} + f_g P_{g2} \right)$$

$$= \sum_{g=1}^{G} (-r_{g2} + f_g P_{g2}) .$$

Similar results can be obtained for the remaining derivatives.

The derivatives with respect to ξ_k will be of the same general form as those for γ_k with the addition of the θ_g multiplier. For example, when $v \neq k + 1$,

$$\frac{\partial Z_v}{\partial \xi_k} = \frac{\theta_g}{m}$$

and, when $v = k + 1$,

$$\frac{\partial Z_{k+1}}{\partial \gamma_k} = \left(\frac{1}{m} - 1 \right) \theta_g .$$

From Equation 9.18, the general derivative with respect to ξ_k can be written as

$$\frac{\partial L}{\partial \xi_k} = \sum_{g=1}^{G} \theta_g \left(-r_{g,k+1} + f_g P_{g,k+1} \right) \tag{9.20}$$

and the vector of derivatives as

$$\left[\frac{\partial L}{\partial \xi_k} \right] = \sum_{g=1}^{G} \theta_g T \left[-R_{gv} + f_g \Omega_{gv} \right] . \tag{9.21}$$

The second derivatives of the log-likelihood with respect to γ_k and ξ_k can be obtained from Equations 9.19 and 9.20. Since these involve many combinations, only the derivatives for a few cases will be shown; then the results due to Bock (1972) will be used. To illustrate the process,

$$\frac{\partial^2 L}{\partial \gamma_1 \partial \gamma_1} = \frac{\partial}{\partial \gamma_1} \left[\sum_{g=1}^{G} (-r_{g2} + f_g P_{g2}) \right] = \frac{\partial}{\partial \gamma_1} \sum_{g=1}^{G} (-r_{g2}) + \frac{\partial}{\partial \gamma_1} \sum_{g=1}^{G} f_g P_{g2}$$

$$= \sum_{g=1}^{G} f_g \frac{\partial P_{g2}}{\partial \gamma_1} = \sum_{g=1}^{G} [-P_{g2} (1 - P_{g2}) f_g]$$

$$\frac{\partial^2 L}{\partial \gamma_1 \partial \gamma_2} = \frac{\partial}{\partial \gamma_1} \left[\sum_{g=1}^{G} (-r_{g3} + f_g P_{g3}) \right] = \sum_{g=1}^{G} f_g \frac{\partial P_{g3}}{\partial \gamma_1} = \sum_{g=1}^{G} P_{g2} P_{g3} f_g$$

$$\vdots$$

$$\frac{\partial^2 L}{\partial \gamma_1 \partial \gamma_{m-1}} = \frac{\partial}{\partial \gamma_1} \left[\sum_{g=1}^{G} (-r_{gm} + f_g P_{gm}) \right] = \sum_{g=1}^{G} f_g \frac{\partial P_{gm}}{\partial \gamma_1} = \sum_{g=1}^{G} P_{g2} P_{gm} f_g .$$

It is of interest to note that the observed number of examinees selecting a given response category does not appear in the final expressions for the second derivative, as was also the case for dichotomous response under a logistic ICC model. Now, these results should be familiar as they are similar to the elements in the variance-covariance matrix used in the dichotomous case. Let us define a matrix of weights:

$$W_g = f_g \cdot \begin{bmatrix} P_{g1}(1 - P_{g1}) & -P_{g1}P_{g2} & -P_{g1}P_{g3} & \cdots & -P_{g1}P_{gm} \\ -P_{g2}P_{g1} & P_{g2}(1 - P_{g2}) & -P_{g2}P_{g3} & \cdots & -P_{g2}P_{gm} \\ \vdots & \vdots & \vdots & \ddots & \vdots \\ -P_{gm}P_{g1} & -P_{gm}P_{g2} & -P_{gm}P_{g3} & \cdots & P_{gm}(1 - P_{gm}) \end{bmatrix} \qquad (9.22)$$

Then, from Bock (1972), we have

$$\left[\frac{\partial^2 L}{\partial \gamma \partial \gamma} \right] = -\sum_{g=1}^{G} T W_g T'. \qquad (9.23)$$

Again, let us use a simple example, $m = 3$, to verify the results:

$$TWT' = f_g \cdot \begin{bmatrix} \frac{1}{m} & \frac{1}{m} - 1 & \frac{1}{m} \\ \frac{1}{m} & \frac{1}{m} & \frac{1}{m} - 1 \end{bmatrix}$$

$$\times \begin{bmatrix} P_{g1}(1 - P_{g1}) & -P_{g1}P_{g2} & -P_{g1}P_{g3} \\ -P_{g2}P_{g1} & P_{g2}(1 - P_{g2}) & -P_{g2}P_{g3} \\ -P_{g3}P_{g1} & -P_{g3}P_{g2} & P_{g3}(1 - P_{g3}) \end{bmatrix}$$

$$\times \begin{bmatrix} \frac{1}{m} & \frac{1}{m} \\ \frac{1}{m} - 1 & \frac{1}{m} \\ \frac{1}{m} & \frac{1}{m} - 1 \end{bmatrix}$$

$$= f_g \cdot \begin{bmatrix} \frac{1}{m} & \frac{1}{m} - 1 & \frac{1}{m} \\ \frac{1}{m} & \frac{1}{m} & \frac{1}{m} - 1 \end{bmatrix} \cdot \begin{bmatrix} P_{g1}P_{g2} & P_{g1}P_{g3} \\ -P_{g2}(1 - P_{g2}) & P_{g2}P_{g3} \\ P_{g3}P_{g2} & -P_{g3}(1 - P_{g3}) \end{bmatrix}$$

$$= f_g \cdot \begin{bmatrix} P_{g2}(1 - P_{g2}) & -P_{g2}P_{g3} \\ -P_{g3}P_{g2} & P_{g3}(1 - P_{g3}) \end{bmatrix} .$$

Then,

$$\frac{\partial^2 L}{\partial \gamma \partial \gamma} = -\sum_{g=1}^{G} f_g \cdot \begin{bmatrix} -P_{g2}(1 - P_{g2}) & P_{g2}P_{g3} \\ P_{g3}P_{g2} & -P_{g3}(1 - P_{g3}) \end{bmatrix} .$$

By analogy,

$$\left[\frac{\partial^2 L}{\partial \gamma \partial \xi} \right] = \sum_{g=1}^{G} \theta_g T W T'$$

and

$$\left[\frac{\partial^2 L}{\partial \xi \partial \xi}\right] = -\sum_{g=1}^{G} \theta_g^2 TWT'.$$

Then, for a single item with m response categories, the $(m-1) \times (m-1)$ matrix of second derivatives Ψ will be of the form shown below and is actually the Kronecker product of

$$\begin{bmatrix} 1 & \theta_g \\ \theta_g & \theta_g^2 \end{bmatrix} \otimes \begin{bmatrix} TWT' \end{bmatrix}.$$

The Newton-Raphson equation to be solved iteratively will be

$$\begin{bmatrix} \hat{\gamma}_1 \\ \hat{\gamma}_2 \\ \vdots \\ \hat{\gamma}_{m-1} \\ \hat{\xi}_1 \\ \hat{\xi}_2 \\ \vdots \\ \hat{\xi}_{m-1} \end{bmatrix}_{(t+1)} = \begin{bmatrix} \hat{\gamma}_1 \\ \hat{\gamma}_2 \\ \vdots \\ \hat{\gamma}_{m-1} \\ \hat{\xi}_1 \\ \hat{\xi}_2 \\ \vdots \\ \hat{\xi}_{m-1} \end{bmatrix}_{(t)} - \hat{\Psi}_{(t)}^{-1} \cdot \begin{bmatrix} \partial L/\partial \gamma_1 \\ \partial L/\partial \gamma_2 \\ \vdots \\ \partial L/\partial \gamma_{m-1} \\ \partial L/\partial \xi_1 \\ \partial L/\partial \xi_2 \\ \vdots \\ \partial L/\partial \xi_{m-1} \end{bmatrix}_{(t)},$$

where Ψ and the vector of first derivatives are evaluated using the estimated values of the parameters at iteration t. Once the estimates of these parameters are obtained, they can be used to find the values of the estimates of original parameters ζ_k, λ_k for the item via

$$\Gamma = BS = \begin{bmatrix} \zeta_1 & \zeta_2 & \cdots & \zeta_m \\ \lambda_1 & \lambda_2 & \cdots & \lambda_m \end{bmatrix} \cdot \begin{bmatrix} 1 & 1 & 1 & \cdots & 1 \\ -1 & 0 & 0 & \cdots & 0 \\ 0 & -1 & 0 & \cdots & 0 \\ 0 & 0 & -1 & \cdots & 0 \\ \vdots & \vdots & \vdots & \ddots & \vdots \\ 0 & 0 & 0 & \cdots & -1 \end{bmatrix}$$

and

$$C = BS$$
$$CS'(SS')^{-1} = B(SS')(SS')^{-1}$$
$$CS'(SS')^{-1} = B$$

which can be solved for the B since the elements in Γ and S are known. In the case in which $m = 3$,

$$\begin{bmatrix} \gamma_1 & \gamma_2 \\ \xi_1 & \xi_2 \end{bmatrix} = \begin{bmatrix} \zeta_1 - \zeta_2 & \zeta_1 - \zeta_3 \\ \lambda_1 - \lambda_2 & \lambda_1 - \lambda_3 \end{bmatrix}.$$

Equating by cells yields

$$\gamma_1 = \zeta_1 - \zeta_2$$
$$\gamma_2 = \zeta_1 - \zeta_3$$

and

$$\zeta_2 = \zeta_1 - \gamma_1$$
$$\zeta_3 = \zeta_1 - \gamma_2.$$

But

$$\sum_{k=1}^{m} \zeta_k = 0 \quad \text{and} \quad \sum_{k=1}^{m} \lambda_k = 0$$

and, then,

$$\zeta_1 + \zeta_2 + \zeta_3 = \zeta_1 + \zeta_1 - \gamma_1 + \zeta_1 - \gamma_2 = 0$$
$$3\zeta_1 = \gamma_1 + \gamma_2$$
$$\zeta_1 = \frac{\gamma_1 + \gamma_2}{3}.$$

Similarly,

$$\zeta_2 = \frac{\gamma_1 + \gamma_2}{3} - \gamma_1 = \frac{\gamma_1 + \gamma_2 - 3\gamma_1}{3} = \frac{\gamma_2 - 2\gamma_1}{3}$$
$$\zeta_3 = \frac{\gamma_1 + \gamma_2}{3} - \gamma_2 = \frac{\gamma_1 + \gamma_2 - 3\gamma_2}{3} = \frac{\gamma_1 - 2\gamma_2}{3}$$

and

$$\lambda_1 = \frac{\xi_1 + \xi_2}{3} \quad \lambda_2 = \frac{\xi_2 - 2\xi_1}{3} \quad \lambda_3 = \frac{\xi_1 - 2\xi_2}{2}.$$

Since these are in the regression form, it would be useful for interpretation purposes to convert them to item parameter form and the following results:

$$\hat{\beta}_1 = -\frac{\hat{\zeta}_1}{\hat{\lambda}_1} \quad \hat{\beta}_2 = -\frac{\hat{\zeta}_2}{\hat{\lambda}_2} \quad \hat{\beta}_3 = -\frac{\hat{\zeta}_3}{\hat{\lambda}_3}$$

$$\hat{\alpha}_1 = \hat{\lambda}_1 \quad \hat{\alpha}_2 = \hat{\lambda}_2 \quad \hat{\alpha}_3 = \hat{\lambda}_3.$$

9.3 Maximum Likelihood Estimation of Ability

In the present context, it has been assumed that the Birnbaum paradigm will be used to estimate both the item and ability parameters when the items are nominally scored. When the item parameters were estimated the items were dealt with one at a time over all examinees. In order to estimate ability, a single ability level over all items is of interest. In order to keep the mathematics consistent with that of Bock (1972), the item parameters γ_{iv}, ξ_{iv} obtained after reparameterization will be used:

$$\text{Prob}([u_{ijv}]|\theta_j, \zeta_k, \lambda_k) = \prod_{j=1}^{N}\prod_{i=1}^{n}\prod_{v=1}^{m} P_{iv}(\theta_j)^{u_{ijv}},$$

where $[u_{ijv}]$ is the $n \times N \times m$ matrix of item response category choices u_{ijv}. When examinee j chooses category k of item i, $u_{ijk} = 1$ and is zero otherwise. The log-likelihood is

$$L = \log\{\text{Prob}([u_{ijv}]|\theta_j, \zeta_k, \lambda_k)\} = \sum_{j=1}^{N}\sum_{i=1}^{n}\sum_{v=1}^{m} u_{ijv} \log P_{iv}(\theta_j). \qquad (9.24)$$

Since the Newton-Raphson procedure is used to obtain the maximum likelihood estimate of ability, the first and second derivatives of the log-likelihood with respect to ability are needed. Following the approach taken in previous chapters, the ability estimates will be obtained with respect to an examinee's ability rather than the ability level of a group as was done by Bock (1972). The first derivative of the log-likelihood with respect to the ability of an arbitrary examinee is

$$\frac{\partial L}{\partial \theta_j} = \sum_{i=1}^{n}\sum_{v=1}^{m} \frac{u_{ijv}}{P_{ijv}} \frac{\partial P_{ijv}}{\partial \theta_j},$$

where $P_{ijv} = P_{iv}(\theta_j)$ for notational simplicity. Because an examinee can select only one item response category per item, the $\sum_{v=1}^{m}$ is not needed; and the log-likelihood can be written as

$$L = \sum_{j=1}^{N}\sum_{i=1}^{n} u_{ijv} \log P_{ijv}. \qquad (9.25)$$

Because of the reparameterization of the item parameters, the algebra of the derivatives is best expressed in terms of vectors and matrices. The derivatives of a single item will be used to show the results in algebraic form for a simple case. From Bock (1972),

$$\frac{\partial L}{\partial \theta_j} = \sum_{i=1}^{n} \xi_i' T_i (U_{iv} - \Omega_{iv}), \qquad (9.26)$$

where $U_{iv} = [u_{1jv}, u_{2jv}, \ldots, u_{njv}]$ is the vector of item response category choice indications of subject j. Upon expansion, Equation 9.26 becomes

$$\frac{\partial L}{\partial \theta_j} = -\sum_{i=1}^{n} [\xi_{i1}(u_{i2} - P_{i2}) + \xi_{i2}(u_{i3}' - P_{i3}) + \cdots + \xi_{i,m-1}(u_{im} - P_{im})] \quad (9.27)$$

which parallels the structure of the derivative with respect to ξ_{iv} presented in Equation 9.21.

The vector of second derivatives with respect to ability is given by

$$\frac{\partial^2 L}{\partial \theta_j \partial \theta_j'} = -\sum_{i=1}^{n} \xi_i' T_i W_{ij} T_i' \xi_i \quad \text{for } j = j'. \qquad (9.28)$$

Since the examinees are independent, this derivative will be zero when $j \neq j'$. In order to illustrate the features of the derivative, the expression for a single item having three response categories will be expanded:

$$
\begin{bmatrix} \xi_{i1} & \xi_{i2} \end{bmatrix} \cdot \begin{bmatrix} \frac{1}{m}\frac{1}{m} - 1 & \frac{1}{m} \\ \frac{1}{m} & \frac{1}{m} & \frac{1}{m} - 1 \end{bmatrix}
$$

$$
\times \begin{bmatrix} u_{ij1}P_{i1}(1 - P_{i1}) & -u_{ij1}P_{i1}P_{i2} & -u_{ij1}P_{i1}P_{i3} \\ -u_{ij2}P_{i2}P_{i1} & u_{ij2}P_{i2}(1 - P_{i2}) & -u_{ij2}P_{i2}P_{i3} \\ -u_{ij3}P_{i3}P_{i1} & -u_{ij3}P_{i3}P_{i2} & u_{ij3}P_{i3}(1 - P_{i3}) \end{bmatrix}
$$

$$
\times \begin{bmatrix} \frac{1}{m} & \frac{1}{m} \\ \frac{1}{m} - 1 & \frac{1}{m} \\ \frac{1}{m} & \frac{1}{m} - 1 \end{bmatrix} \cdot \begin{bmatrix} \xi_{i1} \\ \xi_{i2} \end{bmatrix}.
$$

While this can be evaluated rather easily via a computer program, the expansion will be carried out algebraically to show the internal structure of the result. Once the structure is seen, generalization to a larger number of response categories is direct.

Start by multiplying the T matrix and ξ vectors on the right and left to simplify the expression. This yields

$$
\begin{bmatrix} \frac{1}{m}(\xi_{i1} + \xi_{i2}) & \left(\frac{1}{m} - 1\right)\xi_{i1} + \frac{1}{m}\xi_{i2} & \frac{1}{m}\xi_{i1} + \left(\frac{1}{m} - 1\right)\xi_{i2} \end{bmatrix}
$$

$$
\times W \cdot \begin{bmatrix} \frac{1}{m}(\xi_{i1} + \xi_{i2}) \\ \left(\frac{1}{m} - 1\right)\xi_{i1} + \frac{1}{m}\xi_{i2} \\ \frac{1}{m}\xi_{i1} + \left(\frac{1}{m} - 1\right)\xi_{i2} \end{bmatrix}. \tag{9.29}
$$

Multiplying W by the vector on the right yields

$$
\begin{bmatrix} u_{ij1}(\xi_{i1}P_{i1}P_{i2} + \xi_{i2}P_{i1}P_{i3}) \\ u_{ij2}[-\xi_{i1}P_{i2}(1 - P_{i2}) + \xi_{i2}P_{i2}P_{i3}] \\ u_{ij3}[\xi_{i1}P_{i3}P_{i2} + \xi_{i2}P_{i3}(1 - P_{i3})] \end{bmatrix}.
$$

Then, multiplying the vector by the vector on the left in Equation 9.29 gives

$$
\frac{u_{ij1}}{m}\left(\xi_{i1}^2 P_{i1}P_{i2} + \xi_{i1}\xi_{i2}P_{i1}Q_{i1} + \xi_{i2}^2 P_{i1}P_{i3}\right)
$$

$$
+ u_{ij2}\left[\frac{1}{m}\left(-\xi_{i1}^2 P_{i2}Q_{i2} - \xi_{i1}\xi_{i2}P_{i2}P_{i1} + \xi_{i2}^2 P_{i2}P_{i3}\right) + \xi_{i1}^2 P_{i2}Q_{i2} - \xi_{i1}\xi_{i2}P_{i2}P_{i3}\right]
$$

$$
+ u_{ij3}\left[\frac{1}{m}\left(\xi_{i1}^2 P_{i2}P_{i3} - \xi_{i1}\xi_{i2}P_{i1}P_{i3} - \xi_{i2}^2 P_{i3}Q_{i3}\right) + \xi_{i2}^2 P_{i3}Q_{i3} - \xi_{i1}\xi_{i2}P_{i2}P_{i3}\right]
$$

which is then summed over items to yield the second derivative of the log-likelihood with respect to an examinee's ability. Although this expression is rather lengthy it should be remembered that only one of the u_{ijk} is nonzero so that two terms always drop for a given item.

Since observed data appear in the second derivative, its expectation must be taken, and the u_{ijk} are replaced by the P_{ijk}. The Fisher scoring equation will be

$$\left[\hat{\theta}_j\right]_{(t+1)} = \left[\hat{\theta}_j\right]_{(t)} - \left[\frac{\partial L/\partial\theta_j}{E(\partial^2 L/\partial\theta_j^2)}\right]_{(t)} \tag{9.30}$$

and is identical in structure to that used previously. Again, it would be solved iteratively until $\Delta\hat{\theta}_j$ is sufficiently small. Consequently, it is possible to estimate the ability level of an examinee, even though it is not possible to obtain an observed test score for an examinee. When the P_{igv} is extreme for an examinee and the elements of W_{ig} are small for all items, the numerator in the second term of Equation 9.30 approaches unity while the denominator is small. In this situation, the conditions necessary for obtaining a maximum likelihood estimate may not be met. Bock (1972) recommends removing such examinees from the sample and repeating the overall estimation process.

At this point, the two basic building blocks of the parameter estimation procedure for nominal response items are at hand. Appendix G presents computer programs written in MICROSOFT QUICKBASIC for each of these and an analysis of an example three-response-category data set. Again, enough computational detail will be shown to illustrate the estimation processes.

9.4 The Information Functions

The general approach to the information functions for a nominally scored item closely follows that for the graded response case. The item information function is given by

$$I_i(\theta) = \sum_{v=1}^{m_i} I_v(\theta) P_v(\theta), \tag{9.31}$$

where i denotes the item, $I_v(\theta)$ is the amount of information associated with a particular item response category, and $P_v(\theta)$ is the probability of an examinee of ability θ selecting response category v. As was the case with the graded response case, the quantity $I_k(\theta)P_k(\theta)$ is the amount of information share of category k. The information function of an item response category is given by

$$I_k(\theta) = -\frac{\partial^2 \log P_k(\theta)}{\partial\theta^2} = -\frac{\partial}{\partial\theta}\left[\frac{P_k'(\theta)}{P_k(\theta)}\right], \tag{9.32}$$

where $P_k'(\theta) = \partial P_k(\theta)/\partial\theta$. Then,

$$I_k(\theta) = \frac{[P_k'(\theta)]^2 - P_k(\theta)P_k''(\theta)}{[P_k(\theta)]^2}, \tag{9.33}$$

where $P_k''(\theta) = \partial^2 P_k(\theta)/\partial\theta^2$. The item response information share is

$$I_k(\theta)P_k(\theta) = \frac{[P_k'(\theta)]^2}{P_k(\theta)} - P_k''(\theta). \tag{9.34}$$

When Equation 9.34 is substituted in Equation 9.31, the item information function then becomes

$$I_i(\theta) = \sum_{k=1}^{m_i} \left\{ \frac{[P_k'(\theta)]^2}{P_k(\theta)} - P_k''(\theta) \right\}. \tag{9.35}$$

The necessary derivatives have been provided by Vale and Weiss (1977) and are derived below. From Equation 9.2, we have

$$P_k(\theta) = \frac{e^{Z_k}}{D}.$$

Then, with $m = m_i$ for a simplification purpose,

$$\frac{\partial P_k(\theta)}{\partial\theta} = \frac{D(\partial e^{Z_k}/\partial\theta) - e^{Z_k}(\partial D/\partial\theta)}{D^2}$$

$$= \frac{D\lambda_k e^{Z_k} - e^{Z_k}\sum_{v=1}^{m}\lambda_v e^{Z_v}}{D^2}$$

$$= \frac{e^{Z_k}[(e^{Z_1} + e^{Z_2} + \cdots + e^{Z_m})\lambda_k - (\lambda_1 e^{Z_1} + \lambda_2 e^{Z_2} + \cdots + \lambda_m e^{Z_m})]}{D^2}$$

$$= \frac{e^{Z_k}(e^{Z_1}(\lambda_k - \lambda_1) + e^{Z_k}(e^{Z_2}(\lambda_k - \lambda_2) + \cdots + e^{Z_k}(e^{Z_m}(\lambda_k - \lambda_m)}{D^2},$$

and

$$P_k'(\theta) = \frac{e^{Z_k}\sum_{v=1}^{m}e^{Z_v}(\lambda_k - \lambda_v)}{D^2}. \tag{9.36}$$

Given this result, the second derivative is

$$P_k''(\theta) = \frac{\partial P_k'(\theta)}{\partial\theta}$$

$$= \frac{D^2\dfrac{\partial}{\partial\theta}\left[e^{Z_k}\sum_{v=1}^{m}e^{Z_v}(\lambda_k - \lambda_v)\right] - \left[e^{Z_k}\sum_{v=1}^{m}e^{Z_v}(\lambda_k - \lambda_v)\right]\dfrac{\partial D^2}{\partial\theta}}{D^4}.$$

Then, by parts, for part 1,

$$\frac{\partial}{\partial\theta}\left[e^{Z_k}\sum_{v=1}^{m}e^{Z_v}(\lambda_k - \lambda_v)\right]$$

$$= e^{Z_k} \frac{\partial}{\partial \theta} \left[\sum_{v=1}^{m} e^{Z_v} (\lambda_k - \lambda_v) \right] + \sum_{v=1}^{m} e^{Z_v} (\lambda_k - \lambda_v) \frac{\partial e^{Z_k}}{\partial \theta}$$

$$= e^{Z_k} \left(\sum_{v=1}^{m} \lambda_k \lambda_v e^{Z_v} - \sum_{v=1}^{m} \lambda_v^2 e^{Z_v} \right) + \left[\sum_{v=1}^{m} e^{Z_v} (\lambda_k - \lambda_v) \right] \lambda_k e^{Z_k}$$

$$= e^{Z_k} \left(\sum_{v=1}^{m} \lambda_k \lambda_v e^{Z_v} - \sum_{v=1}^{m} \lambda_v^2 e^{Z_v} + \sum_{v=1}^{m} e^{Z_v} \lambda_k^2 - \sum_{v=1}^{m} \lambda_k \lambda_v e^{Z_v} \right)$$

$$= e^{Z_k} \sum_{v=1}^{m} e^{Z_v} (\lambda_k^2 - \lambda_v^2).$$

For part 2, we need

$$\frac{\partial D^2}{\partial \theta^2} = 2D \frac{\partial D}{\partial \theta}$$

$$= 2D \left(\lambda_1 e^{Z_1} + \lambda_2 e^{Z_2} + \cdots + \lambda_m e^{Z_m} \right)$$

$$= 2D \sum_{v=1}^{m} \lambda_v e^{Z_v}.$$

Then, substituting yields

$$P_k''(\theta) = \frac{D^2 \left[e^{Z_k} \sum_{v=1}^{m} e^{Z_v} (\lambda_k^2 - \lambda_v^2) \right] - \left[e^{Z_k} \sum_{v=1}^{m} e^{Z_v} (\lambda_k - \lambda_v) \right] 2D \sum_{v=1}^{m} \lambda_v e^{Z_v}}{D^4},$$

and

$$P_k''(\theta) = \frac{e^{Z_k} \left\{ D \sum_{v=1}^{m} e^{Z_v} (\lambda_k^2 - \lambda_v^2) - 2 \left[\sum_{v=1}^{m} e^{Z_v} (\lambda_k - \lambda_v) \right] \sum_{v=1}^{m} \lambda_v e^{Z_v} \right\}}{D^3}. \qquad (9.37)$$

While these terms in Equation 9.33 and 9.34 are complex, they are very regular and, hence, it is easy to compute them or at least to write a computer program to compute them.

Bock (1972) has presented the information functions in matrix algebra form. The item information function at ability level θ_j is given by

$$I_i(\theta_j) = \lambda_i' W_i \lambda_i \sum_{v=1}^{m} \Omega_{iv}(\theta_j) = \lambda_i' W_i \lambda_i, \qquad (9.38)$$

where

$$\lambda_i' = \left[\lambda_1 \; \lambda_2 \; \cdots \; \lambda_k \; \cdots \; \lambda_m \right],$$

$$\Omega_{iv}(\theta_j) = \left[P_{j1} \; P_{j2} \; \cdots \; P_{jk} \; \cdots \; P_{jm} \right],$$

and

$$W_i = \begin{bmatrix} P_{j1}(1-P_{j1}) & -P_{j1}P_{j2} & \cdots & -P_{j1}P_{jk} & \cdots & -P_{j1}P_{jm} \\ -P_{j2}P_{j1} & P_{j2}(1-P_{j2}) & \cdots & -P_{j2}P_{jk} & \cdots & -P_{j2}P_{jm} \\ \vdots & \vdots & \ddots & \vdots & \ddots & \vdots \\ -P_{jk}P_{j1} & -P_{jk}P_{j2} & \cdots & P_{jk}(1-P_{jk}) & \cdots & -P_{jk}P_{jm} \\ \vdots & \vdots & \ddots & \vdots & \ddots & \vdots \\ -P_{jm}P_{j1} & -P_{jm}P_{j2} & \cdots & P_{jm}P_{jk} & \cdots & P_{jm}(1-P_{jm}) \end{bmatrix}.$$

The item information share of an item response category is given by

$$I_{iv}(\theta_j) = \lambda_i' W_i \lambda_i \Omega_{iv}(\theta_j) \tag{9.39}$$

which can (after much tedious algebra) be shown to be the same as the expression in Equation 9.34.

Table 9.2 contains the numerical values of the item information function and the amount of item information shares for the item of Table 9.1. Figure 9.2 is a plot of these values to graphically display the information functions.

Table 9.2. Evaluation of the Multivariate Logistic Function with Three Response Categories

θ	Item Information Share			Item Information
	Category 1	Category 2	Category 3	
-3.0	0.0020	0.0339	0.0002	0.0362
-2.5	0.0048	0.0588	0.0008	0.0644
-2.0	0.0117	0.1058	0.0037	0.1212
-1.5	0.0279	0.1865	0.0161	0.2305
-1.0	0.0587	0.2907	0.0617	0.4111
-0.5	0.0940	0.3451	0.1802	0.6193
0.0	0.0981	0.2666	0.3424	0.7071
0.5	0.0624	0.1256	0.3967	0.5847
1.0	0.0262	0.0391	0.3040	0.3694
1.5	0.0085	0.0094	0.1795	0.1974
2.0	0.0024	0.0020	0.0928	0.0972
2.5	0.0006	0.0004	0.0454	0.0465
3.0	0.0002	0.0001	0.0219	0.0222

It can be seen in Figure 9.2 that the information share of each response category is a unimodal curve that is not perfectly symmetrical. Response category 1 reaches a maximum of about 0.11 at roughly $\theta = -0.20$, category 2 has a maximum of about 0.35 at roughly $\theta = -0.60$, and category 3 reaches a maximum of about 0.40 at roughly $\theta = 0.45$. The item information curve is a not quite symmetric curve that is somewhat leptokurtic with a maximum of about 0.71 at an ability level slightly below zero.

The total amount of information yielded by a nominally scored item will be greater than that yielded by the same item when it is dichotomously scored. The item of Table 9.1 and Figure 9.1 has been reanalyzed with category 3 considered the correct response and the remaining two categories

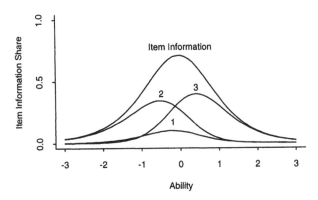

Fig. 9.2. Item information shares and item information function for the items of Table 9.1.

pooled into an incorrect response category. The obtained parameter estimates were $\hat{\beta} = 0.065$ and $\hat{\alpha} = 1.6$. The item information functions corresponding to these two scoring procedures are shown in Figure 9.3. The maximum amount of information under nominal scoring is about 0.71 while the maximum under binary scoring is roughly 0.64. The two information curves cross at about $\theta = 1.0$ and above this ability level the binary item yields slightly more information. For ability levels less than one, the nominally scored item uniformly yields a greater amount of information. The overall similarity of these two item information functions reflects the fact that category 1 has a rather small probability of being selected at any ability level. As a result, in this example the normally scored item is behaving much like a binary item. It should be noted that the item information for category 3 is uniformly larger when dichotomously scored than when treated as one of the nominally scored categories.

9.5 Extensions of Bock's Nominal Response Model

As was the case with dichotomously scored items, guessing can be involved in an examinees choice of nominal response categories. Samejima (1979) introduced an extension of Bock's nominal response model that incorporated a "guessing" component. Under this model, some proportion c_k chose each of the response categories by guessing, and the number of examinees choosing a response category includes these persons. The IRCCC model is

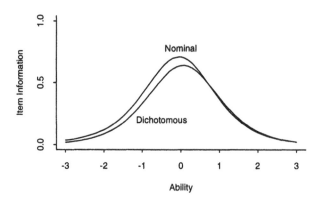

Fig. 9.3. Item information functions for the item of Table 9.1 when scored nominally and dichotomously; category 3 considered as correct.

$$P(u_{ijk} = k|\theta, \lambda_v, \zeta_v) = \frac{\exp(\zeta_k + \lambda_k\theta) + c_k \exp(\zeta_0 + \lambda_0\theta)}{\displaystyle\sum_{v=0}^{m_i} \exp(\zeta_v + \lambda_v\theta)}, \tag{9.40}$$

where c_k was fixed at $1/m_i$ and not estimated. An additional "don't know" category, denoted by 0, was added having parameters λ_0 and ζ_0 to model the "guessing" behavior. The logic of this model is that the low-ability examinees assign their responses at random to the m_i response categories with equal probability. Thissen and Steinberg (1984) found that this assumption was not supported empirically and developed an extension of the model in which the c_k were a function of other parameters c_k^* that were estimated. Under this approach,

$$c_k = \frac{e^{c_k^*}}{\displaystyle\sum_{v=1}^{m_i} e^{c_v^*}}, \qquad \text{where} \qquad \sum_{v=1}^{m_i} c_v^* = 0.$$

In the MULTILOG manual (Thissen, 1991), this extension is called the Bock-Samejima (BS) model, and the IRCCC are defined by

$$P(u_{ijk} = k) = \frac{h^* \exp(\zeta_k + \lambda_k\theta) + hc_k \exp(\zeta_1 + \lambda_1\theta)}{\displaystyle\sum_{v=1}^{m+1} \exp(\zeta_v + \lambda_v\theta)}. \tag{9.41}$$

Here response category 1 denotes the "don't know" response that is distributed over the remaining response categories. While this model appears attractive, there are a large number of parameters to be estimated. Thissen and Steinberg (1984) indicate that estimation of them is fraught with problems. Perhaps this is another case of overparameterization.

The control codes h^* and h are used to particularize this general model and are set by the INFORLOG preprocessor program for the MULTILOG program. When $m > 2$, h^* is always set to unity. If both h^* and h are set to unity Equation 9.41 yields Thissen and Steinberg's (1984) multiple choice model. In addition, if all the c_k are set to $1/m_i$, Samejima's (1979) model is obtained. When $m > 2$, $h^* = 1$ and $h = 0$, Equation 9.41 yields Bock's (1972) nominal response model.

When there are only two response categories, we have the dichotomous response case. Let category 1 denote the incorrect response and category 2 denote the correct response. In this case, $m = 2$, and h^* for category 1 is set to zero and h^* for category 2 is unity. Then Equation 9.41 particularizes to

$$P(u_{ijk} = 2) = \frac{\exp(\zeta_2 + \lambda_2\theta) + c_2\exp(\zeta_1 + \lambda_1\theta)}{\displaystyle\sum_{v=1}^{2}\exp(\zeta_v + \lambda_v\theta)}$$

$$= \frac{\exp(\zeta_2 + \lambda_2\theta) + c_2\exp(\zeta_1 + \lambda_1\theta)}{\exp(\zeta_1 + \lambda_1\theta) + \exp(\zeta_2 + \lambda_2\theta)}.$$

But, under a nominal response model, $\lambda_1 = -\lambda_2$ and $\zeta_1 = -\zeta_2$; hence,

$$P(u_{ijk} = 2) = \frac{\exp(\zeta_2 + \lambda_2\theta) + c_2\exp[-(\zeta_2 + \lambda_2\theta)]}{\exp[-(\zeta_2 + \lambda_2\theta)] + \exp(\zeta_2 + \lambda_2\theta)}$$

$$= \frac{1 + c_2\exp[-2(\zeta_2 + \lambda_2\theta)]}{\exp[-2(\zeta_2 + \lambda_2\theta)] + 1}$$

$$= \frac{c_2\exp[-2(\zeta_2 + \lambda_2\theta)] + 1}{1 + \exp[-2(\zeta_2 + \lambda_2\theta)]}$$

$$= \frac{c_2\{1 + \exp[-2(\zeta_2 + \lambda_2\theta)]\} + 1 - c_2}{1 + \exp[-2(\zeta_2 + \lambda_2\theta)]}$$

and, then,

$$P(u_{ijk} = 2) = c_2 + (1 - c_2)\left\{\frac{1}{1 + \exp[-2(\zeta_2 + \lambda_2\theta)]}\right\}. \tag{9.42}$$

This expression is equivalent to the usual three-parameter ICC model with $c_2 = c$, and the term in the bracket is the two-parameter model based on the multivariate logistic function (see Equation 9.6).

From the above, it can be seen that the Thissen and Steinberg (1984) extension of the Bock-Samejima models provides a general equation for the IRCCC that can encompass the one-, two-, and three-parameter logistic ICC models for dichotomous response, as well as several variants of the nominal response case. The MULTILOG computer program capitalizes on this generality to encompass a range of models within the framework of a single computer program that employs marginal maximum likelihood estimation.

9.6 The Relation of Nominally Scored Items and Logit-Linear Models

In the sections above, the multivariate logistic function and the associated maximum likelihood estimation procedures were presented within the context of IRT. The material presented actually represents a particular application of a more general model for analyzing qualitative data due to Bock (1970, 1975, Chap. 8) and Grizzle, Starmer, and Koch (1969) known as logit-linear models. These, in turn, are equivalent to a recently developed area of statistics known as log-linear models. The book by Bishop, Fienberg, and Holland (1975) contains a comprehensive presentation of the log-linear approach from a survey research orientation. (See also Fienberg, 1977; Everitt, 1977; and Haberman, 1978). Although the log-linear and the logit-linear approaches will yield identical results and can be related algebraically, they do involve somewhat different orientations to the analysis of qualitative data (see Baker, 1981). The interest here is not in comparing the logit-linear and log-linear approaches but in showing that the estimation procedures for nominally scored items are a special case of a more general model.

A basic feature of the logit-linear approach is the separation of explanatory variables from response variables in both the conceptualization of a problem and in the analysis procedures. For example, assume that a population can be divided into r mutually exclusive and exhaustive groups on the basis of a single classificatory variable. Within a sample of size M_j from each group, the responses to an item of a questionnaire can be allocated to c mutually exclusive and exhaustive categories. The result would be an $r \times c$ contingency table such as the one shown in Table 9.3. Since the number of subjects in each group is determined by the sample design, the row marginals are fixed while the column marginals are free to reflect the response data. Bock (1975) assumed the proportion of responses of a given group that falls into each of the response categories is given by the multivariate logistic function, Equation 9.1.

Table 9.3. An $r \times c$ Contingency Table

Stratification Variable Level	Questionnaire Item Response Category					Group Size
	1	2	3	$\cdots$	c	
1	f_{11}	f_{12}	f_{13}	$\cdots$	f_{1c}	M_1
2	f_{21}	f_{22}	f_{23}	$\cdots$	f_{2c}	M_2
3	f_{31}	f_{32}	f_{33}	$\cdots$	f_{3c}	M_3
$\vdots$	$\vdots$	$\vdots$	$\vdots$	$\ddots$	$\vdots$	$\vdots$
r	f_{r1}	f_{r2}	f_{r3}	$\cdots$	f_{rc}	M_r

The row marginal totals are fixed by the sample design; hence, there can be a unique multivariate logistic function for each contingency table. The corresponding multivariate logits are a linear function of the sample design and the response design. This function is structurally identical to the linear models used in fixed-effects ANOVA involving a constant, main effects, and interactions (association) of first and higher orders. This parallelism extends to an identical degrees of freedom breakdown.

The linear model for the complete set of multivariate logits can be expressed as a matrix equation to be solved for the unknown parameters.

$$\underset{r \times c}{Z} = \underset{r \times d}{X} \cdot \underset{d \times g}{B} \cdot \underset{g \times c,}{A} \tag{9.43}$$

where X is the sample design matrix, B is the matrix of parameters, and A is the response design matrix. When ANOVA was formulated in linear model terms, the design matrix involved linear dependencies and was less than full rank. In Equation 9.43, both the sample design and response design matrices involve such dependencies and, hence, must be reparameterized. Bock (1975, Chap. 8) presents this reparameterization in detail, and it is identical to that used in ANOVA (see Finn, 1974). The X matrix will, in general, be of deficient rank $s \leq n$. It can be reparameterized via by setting

$$\underset{r \times d}{X} = \underset{r \times s}{K} \cdot \underset{s \times d}{L} \tag{9.44}$$

and L is chosen by the investigator and contains contrasts among the sample design categories. The "sample basis" matrix K is calculated from $K = XL'(LL')^{-1}$. One consequence of the restriction given in Equation 9.3 is that the logits are invariant under a shift in origin and must be "anchored." This is implemented by setting the first row of the A matrix to zero whereas, in the item analysis setting, projection operators were used in A. In addition, the response design matrix will always be of deficient rank and must be reparameterized. This is done via

$$\underset{g \times c}{A} = \underset{g \times t}{S} \cdot \underset{t \times c,}{T} \tag{9.45}$$

where S is chosen by the investigator and involves contrasts among the response categories. The "response basis" matrix is calculated from $T = (S'S)^{-1}S'A$. Substituting from Equations 9.44 and 9.45 into Equation 9.43 yields

$$Z = KLBST.$$

Let $\Gamma = LBS$ and, then,

$$\underset{r \times c}{Z} = \underset{r \times s}{K} \cdot \underset{s \times t}{\Gamma} \cdot \underset{t \times c}{T} . \tag{9.46}$$

The Newton-Raphson procedure is used to obtain maximum likelihood estimates of the parameters, the γ, in the Γ matrix. The likelihood equation is identical to Equation 9.9, and the log-likelihood is identical to Equation 9.10, and the maximum likelihood procedures are similar to those presented earlier.

With the estimates $\hat{\gamma}$ in hand, Equation 9.46 is used to obtain the values of the estimated logits. The multivariate logistic function is then evaluated for each row in the sample design, and the expected cell frequencies $E_{jk} = \hat{P}_k M_j$ are obtained. The likelihood ratio statistic is then used to index the goodness of fit of the expected frequencies yielded by the model and the observed frequencies of the contingency table:

$$G^2 = 2 \sum_{j=1}^{r} \sum_{k=1}^{c} \log\left(\frac{f_{jk}}{E_{jk}}\right).$$

Under Bock's logit-linear approach, definition of the models of interest is accomplished via the use of one-degree-of-freedom contrasts associated with each of the two design matrices. Thus, a sample basis matrix of a given rank paired with a response basis matrix of a given rank, coupled with the particular types of underlying contrasts in each, specifies a particular linear model. It is important to note that a model of a given sample and response rank assumes a hierarchical ordering and inclusion of all models of lesser rank. Much of the utility and flexibility of Bock's logit-linear approach stems from the capacity for combining particular types of contrasts across the sample and response designs in particular orders.

9.6.1 Maximum Likelihood Estimation of Item parameters Under a Logit-Linear Approach

When the ability scores of a sample of examinees are known, one can plot the observed proportion of correct response P_j to a dichotomously scored item as a function of ability. The classic estimation problem of IRT is to estimate the parameters of the functional equation, the item characteristic curve, relating ability θ_j and the true proportion P_j of correct response. The logistic function can be used as the model for the item characteristic curve. Under the logistic transformation, the estimation problem reduces to one of estimating the regression parameters, ζ and λ, usually via maximum likelihood procedures. The goodness of fit of the item characteristic curve specified by the $\hat{\zeta}$ and $\hat{\lambda}$ and the observed proportions of correct response can be indexed by the usual Pearson χ^2 or G^2 statistic.

Berkson (1944), in the context of quantal response bioassay, recognized that this problem could be formulated in contingency table terms. Dividing the ability score scale into r levels and scoring the item responses dichotomously yields an $r \times 2$ contingency table. The cell entries within a row would be the number of examinees having ability level θ_g answering the item correctly, f_{g2}, and incorrectly, f_{g1}. Dividing each cell entry by its corresponding row marginal f_g yields the observed proportion of correct response, $p_{g2} = p_g$, of incorrect response, $p_{g1} = q_g$, at the given ability level. In the usual contingency table analysis, the interest is in the independence of the row and column classifications. Under the independence assumption, the row and column marginals are used to obtain estimates of the true P_{jk}, and the expected cell frequencies are given by $\hat{E}_{jk} = f\hat{P}_{jk} = f\hat{P}_{j.}\hat{P}_{.k}$. The goodness of fit of the observed and expected frequencies is then indexed by a χ_2 statistic with $(r-1)(c-1)$ degrees of freedom. However, in the present situation, the interest is not in independence but in the functional relationship between P_g and θ_g. Using a logistic model for the relationship and an appropriate estimation procedure will yield the regression parameter estimates $\hat{\zeta}$ and $\hat{\lambda}$. Since $P_j + Q_j = 1$, the row marginals are fixed, and the expected cell frequencies are given by

$$E_{gk} = M_g P_{gk} = M_g \frac{e^{u_k Z_{gk}}}{1 + e^{Z_{gk}}},$$

where

$$u_k = \begin{cases} 1 \text{ when } k = 2 \\ 0 \text{ when } k = 1. \end{cases}$$

The goodness of fit of the observed and expected cell frequencies can be indexed by a χ^2 goodness-of-fit statistic:

$$\chi^2 = \sum_{g=1}^{r} \sum_{k=1}^{2} \frac{(f_{gk} - E_{gk})^2}{E_{gk}}$$

$$= \sum_{g=1}^{r} \left[\frac{(P_{g1} M_g - \hat{P}_{g1} M_g)^2}{\hat{P}_{g1} M_g} + \frac{(P_{g2} M_g - \hat{P}_{g2} M_g)^2}{\hat{P}_{g2} M_g} \right]$$

with $r - 2$ degrees of freedom. Thus, the item parameter problem of IRT can be formulated in terms of a problem in contingency table analysis as was done in Chapter 3.

Given the above results, the item parameter estimation problem can also be formulated in terms of logit-linear models. For didactic purposes, a contrived data set reported by Baker and Subkoviak (1981) will be used. An ability scale was divided into four equally spaced intervals, each containing 1000 examinees. The observed proportions of correct response were the same as the true proportion of correct response yielded by a logistic model for the item characteristic curve ($\alpha = 1.0$, $\beta = 0.0$). Since the observed cell frequencies are also the true cell frequencies, a ready check is provided on the ability of the logit-linear procedures to yield the item parameters. The resulting contingency table is shown in Table 9.4.

In this example, the sample design is a stratified sampling plan having 4 levels of ability and 1000 examinees within each level. The response design employs a single factor having two levels, correct and incorrect response. Under the logistic model, the logit Z_j is a linear function of ability and, because of the equal spacing of the ability scores, the sample design contrast matrix L contains the orthogonal polynomial coefficients for a linear trend and is of rank 2:

Table 9.4. Item Response data Yielded by a Logistic Model ($\alpha = 1.0, \beta = 0.0$)

	Item Response	
Ability	Correct	Incorrect
−3.0	50	950
−2.0	269	731
1.0	731	269
3.0	950	50

$$L' = \begin{bmatrix} 1 & -3 \\ 1 & -1 \\ 1 & 1 \\ 1 & 3 \end{bmatrix} = K \qquad \text{due to the orthogonality.}$$

The logit also is a simple contrast between the logarithms of the proportions of correct response since $Z_g = \log P_g - \log Q_g$. Therefore, the contrast matrix S for the response design is of rank 1 and is given by

$$S = \begin{bmatrix} 1 & -1 \end{bmatrix}.$$

When these contrast matrices are used in the reparameterization process, Equation 9.11 yields

$$\begin{bmatrix} Z_{11} & Z_{12} \\ Z_{21} & Z_{22} \\ Z_{31} & Z_{32} \\ Z_{41} & Z_{42} \end{bmatrix} = \begin{bmatrix} 1 & -3 \\ 1 & -1 \\ 1 & 1 \\ 1 & 3 \end{bmatrix} \cdot \begin{bmatrix} \gamma_{11} \\ \gamma_{21} \end{bmatrix} \cdot \begin{bmatrix} 0.5 & -0.5 \end{bmatrix}.$$

The task then is to estimate γ_{11} and γ_{21} which, in the present example, correspond to ζ and λ via the likelihood equation and the Newton-Raphson technique. This can be accomplished using the MULTIQUAL computer program (Bock & Yates, 1973), which yielded $\hat{\alpha} = 1.0$, $\hat{\beta} = 0.0$. Left to its own devices, the MULTIQUAL program normalizes the column elements by dividing them by the square root of the sum of the squared elements, the Fisher-Techebkov normalization. Thus, the $\hat{\alpha}$ and $\hat{\beta}$ must be rescaled to obtain the actual item parameters. In the present example, this automatic rescaling was bypassed, and the parameter estimates obtained directly. In the usual application of logit-linear models, the interest is in the fit of the model, not the numerical values of the parameters; hence, the rescaling is of little consequence.

The extension of the logit-linear approach to items that are nominally scored into more than two response categories is direct. Again, a simple numerical problem will be used to illustrate the procedures. The same ability levels of the previous example will be employed since they match the orthogonal polynomial coefficients for a linear trend. The item responses will be scored on the basis of three nominally scaled categories. For illustrative purposes, the multivariate logistic function ($\zeta_1 = -0.75$, $\lambda_1 = 0.50$; $\zeta_2 = 0.25$, $\lambda_2 = -1.00$; $\zeta_3 = 0.50$, $\lambda_3 = 0.50$) will be evaluated to obtain the observed frequency of response. The resulting contingency table is shown in Table 9.5.

Table 9.5. Observed Frequency of Category Selection Corresponding to the Item of Table 9.1

Ability	Row	Response Category			Number of Examinees
		1	2	3	
−3.0	1	4	982	14	1000
−2.0	2	60	731	209	1000
1.0	3	196	119	685	1000
3.0	4	221	7	772	1000

The sample design matrix, the sample contrast matrix, and the sample design basis matrix will be the same as in the previous example. In the case of the response design, the single contrast of the S matrix must be extended to involve all three response categories. A reasonable extension is to use simple contrasts against the last category. The S matrix is given by

$$S = \begin{bmatrix} 1 & 0 \\ 0 & 1 \\ -1 & -1 \end{bmatrix}$$

and the response design basis matrix is

$$T = \begin{bmatrix} \frac{2}{3} & -\frac{1}{3} & -\frac{1}{3} \\ -\frac{1}{3} & \frac{2}{3} & -\frac{1}{3} \end{bmatrix}.$$

Given these definitions, Equation 9.21 becomes

$$\begin{bmatrix} Z_{11} & Z_{12} & Z_{13} \\ Z_{21} & Z_{22} & Z_{23} \\ Z_{31} & Z_{32} & Z_{33} \\ Z_{41} & Z_{42} & Z_{43} \end{bmatrix} = \begin{bmatrix} 1 & -3 \\ 1 & -1 \\ 1 & 1 \\ 1 & 3 \end{bmatrix} \cdot \begin{bmatrix} \gamma_{11} & \gamma_{12} \\ \gamma_{21} & \gamma_{22} \end{bmatrix} \cdot \begin{bmatrix} \frac{2}{3} & -\frac{1}{3} & -\frac{1}{3} \\ -\frac{1}{3} & \frac{2}{3} & -\frac{1}{3} \end{bmatrix}. \tag{9.47}$$

Analysis of the frequency data of Table 9.5 via the MULTIQUAL program yielded the estimates

$$\hat{\zeta}_1 = -0.74 \quad \hat{\zeta}_2 = 0.22 \quad \hat{\zeta}_3 = 0.51$$
$$\hat{\lambda}_1 = -0.52 \quad \hat{\lambda}_2 = -1.03 \quad \hat{\lambda}_3 = -0.52$$

which are essentially the values used to establish the table. Because of the automatic rescaling done by the MULTIQUAL program, the maximum likelihood estimates the program reports need to be rescaled by the square root of the sum of the squares of the orthogonal polynomial coefficients used. In the present case, the scale factor is $\sqrt{4}$ for the intercept parameters and $\sqrt{20}$ for the slope parameters.

Equation 9.47 is quite revealing. Although the representational process was started from a contingency table frame of reference, the final equation is structurally identical to Equation 9.12 used for nominally scored items. What has been done here is to make the elements used in a generalized reparameterization process depend on the application at hand. We also "cheated" in that the levels of the ability factor were chosen to correspond to the orthogonal polynomials for a linear trend. This was done only to simplify the presentation and does not impact the underlying logic. Under Bock's general model, variables other than ability and item response can be used to cross-classify the people and obtain a contingency table. Once these have been established, the procedures parallel those presented above. The net result is that there is a direct link between IRT and log-linear models. This linkage is well known in the case of binary items, but the extension to nominally scored items is unique to the work of Bock.

In both of the examples above, the responses to a single item were organized in a two-dimensional contingency table and analyzed via the logit-linear approach. This can be extended to a three-dimensional table in which all n items in a test are represented. Mellenbergh and Vijn (1981) have done this for dichotomously scored items and have shown that the Rasch model is a special case of a log-linear/logit-linear model. They established a three-dimensional (item × raw test score × item response) contingency table containing frequency counts. There were n items, with two response categories, and the score dimension had $n + 1 - p$ categories. The p represents the number of nonobserved raw scores as well as scores of zero and n that were edited out of the data set. Mellenbergh and Vijn formulated the cell frequencies in terms of both a log-linear and a logit-linear model. A data set due to Perline, Wright, and Wainer (1977) was analyzed via the BICAL program using JMLE and under the log-linear approach. Plots of the two sets of item difficulty estimates showed the expected linear relationship, as did the plot of the two sets of ability estimates. The log-linear estimates were also close in value to their CML estimates (Kelderman, 1984). As is typical under the log-linear/logit linear approach, a χ^2 or a likelihood ratio statistic G^2 can be used to index the overall goodness of fit of the Rasch model to the cell frequencies. However, Kelderman (1984) indicates that the test of goodness of fit should be based on a quasi-log-linear model (see Goodman, 1968) since the cell counts are not completely independent. This lack of independence is due to the fact that a test score is the sum of the examinee's item response choices.

9.7 Summary

Although it does not appear to be used in practice very often, the nominal response model plays an important role within item response theory. On its own, it provides a vehicle for the simultaneous estimation of the item parameters of all the response alternatives to a multiple-choice item that is not possible under other models. It also provides a vehicle for estimating an examinee's ability when all response categories to an item are taken into account even though a raw test score is not possible. Through the use of the multivariate logistic function, the one-, two-, and three-parameter logistic ICC models for dichotomous response items can be formulated as special cases of the nominal response model. In addition, there are several variants of the nominal response model itself. Since these models share a common base, so do the estimation procedures. Bock's formulation of the maximum likelihood estimation process in terms of various matrices provides a generalized representation that integrates what has historically been developed as separate techniques. This formulation, in turn, is encompassed within logit-linear models. Because of complexity of the underlying mathematical and computational procedures, accomplishing simultaneous parameter estimation for n items in the nominal response case via logit-linear models is not very attractive. However, computationally efficient IRT parameter estimation procedures can be established using the usual two basic building blocks when they are based on the nominal response model. The Birnbaum JMLE paradigm was used in the LOGOG computer program to estimate item and ability parameters under the nominal response model. A more modern approach is to employ marginal maximum likelihood estimation and the EM algorithm to estimate the item parameters, as is done in the MULTILOG program, and then to estimate ability in a separate stage. For the recovery results of the nominal response model via marginal maximum likelihood estimation, see De Ayala and Sava-Bolesta (1999) and Wollack, Bolt, Cohen, and Lee (2002).

10. Parameter Estimation for Multiple Group Data

10.1 Introduction

Earlier chapters dealt with estimation of item and ability parameters for a single group of examinees. In many practical situations, it is necessary to analyze item response data obtained from two or more (i.e., multiple) groups of examinees. The focus of the present chapter is on the method of obtaining item parameter estimates and the estimates of latent group characteristics for such multiple group data. In this context, the ability parameters for each examinee in a group are not to be estimated.

In a large testing program, several forms of a test are continually developed year after year to measure ability or proficiency of examinees. Unless the test forms are constructed using preequating, it is necessary to have good equating methods in order to effectively scale the test forms. Before equating is performed, several important factors should be considered including, for example, equating design, calibration procedures, and equating methods.

When test data obtained from multiple groups are analyzed to place the calibration results onto the same scale so that examinee's ability and item parameter estimates can be compared, careful attention should be given to the design from which the test data are obtained. If the estimation procedures from previous chapters should be applied to this situation, separate calibration runs can be performed. Such runs may yield comparable ability and item parameter estimates only if groups are equivalent or randomly equivalent. When groups of examinees are not equivalent in terms of their ability, either separate calibration with linking or concurrent calibration can be used to solve the metric issue. Linking for separate calibration runs is required because the estimates that result from parameter estimation procedures for nonequivalent groups are potentially expressed on different ability scales. When the item response models holds, the parameter estimates from different calibration runs are expressed on the scales which are linearly related. Hence, a linear transformation of the estimates should be applied (see Cook & Eignor, 1991; Kolen & Brennan, 1995).

In concurrent calibration all data from multiple groups of examinees and multiple test forms are analyzed to develop an item response theory ability scale on which all the estimates are numerically expressed. When multiple groups are calibrated, there exists in essence an equating problem. Note that there exist many different equating designs and each may require a different calibration strategy. This chapter treats the anchor test design where groups of examinees regardless of being equivalent or nonequivalent take test forms that share some common items.

There may be several different sets of parameters (cf. Mislevy, 1986b) in the multiple group model. The parameters that can be hypothesized are item parameters ξ, item hyperparameters η, ability parameters θ, and ability hyperparameters τ. Again, in this chapter the main concern is the estimation of the item parameters and the ability hyperparameters assuming that the item hyperparameters are known (or do not exist) and the ability parameters are to be estimated subsequently after obtaining the estimates of the item parameters and the latent group characteristics that reflect the ability hyperparameters.

Making inference about the examinee's ability level is undeniably the main task of the analysis of test data. In some research designs, however, the task may lie in obtaining information about a group of examinees instead of or in addition to estimating examinee's ability and item parameters (e.g., Lord & Novick, 1968, pp. 234–260). Classical test theory which is built upon weak true score models does not provide any theoretically viable solution to such research design. Item response theory on the contrary in conjunction with the method of maximum marginal likelihood can provide unique approaches to the problem of estimating group characteristics. Although group characteristics can be depicted in many different ways, it is assumed in this chapter that essentially a latent distribution with population parameters exists for each group of examinees.

First, estimation of properties or characteristics of distributions is presented; this section is not directly related to the item response theory framework, but may provide some background information regarding estimation of distributions in usual statistical applications. Estimation of a latent distribution with known item parameters is presented next. Estimation of both item parameters and latent distributions is presented subsequently. It should be noted that Bock and Zimowski (1997) is the major reference for the current chapter.

10.2 Estimating Properties of Distributions

Scientific investigations are frequently concerned with obtaining information about some population. In order to obtain accurate knowledge about a population, it may be necessary to examine every individual in the population. This is in general not feasible and, hence, a sample is to be used for such a purpose.

In statistical estimation the sample is usually assumed to be a random sample. A variable of interest that can be considered to be a sequence of independent and identically distributed random variables is obtained from the sample for the investigation. A statistic is a function of several random variables, and one of the primary uses of a statistic is to estimate unknown properties of the population. A statistic that is used to estimate is called an estimator.

The true distribution function, or equivalently the true density, of a random variable in practical settings is never known. Sometimes an inference can be drawn as to the form of the distribution function and use the outcome as an approximation of the true distribution function. One way of making a good inference is by obtaining many values of the random variable and constructing an empirical distribution function that may be used as an estimate of the unknown distribution function of the random variable.

The empirical distribution function can be used as an estimator of the population distribution function, and the sample mean, variance, and quantiles can be used as estimators of the respective population mean, variance, and quantiles. It may seem natural to compare the empirical distribution function from the sample with the hypothesized true distribution function. In order to compare, some measure of discrepancy between two functions is required. It is possible to use statistical procedures of the Kolmogrove and Smirnov type that analyze the distance between the functions as a measure of how much the functions discord each other (Conover, 1999). Some parameters of the hypothesized distribution can be estimated from data for the Kolmogrov test (e.g., Lilliefors, 1967, 1969).

The fundamental problem of using the procedures based on the empirical distribution function in the item response theory context is that ability cannot be directly observed. Moreover, when some properties of latent distributions (e.g., mean, variance, shape) are estimated, the distribution of optimal point estimates does not yield proper answers.

10.3 Estimation of a Latent Distribution

In the item response theory field several studies presented how the latent distribution or the characteristics of the latent distribution can be estimated. Andersen and Madsen (1977) and Sanathanan and Blumenthal (1978) investigated the issue of estimating latent distributions in conjunction of the Rasch model (see also de Leeuw & Verhelst, 1986; Lindsay, Clogg, & Grego, 1991). Mislevy (1984) contains general approaches to estimate latent distributions.

In this section, estimation of a latent distribution is presented assuming that the item parameters are known. Equations are presented for both original response and patterned data. Although in a real situation for multiple groups both item parameters and latent distributions are estimated together, estimation of a latent distribution with known item parameters can lay the foundation to later more complicated situations.

Examinees are assumed to be sampled from a population at random with a density function $\pi(\theta|\tau)$, where τ is a population parameter. Although there are many different ways to represent the density, the attention is limited to a case in which the density is a normal distribution and, consequently, τ consists of μ and σ^2.

Assume also that U_i represents a particular dichotomous response to item i and $P(U_i|\theta, \xi_i)$ is the probability of the particular response given the examinee's ability θ value and a possibly vector valued item parameter ξ_i. If ξ is a set of all item parameters and U_j is a response pattern of person j to the n items, $U_j = [U_{1j}, U_{2j}, \ldots, U_{nj}]$, the probability of obtaining U_j is

$$P(U_j|\theta, \xi) = \prod_{i=1}^{n} P(U_{ij}|\theta, \xi). \tag{10.1}$$

This above equation can be expressed as $L(U_j|\theta)$ if ξ is known. When ξ is assumed to be known, by integrating out the incidental parameter θ, the marginal probability of U_j can be written as

$$P(U_j|\tau) = \int_{\Theta} P(U_j|\theta, \xi)\pi(\theta|\tau)d\theta. \tag{10.2}$$

The probability of the response matrix U for a sample of N examinees is

$$P(U|\tau) = \prod_{j=1}^{N} P(U_j|\tau). \tag{10.3}$$

If the probability of the response matrix U is treated as a function of τ, then it is the marginal likelihood function for τ, that is, $L(\tau)$. Maximizing the marginal likelihood function or, equivalently, the log of the function with respect to τ yields the estimates of population parameters. The likelihood equation is

$$\frac{\partial \log L}{\partial \tau} = \sum_{j=1}^{N} \frac{1}{P_j} \frac{\partial P_j}{\partial \tau} = 0, \tag{10.4}$$

where $P_j = P(U_j|\tau)$.

The estimated variance and covariance matrix is the inverse of the information matrix,

$$I = -\frac{\partial^2 \log L}{\partial \tau \partial \tau'}. \tag{10.5}$$

The inverse of the Fisher expected information,

$$I = E\left[\left(\frac{\partial \log L}{\partial \tau}\right)\left(\frac{\partial \log L}{\partial \tau'}\right)\right], \tag{10.6}$$

can also be used to approximate the estimated variance and covariance matrix. Other approximation formulas can be found in Mislevy (1984).

The calculation of the likelihood equation involves evaluation of integrals, and the integration must be approximated by numerical methods. The three methods are available; Gauss-Hermite quadrature, quadrature over fixed points, and Monte Carlo integration (Mislevy, 1984). The preferred method might be Gauss-Hermite quadrature (Stroud & Secrest, 1966), but computationally the use of quadrature over fixed points is simpler. See Bock (1983) for the use of Monte Carlo numerical integration for the triple logistic model.

For the method of quadrature over fixed points, the integrals will be approximated as sums over Q quadrature points (i.e., $X_1, X_2, \ldots, X_Q$). For example, $Q = 20$, and the minimum ability point can be -4.75 and the maximum ability point can be $+4.75$ in increments of 0.5. The initial estimates of the population mean μ and standard deviation σ are needed, and these can be 0 and 1, respectively. The corresponding quadrature weights for the quadrature points can be obtained using

$$A(X_q) = \frac{\exp\left[-\frac{1}{2}\left(\frac{X_q-\mu}{\sigma}\right)^2\right]}{\sum_{q=1}^{Q}\exp\left[-\frac{1}{2}\left(\frac{X_q-\mu}{\sigma}\right)^2\right]}. \tag{10.7}$$

Again, note that the quadrature weights are based on the initial value of τ.

After performing the approximation of the integration by quadrature over fixed points, solving the likelihood equations yields the estimates of the population mean and variance as

$$\hat{\mu} = \frac{1}{N}\sum_{q=1}^{Q}X_q\sum_{j=1}^{N}P_{jq} \tag{10.8}$$

and

$$\hat{\sigma}^2 = \frac{1}{N}\sum_{q=1}^{Q}(X_q - \hat{\mu})^2\sum_{j=1}^{N}P_{jq}, \tag{10.9}$$

where

$$P_{jq} = \frac{L(U_j|X_q)A(X_q)}{\sum_{q=1}^{Q}L(U_j|X_q)A(X_q)}. \tag{10.10}$$

The calculation of P_{jq} depends on the initial values of μ and σ^2 and the solution is necessarily iterative until a set of stable estimates $\hat{\mu}$ and $\hat{\sigma}^2$ are obtained. It can be noted that the Sheppard correction can be applied to the variance so that

$$\hat{\sigma}_s^2 = \hat{\sigma}^2 - \frac{d^2}{12}, \tag{10.11}$$

where d is the increment of the quadrature points (Kendall, Stuart, & Ord, 1987, pp. 93–94).

The elements of the information matrix can be obtained for $\mu\mu$, $\mu\sigma^2$, and $\sigma^2\sigma^2$, respectively, using the following equations:

$$\sum_{j=1}^{N}\left[\sum_{q=1}^{Q}P_{jq}\frac{X_q-\hat{\mu}}{\hat{\sigma}^2}\right]^2 \tag{10.12}$$

$$\sum_{j=1}^{N}\left[\sum_{q=1}^{Q}P_{jq}\frac{X_q-\hat{\mu}}{\hat{\sigma}^2}\right]\left[\sum_{q=1}^{Q}P_{jq}\frac{(X_q-\hat{\mu})^2-\hat{\sigma}^2}{2\hat{\sigma}^4}\right] \tag{10.13}$$

$$\sum_{j=1}^{N}\left[\sum_{q=1}^{Q}P_{jq}\frac{(X_q-\hat{\mu})^2-\hat{\sigma}^2}{2\hat{\sigma}^4}\right]^2 \tag{10.14}$$

Note that there are possibilities that the same response patterns can be observed, and the probability of U can also be written as

$$P(U|\tau) = \frac{N!}{\prod_{l=1}^{L}r_l!}\prod_{l=1}^{L}[P(U_l|\tau)]^{r_l}, \tag{10.15}$$

where l designates the response pattern from 1 to L, L is the minimum of 2^n or N, and r_l is the number of cases for response pattern U_l. The marginal likelihood function of τ is

$$L(\tau) \propto \prod_{l=1}^{L}[P(U_l|\tau)]^{r_l}. \tag{10.16}$$

The likelihood equation is

$$\frac{\partial \log L}{\partial \tau} = \sum_{l=1}^{L}\frac{r_l}{P_l}\frac{\partial P_l}{\partial \tau} = 0, \tag{10.17}$$

where $P_l = P(U_l|\tau)$.

Solving the likelihood equations for the patterned data yields

$$\hat{\mu} = \frac{1}{N}\sum_{q=1}^{Q}X_q\sum_{l=1}^{L}P_{lq}r_l \tag{10.18}$$

and

$$\hat{\sigma}^2 = \frac{1}{N}\sum_{q=1}^{Q}(X_q-\hat{\mu})^2\sum_{l=1}^{L}P_{lq}r_l, \tag{10.19}$$

where

$$P_{lq} = \frac{L(U_l|X_q)A(X_q)}{\sum_{q=1}^{Q}L(U_l|X_q)A(X_q)}. \tag{10.20}$$

The elements of the information matrix can be approximated for $\mu\mu$, $\mu\sigma^2$, and $\sigma^2\sigma^2$, respectively, using the following equations:

$$\sum_{l=1}^{L} r_l \left[\sum_{q=1}^{Q} P_{lq} \frac{X_q - \hat{\mu}}{\hat{\sigma}^2} \right]^2 \tag{10.21}$$

$$\sum_{l=1}^{L} r_l \left[\sum_{q=1}^{Q} P_{lq} \frac{X_q - \hat{\mu}}{\hat{\sigma}^2} \right] \left[\sum_{q=1}^{Q} P_{lq} \frac{(X_q - \hat{\mu})^2 - \hat{\sigma}^2}{2\hat{\sigma}^4} \right] \tag{10.22}$$

$$\sum_{l=1}^{L} r_l \left[\sum_{q=1}^{Q} P_{lq} \frac{(X_q - \hat{\mu})^2 - \hat{\sigma}^2}{2\hat{\sigma}^4} \right]^2 \tag{10.23}$$

In addition to the normal distribution presented in this section, other representations of latent distributions are available. Mislevy (1984) presented the use of a latent mixture of normal distributions with different population compositions and potentially different population means. A nonparametric representation is also possible in which the latent distributions are characterized by a finite set of points on the ability continuum and corresponding probability masses (Bock & Aitkin, 1981; Mislevy, 1984). Mislevy (1984) also presented a beta-binomial solution based on Lord and Novick (1968).

10.4 The Multiple Group Model

When multiple groups are analyzed, the latent distributions of respective groups can be characterized differently. Because it is possible to estimate item parameters together with the parameters of the latent distributions, marginal estimation is used in which ability parameters are removed from the likelihood.

Consider data obtained from person j in group g responding in category k of item i. Suppose there are G groups, N_g persons in group g, n items in a test, and K_i response categories of item i. A response to item i can be denoted by U_i, where U_i takes on any integer value 1 through K_i.

A response function,

$$P_{ik}(\theta) = P(U_i = k | \theta, \xi_i), \tag{10.24}$$

that is common to all groups and persons is hypothesized, where ξ_i is the item parameter. It describes the probability of a response in category k, given the value of a continuous and unbounded person parameter θ and the item parameter ξ_i. The response categories are assumed to be mutually exclusive and exhaustive, and consequently

$$\sum_{k=1}^{K_i} P_{ik}(\theta) = 1. \tag{10.25}$$

The sequence of response of person j to the n items can be expressed as $U_j = [U_{1j}, U_{2j}, \ldots, U_{nj}]$, which is a vector. Assuming conditional independence of the responses given θ, the probability of response sequence U_j can be obtained as a multiplicative product of respective item response functions, so that

$$P(U_j|\theta,\xi) = \prod_{i=1}^{n} P(U_{ij} = k|\theta,\xi_i), \qquad (10.26)$$

where ξ is the set of all item parameters.

Suppose θ to have a continuous distribution with finite mean and variance in the population of persons for specific group g with its density function $\pi_g(\theta)$. The marginal probability of response sequence U_j in group g can be expressed as

$$P_g(U_j) = \int_{\Theta} P(U_j|\theta,\xi)\pi_g(\theta)d\theta, \qquad (10.27)$$

where Θ is the parameter space for θ.

In fact, it is assumed that the population distribution functions depend on some other unknown fixed parameters τ_g for group g. The marginal probability of response sequence with the fixed parameters can be explicitly written as

$$P_g(U_j) = \int_{\Theta} P(U_j|\theta,\xi)\pi(\theta|\tau_g)d\theta. \qquad (10.28)$$

Given the observed item scores U_{gij}, ξ and τ can be estimated simultaneously.

The distributions of θ within groups are assumed to be normal. If not, alternative representations presented in Mislevy (1984) and Bock and Zimowski (1997) can be applied; for example, a mixture of Gaussian components with common variances but different means and proportions (Day, 1969; Mislevy, 1984) and nonparametric representation of the latent density using a finite number of equally spaced points in an interval that includes almost all of the probability.

10.5 Parameter Estimation

It is convenient to employ distinct response patterns instead of the individual response sequences. Let r_{gl} be the number of occurrence of pattern l in group g, and let $L_g \leq \min(N_g, L)$, where $L = \prod_{i=1}^{n} K_i$, be the total number of patterns with r_{gl}. The pattern reflects the ability level and hence the subscript l is used.

Assuming independence of persons as well as groups, the marginal likelihood can be written as

$$L(\xi,\tau) = \prod_{g=1}^{G} \frac{N_g!}{\prod_{l=1}^{L_g} r_{gl}!} \prod_{l=1}^{L_g} [P_g(U_l)]^{r_{gl}}. \qquad (10.29)$$

In order to find maximum likelihood estimates, the log marginal likelihood is maximized with regard to ξ and τ, respectively. Because each item is independent, the log marginal likelihood is maximized with regard to ξ_i. Hence, the likelihood equation for the parameters of item i can be expressed as

$$\frac{\partial \log L}{\partial \xi_i} = \sum_{g=1}^{G} \sum_{l=1}^{L_g} \frac{r_{gl}}{P_{gl}} \frac{\partial P_{gl}}{\partial \xi_i} = 0, \tag{10.30}$$

where $P_{gl} = P_g(U_l)$ for brevity. If we let

$$L_{gl}(\theta) = P(U_{gl}|\theta) = \prod_{i=1}^{n} P(U_{gil} = k|\theta)$$

and

$$u_{gilk} = \begin{cases} 1 & \text{if } U_{gil} = k \\ 0 & \text{otherwise} \end{cases},$$

then

$$\frac{\partial P_{gl}}{\partial \xi_i} = \int \left(\frac{\frac{\partial}{\partial \xi_i}\{\prod_{k=1}^{K_i}[P_{ik}(\theta)]^{u_{gilk}}\}}{\prod_{k=1}^{K_i}[P_{ik}(\theta)]^{u_{gilk}}} \right) L_{gl}(\theta)\pi_g(\theta)d\theta.$$

Because

$$\frac{\partial \prod_{k=1}^{K_i}[P_{ik}(\theta)]^{u_{gilk}}}{\partial \xi_i} = \sum_{k=1}^{K_i} \frac{u_{gilk}}{P_{ik}(\theta)} \frac{\partial P_{ik}(\theta)}{\partial \xi_i} \prod_{k=1}^{K_i}[P_{ik}(\theta)]^{u_{gilk}},$$

it can be shown that

$$\frac{\partial \log L}{\partial \xi_i} = \sum_{g=1}^{G} \sum_{l=1}^{L_g} \frac{r_{gl}}{P_{gl}} \int \left(\sum_{k=1}^{K_i} \frac{u_{gilk}}{P_{ik}(\theta)} \frac{\partial P_{ik}(\theta)}{\partial \xi_i} \right) L_{gl}(\theta)\pi_g(\theta)d\theta. \tag{10.31}$$

The above likelihood equation can be applied to many different item response theory models for both dichotomous and polytomous responses. Let's look at an example to solidify our understanding of the equation. For the two-parameter logistic model the likelihood equation with the restriction $P_{i2}(\theta) = 1 - P_{i1}(\theta)$ becomes

$$\frac{\partial \log L}{\partial \xi_i} = 0$$

$$= \sum_{g=1}^{G} \sum_{l=1}^{L_g} \frac{r_{gl}}{P_{gl}} \int \left(\frac{u_{gil1} - P_{i1}(\theta)}{P_{i1}(\theta)[1 - P_{i1}(\theta)]} \right) \frac{\partial P_{i1}(\theta)}{\partial \xi_i} L_{gl}(\theta)\pi_g(\theta)d\theta. \tag{10.32}$$

Using $z_i(\theta) = \zeta_i + \lambda_i\theta$ which is the logit in θ, the likelihood equations can be expressed as

$$\frac{\partial \log L}{\partial \zeta_i} = \sum_{g=1}^{G} \sum_{l=1}^{L_g} \frac{r_{gl}}{P_{gl}} \int [u_{gil1} - P_{i1}(\theta)] \frac{\partial z_i(\theta)}{\partial \zeta_i} L_{gl}(\theta)\pi_g(\theta)d\theta = 0 \tag{10.33}$$

and

$$\frac{\partial \log L}{\partial \lambda_i} = \sum_{g=1}^{G} \sum_{l=1}^{L_g} \frac{r_{gl}}{P_{gl}} \int [u_{gil1} - P_{i1}(\theta)] \frac{\partial z_i(\theta)}{\partial \lambda_i} L_{gl}(\theta) \pi_g(\theta) d\theta = 0, \quad (10.34)$$

where $\partial z_i(\theta)/\partial \zeta_i = 1$ and $\partial z_i(\theta)/\partial \lambda_i = \theta$. For other polytomous models, we may construct similar likelihood equations (e.g., Bock, 1985).

Now consider the population parameter τ_g. The likelihood equation for τ_g can be written as

$$\frac{\partial \log L}{\partial \tau_g} = \sum_{l=1}^{L_g} \frac{r_{gl}}{P_{gl}} \int L_{gl}(\theta) \frac{\partial \pi_g(\theta)}{\partial \tau_g} d\theta = 0. \quad (10.35)$$

As we mentioned earlier, if the distribution of θ for group g is normal with mean μ_g and variance σ_g^2,

$$\pi_g(\theta) = \frac{1}{\sqrt{2\pi}\sigma_g} \exp\left(-\frac{(\theta - \mu_g)^2}{2\sigma_g^2}\right).$$

The likelihood equations are

$$\frac{\partial \log L}{\partial \mu_g} = \frac{1}{\sigma_g^2} \sum_{l=1}^{L_g} \frac{r_{gl}}{P_{gl}} \int (\theta - \mu_g) L_{gl}(\theta) \pi_g(\theta) d\theta = 0 \quad (10.36)$$

and

$$\frac{\partial \log L}{\partial \sigma_g^2} = -\frac{1}{2} \sum_{l=1}^{L_g} \frac{r_{gl}}{P_{gl}} \int [\sigma_g^{-2} - (\theta - \mu_g)^2 \sigma_g^{-4}] L_{gl}(\theta) \pi_g(\theta) d\theta = 0. \quad (10.37)$$

Because

$$\sum_{l=1}^{L_g} r_{gl} = N_g,$$

the likelihood equation for μ_g yields

$$\hat{\mu}_g = \frac{1}{N_g} \sum_{l=1}^{L_g} r_{gl} \bar{\theta}_{gl}, \quad (10.38)$$

where

$$\bar{\theta}_{gl} = \frac{1}{P_{gl}} \int \theta L_{gl}(\theta) \pi_g(\theta) d\theta$$

is the posterior mean of θ given U_{gl}. Also, because

$$\theta - \mu_g = (\theta - \bar{\theta}_{gl}) + (\bar{\theta}_{gl} - \mu_g),$$

the likelihood equation for σ_g^2 yields

$$\hat{\sigma}_g^2 = \frac{1}{N_g} \sum_{l=1}^{L_g} r_{gl}[\sigma_{\theta|U_{gl}}^2 + (\bar{\theta}_{gl} - \hat{\mu}_g)^2], \tag{10.39}$$

where

$$\sigma_{\theta|U_{gl}}^2 = \frac{1}{P_{gl}} \int (\theta - \bar{\theta}_{gl})^2 L_{gl}(\theta)\pi_g(\theta)d\theta$$

is the posterior variance of θ given U_{gl}. Note that these likelihood equations are in a form suitable for the expectation and maximization (EM) solution (Bock, 1989; Dempster et al., 1981).

As presented in Chapter 6 and the previous section, the quadrature approximation is to be used for integrals. In the expectation step the calculation of $L_{gl}(\theta)$ and P_{gl} are obtained first using the quadrature points and weights. Then r_{gl} terms for the quadrature points as well as the corresponding expected frequencies are obtained. In the maximization step, using the artificial data, the maximum likelihood equations for item parameters are solved using the Newton-Raphson procedure. Subsequently the maximum likelihood equations for the population ability parameters are evaluated. Because the dependence among the quantities in the two steps, the iterative expectation and maximization steps will be repeated until the convergence criterion is met.

Now the two-parameter logistic model is employed to present a detailed illustration of the expectation and maximization steps. In the expectation step artificial values of the correct responses and the corresponding numbers of examinees for quadrature points are obtained. If

$$\bar{r}_{gi} = \sum_{l=1}^{L_g} \frac{r_{gl}u_{gil1}L_{gl}(\theta)\pi_g(\theta)}{P_{gl}} \tag{10.40}$$

and

$$\bar{N}_g = \sum_{l=1}^{L_g} \frac{r_{gl}L_{gl}(\theta)\pi_g(\theta)}{P_{gl}}, \tag{10.41}$$

then

$$\bar{r}_{giq} = \sum_{l=1}^{L_g} \frac{r_{gl}u_{gil1}L_{gl}(X_{gq})A(X_{gq})}{P_{gl}} \tag{10.42}$$

and

$$\bar{N}_{gq} = \sum_{l=1}^{L_g} \frac{r_{gl}L_{gl}(X_{gq})A(X_{gq})}{P_{gl}}, \tag{10.43}$$

for $q = 1, 2, \ldots, Q$. The expectation step is for obtaining sets of $\bar{r}_{giq}$ and $\bar{N}_{gq}$ using initial values of item parameters and population ability parameters.

The maximization step solves the likelihood equations as if the artificial data obtained in the expectation step are the complete data. Because

$$\frac{\partial \log L}{\partial \xi_i} = \sum_{g=1}^{G} \int \frac{\bar{r}_{gi} - \bar{N}_g P_{i1}(\theta)}{P_{i1}(\theta)[1 - P_{i1}(\theta)]} \frac{\partial P_{i1}(\theta)}{\partial \xi_i} d\theta, \tag{10.44}$$

by replacing the definite integral with the quadrature formula,

$$\frac{\partial \log L}{\partial \xi_i} = \sum_{g=1}^{G} \sum_{q=1}^{Q} \frac{\bar{r}_{giq} - \bar{N}_{gq} P_{i1}(X_{gq})}{P_{i1}(X_{gq})[1 - P_{i1}(X_{gq})]} \frac{\partial P_{i1}(X_{gq})}{\partial \xi_i}. \tag{10.45}$$

Using ζ_i and λ_i and such terms as

$$\frac{\partial \log L}{\partial \zeta_i} = \sum_{g=1}^{G} \sum_{q=1}^{Q} [\bar{r}_{giq} - \bar{N}_{gq} P_{i1}(X_{gq})], \tag{10.46}$$

$$\frac{\partial \log L}{\partial \lambda_i} = \sum_{g=1}^{G} \sum_{q=1}^{Q} [\bar{r}_{giq} - \bar{N}_{gq} P_{i1}(X_{gq})] X_{gq}, \tag{10.47}$$

$$\frac{\partial^2 \log L}{\partial \zeta_i^2} = - \sum_{g=1}^{G} \sum_{q=1}^{Q} \bar{N}_{gq} W_{giq}, \tag{10.48}$$

$$\frac{\partial^2 \log L}{\partial \zeta_i \lambda_i} = \frac{\partial^2 \log L}{\partial \lambda_i \zeta_i} = - \sum_{g=1}^{G} \sum_{q=1}^{Q} \bar{N}_{gq} W_{giq} X_{gq}, \tag{10.49}$$

and

$$\frac{\partial^2 \log L}{\partial \lambda_i^2} = - \sum_{g=1}^{G} \sum_{q=1}^{Q} \bar{N}_{gq} W_{giq} X_{gq}^2, \tag{10.50}$$

where $W_{giq} = P_{i1}(X_{gq})[1 - P_{i1}(X_{gq})]$, the Newton-Raphson equation can be written as

$$\begin{bmatrix} \hat{\zeta}_i \\ \hat{\lambda}_i \end{bmatrix}_{(t+1)} = \begin{bmatrix} \hat{\zeta}_i \\ \hat{\lambda}_i \end{bmatrix}_{(t)} - \begin{bmatrix} \frac{\partial^2 \log L}{\partial \zeta_i^2} & \frac{\partial^2 \log L}{\partial \zeta_i \partial \lambda_i} \\ \frac{\partial^2 \log L}{\partial \lambda_i \partial \zeta_i} & \frac{\partial^2 \log L}{\partial \lambda_i^2} \end{bmatrix}_{(t)}^{-1} \begin{bmatrix} \frac{\partial \log L}{\partial \zeta_i} \\ \frac{\partial \log L}{\partial \lambda_i} \end{bmatrix}_{(t)}. \tag{10.51}$$

Let D be the determinant of the Hessian, where

$$D_{(t)} = \left(\frac{\partial^2 \log L}{\partial \zeta_i^2} \frac{\partial^2 \log L}{\partial \lambda_i^2} - \frac{\partial^2 \log L}{\partial \zeta_i \partial \lambda_i} \frac{\partial^2 \log L}{\partial \lambda_i \partial \zeta_i} \right)_{(t)},$$

then

$$\begin{bmatrix} \hat{\zeta}_i \\ \hat{\lambda}_i \end{bmatrix}_{(t+1)} = \begin{bmatrix} \hat{\zeta}_i \\ \hat{\lambda}_i \end{bmatrix}_{(t)} - \frac{1}{D_{(t)}} \begin{bmatrix} \frac{\partial^2 \log L}{\partial \lambda_i^2} & -\frac{\partial^2 \log L}{\partial \zeta_i \partial \lambda_i} \\ -\frac{\partial^2 \log L}{\partial \lambda_i \partial \zeta_i} & \frac{\partial^2 \log L}{\partial \zeta_i^2} \end{bmatrix}_{(t)} \begin{bmatrix} \frac{\partial \log L}{\partial \zeta_i} \\ \frac{\partial \log L}{\partial \lambda_i} \end{bmatrix}_{(t)}. \tag{10.52}$$

Let

$$\Delta\hat\zeta_{i(t)} = \frac{1}{D_{(t)}} \left(\frac{\partial^2 \log L}{\partial \lambda_i^2} \frac{\partial \log L}{\partial \zeta_i} - \frac{\partial^2 \log L}{\partial \zeta_i \partial \lambda_i} \frac{\partial \log L}{\partial \lambda_i} \right)_{(t)} \tag{10.53}$$

and

$$\Delta\hat\lambda_{i(t)} = \frac{1}{D_{(t)}} \left(-\frac{\partial^2 \log L}{\partial \lambda_i \partial \zeta_i} \frac{\partial \log L}{\partial \zeta_i} + \frac{\partial^2 \log L}{\partial \zeta_i^2} \frac{\partial \log L}{\partial \lambda_i} \right)_{(t)}, \tag{10.54}$$

then

$$\hat\zeta_{i(t+1)} = \hat\zeta_{i(t)} - \Delta\hat\zeta_{i(t)} \tag{10.55}$$

and

$$\hat\lambda_{i(t+1)} = \hat\lambda_{i(t)} - \Delta\hat\lambda_{i(t)}. \tag{10.56}$$

Because the Newton-Raphson procedure was presented several times in the earlier chapters, a detailed description is not presented again.

The likelihood equations for the population ability parameters yield

$$\hat\mu_g = \frac{1}{N_g} \sum_{l=1}^{L_g} r_{gl} \bar X_{gl}, \tag{10.57}$$

where

$$\bar X_{gl} = \frac{1}{P_{gl}} \sum_{q=1}^{Q} X_{gq} L_{gl}(X_{gq}) A(X_{gq}), \tag{10.58}$$

and

$$\hat\sigma_g^2 = \frac{1}{N_g} \sum_{l=1}^{L_g} r_{gl} [\sigma_{X|U_{gl}}^2 + (\bar X_{gl} - \hat\mu_g)^2], \tag{10.59}$$

where

$$\sigma_{X|U_{gl}}^2 = \frac{1}{P_{gl}} \sum_{q=1}^{Q} (X_{gq} - \bar X_{gl})^2 L_{gl}(X_{gq}) A(X_{gq}). \tag{10.60}$$

Standard errors can be estimated by different ways. For example, we may use

$$\frac{\partial^2 \log L}{\partial \mu_g^2} = -\frac{1}{\sigma_g^2} \sum_{l=1}^{L_g} \frac{r_{gl}}{P_{gl}} \int L_{gl}(\theta)\pi_g(\theta)d\theta$$

$$+\frac{1}{\sigma_g^4} \sum_{l=1}^{L_g} \frac{r_{gl}}{P_{gl}} \int (\theta - \mu_g)^2 L_{gl}(\theta)\pi_g(\theta)d\theta \tag{10.61}$$

and other similar equations. For the above equation, in terms of the quadrature formula and with the terms defined earlier, it can be shown that

$$\frac{\partial^2 \log L}{\partial \mu_g^2} = -\frac{1}{\sigma_g^2} \sum_{q=1}^{Q} \bar N_{gq} + \frac{1}{\sigma_g^4} \sum_{l=1}^{L_g} r_{gl} [\sigma_{X|U_{gl}}^2 + (\bar X_{gl} - \mu_g)^2]. \tag{10.62}$$

10.6 Summary

For a purpose of test validation, test data obtained initially from a group of examinees can be treated as if these are obtained from two or more groups of examinees. A typical example of this type is the detection of differential item functioning (Muraki, 1999; Thissen, Steinberg, & Gerrard, 1986; Thissen, Steinberg, & Wainer, 1988, 1993), and the initial data may be regrouped based on gender, race or ethnicity, school or district, and geographical region. In order to compare item parameter estimates from multiple groups, the estimates should be placed on a same ability metric. Estimation for multiple group data obviously arises in this differential item functioning situation.

When there are several forms of a test that share some common items and each form is to be administered a group of examinees, equating is required to yield comparable scales. It is possible to apply the framework of multiple group estimation in this situation. Note that the multiple group estimation in this case requires response data for all examinees from different groups. Sometimes the full data may not be available as well as obtaining estimates from the full combined data may not be practical.

In some other applications, the focus of testing is not obtaining ability estimates of individual examinees but assessing the proficiency of examinee groups themselves (Bock & Mislevy, 1981; Mislevy, 1983). In this matrix sampling case, typically a small subset of items will be administered to an examinee and consequently scores or ability estimates are not in general reported for each examinee. The usual item and test analysis techniques derived from only one group of examinees may not be directly applicable to this situation. The multiple group estimation based on item response theory can be used, instead.

There are still other applications where the multiple group estimation can be used, according to Bock and Zimowski (1997), including item parameter drift (Bock, Muraki, & Pfiffenberger, 1988), vertical equating (Lord, 1980, pp. 193–211; Stocking & Lord, 1983), and two-stage testing (Bock & Zimowski, 1990; Lord, 1971, 1980; Zimowski et al., 1996, pp. 179–196). For all these applied situations where response data are to be analyzed under some models based on item response theory, the multiple group estimation techniques in this chapter can present a unified framework to estimate both item parameters and latent ability distributions.

In case of estimating latent distributions only handful papers have been presented beyond the original contribution papers by Mislevy (1984, 1987). Thissen (1990, 1991) presented the use of the Johnson (1949) family of distributions to characterize the latent distribution in conjunction with marginal maximum likelihood estimation of item parameters. Wang and Chen (1999) recently presented comparison study for estimating latent distributions. The Markov chain Monte Carlo (i.e., Metropolis-Hasting within Gibbs) algorithm and the expectation and maximization algorithm are compared. They concluded that both yielded similar results and that Markov chain Monte Carlo algorithm seemed promising due to its potential flexibility to handle data with complicated structure.

It should be noted that the estimation of the examinee ability was not presented in this chapter because the ability parameters are integrated out from the likelihood functions when the item parameters and the parameters that characterize the latent distributions are estimated. If the ability parameters are required, these can be estimated after obtaining the item and latent distribution parameters and assuming that the obtained parameters are fixed and true values. Methods of maximum likelihood and Bayes can be used to obtain the ability parameters. When the Bayes methods is used either a separate group distribution or the combined population distribution can be employed as the prior distribution depending on the specific purpose of estimation (Bock & Zimowski, 1997).

The computer programs, MULTILOG (Thissen, Chen, & Bock, 2003), PARSCALE (Muraki & Bock, 2003), and BILOG-MG (Zimowski, Muraki, Mislevy, & Bock, 1996, 2003), have capabilities of handing parameter estimation for multiple group data. Appendix I contains BASIC code that implements multiple group estimation presented above, especially for detection of differential item functioning.

11. Estimation of Item Parameters of Mixed Models

11.1 Introduction

In modern assessment, different item types are frequently used to improve validity of test scores. Two classes of models in item response theory are available for situations when different item types are employed in testing; dichotomously-scored items and polytomously-scored items.

Although data arise from mixed item types can be analyzed any combination of dichotomous and polytomous item response models, some combinations may be more natural than others. For example, the combination of the one-parameter logistic model and the partial credit model, the combination of the one-parameter logistic model and the rating scale model, and the combination of the two-parameter model and the graded response model seem to be more theoretically viable than other awkward combinations of models.

Several researchers have presented papers dealt with some practical aspects of the use of item response theory models to analyze test data obtained with various types of items and to report examinees's ability (e.g., Billeaud, Swygert, Nelson, & Thissen, 1998; Erickan, Schwarz, Julian, Burket, Weber, & Link, 1998; Grima & Weichun, 2002; Rosa, Swygert, Nelson, & Thissen, 2001; Thissen, Nelson, & Swygert, 2001; Thissen & Wainer, 2001; Wainer & Thissen, 1993). The main focus of these papers was on the scoring procedures when mixed item types are used. Especially how to combine scores (or estimate ability) from multiple-choice items and constructed-response items was the main issue. In these papers, it is presumed that the multiple-choice items are dichotomously scored and the constructed-response or open-ended items are polytomously scored.

In this chapter how to estimate item parameters when mixed item types are used in testing will be presented. For the derivation of detailed parameter estimation, only a simple case is used for which the dichotomously-scored items are analyzed under the two-parameter logistic model and the polytomously-scored items are modeled with the logistic form of the graded response model. In a real and practical situation, however, other combination of items may be used, and modeling may require somewhat different and possibly more complicated models.

11.2 A General Model

The fundamental assumption required to calibrate the mixed item types simultaneously using different item response models is that the two or more mixed types of items measure the same unidimensional construct. It is important to note that the mixed types of items are reasonably one-dimensional in a practical and factor-analytic sense.

Now consider general models for the mixed item types in the context of marginal maximum likelihood estimation. Consider data obtained from examinee j responding in category k of item i. Suppose there are N examinees in a calibration group, n items in a test, and K_i response categories of item i. A response to item i can be denoted by U_i, where U_i takes on any integral value 1 through K_i.

A response function,

$$P_{ik}(\theta) = P(U_i = k|\theta), \tag{11.1}$$

that is common to all examinees is hypothesized. It describes the probability of a response in category k, given the value of a continuous and unbounded person parameter θ. The response categories are assumed to be mutually exclusive and exhaustive, and consequently

$$\sum_{k=1}^{K_i} P_{ik}(\theta) = 1. \tag{11.2}$$

The response vector of examinee j to the n items can be expressed as $U_j = [U_{j1}, U_{j2}, \ldots, U_{jn}]$. It can be assumed that U_j contains a set of dichotomously-scored items and a set of polytomously-scored items. Assuming conditional independence of the responses given θ, the probability of response vector U_j can be obtained as a multiplicative product of respective item response functions, so that

$$P(U_j|\theta) = P(U_{j1}|\theta) \times P(U_{j2}|\theta) \times \cdots \times P(U_{jn}|\theta). \tag{11.3}$$

Assume that θ has a continuous distribution with finite mean and variance in the population of persons with its density function $\pi(\theta)$. The marginal probability of pattern U_j can be expressed as

$$P(U_j) = \int_\Theta P(U_j|\theta)\pi(\theta)d\theta, \tag{11.4}$$

where Θ is the parameter space for θ. This probability can also be denoted as $\bar{P}(U_j)$. It is assumed that the response functions depend on some unknown fixed parameters ξ and that the population distribution function depends on some other fixed parameters τ.

The marginal probability of pattern with the fixed parameters can be explicitly written as

$$P(U_j) = \int P(U_j|\theta, \xi)\pi(\theta|\tau)d\theta. \tag{11.5}$$

For a scaling purpose, it is assumed that $\theta \sim N(0,1)$ and $\pi(\theta|\tau) = N(0,1)$, where $\tau = (\mu, \sigma^2)$. There can possibly be many different mixed item response models that can be applied to data obtained from the mixture of dichotomously-scored and polytomously-scored items. It is assumed in this chapter that the two-parameter logistic model and the logistic form of the graded response model are employed.

11.3 Parameter Estimation via Marginal Maximum Likelihood

11.3.1 Estimation of Item Parameters

Because the two-parameter logistic model can be formulated as the simplest case of the graded response model, all derivation for marginal maximum likelihood estimation will be first presented in terms of the graded response model in which

$$P_{ik}(\theta) = P^*_{i,k-1}(\theta) - P^*_{ik}(\theta), \tag{11.6}$$

where

$$P^*_{i,k-1}(\theta) = \frac{1}{1 + e^{-(\zeta_{i,k-1} + \lambda_i \theta)}}, \tag{11.7}$$

$$P^*_{ik}(\theta) = \frac{1}{1 + e^{-(\zeta_{ik} + \lambda_i \theta)}}, \tag{11.8}$$

$$P^*_{i0}(\theta) = 1, \tag{11.9}$$

and

$$P^*_{iK_i}(\theta) = 0. \tag{11.10}$$

It is convenient to employ distinct response patterns instead of individual response vectors with subscript j. Let r_l be the number of occurrence of pattern l, and let $L \le \min(N, \prod_{i=1}^{n} K_i)$, be the total number of patterns of r_l.

Assuming independence of examinees, the marginal likelihood can be written as

$$L(\xi) = \frac{N!}{\prod_{l=1}^{L} r_l!} \prod_{l=1}^{L} [P(U_l)]^{r_l}. \tag{11.11}$$

The likelihood equation for the parameters of item i can be expressed as

$$\frac{\partial \log L}{\partial \xi_i} = \sum_{l=1}^{L} \frac{r_l}{P_l} \frac{\partial P_l}{\partial \xi_i} = 0, \tag{11.12}$$

where $P_l = P(U_l)$. Let

$$L_l(\theta) = P(U_l|\theta) = \prod_{i=1}^{n} P(U_{li}|\theta)$$

and

$$u_{lik} = \begin{cases} 1 & \text{if } U_{li} = k \\ 0 & \text{otherwise} \end{cases},$$

then

$$\frac{\partial P_l}{\partial \xi_i} = \int \left(\frac{\frac{\partial}{\partial \xi_i} \{\prod_{k=1}^{K_i}[P_{ik}(\theta)]^{u_{lik}}\}}{\prod_{k=1}^{K_i}[P_{ik}(\theta)]^{u_{lik}}} \right) L_l(\theta)\pi(\theta)d\theta.$$

Because

$$\frac{\partial \prod_{k=1}^{K_i}[P_{ik}(\theta)]^{u_{lik}}}{\partial \xi_i} = \sum_{k=1}^{K_i} \frac{u_{lik}}{P_{ik}(\theta)} \frac{\partial P_{ik}(\theta)}{\partial \xi_i} \prod_{k=1}^{K_i}[P_{ik}(\theta)]^{u_{lik}},$$

it can be shown that

$$\frac{\partial \log L}{\partial \xi_i} = \sum_{l=1}^{L} \frac{r_l}{P_l} \int \left(\sum_{k=1}^{K_i} \frac{u_{lik}}{P_{ik}(\theta)} \frac{\partial P_{ik}(\theta)}{\partial \xi_i} \right) L_l(\theta)\pi(\theta)d\theta. \tag{11.13}$$

The likelihood equations for the graded response model item can be written as

$$\frac{\partial \log L}{\partial \zeta_{ik}} =$$

$$\sum_{l=1}^{L} \frac{r_l}{P_l} \int \left(-\frac{u_{lik}}{P_{ik}(\theta)} + \frac{u_{li,k+1}}{P_{i,k+1}(\theta)} \right) W_{ik}^*(\theta) L_l(\theta)\pi(\theta)d\theta = 0 \tag{11.14}$$

and

$$\frac{\partial \log L}{\partial \lambda_i} =$$

$$\sum_{l=1}^{L} \frac{r_l}{P_l} \int \left(\sum_{k=1}^{K_i} \frac{u_{lik}}{P_{ik}(\theta)} [W_{i,k-1}^*(\theta) - W_{ik}^*(\theta)]\theta \right) L_l(\theta)\pi(\theta)d\theta = 0, \tag{11.15}$$

where $W_{ik}^*(\theta) = P_{ik}^*(\theta)[1 - P_{ik}^*(\theta)]$ and the subscript k in ζ_i varies 1 to $K_i - 1$. Because θ is treated to be unobserved, we can apply the expectation and maximization (i.e., EM) solution (Bock & Aitkin, 1981; Dempster, Laird, & Rubin, 1977; Dempster, Rubin, & Tsutakawa, 1981). Because the expectation and maximization algorithm and the estimation of item parameters have been presented in earlier chapters, only important results are presented below.

The expectation step involves obtaining artificial data, $\bar{r}_{ik}$ and $\bar{N}$. Let

$$\bar{r}_{ik} = \sum_{l=1}^{L} \frac{r_l u_{lik} L_l(\theta)\pi(\theta)}{P_l} \tag{11.16}$$

and

$$\bar{N} = \sum_{l=1}^{L} \frac{r_l L_l(\theta) \pi(\theta)}{P_l}, \tag{11.17}$$

then

$$\frac{\partial \log L}{\partial \xi_i} = \int \sum_{k=1}^{K_i} \frac{\bar{r}_{ik}}{P_{ik}(\theta)} \frac{\partial P_{ik}(\theta)}{\partial \xi_i} d\theta. \tag{11.18}$$

Because the above equation contains integral that should be evaluated numerically, replacing the definite integral with the quadrature formula yields

$$\frac{\partial \log L}{\partial \xi_i} = \sum_{q=1}^{Q} \sum_{k=1}^{K_i} \frac{\bar{r}_{ikq}}{P_{ik}(X_q)} \frac{\partial P_{ik}(X_q)}{\partial \xi_i}, \tag{11.19}$$

where

$$\bar{r}_{ikq} = \sum_{l=1}^{L} \frac{r_l u_{lik} L_l(X_q) A(X_q)}{P_l} \tag{11.20}$$

and

$$P_l = \sum_{q=1}^{Q} L_l(X_q) A(X_q). \tag{11.21}$$

Note also that

$$\bar{N}_q = \sum_{l=1}^{L} \frac{r_l L_l(X_q) A(X_q)}{P_l}. \tag{11.22}$$

The maximization step tries to solve the Newton-Raphson equation for each item. With such terms as

$$\frac{\partial \log L}{\partial \zeta_{ik}} = \sum_{q=1}^{Q} \left(-\frac{\bar{r}_{ikq}}{P_{ik}(X_q)} + \frac{\bar{r}_{i,k+1,q}}{P_{i,k+1}(X_q)} \right) W_{ik}^*(X_q), \tag{11.23}$$

$$\frac{\partial \log L}{\partial \lambda_i} = \sum_{q=1}^{Q} \sum_{k=1}^{K_i} \frac{\bar{r}_{ikq}}{P_{ik}(X_q)} [W_{i,k-1}^*(X_q) - W_{ik}^*(X_q)] X_q, \tag{11.24}$$

$$E\left(\frac{\partial^2 \log L}{\partial \zeta_{ik}^2} \right) = -\sum_{q=1}^{Q} \bar{N}_q [W_{ik}^*(X_q)]^2 \left(\frac{1}{P_{ik}(X_q)} + \frac{1}{P_{i,k+1}(X_q)} \right), \tag{11.25}$$

$$E\left(\frac{\partial^2 \log L}{\partial \zeta_{ik} \partial \zeta_{i,k-1}} \right) = \sum_{q=1}^{Q} \bar{N}_q W_{ik}^*(X_q) W_{i,k-1}^*(X_q) \frac{1}{P_{ik}(X_q)}, \tag{11.26}$$

$$E\left(\frac{\partial^2 \log L}{\partial \zeta_{ik}\partial \zeta_{i,k+1}}\right) = \sum_{q=1}^{Q} \bar{N}_q W_{ik}^*(X_q) W_{i,k+1}^*(X_q) \frac{1}{P_{i,k+1}(X_q)}, \qquad (11.27)$$

$$E\left(\frac{\partial^2 \log L}{\partial \lambda_i^2}\right) = -\sum_{q=1}^{Q} \bar{N}_q X_q^2 \sum_{k=1}^{K_i} [W_{i,k-1}^*(X_q) - W_{ik}^*(X_q)]^2 \frac{1}{P_{ik}(X_q)}, (11.28)$$

$$E\left(\frac{\partial^2 \log L}{\partial \zeta_{ik}\partial \lambda_i}\right) = \sum_{q=1}^{Q} \bar{N}_q W_{ik}^*(X_q) X_q \times$$

$$\left(\frac{W_{i,k-1}^*(X_q) - W_{ik}^*(X_q)}{P_{ik}(X_q)} - \frac{W_{ik}^*(X_q) - W_{i,k+1}^*(X_q)}{P_{i,k+1}(X_q)}\right), \qquad (11.29)$$

$$E\left(\frac{\partial^2 \log L}{\partial \zeta_{ik}\partial \zeta_{ik'}}\right) = 0, \qquad (11.30)$$

where $|k-k'| > 1$ and $W_{ik}^*(X_q) = P_{ik}^*(X_q)[1 - P_{ik}^*(X_q)]$, the Newton-Raphson equation can be written as

$$\left[\hat{\xi}_i\right]_{(t+1)} = \left[\hat{\xi}_i\right]_{(t)} - \left[E\left(\frac{\partial^2 \log L}{\partial \xi_i \partial \xi_i'}\right)\right]_{(t)}^{-1} \left[\frac{\partial \log L}{\partial \xi_i}\right]_{(t)}, \qquad (11.31)$$

where t indexes the iteration. Equivalently,

$$\begin{bmatrix} \hat{\zeta}_{i1} \\ \hat{\zeta}_{i2} \\ \hat{\zeta}_{i3} \\ \hat{\zeta}_{i4} \\ \vdots \\ \hat{\zeta}_{ik} \\ \vdots \\ \hat{\zeta}_{i,K_i-1} \\ \hat{\lambda}_i \end{bmatrix}_{(t+1)} = \begin{bmatrix} \hat{\zeta}_{i1} \\ \hat{\zeta}_{i2} \\ \hat{\zeta}_{i3} \\ \hat{\zeta}_{i4} \\ \vdots \\ \hat{\zeta}_{ik} \\ \vdots \\ \hat{\zeta}_{i,K_i-1} \\ \hat{\lambda}_i \end{bmatrix}_{(t)} - \left[E\left(\frac{\partial^2 \log L}{\partial \xi_i \partial \xi_i'}\right)\right]_{(t)}^{-1} \begin{bmatrix} \frac{\partial \log L}{\partial \zeta_{i1}} \\ \frac{\partial \log L}{\partial \zeta_{i2}} \\ \frac{\partial \log L}{\partial \zeta_{i3}} \\ \frac{\partial \log L}{\partial \zeta_{i4}} \\ \vdots \\ \frac{\partial \log L}{\partial \zeta_{ik}} \\ \vdots \\ \frac{\partial \log L}{\partial \zeta_{i,K_i-1}} \\ \frac{\partial \log L}{\partial \lambda_i} \end{bmatrix}_{(t)} ,(11.32)$$

where the Hessian, before taking expectation, is defined as

$$\begin{bmatrix} \dfrac{\partial^2 \log L}{\partial \zeta_{i1}^2} & \dfrac{\partial^2 \log L}{\partial \zeta_{i1}\partial \zeta_{i2}} & 0 & 0 & \cdots & 0 & \cdots & 0 & \dfrac{\partial^2 \log L}{\partial \zeta_{i1}\partial \lambda_i} \\[2mm] \dfrac{\partial^2 \log L}{\partial \zeta_{i2}\partial \zeta_{i1}} & \dfrac{\partial^2 \log L}{\partial \zeta_{i2}^2} & \dfrac{\partial^2 \log L}{\partial \zeta_{i2}\partial \zeta_{i3}} & 0 & \cdots & 0 & \cdots & 0 & \dfrac{\partial^2 \log L}{\partial \zeta_{i2}\partial \lambda_i} \\[2mm] 0 & \dfrac{\partial^2 \log L}{\partial \zeta_{i3}\partial \zeta_{i2}} & \dfrac{\partial^2 \log L}{\partial \zeta_{i3}^2} & \dfrac{\partial^2 \log L}{\partial \zeta_{i3}\partial \zeta_{i4}} & \cdots & 0 & \cdots & 0 & \dfrac{\partial^2 \log L}{\partial \zeta_{i3}\partial \lambda_i} \\[2mm] 0 & 0 & \dfrac{\partial^2 \log L}{\partial \zeta_{i4}\partial \zeta_{i3}} & \dfrac{\partial^2 \log L}{\partial \zeta_{i4}^2} & \cdots & 0 & \cdots & 0 & \dfrac{\partial^2 \log L}{\partial \zeta_{i4}\partial \lambda_i} \\[2mm] \vdots & \vdots & \vdots & \vdots & \ddots & \vdots & \ddots & \vdots & \vdots \\[2mm] 0 & 0 & 0 & 0 & \cdots & \dfrac{\partial^2 \log L}{\partial \zeta_{ik}^2} & \cdots & 0 & \dfrac{\partial^2 \log L}{\partial \zeta_{ik}\partial \lambda_i} \\[2mm] \vdots & \vdots & \vdots & \vdots & \ddots & \vdots & \ddots & \vdots & \vdots \\[2mm] 0 & 0 & 0 & 0 & \cdots & 0 & \cdots & \dfrac{\partial^2 \log L}{\partial \zeta_{i,K_i-1}^2} & \dfrac{\partial^2 \log L}{\partial \zeta_{i,K_i-1}\partial \lambda_i} \\[2mm] \dfrac{\partial^2 \log L}{\partial \lambda_i \partial \zeta_{i1}} & \dfrac{\partial^2 \log L}{\partial \lambda_i \partial \zeta_{i2}} & \dfrac{\partial^2 \log L}{\partial \lambda_i \partial \zeta_{i3}} & \dfrac{\partial^2 \log L}{\partial \lambda_i \partial \zeta_{i4}} & \cdots & \dfrac{\partial^2 \log L}{\partial \lambda_i \partial \zeta_{ik}} & \cdots & \dfrac{\partial^2 \log L}{\partial \lambda_i \partial \zeta_{i,K_i-1}} & \dfrac{\partial^2 \log L}{\partial \lambda_i^2} \end{bmatrix}$$

Now, consider that the two-parameter logistic model is used for some items. The two-parameter logistic model can be considered as the simplest case of the logistic graded response model with two categories of response, $k = 1$ for incorrect response and $k = 2$ for correct response,

$$P_{i1}(\theta) = 1 - P_{i1}^*(\theta) \tag{11.33}$$

and

$$P_{i2}(\theta) = P_{i1}^*(\theta) - 0, \tag{11.34}$$

where

$$P_{i1}^*(\theta) = \frac{1}{1 + e^{-(\zeta_{i1}+\lambda_i\theta)}}. \tag{11.35}$$

Because there exists only one intercept parameter, let us use ζ_i for ζ_{i1}. In addition, let $P_i(\theta) = P_{i2}(\theta)$, $Q_i(\theta) = P_{i1}(\theta)$, $u_{li} = u_{li2}$, and $1 - u_{li} = u_{li1}$. The likelihood equation of the two-parameter logistic model can be written as

$$\frac{\partial \log L}{\partial \xi_i} = \sum_{l=1}^{L} \frac{r_l}{P_l} \int \left(\frac{u_{li} - P_i(\theta)}{W_i(\theta)} \right) \frac{\partial P_i(\theta)}{\partial \xi_i} L_l(\theta)\pi(\theta)d\theta = 0, \tag{11.36}$$

where $W_i(\theta) = P_i(\theta)Q_i(\theta)$.

Let

$$\bar{r}_i = \sum_{l=1}^{L} \frac{r_l u_{li} L_l(\theta)\pi(\theta)}{P_l} \tag{11.37}$$

and

$$\bar{N} = \sum_{l=1}^{L} \frac{r_l L_l(\theta)\pi(\theta)}{P_l} \tag{11.38}$$

(i.e., $\bar{r}_{i1} = \bar{N} - \bar{r}_i$ and $\bar{r}_{i2} = \bar{r}_i$), then

$$\frac{\partial \log L}{\partial \xi_i} = \int \left(\frac{\bar{r}_i - \bar{N} P_i(\theta)}{W_i(\theta)} \right) \frac{\partial P_i(\theta)}{\partial \xi_i} d\theta. \tag{11.39}$$

Replacing the definite integral with the quadrature formula yields

$$\frac{\partial \log L}{\partial \xi_i} = \sum_{q=1}^{Q} \left(\frac{\bar{r}_{iq} - \bar{N}_q P_i(X_q)}{W_i(X_q)} \right) \frac{\partial P_i(X_q)}{\partial \xi_i}, \tag{11.40}$$

where

$$\bar{r}_{iq} = \sum_{l=1}^{L} \frac{r_l u_{li} L_l(X_q) A(X_q)}{P_l}, \tag{11.41}$$

$$\bar{N}_q = \sum_{l=1}^{L} \frac{r_l L_l(X_q) A(X_q)}{P_l}, \tag{11.42}$$

$$P_l = \sum_{q=1}^{Q} L_l(X_q) A(X_q), \tag{11.43}$$

and

$$W_i(X_q) = P_i(X_q) Q_i(X_q). \tag{11.44}$$

With such terms as

$$\frac{\partial \log L}{\partial \zeta_i} = \sum_{q=1}^{Q} [\bar{r}_{iq} - \bar{N}_q P_i(X_q)], \tag{11.45}$$

$$\frac{\partial \log L}{\partial \lambda_i} = \sum_{q=1}^{Q} [\bar{r}_{iq} - \bar{N}_q P_i(X_q)] X_q, \tag{11.46}$$

$$\frac{\partial^2 \log L}{\partial \zeta_i^2} = -\sum_{q=1}^{Q} \bar{N}_q W_i(X_q), \tag{11.47}$$

$$\frac{\partial^2 \log L}{\partial \lambda_i^2} = -\sum_{q=1}^{Q} \bar{N}_q W_i(X_q) X_q^2, \tag{11.48}$$

and

$$\frac{\partial^2 \log L}{\partial \zeta_i \partial \lambda_i} = -\sum_{q=1}^{Q} \bar{N}_q W_i(X_q) X_q, \tag{11.49}$$

the Newton-Raphson equation can be written as

$$\left[\hat{\xi}_i \right]_{(t+1)} = \left[\hat{\xi}_i \right]_{(t)} - \left[\frac{\partial^2 \log L}{\partial \xi_i \partial \xi_i'} \right]_{(t)}^{-1} \left[\frac{\partial \log L}{\partial \xi_i} \right]_{(t)}, \tag{11.50}$$

where t indexes the iteration. Equivalently,

$$
\begin{bmatrix} \hat{\zeta}_i \\ \hat{\lambda}_i \end{bmatrix}_{(t+1)} = \begin{bmatrix} \hat{\zeta}_i \\ \hat{\lambda}_i \end{bmatrix}_{(t)} - \begin{bmatrix} \frac{\partial^2 \log L}{\partial \zeta_i^2} & \frac{\partial^2 \log L}{\partial \zeta_i \partial \lambda_i} \\ \frac{\partial^2 \log L}{\partial \lambda_i \partial \zeta_i} & \frac{\partial^2 \log L}{\partial \lambda_i^2} \end{bmatrix}_{(t)}^{-1} \begin{bmatrix} \frac{\partial \log L}{\partial \zeta_i} \\ \frac{\partial \log L}{\partial \lambda_i} \end{bmatrix}_{(t)}. \tag{11.51}
$$

The expectation and maximization steps are repeated until the convergence criterion is met. It can be noted that the equations presented for the two-parameter logistic model are entirely parallel to those for the graded response model. The main difference between two sets of equations is the size of the Newton-Raphson equations. Thus it is possible to use the marginal maximum likelihood estimation procedure for the graded response model to estimate item parameters in this specific mixed model case of the two-parameter logistic model and the graded response model. Appendix J shows how it is implemented.

11.3.2 Estimation of Ability Parameters

The marginal maximum likelihood estimation procedure using the expectation and maximization algorithm yields only item parameter estimates. The ability estimation procedure assumes that the item parameters are the true values.

Ability parameters can be estimated using the method of maximum likelihood, the expected a posteriori estimation, or the maximum a posteriori estimation. For example, the expected a posteriori estimate of ability for response pattern l, $\bar{\theta}_l$ (i.e., mean of the posterior distribution), can be obtained as

$$
\bar{\theta}_l = \frac{\int \theta L_l(\theta) \pi(\theta) d\theta}{P_l}. \tag{11.52}
$$

Note that l can be seen as the combination of l_d and l_p, where l_d is the response pattern for the dichotomously-scored items and l_p is the response pattern for the polytomously-scored items. Also note that the likelihood can be separated into two pieces,

$$
L_l(\theta) = L_{l_d}(\theta) L_{l_p}(\theta). \tag{11.53}
$$

The estimated variance for $\bar{\theta}_l$ is

$$
\hat{\sigma}^2(\bar{\theta}_l) = \frac{\int (\theta - \bar{\theta}_l)^2 L_l(\theta) \pi(\theta) d\theta}{P_l} \tag{11.54}
$$

(cf. Billeaud et al., 1998).

11.4 Research on Mixed Item Types

Several papers that analyzed data obtained with mixed item types have been published. Because any combination of the dichotomous models and the polytomous models can be used to analyze such data, it is of interest to look into the models used in these papers. In Billeaud et al. (1997), the Birnbaum's three-parameter model and the graded response were used. In Ercikan et al. (1998), the Birnbaum's three-parameter model and the two parameter partial credit model (Yen, 1993), that is equivalent to the generalized partial credit model, were used. In Grima and Weichun (2002), the Birnbaum's three-parameter model and the generalized partial credit model were used. In these studies, the marginal maximum likelihood estimation procedures were used for the estimation of item parameters, and the use of mixed item types was discussed especially for the combination of the multiple-choice items and the constructed-response items. When data with mixed item types are analyzed under the Rasch framework, the usual choice seems to be the combination of the one-parameter logistic model and the partial credit model.

Ercikan et al. (1998) presented calibration of mixed item types and scoring of tests with multiple-choice items and constructed-response items. According to Ercikan et al., the reason why we use different item types is to enhance psychometric characteristics of the test. It is, hence, believed that mixed item types enhance test reliability, score validity, and cost efficiency. Ercikan et al. indicated that both types of items could be calibrated together, but there seemed to be some loss of information for the constructed-response items from the simultaneous calibration. Marginal maximum likelihood estimation (Bock & Aitkin, 1981) was used to obtain item parameter estimates, and the method of maximum likelihood was used to obtain ability estimates for respective response patterns. It can be noted that, in Ercikan et al. (1998), the item response theory model assumptions were evaluated by examining fit statistics and local item dependence statistics.

An empirical comparison among various scoring methods of assessment data consists of multiple-choice items and constructed-response items was reported by Grima and Weichun (2002) in conjunction with the marginal maximum likelihood estimation of item parameters. In their study, six scoring procedures, including procedures that utilized testlets and subtests, were compared. Scaling of the multiple-choice items and the constructed-response items simultaneously yielded the best result based on the fit analysis. Note that calibration of items from mixed models simultaneously may produce statistically optimal scale that ideally combine the information from respective models. The ability estimates for respective response patterns are based on the implicit weighting of the multiple-choice items and constructed-response items.

Calibration of the data using different item response theory models is based on several assumptions. All items regardless of types measure the same latent trait (i.e., unidimensionality) can be seen as the main assumption. Factor analysis of item data, assessing local independence, and the examination of model fit also are important considerations (see e.g., Bennett, Rock, & Wang, 1991; Bridgeman & Rock, 1993; Thissen, Wainer, & Wang, 1994).

11.5 Summary

When examinees are assessed using different types of items that are scored dichotomously and polytomously, how to combine information from different types of items is an important practical issue. Although mixed item types in general require different item response theory models to produce a common scale, it is possible to use just one inclusive model to analyze item response data. In this chapter, estimation of item parameters is presented for the case where the dichotomously-scored items and the polytomously-scored items are used. The two-parameter logistic model and the logistic form of the graded response model were used, and equations were presented for marginal maximum likelihood estimation of item parameters. Ability estimation is only briefly mentioned.

Several scoring procedures in the context of mixed item response theory models were compared in various studies. The response pattern scores (e.g., the expected a posteriori estimates) seem to provide an optimal result assuming that the item response models fits data. Ability estimation under item response theory uses implicit weights of items whereas the scoring method under classical test theory uses somewhat explicit weights to maximize the reliability of the resulting scores (cf. Billeaud et al., 1998; Rosa et al., 2001; Rudner, 2001, Wainer & Thissen, 1993).

Computer programs, PARSCALE (Muraki & Bock, 2003), MULTILOG (Thissen, Chen, & Bock, 2003), WINSTEPS (Linacre, 2003), and OPLM (Verhelst, Glas, & Verstralen, 1995), can be used to estimate item and ability parameters under the mixed item response models. Appendix J contains BASIC code for the estimation of item parameters in the mixed models, as a special case presented in this chapter.

12. Parameter Estimation via Gibbs Sampler

12.1 Introduction

This chapter presents a relatively new estimation method in item response theory, namely, Gibbs sampler. Gibbs sampler is a member of a new class of procedures known as Markov chain Monte Carlo techniques.

An introduction to Gibbs sampler is provided by Casella and George (1992). Geman and Geman (1984), in a seminal paper in this area, developed Gibbs sampler in the context of image processing. However, the origins of Markov chain Monte Carlo methods and Gibbs sampler can be traced to the work of Metropolis and Ulam (1949) and Metropolis, Rosenbluth, Rosenbluth, Teller, and Teller (1953). Although Gibbs sampler had primarily been used within Bayesian statistics (Smith & Roberts, 1993), Gelfand and Smith (1990) showed that Gibbs sampler can also be used in a variety of non-Bayesian statistical problems. Note that Markov chain Monte Carlo methods are computationally intensive, which has limited their practicality in the past. Fortunately, the widespread availability of powerful personal computers has made the computationally intensive Monte Carlo methods feasible.

Before examining Gibbs sampler, let us mention the basic structure of the Markov chain Monte Carlo techniques. Let W be a vector of the random variables. Suppose it is feasible to generate a sequence of W,

$$\{W^{(0)}, W^{(1)}, W^{(2)}, \ldots, W^{(t)}, W^{(t+1)}, \ldots\}, \tag{12.1}$$

such that each time t, the next state $W^{(t+1)}$ is sampled from a distribution $p(W^{(t+1)}|W^{(t)})$ that depends on the current state of the chain. This sequence is called a Markov chain, and $p(W^{(t+1)}|W^{(t)})$ is the transition kernel of the Markov chain which can also be denoted as $k(W^{(t)}, W^{(t+1)})$. Markov chains are constructed from a transition kernel k, a conditional probability density that

$$W^{(t+1)} \sim k(W^{(t)}, W^{(t+1)}). \tag{12.2}$$

The chains encountered in Markov chain Monte Carlo settings have a very strong stability property. A stationary probability distribution exists by construction; that is, a distribution π such that if $W^{(t)} \sim \pi$, the $W^{(t+1)} \sim \pi$, provided that the kernel k allows for free moves all over the state space. Gelfand and Smith (1990), Gelman, Carlin, Stern, and Rubin (1995), Robert and

Casella (1999), and Tanner (1991, 1996) wrote excellent overviews of Markov chain Monte Carlo methods. For Bayesian inference the Markov chain can be defined in a way that the stationary distribution π is the posterior distribution. For example, $\pi(W)$ can be seen as $p(\theta, \xi|Y)$, where θ is the vector of ability parameters, ξ is the vector of item parameters, and Y is the item response matrix.

The transition kernel of Gibbs sampler in the context of item response theory can be defined as

$$k[(\theta^{(t)}, \xi^{(t)}), (\theta^{(t+1)}, \xi^{(t+1)})] = p(\theta^{(t+1)}|\xi^{(t)}, Y)p(\xi^{(t+1)}|\theta^{(t+1)}, Y) \quad (12.3)$$

(Patz & Junker, 1999a). Because

$$p(\theta|\xi, Y) = \frac{p(Y|\theta, \xi)p(\theta, \xi)}{\int p(Y|\theta, \xi)p(\theta, \xi)d\theta} \quad (12.4)$$

and

$$p(\xi|\theta, Y) = \frac{p(Y|\theta, \xi)p(\theta, \xi)}{\int p(Y|\theta, \xi)p(\theta, \xi)d\xi}, \quad (12.5)$$

implementing Gibbs sampler may require computing of the normalizing constants appeared in the denominators. According to Patz and Junker (1999a), recent Gibbs sampler methods have been developed to simplify or circumvent the calculations of the normalizing constants.

The following sections presents two Gibbs sampler methods applied to the estimation of parameters in item response models; Albert's (1992) Gibbs sampler for the two-parameter normal ogive model and Gibbs sampler using rejection sampling (Ripley, 1987) for the one-parameter logistic model. Albert's Gibbs sampler can be seen as a non-Bayesian version of Gibbs sampler, particularly for the estimation of item parameters because no priors are imposed on item parameters. Gibbs sampler using rejection sampling is of interest because the computer program BUGS (Spiegelhalter, Thomas, Best, & Gilks, 1997) is readily available to analyze item response data. In addition to these two methods, a Bayesian version of Albert's Gibbs sampler (Johnson & Albert, 1999) is also presented subsequently; it can be considered as an extension of the earlier non-Bayesian version of Gibbs sampler. The actual implementation in BASIC is presented in Appendix K only for Albert's Gibbs sampler. The detailed explications of Gibbs sampler using rejection sampling and the Bayesian version of Albert's Gibbs sampler, however, are necessary because different parameterizations of item response theory models are used by various authors in their publication. Note that there are other Markov chain Monte Carlo methods suggested in item response theory, including the Metropolis-Hastings algorithm and the Metropolis-Hastings within Gibbs algorithm (see Patz & Junker, 1999a).

12.2 Albert's Gibbs Sampler

12.2.1 The Logic of Gibbs Sampler

Assume that item response data of N examinees are modeled using the two-parameter normal ogive model. Within the general rubric of the Markov chain Monte Carlo method, the Gibbs sampler approach to the joint estimation of the $2n$ item parameters of a test and the N examinee ability parameters involves a Markov chain. At each stage t, or cycle t, of the Markov chain, the sampling of parameters from the joint posterior distribution of the $2n + N$ parameters is simulated by sampling from the conditional posterior distribution of each parameter (i.e., the Monte Carlo component of the Markov chain Monte Carlo method). Thus, the difficult task of sampling from a multidimensional joint posterior distribution is reduced to sampling from each of $2n + N$ one-dimensional distributions.

For Gibbs sampler, as employed in item response theory, each examinee's ability parameter is sampled from its posterior distribution, conditional on the values of all the remaining item and examinee parameters. Then each pair of item parameters is sampled from their posterior distribution conditional on the values of all of the remaining item and examinee parameters. The resulting $2n + N$ parameter values are then saved as if they were a sample from their joint posterior distribution. The parameter values contained in the sample or draw at stage t serve as the conditional values in the next stage $t + 1$ of the Markov chain. Over a sequential series of such stages, the sampled parameters should approach those obtained from the underlying joint posterior distribution, and the Markov chain will have reached a stable stage or converged. Once convergence is reached, the sampling process is continued over a large number of such stages to empirically determine the joint posterior distribution of the $2n + N$ parameters. A measure of central tendency, such as the mean of each parameter's marginal posterior distribution, is used as an estimate of that parameter.

12.2.2 Albert's Implementation of Gibbs Sampler

The details of this Gibbs sampler can be seen by examining its implementation by Albert (1992), which was done using MATLAB (The MathWorks, Inc., 1996). The two-parameter normal ogive model was used for the item response function. Other than the constraint that the item discriminations must be greater than 0, no priors were employed for the item parameters. The probit was given by

$$Z_{ij} = \alpha_i \theta_j - \gamma_i, \tag{12.6}$$

where α_i is the item discrimination parameter, $i = 1, 2, \ldots, n$, γ_i is the item difficulty parameter, and θ_j is the trait level for examinee j, $j = 1, 2, \ldots, N$.

However, this is an atypical parameterization of item difficulty. The usual item response theory parameterization is

$$Z_{ij} = \alpha_i(\theta_j - \beta_i) = \alpha_i\theta_j + \zeta_i, \qquad (12.7)$$

where ζ_i is the intercept, and $\beta_i = -\zeta_i/\alpha_i$ is the item difficulty. Thus, Albert's item difficulty parameter is $\gamma_i = \alpha_i\beta_i = \alpha(-\zeta_i/\alpha_i) = -\zeta_i$. The item difficulty is the negative of the usual intercept and $\beta_i = \gamma_i/\alpha_i$ is the usual item response theory item difficultly. It is also possible to use λ_i instead of α_i.

The dichotomously scored item responses of the examinees are stored in the $n \times N$ matrix Y. An item's parameters are given by $\xi_i = (\alpha_i, \gamma_i)$ and the matrix of all item parameters is given by $\xi = (\xi_1, \xi_2, \ldots, \xi_n)$. The vector of ability parameters is $\theta = (\theta_1, \theta_2, \ldots, \theta_N)$. Let Z be the vector of independent random variables $Z = (Z_{11}, Z_{12}, \ldots, Z_{ij}, \ldots, Z_{nN})$. Note that Z can be equivalently defined by a matrix. The implementation of the Monte Carlo component in this Gibbs sampler involves three sampling processes: (1) a sampling of probits that is dependent on the examinee's item responses, (2) a sampling of ability parameters, and (3) a sampling of $2n$ item parameters. Because the sampling of probits, the transition kernel can be written as

$$k[(Z^{(t)}, \theta^{(t)}, \xi^{(t)}), (Z^{(t+1)}, \theta^{(t+1)}, \xi^{(t+1)})] = \qquad (12.8)$$

$$p(Z^{(t+1)}|\theta^{(t)}, \xi^{(t)}, Y)p(\theta^{(t+1)}|\xi^{(t)}, Z^{(t+1)}, Y)p(\xi^{(t+1)}|Z^{(t+1)}, \theta^{(t+1)}, Y),$$

and this implementation can be seen as Gibbs sampler with data augmentation. Initial values of the parameters are set to take $\alpha_i = 2$, $\gamma_i = \sqrt{5}$ times the unit normal deviate corresponding to the classical test theory item difficulty, and $\theta_j = 0$ (Albert, 1992). The three sampling processes are explained below.

First, let us discuss the sampling of probits. Using the current values of all the parameters, the conditional distributions of the Z_{ij}, with means $\eta_{ij} = \alpha_i\theta_j - \gamma_i$ and unit variance, are established and each is divided into two parts. The area above $Z_{ij} = 0$ corresponds to the probability of a correct response P_{ij}, and the area below to the probability of an incorrect response Q_{ij}. When an examinee responds correctly to the item, $Y_{ij} = 1$ and Z_{ij} is randomly sampled from the part of the normal distribution above 0; an incorrect response results in a Z_{ij} that is randomly sampled from the part below 0. This process is performed for all elements in the vector Z. Although complex, the process results in a matrix of probits that are dependent on the current values of ξ, θ, and the item response matrix Y.

Second process is sampling from an examinee's conditional posterior ability distribution. It is assumed that θ_j is independently distributed. The likelihood function for θ_j is the normal form with mean

$$\tilde{\theta}_j = \frac{\sum_{i=1}^n \alpha_i(Z_{ij} + \gamma_i)}{\sum_{i=1}^n \alpha_i^2} \qquad (12.9)$$

with variance

$$v = \frac{1}{\sum_{i=1}^{n} \alpha_i^2}. \tag{12.10}$$

This is combined with a normal prior distribution to yield a normally distributed posterior distribution of θ_j with mean

$$\bar{\theta}_j = \frac{\tilde{\theta}_j/v + \mu/\sigma^2}{1/v + 1/\sigma^2} \tag{12.11}$$

and variance

$$v_p = \frac{1}{1/v + 1/\sigma^2}, \tag{12.12}$$

where the parameters of the normal prior are fixed at $\mu = 0$ and $\sigma^2 = 1$. The calculations proceeds computing $\tilde{\theta}$ for an examinee and then obtaining $\bar{\theta}$ and v_p (i.e., the variance of the posterior distribution). Next, a random normal deviate is generated and a sampled θ_j value is given by

$$\theta_j^{(t)} = RN\sqrt{v_p} + \bar{\theta}_j, \tag{12.13}$$

where t designates the sampling cycle and RN is the random normal deviate. The above steps are performed for all N examinees, yielding a new vector of ability parameters.

The last process is sampling from the item parameter conditional posterior distributions. The item parameters in the vector ξ are sampled, conditional on the current values in Z and θ. There are four steps as follows:

1. The matrix X is defined as

$$X = \begin{bmatrix} \hat{\theta}_1 & -1 \\ \hat{\theta}_2 & -1 \\ \hat{\theta}_3 & -1 \\ \vdots & \vdots \\ \hat{\theta}_N & -1 \end{bmatrix} \tag{12.14}$$

and the Hessian is given by

$$X'X = \begin{bmatrix} \sum_{i=1}^{N} \hat{\theta}_j^2 & -\sum_{i=1}^{N} \hat{\theta}_j \\ -\sum_{i=1}^{N} \hat{\theta}_j & N \end{bmatrix}. \tag{12.15}$$

2. An ordinary least-squares solution for the simultaneous estimation of the parameters in ξ is performed using

$$\tilde{\xi} = (X'X)^{-1}(X'Z), \tag{12.16}$$

yielding a $2 \times n$ matrix of item parameter values,

$$\tilde{\xi} = \begin{bmatrix} \tilde{\alpha}_1 & \tilde{\alpha}_2 & \cdots & \tilde{\alpha}_n \\ \tilde{\gamma}_1 & \tilde{\gamma}_2 & \cdots & \tilde{\gamma}_n \end{bmatrix}. \tag{12.17}$$

3. A Cholesky factorization,

$$A = chol(X'X)^{-1}, \tag{12.18}$$

is performed where A is an upper triangular matrix and A' is its transpose, and $(X'X)^{-1}$ is the posterior variance of ξ_i for each of the n items. Note that the matrix A is the same for all items in the test.

4. Two random normal deviates, RN_1 and RN_2, are generated and the item parameter values for an item of the tth sampling stage are given by

$$\begin{bmatrix} \alpha_i^{(t)} \\ \gamma_i^{(t)} \end{bmatrix} = A' \begin{bmatrix} RN_1 \\ RN_2 \end{bmatrix} + \begin{bmatrix} \tilde{\alpha}_i \\ \tilde{\gamma}_i \end{bmatrix}. \tag{12.19}$$

The net effect is that $\alpha_i^{(t)}$ and $\gamma_i^{(t)}$ have been sampled from a bivariate normal distribution with a variance-covariance matrix defined by A. However, the constraint $\alpha_i > 0$ is imposed on an item's $\alpha_i^{(t)}$ and Step 4 is repeated until the constraint is met. Step 4 is performed for all n items in the test.

At this point, a complete set of new parameter values in θ and ξ has been obtained and stored as a draw from the joint posterior distribution of the parameters. Note that the Monte Carlo component of the Markov chain Monte Carlo process bears greater similarity to Birnbaum's (1968) joint maximum likelihood estimation paradigm than it does to the expectation and maximization algorithm. The former goes back and forth between the item parameter estimation and the ability estimation, as does the Albert's Gibbs sampler. The latter estimates only the item parameters.

12.2.3 Continuing the Markov Chain

The Monte Carlo process described above is repeated at each cycle of the Markov chain until the number of samples needed to form the empirical marginal posterior distribution of each parameter has been obtained. Under the Markov chain Monte Carlo approach, it is necessary for the Markov chain to reach a stable state before samples are retained. When this point in the chain of samples is reaches, the Markov chain is said to have reached convergence (i.e., the sampling is approximating the same joint posterior distribution of the $2n + N$ parameters at each stage). There is considerable discussion as to the number of samples required to reach convergence (see Cowles & Carlin, 1996, for a review of this issue). In Albert's (1992), at the 10th cycle the samples values are assumed to reach convergence, and the values are retained.

One problem with the Markov chain process is that, because successive draws are not independent, there is an autocorrelation between the parameter values in the stages. One solution to this problem is to retain only a spaced subset of samples, for example, saving every 5th sample. The values of the interval of stages is selected such that the samples retained are separated far enough so that the autocorrelation between them is negligible and they can be considered independent. After convergence has been retained, it is reasonable to set the value of the interval to 5 and the total number of the recorded sample values to 200 (Albert, 1992). In such a setting, the total number of samples generated is $10 + (200 - 1) \times 5$. Thus, for each run of the Markov chain, 1,005 draws would be generated, but only 200 would be retained. Previous research suggests that choice of the length of the Markov chain and the number of draws to retain depend on the situation (see Gelman et al., 1995, for a discussion of approaches to this issue).

At this point in the Gibbs sampler procedure, a separate marginal posterior distribution for the parameters of each item and for every examinee's ability parameter exist, each with the total number of the recorded sample values. The mean of each distribution is computed and reported as the Gibbs sampler estimate of the given parameter. This results in n pairs of item parameter estimates and N examinee ability estimates.

12.3 Gibbs Sampler: Another Approach

This section presents Gibbs sampler for the Rasch model using rejection sampling (Ripley, 1987) implemented in the computer program BUGS (Spiegelhalter et al., 1997). Gibbs sampler for a general case is presented first and detailed steps of Gibbs sampling for the Rasch model are presented subsequently. Note that some concepts and formulas from the previous sections are modified and reiterated again in this section.

12.3.1 Gibbs Sampler for Item Response Models

The Gibbs sampler algorithm is as follows (Gelfand & Smith, 1990; Tanner, 1996). First, instead of using θ and ξ, let $\omega = (\theta, \xi)$ be a vector of parameters with k elements. Suppose that the full or complete conditional distributions, $p(\omega_i|\omega_j, Y)$, where $i = 1, \ldots, k$ and $j \neq i$, are available for sampling. That is, samples may be generated by some method given values of the appropriate conditioning random variables. Then given an arbitrary set of starting values with the superscript (0), the algorithm starts. Using the ω notation again, the vectors $\omega^{(0)}, \ldots, \omega^{(t)}, \ldots$ are a realization of a Markov chain with a transition probability from $\omega^{(t)}$ to $\omega^{(t+1)}$ given by

$$p(\omega^{(t)}, \omega^{(t+1)}) = \prod_{l=1}^{k} p(\omega_l^{(t+1)}|\omega_j^{(t)}, j > l, \omega_j^{(t+1)}, j < l, Y). \qquad (12.20)$$

Using θ and ξ, the $(t + 1)$th iteration performs the following:

Draw $\theta_1^{(t+1)}$ from $p(\theta_1|\theta_2^{(t)},\ldots,\theta_N^{(t)},\xi_1^{(t)},\ldots,\xi_n^{(t)},Y)$.
Draw $\theta_2^{(t+1)}$ from $p(\theta_2|\theta_1^{(t+1)},\theta_3^{(t)},\ldots,\theta_N^{(t)},\xi_1^{(t)},\ldots,\xi_n^{(t)},Y)$.

$\vdots$

Draw $\theta_N^{(t+1)}$ from $p(\theta_N|\theta_1^{(t+1)},\ldots,\theta_{N-1}^{(t+1)},\xi_1^{(t)},\ldots,\xi_n^{(t)},Y)$.
Draw $\xi_1^{(t+1)}$ from $p(\xi_1|\theta_1^{(t+1)},\ldots,\theta_N^{(t+1)},\xi_2^{(t)},\ldots,\xi_n^{(t)},Y)$.
Draw $\xi_2^{(t+1)}$ from $p(\xi_2|\theta_1^{(t+1)},\ldots,\theta_N^{(t+1)},\xi_1^{(t+1)},\xi_3^{(t)},\ldots,\xi_n^{t)},Y)$.

$\vdots$

Draw $\xi_n^{(t+1)}$ from $p(\xi_n|\theta_1^{(t+1)},\ldots,\theta_N^{(t+1)},\xi_1^{(t+1)},\ldots,\xi_{n-1}^{(t+1)},Y)$.

Using the ω notation again, the joint distribution of $\omega^{(t)}$ converges geometrically to the posterior distribution $p(\omega|Y)$ as $t \to \infty$ (Bernardo & Smith, 1994; Geman & Geman, 1984). In particular, $\omega_i^{(t)}$ tends to be distributed as a random quantity whose density is $p(\omega_i|Y)$. Now suppose that there exist m replications of the t iterations. For large t, the replicates $\omega_{i1}^{(t)},\ldots,\omega_{im}^{(t)}$ are approximately a random sample from $p(\omega_i|Y)$. If we make m reasonably large, then an estimate, $\hat{p}(\omega_i|Y)$, can be obtained either as a kernel density estimate derived from the replicates or as

$$\hat{p}(\omega_i|Y) = \frac{1}{m}\sum_{l=1}^{m}p(\omega_i|\omega_{jl}^{(t)},j \neq i,Y). \tag{12.21}$$

In the context of item response theory, Gibbs sampler tries to obtain or sample sets of parameters using the joint posterior density $p(\theta,\xi|Y)$. Because the sampling of a particular parameter is conditioned upon all other parameters and the data, the simulated sample of parameters represents a sample from the marginalized posterior. Inferences with regard to parameters can then be made using the sampled parameters. Note that inference for both θ and ξ can be made from the Gibbs sampler procedure.

12.3.2 Steps of Gibbs Sampler

Gibbs sampler uses the following four basic steps (cf. Spiegelhalter, Best, Gilks, & Inskip, 1996):

1. Full conditional distributions and sampling methods for unobserved parameters must be specified.
2. Starting values must be provided.
3. Output must be monitored.
4. Summary statistics (e.g., estimates and standard errors) for quantities of interest must be calculated.

Discussion of the four steps involved are presented in detail below. As mentioned earlier, the main feature of Markov chain Monte Carlo methods is to obtain a sample of parameter values from the posterior density (Tanner,

1996). The sample of parameter values then can be used to estimate some functions or moments (e.g., mean and variance) of the posterior density of the parameter of interest. In other item response theory estimation procedures, the task is generally to obtain modes of the likelihood function.

12.3.3 Model Specifications

The model specifications are used as input to BUGS. In an item response data set, the item responses Y_{ij} are independent, conditional on their parameters in P_{ij}. For item i and examinee j, each P_{ij} is a function of the ability parameter θ_j, the location parameter β_i, and the slope parameter α under the Rasch model (cf. Thissen, 1982). The θ_j are assumed to be independently drawn from a standard normal distribution for scaling purposes. Figure 12.1 is adopted from Spiegelhalter, Thomas, Best, & Gilks (1996) and shows a directed acyclic graph based on these assumptions. It is only possible to proceed by following the directions of the arrows. Each variable or quantity in the model appears as a node in the graph, and directed links correspond to direct dependencies as specified above. The solid arrow denotes the probabilistic dependency, while dashed arrows indicate functional or deterministic relationships. The model can be seen as directed because each link between nodes is represented as an arrow. The model can also be seen as acyclic because it is impossible to return to a node after leaving.

It may be helpful to use the following definitions: Let v be a node in the graph, and V be the set of all nodes. A parent of v is defined as any node with an arrow extending from it and pointing to v, and a descendant of v is defined as any node on a direct path beginning from v. For identifying parents and descendants, deterministic links should be combined so that, for example, the parent of Y_{ij} is P_{ij}. It is assumed in Figure 12.1 for any node v, if we know the value of its parents, then no other nodes would be informative concerning v except descendants of v.

Lauritzen, Dawid, Larsen, and Leimer (1990) indicated that the directed acyclic graph model is equivalent to assuming that the joint distribution of all the random quantities is fully specified in terms of the conditional distribution of each node given its parents:

$$P(V) = \prod_{v \in V} P(v|\text{parents}[v]), \tag{12.22}$$

where $P(\cdot)$ denotes a probability distribution. This factorization not only allows extremely complex models to be built up from local components, but also provides an efficient basis for the implementation of Markov chain Monte Carlo methods (Spiegelhalter, Best, et al., 1996). For any node v, the remaining nodes are denoted by $V-v$. It follows that the full conditional distribution, $P(v|V-v)$, has the form

$$P(v|V-v) \propto P(v, V-v)$$

Fig. 12.1. A directed acyclic graph for the Rasch model

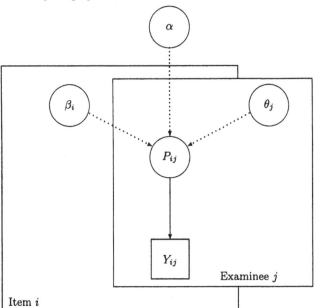

$$\propto P(v|\text{parent}[v]) \prod_{w\in\text{children}[v]} P(w|\text{parents}[w]). \qquad (12.23)$$

The proportionality constant, which is a function of the remaining nodes, ensures that the distribution is a probability function that integrates to unity.

Gibbs sampler implemented in BUGS works by iteratively drawing samples from the full conditional distributions of the model parameters using the adaptive rejection sampling algorithm (Gilks, 1996; Gilks & Wild, 1992; Ripley, 1987) because the full conditional distributions for the Rasch model are log-concave (Ghosh, Ghosh, Chen, & Agresti, 1999). For example, to sample θ from $p(\theta|\xi, Y) \propto p(Y|\theta, \xi)p(\theta, \xi)$, rejection sampling requires an envelope function $P(\theta)$. The envelope function is a convenient, proposal distribution that can be easily defined. Samples, say θ^*, are drawn from the density proportional to $P(\theta)$ and each θ^* is accepted with probability $p(\theta^*|\xi, Y)/P(\theta^*)$; that is, a uniform random variable u is generated and the θ^* will be accepted only if $u \le p(\theta^*|\xi, Y)/P(\theta^*)$ or rejected otherwise. It is essential that the proposal distribution is close to $p(\theta|\xi, Y)$. Note that ξ can be sampled similarly from $p(\xi|\theta, Y)$ using $P(\xi)$. In BUGS the proposal distributions are selected automatically when the target densities are log-concave.

To analyze the item response data, the forms of the parent and child relationships in Figure 12.1 should be specified. Under the one-parameter logistic model (cf. Thissen, 1982; Spiegelhalter, Best, et al. 1996), the probability that examinee j responds correctly to item i is assumed to follow a logistic function

$$P_{ij} = \frac{1}{1 + \exp[-(\alpha\theta_j - \beta_i)]}. \tag{12.24}$$

For scaling purposes, we may use the form

$$\theta'_j - b_i = \alpha\theta_j - \beta_i, \tag{12.25}$$

where θ'_j is the usual Rasch ability parameter and b_i is the Rasch item difficulty parameter defined as $\theta'_j = \alpha\theta_j - \bar{\beta}$ and $b_i = \beta_i - \bar{\beta}$, where $\bar{\beta}$ is the mean of the location parameters, $\bar{\beta} = \sum_j \beta_i/n$. Since Y_{ij} are Bernoulli with parameter P_{ij}, we can define

$$Y_{ij} \sim \text{Bernoulli}(P_{ij}) \tag{12.26}$$

and

$$\text{logit}(P_{ij}) = \alpha\theta_j - \beta_i. \tag{12.27}$$

To complete the specification of a full probability model in for BUGS, prior distributions of the nodes without parents (i.e., θ_j, β_i, and α) also need to be specified. We can define these priors in several different ways. We can impose priors on β_i and α using a hierarchical Bayes approach (e.g., Swaminathan & Gifford, 1982, 1985; Kim, Cohen, Baker, Subkoviak, & Leonard, 1994). If it is preferred that the priors not be too influential, uninformative priors could be imposed. Alternatively, it may also be useful to include external information in the form of fairly informative prior distributions. According to Spiegelhalter, Best, et al. (1996), it is important to avoid causal use of standard improper priors in Markov chain Monte Carlo modeling, since these may result in improper posterior distributions.

12.3.4 Starting Values

The choice of starting values (e.g., $\omega^{(0)}$) is not generally that critical as the Gibbs sampler should be run long enough to be sufficiently updated from its initial states. It is useful, however, to perform a number of runs using different starting values to verify that the final results are not sensitive to the choice of starting values (Gelman, 1996). Raftery (1996) indicated that extreme starting values could lead to a very long burn-in or stabilization process.

12.3.5 Output Monitoring

A critical issue for Markov chain Monte Carlo methods is how to determine when one can safely stop sampling and use the results to estimate characteristics of the distributions of the parameters of interest. For this purpose, the values for the unknown quantities generated by Gibbs sampler can be graphically and statistically summarized to check mixing and convergence. Cowles and Carlin (1996) presented a comparative review of convergence diagnostics for the Markov chain Monte Carlo algorithms. The method proposed by Gelman and Rubin (1992) is one of the most popular for monitoring Gibbs sampler, and used in the analyses presented in the subsequent sections.

Details of the Gelman and Rubin method are also given in Gelman (1996). For each parameter, the Gelman-Rubin statistics estimate the reduction in the pooled estimate of variance if the runs were continued indefinitely. The Gelman-Rubin statistics can be calculated sequentially as the runs proceed. The Gelman-Rubin statistics should be near 1 in order to be reasonably assured that convergence has occurred.

12.3.6 Summary Statistics

The last step of Gibbs sampler is to obtain summary statistics for the quantities of interest. The posterior mean of Gibbs sampler can be obtained for each parameter. The posterior interval as well as the posterior standard deviation can also be obtained for each parameter from the results of Gibbs sampler. The parameter estimates and the posterior intervals from Gibbs sampler can be compared with those from other estimation procedures that were based on the normality assumption.

12.4 Extensions of Gibbs Sampler within Item Response Theory

Albert's (1992) Gibbs sampler has been extended to various, complicated item response theory models. For example, Bradlow, Wainer, and Wang (1999) and Wainer, Bradlow, and Du (2000) presented Gibbs sampler for models with testlet structures; Fox and Glas (2001) presented Gibbs sampler for models for multilevel item response theory; Béguin and Glas (2001) presented Gibbs sampler for multidimensional models; Johnson (1996, 1997) presented Gibbs sampler applications of the graded response model and the multirater ordinal model. Also Patz and Junker (1999b) extended the Metropolis-Hastings within Gibbs method to models with multiple item types, missing data, and rated responses. All of these papers dealt with Gibbs sampler in the context of the Bayesian framework. Because Albert's Gibbs sampler that did not employ priors for item parameters was presented earlier, one specific extension that is based on the Bayesian framework is presented below,

Johnson and Albert (1999) in fact contained several extensions of Albert's (1992) Gibbs sampler. Gibbs sampler for the Bayesian framework of the two-parameter normal ogive model is now briefly discussed using similar equations presented earlier in this chapter. It can be noted that, in order to differentiate the sets of equations of the two approaches of Gibbs sampler, slightly different equations are used in this section. Johnson and Albert (1999) presented programs for the modified Gibbs sampler procedure using MATLAB (The MathWorks, Inc., 1996).

The two-parameter normal ogive model was used for the item response function. The probit, without the addition of five (see Finney, 1971; i.e., the normal equivalent deviate), was given by

$$Z_{ij} = a_i \theta_j - b_i, \tag{12.28}$$

where a_i is the item discrimination parameter, and b_i is the negative intercept parameter for item i, that is $\xi_i = (a_i, b_i)$. Using the usual item response theory parameterization, we have

$$Z_{ij} = a_i \theta_j - b_i = \lambda_i \theta_j + \zeta_i = \alpha_i(\theta_j - \beta_i), \tag{12.29}$$

where $a_i = \lambda_i = \alpha_i$ is the slope parameter, ζ_i is the intercept parameter, and β_i is the item difficulty or threshold parameter (see Baker, 1992). Note that under the two-parameter logistic item response theory model,

$$Z_{ij} = \alpha_i D(\theta_j - \beta_i), \tag{12.30}$$

where $D = 1.7$ is the scaling factor (e.g., Mislevy & Bock, 1990).

Let $Z = \{Z_{ij}\}$ be the matrix of independent random probits, that represents the matrix of the augmented data (Patz & Junker, 1999). A Gibbs sampling procedure can be used to sample from the joint posterior distribution over the entire collection of unknown parameters and latent data. The sampler is based on iteratively drawing values from three sets of conditional probability distributions, $p(Z|\theta, \xi, Y)$, $p(\theta|Z, \xi, Y)$, and $p(\xi|Z, \theta, Y)$ (Johnson & Albert, 1999).

To implement the Gibbs sampler, suppose at iteration $(t-1)$ the current values of the model parameters are denoted by $\{Z_{ij}^{(t-1)}\}$, $\{\theta_j^{(t-1)}\}$, $\{a_i^{(t-1)}\}$, and $\{b_i^{(t-1)}\}$. Then one complete cycle of the Gibbs sampler can be described as follows:

First, values of the latent data $\{Z_{ij}^{(t)}\}$ are simulated conditional on the current values of the latent traits and item parameters and on the item response data. In other words, Z_{ij} is randomly sampled from $p(Z^{(t)}|\theta^{(t-1)}, \xi^{(t-1)}, Y)$. The conditional posterior distribution of Z_{ij} is a truncated normal distribution with mean

$$m_{ij} = a_i^{(t-1)} \theta_j^{(t-1)} - b_i^{(t-1)} \tag{12.31}$$

and variance 1. The truncation of the posterior distribution depends on the value of the corresponding observation y_{ij}. If $y_{ij} = 1$, the truncation of Z_{ij} is from the left at 0 and Z_{ij} is sampled from the part of the conditional posterior distribution above 0. If $y_{ij} = 0$, the truncation of Z_{ij} is from the right at 0 and Z_{ij} is sampled from the part below 0. Let the new latent data value simulated from this truncated normal distribution be denoted by $\{Z_{ij}^{(t)}\}$.

Second, latent traits $\{\theta_j^{(t)}\}$ are simulated from their posterior distribution conditional on current values of the latent data and item parameters, where the posterior can be denoted by $p(\theta^{(t)}|Z^{(t)}, \xi^{(t-1)}, Y)$. Using the latent data representation, the item response model can be written as

$$Z_{ij}^{(t)} + b_i^{(t-1)} = a_i^{(t-1)}\theta_j + \epsilon_{ij}, \tag{12.32}$$

where the error term ϵ_{ij} are independent normal with mean 0 and variance 1. For a given value of j, this is a special case of the linear regression model with unknown parameter θ_j. The likelihood function for θ_j is of the normal form with mean

$$\bar{\theta}_j = \frac{\sum_{i=1}^{n} a_i^{(t-1)}(Z_{ij}^{(t)} + b_i^{(t-1)})}{\sum_{i=1}^{n}(a_i^{(t-1)})^2} \tag{12.33}$$

and variance

$$\sigma_{\theta_j}^2 = \frac{1}{\sum_{i=1}^{n}(a_i^{(t-1)})^2}. \tag{12.34}$$

Combining the sampling model with the $N(\mu_\theta, \tau_\theta^2)$ prior, it follows that the conditional posterior density of θ_i is normally distributed with mean

$$m_{\theta_j} = \frac{\bar{\theta}_j/\sigma_{\theta_j}^2 + \mu_\theta/\tau_\theta^2}{1/\sigma_{\theta_j}^2 + 1/\tau_\theta^2} \tag{12.35}$$

and variance

$$v_{\theta_j} = \frac{1}{1/\sigma_{\theta_j}^2 + 1/\tau_\theta^2}. \tag{12.36}$$

Let $\{\theta_j^{(t)}\}$ denote the vector of latent traits randomly drawn from the conditional posterior density. Specifically,

$$\theta_j^{(t)} = R\sqrt{v_{\theta_j}} + m_{\theta_j}, \tag{12.37}$$

where R is a random Gaussian deviate (Press, Teukolsky, Vetterling, & Flannery, 1992, p. 280).

Third, the item parameters $\{a_i, b_i\}$ are simulated from their joint posterior density, $p(\xi^{(t)}|Z^{(t)}, \theta^{(t)}, Y)$, conditionally on the current values of the latent data and the latent traits. To determine the conditional distribution, the latent data model can be written as

$$Z_{ij}^{(t)} = a_i\theta_j^{(t)} - b_i + \epsilon_{ij}. \tag{12.38}$$

Since the values of the latent data $\{Z_{ij}^{(t)}\}$ and the latent traits $\{\theta_j^{(t)}\}$ are fixed, the model can be seen as a linear regression model with unknown parameters a_i and b_i for a fixed value of i. Using matrix notations, the model can be written as

$$z_i = X\xi_i + \epsilon_i \tag{12.39}$$

or

$$
\begin{bmatrix} Z_{i1}^{(t)} \\ Z_{i2}^{(t)} \\ \vdots \\ Z_{iN}^{(t)} \end{bmatrix} = \begin{bmatrix} \theta_1^{(t)} & -1 \\ \theta_2^{(t)} & -1 \\ \vdots & \vdots \\ \theta_N^{(t)} & -1 \end{bmatrix} \begin{bmatrix} a_i \\ b_i \end{bmatrix} + \begin{bmatrix} \epsilon_{i1}^{(t)} \\ \epsilon_{i2}^{(t)} \\ \vdots \\ \epsilon_{iN}^{(t)} \end{bmatrix}.
\tag{12.40}
$$

Let μ_ξ denote the prior mean vector

$$
\mu_\xi = \begin{bmatrix} \mu_a \\ \mu_b \end{bmatrix}
\tag{12.41}
$$

and let Σ_ξ denote the prior covariance matrix

$$
\Sigma_\xi = \begin{bmatrix} \sigma_a^2 & 0 \\ 0 & \sigma_b^2 \end{bmatrix}.
\tag{12.42}
$$

It follows that the conditional posterior density of ξ_i is multivariate normal with mean vector

$$
m_i = (X'X + \Sigma_\xi^{-1})^{-1}(X'z_i + \Sigma_\xi^{-1}\mu_\xi)
\tag{12.43}
$$

and covariance matrix

$$
V_i = (X'X + \Sigma_\xi^{-1})^{-1}.
\tag{12.44}
$$

The Cholesky decomposition of V_i is performed and let

$$
A = chol(V_i) = chol(X'X + \Sigma_\xi^{-1})^{-1}.
\tag{12.45}
$$

Two random normal deviates R_1 and R_2 are generated and the item parameter values for item i are given by

$$
\xi_i^{(t)} = A' \begin{bmatrix} R_1 \\ R_2 \end{bmatrix} + m_i.
\tag{12.46}
$$

According to Johnson and Albert (1999), given suitable starting values for the parameter values, these steps define one cycle in Gibbs sampling scheme that can be used to obtain samples from the posterior distribution over all model parameters. Convergence of the algorithm is typically obtained within several hundred observations and is usually not very sensitive to the choice of starting values.

12.5 Empirical Studies of Gibbs Sampler

Baker (1998) presented a simulation study of the item parameter recovery characteristics of a Gibbs sampling method (Albert, 1992) for item parameter estimation under the two-parameter model. The item parameters were estimated, under a normal ogive item response function model, using both a FORTRAN version of Albert's (1992) Gibbs sampler and BILOG (Mislevy &

Bock, 1989). The item parameter estimates were then equated to the metric of the underlying item parameters using the test characteristic curve method (Stocking & Lord, 1983) for tests with 10, 20, 30, and 50 items, and samples of 30, 60, 120, and 500 examinees. Summary statistics of the equating coefficients showed that Gibbs sampler and BILOG both produced trait scale metrics with units of measurement that were too small, but yielding a proper midpoint of the metric. When expressed in a common metric, the biases of the BILOG estimates of the item discriminations were uniformly smaller and less variable than those from Gibbs sampling. The biases of the item difficulty estimates yielded by the two estimation procedures were small and similar to each other. In addition, the item parameter recovery characteristics were comparable for the largest dataset of 50 items and 500 examinees. However, for short tests and sample sizes the item parameter recovery characteristics of BILOG were superior to those of the Gibbs sampling approach.

Using the Bayesian framework of Gibbs sampler (Johnson & Albert, 1999), Kim and Cohen (2000) investigated the ability estimates of Gibbs sampler and the magnitudes of the posterior standard deviations. Item parameters of the Q-E intelligence test with 10 items and 44 examinees were obtained using a FORTRAN version of Gibbs sampler and marginal Bayesian estimation with BILOG (Mislevy & Bock, 1990). Two normal priors were used in item parameter estimation. Ability estimates were obtained using Gibbs sampler (i.e., jointly with item parameter estimates) and compared with estimates from the expected a posteriori method employing item parameter estimates obtained from Gibbs sampler and marginal Bayesian estimation, respectively. Item parameter estimates were very similar as were ability estimates, but the patterns of the magnitudes of the posterior standard deviations of ability estimates from Gibbs sampler were different from those based on the expected a posteriori method.

Kim (2001) investigated the accuracy of the Markov chain Monte Carlo procedure, Gibbs sampler implemented in BUGS (Spiegelhalter et al., 1997) under the one-parameter logistic Rasch model for estimation of item and ability parameters. Four empirical data sets were analyzed to evaluate the Gibbs sampling procedure. Data sets were also analyzed using methods of conditional maximum likelihood, marginal maximum likelihood, and joint maximum likelihood. Two different ability estimation methods, maximum likelihood and expected a posteriori, were employed under the marginal maximum likelihood estimation of item parameters. Item parameter estimates from the four methods were almost identical. Ability estimates from Gibbs sampler were similar to those obtained from the expected a posteriori method.

Kim and Cohen (1999) investigated the accuracy of Gibbs sampler implemented in BUGS (Spiegelhalter et al., 1997) for estimation of item and ability parameters under the two-parameter logistic model. Memory test data (Thissen, 1982) were analyzed to illustrate the Gibbs sampler procedure using BUGS. In addition simulated data sets for 50, 100, and 200 examinees and for 10, 20, and 40 items (i.e., nine conditions with 100 replications for each condition) were analyzed using Gibbs sampler and the marginal Bayesian method using the computer program BILOG (Mislevy & Bock, 1989). The marginal Bayesian method combined with the expected a posteriori estimation of ability yielded consistently smaller root mean square errors and better bias results than Gibbs sampler.

12.6 Summary

Gibbs sampler is a new addition to the many estimation procedures in item response theory. As a member of Markov chain Monte Carlo methods, it has a great potential to be an efficient and versatile estimation procedure in item response theory. In this chapter two different methods of Gibbs sampler were presented; Albert's (1992) Gibbs sampler and Gibbs sampler using rejection sampling (Ripley, 1987). In addition, a fully Bayesian version of Albert's Gibbs sampler (Johnson & Albert, 1999) is also presented. In the Bayesian framework, a sample of parameter values from the posterior density can be obtained and used to estimate posterior means and variances.

Gibbs sampler in general involves four basic steps: Full conditional distributions and sampling methods for unobserved parameters must be specified; starting values must be provided; output must be monitored; and summary statistics (e.g., estimates and standard errors) for quantities of interest must be calculated. Discussion of the four steps involved were presented in detail in the context of BUGS (Spiegelhalter, Best, et al., 1996).

It should be noted that recently Gibbs sampler has been extended to more complicated models that try to estimate parameters under the Bayesian framework. Several empirical evaluation studies are summarized (e.g., Baker, 1998; Kim & Cohen, 1999, 2000; Kim, 2001). Gibbs sampler may provide a useful alternative method for item and ability parameter estimation when small sample sizes and small numbers of items are used. Even though implementation of Gibbs sampler is available in several computer programs, the accuracy of the resulting estimates have not been thoroughly studied. More simulation results should be reported.

The main difference between Gibbs sampler and the other item response theory estimation methods lies in the way these methods obtain parameter estimates. Basically, Gibbs sampler is a tool for obtaining Bayesian estimates although Albert (1992) initially presented the procedure under a non-Bayesian context. Gibbs sampler in general uses the sample of parameter values to estimate the mean and variance of the posterior density of the parameter. Note that, for Gibbs sampler, ability parameters can be estimated either jointly with item parameters or sequentially after obtaining item parameters and assuming the obtained values are true parameters.

It can be noted that the estimation of item and ability parameters using Gibbs sampler requires a considerable amount of computing time. Gibbs sampler and general Markov chain Monte Carlo methods, however, are likely to be more useful for situations where complicated models are required.

In this chapter, Gibbs sampler was presented without addressing the problem of model selection and criticism (e.g., the choice of the linking function, model fit). The model criticism for Gibbs sampler seems to be an important topic to investigate in future research. Also the evaluation of the Gibbs sampler procedures to other item response theory models, for example, other logistic or probit models for binary items, the partial credit model, the graded response model, and the linear logistic test model, may provide guidelines for using the method under item response theory.

A. Implementation of Maximum Likelihood Estimation of Item Parameters

A.1 Introduction

A BASIC program will be used to illustrate the implementation of the maximum likelihood estimation of an item's parameters. In order to keep the size of the program to a minimum, many of the initial data processing tasks are assumed to have taken place. For example, the examinees have been grouped into 10 intervals with known ability levels, with the number of examinees and of correct responses in each group obtained. In a regular production computer program, the code to accomplish these tasks also would be present. The program has been structured with a short driver (kernel) section to control the overall flow of the program. The actual maximum likelihood estimation of the intercept and slope item parameters under a two-parameter logistic item characteristic curve model is implemented in a subroutine. Since this subroutine is one of the two building blocks of most IRT test analysis computer programs, it will be described in detail in this appendix and in less detail where it is used subsequently. Because computer programming languages do not allow Greek symbols, variable names were chosen to represent terms expressed in mathematics in order to convey the appropriate meaning. For example, CPT represents intercept, and an initial letter S in a variable name indicates a sum.

A.2 Implementation

At line 140, the subroutine beginning at line 1500 is called that establishes the context of the estimation process. The number of ability levels is specified, and the initial values of the intercept and slope are set to zero and one. The prestored values of the ability levels of the 10 examinee groups are those used in the computer program BILOG. They are read from the data statements at lines 9410 and 9420. In addition, the number of correct responses and examinees at each ability level are read via the data statements in lines 9440 and 9460. In lines 1540–1590, two variables are set that control the output of the computational aspects of the program. When set to zero, the variable TRALL causes every term that is computed to be printed at the computer's printer terminal. When only the variable TRMLE is set to zero, the current values of the intercept and slope and the change in their values are printed after each iteration. These two trace variables allow the user to control the amount of computer output. The user is also asked to specify the maximum number of Newton-Raphson (i.e., Fisher scoring method) iterations to be allowed before automatic termination. This limitation comes into effect when convergence of the solution equations is not reached—a rare occurrence under the two-parameter logistic item characteristic curve model.

Returning to the driver section at line 150, the item parameter estimation subroutine beginning at line 1000 is called. The primary iterative loop begins at line 1030 and ends at line 1330, and each iteration is identified by the message ITERATION=. The nested loop defined by lines 1050–1180 implements the computation of the terms in the Newton-Raphson procedure in Equation 2.20 that are performed at each ability level X(K). The observed proportion of correct response PI at ability level X(K) is computed at line 1070 using the prestored data values R(K) and F(K), which were read into the arrays earlier. The ability level X(K) and the initial values of the intercept, CPT=0, and slope, A=1, are used at lines 1040 and 1042 to obtain the values of the probability of correct response PH and the weighting coefficient W, which depend on the item characteristic curve. Given these values, the variables P1–P6 that correspond to the elements in the terms of Equation 2.20 are computed. Since all these elements contain the $f_j W_j$ of P1, P2–P6 are obtained by multiplying P1 by the appropriate variables. These elements are then accumulated in the summation terms which correspond directly to the terms in Equation 2.20. For example,

$$ \text{SFW} = \sum_{j=1}^{k} f_j W_j \quad \text{and} \quad \text{SFWVX} = \sum_{j=1}^{k} f_j W_j v_j \theta_j. $$

When TRALL is set to zero, the information, given here for only the first and last ability level, is printed for each ability level:

```
ITERATION=  1
X( 1 )= -4      PI= .06      P HAT=  1.798621E-02
                W=  1.766271E-02       V=  2.378672
     P1=  1.766271      P2=  4.201379      P3=  9.993705
     P4= -7.065083      P5=  28.26033     P6= -16.80552
     SFW=  1.766271     SFWV=  4.201379   SFWX= -7.065083
     SFWVX= -16.80552   SFWX2=  28.26033  SFWV2=  9.993705

    ⋮

X( 10 )=  4      PI= .97      P HAT=  .9820138
                W=  1.766273E-02       V= -.6801741
     P1=  1.766273      P2= -1.201373     P3=  .817143
     P4=  7.065094      P5=  28.26037     P6= -4.805493
     SFW=  109.9712     SFWV=  59         SFWX=  1.382828E-05
     SFWVX= -103.9238   SFWX2=  290.7876  SFWV2=  81.25681
```

Upon completion of the last ability level, the loop is exited at line 1180, and the denominator common to Equations 2.21 and 2.22 is computed and its value printed:

```
DENOMINATOR= 31978.28
```

This value is checked at line 1230 and, if it is unreasonably small (say, < .000099), the error alarm "OUT OF BOUNDS ERROR ITERATION #" is printed at line 1360 with the iteration number # and the subroutine is exited. If the value is proper, Equation 2.22 is evaluated at line 1240 yielding the increment DCPT to the intercept. At line 1250, Equation 2.21 is evaluated, yielding the increment to the slope DA. The new values of the intercept and slope are obtained at line 1270. If either TRALL or TRMLE is set, the following information is printed:

```
NIT=  1 INTERCEPT=  .536504    CHANGE=  .536504
         SLOPE=  .6426126       CHANGE= -.3573874
```

At this point, the absolute values of the new slope and intercept are checked to determine if they have exceeded an arbitrary bound, 30 in case of the intercept and 20 in the case of the slope. If either bound is exceeded, the "OUT OF BOUNDS ERROR ITERATION #" alarm is printed and the subroutine is exited. If the values are acceptable, the solution of the Newton-Raphson procedure in Equation 2.20 is tested for convergence. While many different convergence criteria are possible, the one employed at line 1300 is the simplest. If the absolute values of the changes to the intercept and slope are both less than .05, the iterative process is complete, and the normal subroutine return at line 1350 is executed. If either value is greater than .05, the program goes to line 1330, which is the last line of the Newton-Raphson procedure. The iteration index NIT is incremented by one, and control goes to line 1035, and ITERATION= 2 is printed. At line 1040, all the sums are cleared to zero, and a new solution of the Newton-Raphson equation is performed using .5365 as the intercept and .6426 as the slope. The computations for the first and last ability level, in the second iteration, are as follows:

```
ITERATION=  2
X( 1 )= -4      PI=  .06      P HAT=  .1156846
                W=  .1023017           V= -.5443176
     P1=  10.23017       P2= -5.56846       P3=  3.031011
     P4= -40.92067       P5=  163.6827      P6=  22.27384
     SFW=  10.23017      SFWV= -5.56846     SFWX= -40.92067
     SFWVX=  22.27384    SFWX2=  163.6827   SFWV2=  3.031011

  .
  .
  .

X( 10 )=  4     PI=  .97      P HAT=  .9571787
                W=  4.098763E-02       V=  .3128094
     P1=  4.098763       P2=  1.282132     P3=  .4010628
     P4=  16.39505       P5=  65.5802      P6=  5.128526
     SFW=  153.9301      SFWV= -24.41483   SFWX= -85.49332
     SFWVX=  79.18647    SFWX2=  678.6666  SFWV2=  12.87017
```

Upon completion of the last ability level, the changes to the intercept and the slope and their new values are printed:

```
NIT=  2 INTERCEPT=  .4356415    CHANGE= -.1008625
         SLOPE=  .7465861        CHANGE=  .1039736
```

Since the absolute values of the increments to both parameters are larger than .05, a third iteration is performed. Again, the computations at each ability level are printed if the traces are set. Those for only the first and last ability level are given below:

```
ITERATION=  3
X( 1 )= -4      PI=  .06      P HAT=  7.237927E-02
                W=  6.714051E-02       V= -.1843785
    P1=  6.714051       P2= -1.237927       P3=  .2282472
    P4= -26.8562        P5=  107.4248       P6=  4.951708
    SFW=  6.714051      SFWV= -1.237927     SFWX= -26.8562
    SFWVX=  4.951708    SFWX2=  107.4248    SFWV2=  .2282472
```

$\vdots$

```
X( 10 )=  4      PI=  .97      P HAT=  .9683846
                 W=  3.061584E-02       V=  .0527637
    P1=  3.061584       P2=  .1615405       P3=  8.523474E-03
    P4=  12.24634       P5=  48.98535       P6=  .646162
    SFW=  139.4293      SFWV= -2.001873     SFWX= -61.61202
    SFWVX=  8.961403    SFWX2=  534.2224    SFWV2=  2.565771
```

The results of the third iteration are:

```
NIT=  3 INTERCEPT=  .4283234    CHANGE= -7.318063E-03
        SLOPE=  .7625168        CHANGE=  1.593067E-02
```

Since the absolute values of the increments to both parameters are less than .05, convergence of the Newton-Raphson equation has been attained. The subroutine exits normally via line 1350 and returns to line 160 in the driver section. The subroutine called at line 160 computes the item difficulty DIFF from the ratio of the negative intercept to the slope. The value of the chi-square goodness-of-fit statistic is given by the final value of SFWV2= 2.5572. The degrees of freedom are the number of ability levels less the number of parameters estimated and has a value of 8. Thus, the item characteristic curve defined by the item parameter estimates is a good fit to the observed proportions of correct response at the 10 ability levels. The final results are printed as follows:

```
INTERCEPT=  .4283234    SLOPE=  .7625168
DIFFICULTY= -.5617232   DISCRIMINATION=  .7625168
CHI-SQUARE=  2.565771   D.F.=  8
```

A point of interest is the improvement of the goodness of fit of the item characteristic curve defined by the item parameter estimates to the observed proportions of correct response over the three iterations. The three iterations yielded 81.25681, 12.87017, and 2.565771, respectively, a dramatic improvement in fit. However, because of sampling variation in the observed proportions of correct response, a perfect fit is rarely obtained, and the present value is a very good one.

While the present BASIC program was particularized to 10 ability levels, the program is easily modified to analyze other data sets. The only terms that need to be changed are the variable limits in the dimension statements, the number of ability levels, and the information contained in the DATA statements.

A.3 BASIC Computer Program

```
100 REM PROGRAM TO GET MAXIMUM LIKELIHOOD ESTIMATES OF THE
110 REM SLOPE AND INTERCEPT PARAMETERS FOR A SINGLE ITEM
120 REM UNDER A TWO PARAMETER LOGISTIC ICC MODEL
130 DIM X(10),R(10),F(10),PQ(10)
140 GOSUB 1500:REM READ CANNED DATA AND ESTABLISH TRACING
150 GOSUB 1000:REM MLE OF ITEM PARAMETERS
160 GOSUB 1700:REM PRINT ITEM PARAMETER ESTIMATES
170 END
1000 REM ITEMBIO SUBROUTINE
1030 FOR NIT=1 TO MAXIT
1035 PRINT:PRINT "ITERATION= ";NIT
1040 SFW=0:SFWV=0:SFWV2=0:SFWX=0:SFWVX=0:SFWX2=0
1050 FOR K=1 TO NXL:REM THETA LOOP
1060 IF F(K)=0 GOTO 1180
1070 PI=R(K)/F(K)
1140 DEV=CPT+A*X(K)
1142 PH=1/(1+EXP(-DEV)):W=PH*(1-PH):IF W<.0000009 GOTO 1180
1145 V=(PI-PH)/W:IF TRALL<>0 GOTO 1160
1150 PRINT "X(";K;")= ";X(K);TAB(17);"PI= ";PI;TAB(30);"P HAT= ";PH
1155 PRINT TAB(17);"W= ";W;TAB(40);"V= ";V
1160 P1= F(K)*W:P2=P1*V:P3=P2*V:P4=P1*X(K):P5=P4*X(K):P6=P4*V
1165 SFW=SFW+P1:SFWV=SFWV+P2:SFWX=SFWX+P4:SFWVX=SFWVX+P6
1170 SFWX2=SFWX2+P5:SFWV2=SFWV2+P3
1171 IF TRALL<>0 GOTO 1180
1172 PRINT TAB(5);"P1= ";P1;TAB(25);"P2= ";P2;TAB(45);"P3= ";P3
1173 PRINT TAB(5);"P4= ";P4;TAB(25);"P5= ";P5;TAB(45);"P6= ";P6
1174 PRINT TAB(5);"SFW= ";SFW;TAB(25);"SFWV= ";SFWV;
1175 PRINT TAB(45);"SFWX= ";SFWX
1176 PRINT TAB(5);"SFWVX= ";SFWVX;TAB(25);"SFWX2= ";SFWX2;
1177 PRINT TAB(45);"SFWV2= ";SFWV2
1180 NEXT K
1190 IF SFW<=0 GOTO 1360
1200 DM=SFW*SFWX2-SFWX*SFWX:IF TRALL<>0 GOTO 1230
1210 PRINT TAB(5);"DENOMINATOR= ";DM
1230 IF DM<=.000099 GOTO 1350
1240 DCPT=(SFWV*SFWX2-SFWVX*SFWX)/DM
1250 DA=(SFW*SFWVX-SFWX*SFWV)/DM
1270 CPT=CPT+DCPT:A=A+DA
1275 IF (TRALL=0) OR (TRMLE=0) THEN
     PRINT "NIT= ";NIT;TAB(9);"INTERCEPT= ";CPT;TAB(32);"CHANGE= ";DCPT
     PRINT TAB(9);"SLOPE= ";A;TAB(32);"CHANGE= ";DA
     PRINT
     ELSE
     END IF
1290 IF(ABS(CPT)>30)OR(ABS(A)>20) GOTO 1360
1300 IF(ABS(DCPT)<=.05)AND(ABS(DA)<=.05) GOTO 1350
1330 NEXT NIT
1340 PRINT "REACHED MAXIMUM NUMBER OF ITERATIONS"
1350 RETURN
1360 PRINT "  OUT OF BOUNDS ERROR ";"ITERATION ";NIT:GOTO 1350
REM **********CANNED DATA AND CONSTANTS**********
```

```
1500 NXL=10:CPT=0:A=1:REM INITIAL VALUES
1510 FOR K=1 TO NXL:READ X(K):NEXT K:REM READ ABILITY LEVELS
1520 FOR K=1 TO NXL:READ R(K):NEXT K:REM READ NUMBER CORRECT RESPONSES
1530 FOR K=1 TO NXL:READ F(K):NEXT K:REM READ NUMBER AT ABILITY LEVEL
1540 REM SET UP TRACE OF COMPUTATIONS
1550 TRALL=-1:TRMLE=-1
1560 INPUT "TRACE ALL? Y/N ";YN$
1570 IF (YN$="Y") OR (YN$="y") THEN TRALL=0:TRMLE=0:GOTO 1600
1580 INPUT "TRACE MLE? Y/N ";YN$
1590 IF (YN$="Y") OR (YN$="y") THEN TRMLE=0
1600 INPUT "ENTER NUMBER OF ITERATIONS TO DO ";MAXIT
1610 CLS:PRINT " ALL    MLE    MAX ITERATIONS"
1620 PRINT TAB(3);TRALL;TAB(9);TRMLE;TAB(17);MAXIT
1630 RETURN
REM***********PRINT ITEM PARAMETER ESTIMATES***************
1700 DIFF=-CPT/A:DF=NXL-2
1710 PRINT "INTERCEPT= ";CPT;TAB(25);"SLOPE= ";A
1720 PRINT "DIFFICULTY= ";DIFF;TAB(25);"DISCRIMINATION= ";A
1730 PRINT "CHI-SQUARE= ";SFWV2;TAB(25);"D.F.= ";DF
1740 RETURN
REM *********CANNED DATA*******************************
9400 REM ABILITY(X) LEVELS USED
9410 DATA -4.0000,-3.1111,-2.2222,-1.3333,-.4444,.4444,1.3333
9420 DATA 2.2222,3.1111,4.0000
9430 REM NUMBER OF CORRECT RESPONSES R(K) AT EACH ABILITY LEVEL
9440 DATA 6,17,20,34,51,68,81,90,95,97
9450 REM NUMBER OF EXAMINEES F(K) AT EACH ABILITY LEVEL
9460 DATA 100,100,100,100,100,100,100,100,100,100
```

B. Implementation of Maximum Likelihood Estimation of Examinee's Ability

B.1 Introduction

The second building block of IRT test analysis is the estimation of an examinee's ability. This process assumes that the parameters of the test items and the examinee's item responses, scored right-wrong, are available. Thus, the BASIC program given below assumes that these quantities exist. The program can estimate an examinee's ability using either a one-, two-, or three-parameter logistic item characteristic curve model. Again, the program is structured with a short driver to control the overall flow of the program. The actual ability estimation is implemented in a subroutine.

B.2 Implementation

At line 130, a subroutine is called that establishes the context of the estimation process. The number of items is set to 10, and the bound on the size of an increment to an ability estimate is set to 0.5. The examinee's item score vector, the intercepts, the slopes and the guessing parameters are read in from the data statements that begin at line 9500. It should be noted that data values must be present for the slope and guessing parameters even if the model does not employ them. The program code will handle the situation appropriately after the full set item parameters has been read into the arrays. Again, the TRALL and TRMLE variables are used to control the amount of computational detail that the user desires to see. The user is also asked to specify the maximum number of iterations of the Newton-Raphson (Fisher-scoring for parameters) equation that are to be performed when the process does not converge. Finally, the number of parameters in the item characteristic curve model are specified. In the present example, the two-parameter logistic item characteristic curve model will be employed. At line 160, the ability estimation subroutine beginning at line 2000 is called, using an initial ability estimate value of THETA=0. The primary iterative loop of the Newton-Raphson procedure begins at line 2000 and ends at line 2460. The nested loop over the test items begins at line 2030 and ends at line 2370. Since all three item characteristic curve models need the value of PHAT, based upon the logistic ogive, for an item, it is calculated in lines 2050 and 2060. If the three-parameter model is used, the program branches at line 2070 to the segment beginning at line 2210. However, our present interest is in the two-parameter case that starts at line 2080. The weighting coefficient W_{ij}, WIJ, and v_{ij}, VIJ, are computed, and these are used to compute the terms added to the numerator and denominator of Equation 3.13. Note that the W_{ij} and the $P_{ij}Q_{ij}$ in the denominator of v_{ij} in Equation 3.13 were canceled algebraically and do not appear in the code. If TRALL is set to zero, the details of these computations are printed. To illustrate this, the detailed print for the first and last items are shown below:

```
ITERATION=  1
         ITEM=  1
INTERCEPT=  .489           SLOPE=  .997         THETA= 0
UIJ=  1            DEV=  .489        P HAT=  .6198708
WIJ=  .235631      VIJ=  .3801292
NUMERATOR SUM=  .3789888              DENOMINATOR SUM=  .2342193
```

⋮

```
         ITEM=  10
INTERCEPT= -.49            SLOPE=  1.146        THETA= 0
UIJ=  1            DEV= -.49         P HAT=  .3798936
WIJ=  .2355744     VIJ=  .6201065
NUMERATOR SUM=  1.832667              DENOMINATOR SUM=  4.119302
```

Upon completion of the last item, the increment DELTA to the initial ability estimate is computed at line 2380 and the following is printed if the trace variables are set to zero:

CHANGE IN THETA= .4448976

At line 2400, the absolute magnitude of the change is compared to the bound (BIGT). If greater than the bound, the change is limited to the appropriately signed value of BIGT at line 2420. The new value of the ability estimate THETA is obtained at line 2430 by adding the change to the initial estimate. If either of the trace variables is set to zero, the following is printed:

THETA= .4448976   CHANGE= .4448976

Next, the convergence of the overall Newton-Raphson procedure is determined. If the absolute value of DELTA is greater than 0.05, another iteration is initiated at lines 2000 and 2010, which sets SUMNUM and SUMDEM to zero. The new value of the examinee's ability estimate will now be used in the calculations. To illustrate these calculations, the detailed results for the first and tenth item in the second iteration are shown below:

```
ITERATION=  2
         ITEM=  1
INTERCEPT=  .489           SLOPE=  .997         THETA= .4448976
UIJ=  1            DEV=  .9325629    P HAT=  .7175949
WIJ=  .2026525     VIJ=  .2824051
NUMERATOR SUM=  .2815579              DENOMINATOR SUM=  .2014384
```

⋮

```
         ITEM=  10
INTERCEPT= -.49            SLOPE=  1.146        THETA= .4448976
UIJ=  1            DEV=  1.985264E-02 P HAT=  .504963
WIJ=  .2499754     VIJ=  .495037
NUMERATOR SUM=  1.648398E-02          DENOMINATOR SUM=  3.936426
```

Upon completion of the last item, the increment DELTA to the initial ability estimate is computed at line 2380, and the following is printed if the traces are set to zero:

    CHANGE IN THETA= 4.187549E-03

Since the absolute value of the change is less than 0.05, the Newton-Raphson procedure has converged, and the normal return from the program is taken at line 2480. If convergence had not been reached by the last iteration, the "REACHED MAX ITERATIONS" message would have been printed before the subroutine return was taken. At line 170 in the driver section, the final value of the ability estimate is printed:

    ESTIMATED ABILITY= .4490851

While the present example used 10 items and the two-parameter item characteristic curve model, different numbers of items and models can be used by changing the values established in the subroutine beginning at line 3000.

B.3 BASIC Computer Program

```
100 REM PROGRAM TO ESTIMATE AN EXAMINEES ABILITY UNDER
110 REM THE ONE-, TWO-, OR THREE-PARAMETER LOGISTIC ICC MODEL
120 DIM CPT(10),A(10),C(10),UIJ(10)
130 GOSUB 3000:REM READ CANNED DATA AND SET TRACES
140 IF(NPARA=1) THEN FOR I=1 TO 10:A(I)=1.0:NEXT I
150 THETA=0
160 GOSUB 2000:REM ESTIMATE ABILITY
170 PRINT "ESTIMATED ABILITY= ";THETA
180 STOP
REM ********END OF DRIVER SECTION
REM ********ABILITY ESTIMATION SUBROUTINE*******
2000 FOR NIT=1 TO MAXIT
2010 SUMNUM=0:SUMDEM=0:REM CLEAR SUMS
2020 PRINT:PRINT "ITERATION= ";NIT
2030 FOR I=1 TO NITEM:REM ITEM LOOP
2040 REM CALCULATE PHAT
2050 DEV=CPT(I)+A(I)*THETA
2060 PHAT=1.0/(1.0+EXP(-DEV))
2070 IF(NPARA=3) GOTO 2210
2080 REM ONE- AND TWO-PARAMETER ICC MODELS
2090 WIJ=PHAT*(1.0-PHAT)
2100 VIJ=(UIJ(I)-PHAT)
2110 SUMNUM=SUMNUM+A(I)*VIJ
2120 SUMDEM=SUMDEM+A(I)^2*WIJ
2130 IF(TRALL=-1) THEN GOTO 2370
2140 PRINT TAB(10);"ITEM= ";I
2150 PRINT TAB(5);"INTERCEPT= ";CPT(I);TAB(30);"SLOPE= ";A(I);
2151 PRINT TAB(50);"THETA=";THETA
2160 PRINT TAB(5);"UIJ= ";UIJ(I);TAB(25);"DEV= ";DEV;
2161 PRINT TAB(45);"P HAT= ";PHAT
```

```
2170 PRINT TAB(5);"WIJ= ";WIJ;TAB(25);"VIJ= ";VIJ
2180 PRINT TAB(5);"NUMERATOR SUM= ";SUMNUM;
2181 PRINT TAB(40);"DENOMINATOR SUM= ";SUMDEM:PRINT
2190 GOTO 2370
2200 REM THREE-PARAMETER ICC MODEL
2210 PT=C(I)+(1.0-C(I))*PHAT
2220 REM PROTECT AGAINST DIVIDE BY ZERO
2230 IF(PT < .00001) THEN PT=.00001
2240 IF(PT > .99999) THEN PT=.99999
2250 WIJ=PT*(1.0-PT)
2260 VIJ=(UIJ(I)-PT)
2270 PSP=PHAT/PT
2280 SUMNUM=SUMNUM+A(I)*VIJ*PSP
2290 SUMDEM=SUMDEM+A(I)^2*WIJ*PSP*PSP
2300 IF(TRALL=-1) THEN GOTO 2370
2310 PRINT TAB(10);"ITEM= ";I
2320 PRINT TAB(5);"C= ";C(I)
2321 PRINT TAB(5);"INTERCEPT= ";CPT(I);TAB(30);"SLOPE= ";A(I);
2322 PRINT TAB(50);"THETA=";THETA
2330 PRINT TAB(5);"UIJ= ";UIJ(I);TAB(25);"DEV= ";DEV;
2331 PRINT TAB(45);"P HAT= ";PHAT
2340 PRINT TAB(5);"PT= ";PT;TAB(25);"PSP= ";PSP
2350 PRINT TAB(5);"WIJ= ";WIJ;TAB(25);"VIJ= ";VIJ
2360 PRINT TAB(5);"NUMERATOR SUM= ";SUMNUM;
2361 PRINT TAB(40);"DENOMINATOR SUM= ";SUMDEM:PRINT
2370 NEXT I
2380 DELTA=SUMNUM/SUMDEM
2390 IF (TRALL=0)OR(TRMLE=0) THEN
         PRINT TAB(5);"CHANGE IN THETA= ";DELTA
      END IF
2400 IF(ABS(DELTA)<BIGT) GOTO 2430
2410 REM PROTECT AGAINST BIG CHANGE IN THETA
2420 IF(DELTA > 0.0) THEN
         DELTA=BIGT
      ELSE
         DELTA= -BIGT
      END IF
2430 THETA=THETA+DELTA
2440 IF(TRALL=0) OR (TRMLE=0) THEN
         PRINT "THETA= ";THETA;TAB(20);"CHANGE= ";DELTA
      END IF
2450 IF(ABS(DELTA) < .05) GOTO 2480
2460 NEXT NIT
2470 PRINT " REACHED MAX ITERATIONS"
2480 RETURN
REM *******READ CANNED DATA AND ESTABLISH TRACES*********
3000 NITEM=10:BIGT=0.5
3010 FOR I=1 TO NITEM:READ UIJ(I):NEXT I
3020 FOR I=1 TO NITEM:READ CPT(I):NEXT I
3030 FOR I=1 TO NITEM:READ A(I):NEXT I
3040 FOR I=1 TO NITEM:READ C(I):NEXT I
3050 TRALL=-1:TRMLE=-1
3060 INPUT "TRACE ALL? Y/N ";YN$
```

```
3070 IF(YN$="Y")OR(YN$="y") THEN TRALL=0:TRMLE=0:GOTO 3100
3080 INPUT "TRACE MLE? Y/N ";YN$
3090 IF(YN$="Y")OR(YN$="y") THEN TRMLE=0
3100 INPUT "ENTER MAXIMUM NUMBER OF ITERATIONS TO DO ";MAXIT
3110 INPUT "ENTER NUMBER OF PARAMETERS IN ICC MODEL ";NPARA
3120 CLS:PRINT "ALL   MLE   MAX ITERATIONS   ICC MODEL"
3130 PRINT TAB(2);TRALL;TAB(6);TRMLE;TAB(14);MAXIT;TAB(28);NPARA
3140 RETURN
9500 DATA 1,1,1,0,0,0,1,0,1,1
9510 DATA .489,-.869,-1.247,.595,.126,.469,.058,.665,-1.292,-.490
9520 DATA .997,.874,1.100,1.435,1.351,1.501,1.955,.906,1.897,1.146
9530 DATA .15,.21,.23,.01,.21,.02,.27,.05,.31,.27
```

C. Implementation of JMLE Procedure for the Rasch Model

C.1 Introduction

Birnbaum's joint maximum likelihood estimation (JMLE) paradigm will be implemented in a BASIC program using the Rasch one-parameter logistic item characteristic curve model. This model was employed here in order to have a compact program illustrating the JMLE paradigm. Again, it is assumed that the initial data processing tasks that produce the vectors of number of examinees with each possible test score and number of examinees answering each item correctly have been performed beforehand. Since the item parameter and ability estimation stages are performed in an alternating manner until convergence is reached, the driver (kernel) section of the program is more involved than those of the preceding appendices.

C.2 Implementation

The procedures described in Section 5.6 will be the basis of the implementation of the JMLE paradigm. At line 130 of the driver section, a call is made to the subroutine, beginning at line 3000, that establishes the environment of the estimation process. The number of items is set to 10 (NITEM=10) and, since perfect and null raw scores are not allowed, the maximum raw score is set to 9 (MAXSCORE=9). The maximum number of cycles of the JMLE paradigm is set arbitrarily to 10 (MAXCYCLE=10), and the maximum number of iterations within the item and ability estimation subroutines are each set to 5 (MAXIT=5). The vectors of raw score frequencies FDG(J) and the item scores S(I) are read into their arrays via the DATA statements at lines 9000 and 9020. All the computations can be traced via the TRALL variable, and this results in about 16 printed pages of output. The TRITM and TRABL trace variables are used to trace just the maximum likelihood estimation results within the corresponding subroutines. Beginning at line 3110, the PROX procedure due to Cohen (1979) is used to compute the initial estimates of the item difficulties and the raw score group ability estimate. The TRALL variable can be used to print the following values:

```
INITIAL B( 1 )= -3.355445E-02
INITIAL B( 2 )= -.5603082
INITIAL B( 3 )=  .5141606
INITIAL B( 4 )= -.1362354
```

```
INITIAL B( 5 )= -7.867818E-02
INITIAL B( 6 )=  .449097
INITIAL B( 7 )=  .2172619
INITIAL B( 8 )= -.1527187
INITIAL B( 9 )=  .3679781
INITIAL B( 10 )= -.5870027
INITIAL THETA( 1 )= -2.197225
INITIAL THETA( 2 )= -1.386294
INITIAL THETA( 3 )= -.8472978
INITIAL THETA( 4 )= -.4054651
INITIAL THETA( 5 )=  0
INITIAL THETA( 6 )=  .4054651
INITIAL THETA( 7 )=  .8472978
INITIAL THETA( 8 )= 1.386294
INITIAL THETA( 9 )= 2.197225
```

Upon returning to the driver section, the variable named FIRST$ is set to Y to tell the program that the first iteration of the JMLE procedure is being performed. Since the difference between two successive values of the average item difficulty is used to test for convergence, this step is necessary to insure that at least two cycles of the JMLE procedure are performed. The variable named LAST$ is set to N to ensure that the ability estimates are not corrected for bias until the overall convergence criterion is met and a final set of ability estimates has been obtained using the corrected-for-bias item difficulty estimates. The primary cycle of the JMLE procedure is defined by lines 150–390. At line 180, the subroutine for estimation the item difficulty parameters is called. Within this subroutine, lines 1000–1190 are the loop over the 10 items of the test. Lines 1010–1170 define the iterations of the Newton-Raphson procedure for an item and implement Equation 5.62. Lines 1030–1120 contain the code for the terms computed at each raw score for an item. If the TRALL variable is set, the information shown below, for only the first item, will be printed:

```
JMLE CYCLE= 1
ITEM= 1  NR ITERATION= 1  RAW SCORE= 1
DEV= -2.16367  PIJ= .1030607  SUMFDGP= 5.462217  SUMFDGPQ= 4.899277
ITEM= 1  NR ITERATION= 1  RAW SCORE= 2
DEV= -1.35274  PIJ= .2054228  SUMFDGP= 22.92315  SUMFDGPQ= 18.77334
ITEM= 1  NR ITERATION= 1  RAW SCORE= 3
DEV= -.8137434  PIJ= .3070934  SUMFDGP= 61.92402  SUMFDGPQ= 45.79729
ITEM= 1  NR ITERATION= 1  RAW SCORE= 4
DEV= -.3719106  PIJ= .4080794  SUMFDGP= 122.3198  SUMFDGPQ= 81.54678
ITEM= 1  NR ITERATION= 1  RAW SCORE= 5
DEV= 3.355445E-02
              PIJ= .5083878  SUMFDGP= 205.187   SUMFDGPQ= 122.2853
ITEM= 1  NR ITERATION= 1  RAW SCORE= 6
DEV= .4390196  PIJ= .6080254  SUMFDGP= 278.15    SUMFDGPQ= 150.885
ITEM= 1  NR ITERATION= 1  RAW SCORE= 7
DEV= .8808523  PIJ= .7069988  SUMFDGP= 373.5949  SUMFDGPQ= 178.8504
ITEM= 1  NR ITERATION= 1  RAW SCORE= 8
DEV= 1.419849  PIJ= .8053147  SUMFDGP= 450.0998  SUMFDGPQ= 193.7448
ITEM= 1  NR ITERATION= 1  RAW SCORE= 9
DEV= 2.230779  PIJ= .9029796  SUMFDGP= 495.2488  SUMFDGPQ= 198.1252
```

Upon completion of the last raw score level, the increment to the initial difficulty estimate is computed at line 1130 and the new value of item difficulty B(I) = -.03355-(.00838) = -.04239 is obtained. If the trace variables are set, the following is printed for an item:

```
DELTAB=  8.838961E-03    B( 1 )= -4.239342E-02
```

This information will be printed for each iteration of the Newton-Raphson procedure for an item and for all items in the test.

Since the increment for this first item difficulty is less than the convergence criterion of 0.05, the Newton-Raphson procedure for this item is exited. The sum of the item difficulties, SUMB, is accumulated at line 1180 and the estimation procedure for the next item is initiated at line 1010. When all items have been completed, the average item difficulty is computed at line 1200. If TRALL or TRITM is set, the following information is printed:

```
SUMB= -9.053469E-03  BBAR= -9.053469E-04  NITEM=  10
B( 1 )= -4.148807E-02
B( 2 )= -.687847
B( 3 )=  .631401
B( 4 )= -.1678604
B( 5 )= -.0970379
B( 6 )=  .5518064
B( 7 )=  .2671642
B( 8 )= -.1881324
B( 9 )=  .4523866
B( 10 )= -.7203926
```

Returning to the driver section at line 190, a check of the value of FIRST$ is made to see if this is the first cycle. If so, there is no need to check for convergence; the variable FIRST$ is set to N and the average item difficulty saved as BBOLD.

At line 220, the ability estimation subroutine beginning at line 2000 is called. Within this subroutine, the primary loop over the raw scores is defined by lines 2000–2180. The Newton-Raphson iterations for each ability level (i.e., raw score) are defined by lines 2010–2160 which implement Equation 5.60. The computations performed on an item by item basis are accomplished in lines 2030–2110. The proportion of correct response PIJ to item I at the ability level THETA(J) is computed at line 2050 under the Rasch model. The sum of PIJ and the sum of PIJ × (1-PIJ) are accumulated at line 2070. If the TRALL is set, the information shown below, for only the first and last items, will be printed for each item:

```
RAW SCORE=  1  NR ITERATION=  1  ITEM=  1
DEV= -2.155736  PIJ=  .1037964  SUMP=  .1037964  SUMPQ=  .0930227

 .
 .
 .

RAW SCORE=  1  NR ITERATION=  1  ITEM=  10
DEV= -1.476832  PIJ=  .1859064  SUMP=  1.074562  SUMPQ=  .9398482
```

Upon completion of the last item, the increment DELTA to the ability estimate is computed at line 2120 and added to the existing ability THETA(J) at line 2130. If either TRALL or TRABL is set, the following is printed at an ability level:

```
DELTA= -7.933417E-02 THETA( 1 )= -2.276559
```

Then the absolute value of DELTA is checked at line 2150 to see if convergence has been reached. In the present case, it has not, and a new iteration is begun at line 2020. If convergence has been reached, the new value of THETA for the raw score can be printed when either TRALL or TRABL is set:

```
THETA( 1 )= -2.279047
```

In the present case, two iterations were required before the absolute value of DELTA was less than 0.05. This process is repeated for all nine raw score levels, yielding a vector THETA(J) of new ability estimates. Control then returns to the driver section at line 220. Since this is the first time ability estimates have been computed, the program branches to line 320, where the value of the variable LAST$ is checked. However, this is not the last cycle, and the program branches to line 390, where the number of JMLE cycles completed is incremented. Then the next cycle of the JMLE procedure is begun at line 160.

After the new item difficulty parameters are obtained in a second or sub-sequent cycle, the difference between the average difficulty in two successive cycles is used to determine if convergence is reached. This is accomplished via the value of the variable FIRST$ tested at line 190 and the IF statement at line 230, which determines if the absolute difference between two mean item difficulties is less than 0.05. If so, the item difficulty estimates are corrected for bias via the subroutine, beginning at line 4000, that is called at line 250, and the final values are printed:

```
ITEM DIFFICULTY CORRECTED FOR BIAS
B( 1 )= -3.787961E-02
B( 2 )= -.6277743
B( 3 )=  .5762585
B( 4 )= -.1532521
B( 5 )= -8.858965E-02
B( 6 )=  .503632
B( 7 )=  .2439306
B( 8 )= -.1717645
B( 9 )=  .4129056
B( 10 )= -.6574666
```

The variable LAST$ is set to Y, and the ability estimation subroutine is called at line 310. The ability estimates are computed using the corrected-for-bias item difficulty estimates. Upon completion, the test at line 320 is not met, and the ability estimates are corrected for bias at lines 340–370. Since this is the last cycle, the program stops at line 380. If the maximum number of cycles is reached before convergence is achieved, the message "MAX CYCLES REACHED" is printed. The item difficulty estimates are corrected for bias via the subroutine called at line 410, and a branch to line 300 is made where LAST$ is set to Y. The final ability estimates are computed, corrected for bias, and printed. The obtained ability estimates corresponding to each raw score are as follows:

```
ABILITY CORRECTED FOR BIAS
THETA( 1 )= -1.812325
THETA( 2 )= -1.149475
THETA( 3 )= -.7043234
THETA( 4 )= -.3370777
```

```
THETA( 5 )=   8.782735E-04
THETA( 6 )=   .3386227
THETA( 7 )=   .705235
THETA( 8 )=   1.149332
THETA( 9 )=   1.810713
```

At this point, Birnbaum's JMLE estimation paradigm has been completed, and the program stops at line 380.

From the above, the simplicity of the JMLE procedure when the Rasch model is used should be clear. However, the present program omitted much of the data input and initial processing necessary to read in actual test data and report on other features of the test analysis process. For a production program, the reader is referred to the WINSTEPS computer program (Linacre, 2003).

C.3 BASIC Program

```
100 REM PROGRAM TO IMPLEMENT THE BIRNBAUM JOINT MAXIMUM LIKELIHOOD
110 REM ESTIMATION PARADIGM FOR THE RASCH MODEL.
120 DIM FDG(9),S(10),B(10),THETA(9)
130 GOSUB 3000:REM READ CANNED DATA,SET TRACES, GET INITIAL ESTIMATES
140 FIRST$="Y":LAST$="N"
150 FOR K=1 TO MAXCYCLE
160 PRINT "JMLE CYCLE= ";K
170 SUMB=0
180 GOSUB 1000:REM ESTIMATE ITEM DIFFICULTIES
190 IF(FIRST$ = "N") THEN GOTO 230
200 FIRST$="N"
210 BBOLD=BBAR
220 GOSUB 2000:GOTO 320
230 IF ABS(BBAR-BBOLD) > .05  THEN GOTO 210
240 REM REACHED CONVERGENCE
250 GOSUB 4000:REM CORRECT ITEM DIFFICULTY FOR BIAS
300 LAST$="Y"
310 GOSUB 2000
320 IF(LAST$="N") THEN GOTO 390
330 PRINT "ABILITY CORRECTED FOR BIAS"
340 FOR J=1 TO MAXSCORE
350 THETA(J)=THETA(J)*(NITEM-2)/NITEM
360 PRINT "THETA(";J;")= ";THETA(J)
370 NEXT J
380 IF LAST$="Y" THEN STOP
390 NEXT K
400 PRINT "MAX CYCLES REACHED"
410 GOSUB 4000: REM CORRECT ITEM DIFFICULTY FOR BIAS
420 GOTO 300
REM *******ITEM PARAMETER ESTIMATION SUBROUTINE******
1000 FOR I=1 TO NITEM
1010 FOR KK=1 TO MAXIT
1020 SUMFDGP=0:SUMFDGPQ=0
```

```
1030 FOR J=1 TO MAXSCORE
1040 IF TRALL <>0 GOTO 1060
1050 PRINT "ITEM= ";I;" NR ITERATION= ";KK;" RAW SCORE= ";J
1060 DEV=(THETA(J)-B(I))
1070 PIJ=1.0/(1+EXP(-DEV))
1080 SUMFDGP=SUMFDGP+FDG(J)*PIJ
1090 SUMFDGPQ=SUMFDGPQ+FDG(J)*PIJ*(1.0-PIJ)
1100 IF TRALL<>0 THEN GOTO 1120
1110 PRINT "DEV= ";DEV;TAB(17);"PIJ= ";PIJ;TAB(33);
1111 PRINT "SUMFDGP= ";SUMFDGP;TAB(53);"SUMFDGPQ= ";SUMFDGPQ
1120 NEXT J
1130 DELTAB=(S(I)-SUMFDGP)/SUMFDGPQ
1140 B(I)=B(I)-DELTAB
1150 IF (TRALL = 0) OR (TRITM= 0) THEN
     PRINT "DELTAB= ";DELTAB;TAB(25);"B(";I;")= ";B(I)
     END IF
1160 IF ABS(DELTAB) < .05 THEN GOTO 1180
1170 NEXT KK
1180 SUMB=SUMB+B(I)
1190 NEXT I
1200 BBAR=SUMB/NITEM
1201 REM CORRECTION 2/16/98
1202 FOR I=1 TO NITEM:B(I)=B(I)-BBAR:NEXT I
1210 IF (TRALL=0) OR (TRITM=0) THEN
     PRINT "SUMB= ";SUMB;" BBAR= ";BBAR;" NITEM= ";NITEM
     FOR I=1 TO NITEM:PRINT "B(";I;")= ";B(I):NEXT I
     ELSE
     END IF
1220 RETURN
REM *******ABILITY ESTIMATION BY RAW SCORE **********
2000 FOR J=1 TO MAXSCORE
2010 FOR KK=1 TO MAXIT
2020 SUMP=0.0:SUMPQ=0.0
2030 FOR I=1 TO NITEM
2040 DEV=(THETA(J)-B(I))
2050 PIJ=1.0/(1+EXP(-DEV))
2060 SUMP=SUMP+PIJ
2070 SUMPQ=SUMPQ+PIJ*(1.0-PIJ)
2080 IF TRALL <>0 GOTO 2110
2090 PRINT "RAW SCORE= ";J;" NR ITERATION= ";KK;" ITEM= ";I
2100 PRINT "DEV= ";DEV;" PIJ= ";PIJ;" SUMP= ";SUMP;" SUMPQ= ";SUMPQ
2110 NEXT I
2120 DELTA=(J-SUMP)/SUMPQ
2130 THETA(J)=THETA(J)+DELTA
2140 IF (TRALL=0) OR (TRABL=0) THEN
     PRINT "DELTA= ";DELTA;" THETA(";J;")= ";THETA(J)
     END IF
2150 IF(ABS(DELTA)<.05) GOTO 2170
2160 NEXT KK
2170 IF (TRALL=0) OR (TRABL=0) THEN PRINT "THETA(";J;")= ";THETA(J)
2180 NEXT J
2190 RETURN
REM*********READ CANNED DATA, SET TRACES, GET INITAL ESTIMATES******
```

```
3000 NITEM=10:MAXSCORE=9:MAXCYCLE=10:MAXIT=5
3010 FOR J=1 TO MAXSCORE:READ FDG(J):NEXT J
3020 FOR I=1 TO NITEM:READ S(I):NEXT I
3030 TRALL=-1:TRITM=-1:TRABL=-1
3040 INPUT "TRACE ALL? Y/N ";YN$
3050 IF YN$="Y" OR YN$="y" THEN TRALL=0:TRITM=0:TRABL=0:GOTO 3110
3060 INPUT "TRACE ITEM MLE? Y/N ";YN$
3070 IF YN$="Y" OR YN$="y" THEN TRITM=0
3080 INPUT "TRACE ABILITY MLE? Y/N ";YN$
3090 IF YN$="Y" OR YN$="y" THEN TRABL=0
3100 REM CALCULATE INTIAL ESTIMATES
3110 N=0:FOR J=1 TO MAXSCORE:N=N+FDG(J):NEXT J
3111 IF TRALL=0 THEN PRINT "N= ";N
3120 SUMI=0.0
3130 FOR I=1 TO NITEM
3140 B(I)=LOG((N-S(I))/S(I))
3150 SUMI=SUMI+B(I)
3151 IF TRALL=0 THEN PRINT "SUMI= ";SUMI;" B(";I";)= ";B(I)
3160 NEXT I
3170 MEANL=SUMI/10:IF TRALL=0 THEN PRINT "MEAN OF LOG TERMS= ";MEANL
3180 FOR I=1 TO NITEM
3190 B(I)=B(I)-MEANL:IF TRALL=0 THEN PRINT "INITIAL B(";I;")= ";B(I)
3200 NEXT I
3210 REM GET ABILITY ESTIMATES
3220 FOR J=1 TO MAXSCORE
3230 THETA(J)=LOG(J/(NITEM-J))
3240 IF TRALL=0 THEN PRINT "INITIAL THETA(";J;")= ";THETA(J)
3250 NEXT J
3260 RETURN
REM *******CORRECT ITEM DIFFFICULTY FOR BIAS***********
4000 PRINT "ITEM DIFFICULTY CORRECTED FOR BIAS"
4010 FOR I=1 TO NITEM
4020 B(I)=B(I)*((NITEM-1)/NITEM)
4030 PRINT "B(";I;")= ";B(I)
4040 NEXT I
4050 RETURN
REM *********CANNED DATA*********
8999 REM FREQUENCY OF RAW SCORES FOR 1 TO MAXSCORE
9000 DATA 53,85,127,148,163,120,135,95,50
9010 REM ITEM SCORE FREQUENCY
9020 DATA 497,622,366,522,508,381,436,526,400,628
```

D. Implementation of Item Parameter Estimation via MMLE/EM

D.1 Introduction

In order to provide a concrete illustration of the Bock and Atkin's (1981) marginal maximum likelihood estimation (MMLE) / expectation-maximization (EM) approach, a BASIC computer program was developed. This computer program, based on programs due to Zwarts (1986), has been written specifically to handle the LSAT-6 data set reported by Bock and Lieberman (1970). This data set is composed of the responses of 1000 examinees to five dichotomous items. The current computer program was designed for didactic purposes only, and the reader is referred to BILOG (Mislevy & Bock, 1986) for a production program. It is assumed that the computer is equipped with a math-coprocessor chip since without it, the numerical calculations will be very time-consuming.

Because of the well-documented computational problems associated with item parameter estimation under the three-parameter model (see Baker, 1987a), the two-parameter logistic item characteristic curve model will be used. For convenience, the parameterization $Z_{ik} = \zeta_i + \lambda_i \theta_k = d_i + a_i \theta_k$ is employed, where d_i is the intercept and a_i the slope rather than the difficulty (b_i) and discrimination (a_i) parameterization. The usual difficulty parameter is given by $b_i = -d_i/a_i$. The MMLE/EM equations for the estimation of d_i and a_i are

$$\sum_{k=1}^{q} [\bar{r}_{ik} - \bar{n}_{ik} P_i(X_k)] = 0$$

and

$$\sum_{k=1}^{q} [\bar{r}_{ik} - \bar{n}_{ik} P_i(X_k)] X_k = 0. \tag{D.1}$$

As emphasized throughout Chapter 6, these equations have the same form as the corresponding bioassay solution.

The expressions for $\bar{n}_{ik}$ and $\bar{r}_{ik}$, Equations 6.23 and 6.24, involved a summation over the N examinees. In a given data set, many examinees can have the same vector of correct and incorrect responses to the I items[1], and the same computations are often replicated. In order to reduce the computational demands, the N possible examinee item score vectors are grouped into 2^I possible item score patterns. These item score patterns are indexed by l, where $l = 1, 2, \ldots, s$. The number of examinees possessing a given item score pattern is given by f_l. Because of the computational efficiencies, Bock and Aitkin (1981) formulated $\bar{n}_{ik}$ and $\bar{r}_{ik}$ in terms of item score patterns rather than examinee item score vectors. The resulting expressions are

$$\bar{n}_{ik} = \frac{\sum_{l=1}^{s} f_l L_l(X_k) A(X_k)}{\sum_{k=1}^{q} \sum_{l=1}^{s} L_l(X_k) A(X_k)} \tag{D.2}$$

and

$$\bar{r}_{ik} = \frac{\sum_{l=1}^{s} u_{ij} f_l L_l(X_k) A(X_k)}{\sum_{k=1}^{q} \sum_{l=1}^{s} L_l(X_k) A(X_k)}, \tag{D.3}$$

where

$$L_l(X_k) = \prod_{i=1}^{I} P_i(X_k)^{u_{lj}} Q_i(X_k)^{1-u_{lj}} \tag{D.4}$$

and where $u_{li} = 1$ if item i is correct in item score patterns l and $u_{li} = 0$ otherwise. Let

$$P_l = \sum_{k=1}^{q} L_l(X_k) A(X_k). \tag{D.5}$$

It was decided a priori to use 10 values of X_k along the ability scale in the Gaussian quadrature procedures. In BILOG a unit normal distribution is divided into the number of categories specified by the user and the mid-points of the categories become the quadrature nodes X_k. The area of the normal curve encompassed by an interval becomes the weights $A(X_k)$. The following values employed by the BILOG computer program were used in the BASIC program:

---

[1] Because of the potential confusion between n_{ik} and n, the number of items will be designated by I rather than by the usual n in this appendix.

Node	X_k	$A(X_k)$
1	-4.0000	.000119
2	-3.1111	.002805
3	-2.2222	.03002
4	-1.3333	.1458
5	-0.4444	.3213
6	0.4444	.3213
7	1.3333	.1458
8	2.2222	.03002
9	3.1111	.002805
10	4.0000	.000119

Note that

$$\sum_{k=1}^{10} A(X_k) = 1, \qquad \sum_{k=1}^{10} X_k A(X_k) = 0, \qquad \text{and} \qquad \sum_{k=1}^{10} X_k^2 A(X_k) = 1.$$

D.2 Implementation

The overall description of the MMLE/EM procedure presented in Chapter 6 was used as the basis for the structure of the BASIC computer program and is reflected in the driver section. A subroutine beginning at line 1500 is called at line 130 to establish the environment of the estimation process. The 32 possible item score patterns are read from the data statements at lines 9000–9310 and stored in the array U(L,I). The predefined quadrature nodes and weights are read from the data statements at lines 9410–9450 into the vectors X(K) and A(K), respectively. The number of examinees possessing each of the 32 item score patterns, FPT(L), was obtained from the LSAT-6 data and is read from the DATA statements in lines 9500–9510. The initial item parameter estimates for all five items were set to $d_i = 0$ and $a_i = 1$, via line 1540 and the data statement at line 10000. Beginning at line 1550, a number of queries are made as to the computations the user desires to trace. At line 1650, the number of E-step, M-step cycles (MNC), is set by the user. A summary of the traces set and the number of cycles is then printed.

Upon completion of this initialization subroutine, the overall EM algorithm is initiated at line 140 of the driver section, and the current cycle number is printed. The E-step of the algorithm, which computes the "artificial data," is implemented via the subroutine contained in lines 300–470. It consists of a nested set of FOR-NEXT loops. The outer loop, lines 300–448, is over the 32 item score patterns; the next loop, lines 310–380, is over the 10 quadrature points; and the inner loop, lines 320–370, is over the five items. The probability of a correct response, P, $P_i(X_k)$, to item i at node X_k is obtained using the current values of the item parameter estimates D(I) and A(I). This probability is calculated via the subroutine (lines 600–620), for

the two-parameter logistic item characteristic curve model, that is called at
line 330. If the item is correct in the lth pattern, then U(L,I)=1 and PQ(I)=P;
and if it is incorrect, then U(L,I)=0 and PQ(I)=1-P. If the TRALL variable is
set, the PQ(I) values at a quadrature node can be displayed, via line 345,
for each of the five items. At line 360, the PQ(I) are sequentially multiplied
across the I items to create the value of $L_l(X_k)$, LXK(L,K), for pattern l at
that quadrature node. If either the TRALL or TRLP variables are set, the value
of LXK(L,K) is displayed via line 375. This group of six lines is reported for
each of the 10 quadrature nodes, but only the results for the first two nodes
of the twenty-second of the 32 item score patterns are shown below:

:

```
345   22   1   1     1.798621E-02        1
345   22   1   2     .9820138            0
345   22   1   3     1.798621E-02        1
345   22   1   4     .9820138            0
345   22   1   5     1.798621E-02        1
L=  22   K=  1       L(X)=  5.611179E-06
345   22   2   1     4.265171E-02        1
345   22   2   2     .9573483            0
345   22   2   3     4.265171E-02        1
345   22   2   4     .9573483            0
345   22   2   5     4.265171E-02        1
```

At line 390, each of the values of $L_l(X_k)$ are multiplied by the corresponding
value of $A(X_k)$ and summed over the quadrature nodes to create P_l, variable
PL(L), for that pattern. This aggregation process is performed for all 32 patterns. The TRALL variable is used, via line 392, to display the values for the 10
quadrature nodes. Again, the results for only the twenty-second item score
pattern are shown below:

:

```
392   22   6.677303E-10        5.611179E-06   .000119
392   22   2.001398E-07        7.111303E-05   .002805
392   22   2.304132E-05        7.608654E-04   .03002
392   22   8.520607E-04        5.686005E-03   .1458
392   22   7.96567E-03         2.214009E-02   .3213
392   22   1.905973E-02        3.452867E-02   .3213
392   22   2.220465E-02        .0215701       .1458
392   22   2.241542E-02        7.020968E-03   .03002
392   22   .0224199            1.596184E-03   .002805
392   22   2.241994E-02        3.063609E-04   .000119
            PL( 22 )=  2.241994E-02
```

The overall process is repeated for each of the 32 patterns. The details of
these computations can be displayed by responding Y to the "TRACE ALL? Y/N"
question at the beginning of the program. Be prepared for lengthy results
appearing on the computer's screen! If you wish to see only the values of
$L_l(X_k)$ and P_l, respond N to the above prompt and Y to the question "TRACE
L,P? Y/N" in the program initialization.

At this point, a two-dimensional array (patterns by quadrature nodes) of the values of $L_l(X_k)$ and vector of values of P_l of length s have been created. Lines 410–460 implement the computation of the $\bar{r}_{ik}$ and $\bar{n}_{ik}$ terms, as defined in Equations D.2 and D.3 for a given item. This consists of a nested set of FOR-NEXT loops. The outer loop, lines 420–460, is over items; the next inner loop, lines 430–450, is over the 10 quadrature nodes; and the innermost loop, lines 439–448, is over the 32 patterns. At line 435 the values of $\bar{r}_{ik}$ and $\bar{n}_{ik}$ are prestored to zero. Line 439 begins a loop over the 32 patterns in which Equations D.2 and D.3 are implemented. The number of examinees having a given pattern FPT(L) is multiplied by $L_l(X_k)$ and $A(X_k)$ and then divided by P_l. The resulting term (NT) is multiplied by the response for that item U(L,I) in the item score pattern, yielding RT. Given these terms, $\bar{r}_{ik}$ (RIK(I,K)) is the sum of the RT and $\bar{n}_{ik}$ (NK(I,K)) is the sum of the NT over the item score patterns. Lines 442–446 provide a detailed display of this calculation that is activated by setting the TRALL variable and again is a voluminous display. The print at line 449 shows the values of $\bar{r}_{ik}$ and $\bar{n}_{ik}$ at each of the 10 quadrature nodes for each item. Only the results for the first item from the first cycle are shown below:

$\vdots$

```
449    X( 1 )= -4      R=  6.592095E-04   N=  4.661977E-03
449    X( 2 )= -3.1111  R=  4.222087E-02   N=  .1457416
449    X( 3 )= -2.2222  R=  1.362479       N=  2.708088
449    X( 4 )= -1.3333  R=  22.09657       N=  31.06769
449    X( 5 )= -.4444   R=  155.1238       N=  181.9796
449    X( 6 )=  .4444   R=  374.0932       N=  402.315
449    X( 7 )=  1.3333  R=  286.4083       N=  295.7905
449    X( 8 )=  2.2222  R=  76.48243       N=  77.55056
449    X( 9 )=  3.1111  R=  8.03212        N=  8.079121
449    X( 10 )= 4       R=  .3582141       N=  .3590834
```

When this inner loop has been repeated over the 10 quadrature nodes within each of the five items, the E-step artificial data has been created for use in the maximization of the log-likelihood. It should be noted that the value of the $\bar{n}_{ik}$ at the 10 quadrature nodes will be the same for all items. In addition, the sum of the $\bar{n}_{ik}$ values for each item equals the sample size of 1000.

The M-step is a straightforward implementation of the usual maximum likelihood estimation technique used in the bioassay and JMLE procedures for a single item as presented in Appendix A. Lines 190–210 in the driver section form a loop in which the item parameters for the five items are computed via the ITEMBIO subroutine called at line 200. In the first cycle of the EM algorithm, the prestored values of d_i and a_i are used to initiate the estimation process for each item.

The ITEMBIO subroutine beginning at line 1000 executes up to six iterations of the maximum likelihood estimation process, with each iteration using the same 10 quadrature nodes on the ability scale. Line 1050 initiates the FOR-NEXT loop, lines 1050–1330, over the 10 quadrature nodes. At line 1070 the artificial data $\bar{r}_{ik}$ and $\bar{n}_{ik}$ are used to compute the proportion of correct response $\bar{r}_{ik}/\bar{n}_{ik}$ (PI=RIK(I,K)/NK(I,K)) at X_k. Lines 1140 and 1142 obtain the "true" proportion of correct response (PH) by evaluating the two-parameter logistic item characteristic curve function using the current values of d_i and a_i. The weighting term (W=PH(1-PH)) is also computed. In lines 1160 and 1170, the elements of the six different sums in the Newton-Raphson equations are computed for each node and then aggregated over the quadrature nodes. These sums can be displayed via lines 1173–1178, and the screen display is controlled by the value of the TRALL variable. Line 1200 computes the denominator of the equations. At line 1210, the value of the denominator is displayed if TRALL is set. The magnitude of the denominator is checked and, if it is too small, the estimation process is terminated, via the error message at line 1350, as a division by a near-zero term would occur. Lines 1240 and 1250 compute the increments Δ_a (DA) and Δ_d (DCPT), and line 1270 the new values of d_i and a_i (i.e., CPT(I)+DCPT, A(I)+DA). The responding Y to the "TRACE MLE? Y/N" query in the initialization subroutine will activate the code in line 1275 that prints these values at each iteration of the maximum likelihood estimation process. The values for the first iteration of the first item are reported below:

```
NIT=  1  ITEM=  1
         INTERCEPT=  1.688435    CHANGE=  1.688435
         SLOPE=   .4899893       CHANGE= -.5100107
```

Line 1290 is a check to determine if d_i and a_i have gone out of bounds, that is, are no longer within a reasonable range of values. If so, the error message at line 1350 is executed and the subroutine exited. Line 1300 determines if convergence has been achieved. If the increments DCPT and DA are both less than 0.05, convergence has been reached, and the subroutine exits through the normal return at line 1340. If not, another iteration is performed via the NEXT statement at line 1330. In general, only two to three iterations are needed under the two-parameter logistic item characteristic curve model. This process is repeated for each of the remaining items. Recalling that $b_i = -d_i/a_i$, the results of the M-step for the five items at the end of the first cycle were:

ITEM	INTERCEPT	SLOPE	DIFFICULTY
1	2.165932	.9438034	-2.294898
2	.4102187	.9409582	-.4359584
3	-.3786524	.9697451	.3904659
4	.7262625	.9321768	-.7791039
5	1.53213	.919755	-1.665802

Note that, in the MMLE/EM solution, the likelihood equations for each item are solved separately; the results for the five items are presented together here only for convenience.

At this point, a new set of values of d_i and a_i (i.e., the ϵ^{p+1}) are available, and a MMLE/EM cycle has been completed. The program returns to line 150 and prints the new cycle number of the MMLE/EM process. The next set of values of the $\bar{r}_{ik}$ and $\bar{n}_{ik}$ are obtained via the E-step and the d_i and a_i via the M-step. This process is repeated until the maximum number of cycles is reached. This value was previously set by the response to the "ENTER NUMBER OF CYCLES TO DO" message at line 1650 of the initialization subroutine. In a production program, a convergence criterion, such as the change in the value of the likelihood function, would be used. The overall process appears to converge very slowly. It was run for 100 cycles, and the final values obtained were:

ITEM	INTERCEPT	SLOPE	DIFFICULTY
1	2.773251	.8257672	-3.358393
2	.9902008	.7228056	-1.369941
3	.2491467	.8909402	-.2796447
4	1.284758	.6884309	-1.866211
5	2.053275	.6569303	-3.125561

These values are different from those of the first cycle and are very close to those reported by Mislevy and Bock (1985); note that they reported the a_i in a normal ogive metric. Although the overall process converges very slowly, each step of the EM algorithm is performed in a few seconds of computer time. The amount of computing would increase dramatically as the number of items in the test increases. Acceleration techniques, such as the method due to Ramsey (1975) and implemented in the BILOG program, may improve the convergence rate within the M-step.

Recall that the present program is for didactic purposes only. The BILOG program has the code necessary to read in the examinees' item score vectors and provides more extensive reporting of results. There also are two major procedural differences: First, the BILOG program recomputes the quadrature weights for the fixed values of the nodes at the beginning of each EM cycle. Second, when the number of items is small ($I \leq 20$) and a number of EM cycles have been completed, the program performs a simultaneous estimation of the item parameters for all I items via a Newton-Raphson equation. Since the dimensionality of the information matrix is the number of items times the number of parameters in the item characteristic curve model, this simultaneous solution is limited to a small number of items.

D.3 BASIC Computer Program

```
100 REM PROGRAM TO IMPLEMENT MMLE/EM FOR LSAT-6 DATA SET
110 DIM U(32,5),X(10),AK(10),FPT(32),LXK(32,10),PL(32)
120 DIM RIK(5,10),NK(5,10),PQ(10),CPT(5),A(5),B(5)
130 GOSUB 1500:REM READ CANNED DATA AND SET TRACES
140 FOR NC=1 TO MNC
150 PRINT:PRINT "CYCLE= ";NC
155 FOR L=1 TO 32:PL(L)=0:FOR K=1 TO 10:LXK(L,K)=1:NEXT K:NEXT L
160 REM E-STEP    GET EXPECTED R AND N
170 GOSUB 300
180 REM M-STEP    MLE OF ITEM PARAMETERS
190 FOR I=1 TO 5
200 GOSUB 1000
210 NEXT I
220 NEXT NC
230 REM DONE MAX CYCLES
240 GOSUB 1700
250 STOP
```

```
REM *******END OF DRIVER SECTION********
299 REM E STEP SUBROUTINE
300 FOR L=1 TO 32
310 FOR K=1 TO 10
320 FOR I=1 TO 5
330 GOSUB 600
340 PQ(I)=P:IF U(L,I)=0 THEN PQ(I)=Q
342 IF TRALL<>0 GOTO 360
345 PRINT "345";TAB(5);L;TAB(10);K;TAB(15);I;TAB(20);PQ(I);
346 PRINT TAB(40);U(L,I)
360 LXK(L,K)=LXK(L,K)*PQ(I)
370 NEXT I
375 IF (TRALL=0)OR(TRLP=0) THEN
    PRINT "L= ";L;TAB(10);"K= ";K;TAB(20);"L(X)= ";LXK(L,K)
    END IF
380 NEXT K
390 PL(L)=0:FOR K=1 TO 10:PL(L)=PL(L)+LXK(L,K)*AK(K)
391 IF TRALL<>0 GOTO 395
392 PRINT "392";TAB(5);L;TAB(10);PL(L);TAB(35);LXK(L,K);
393 PRINT TAB(50);AK(K)
395 NEXT K
398 IF(TRALL=0)OR(TRLP=0) THEN PRINT TAB(10);"PL(";L;")= ";PL(L)
400 NEXT L
410 REM R BAR AND N BAR LOOP
420 FOR I=1 TO 5
425 IF (TRALL=0) OR (TRRN=0) THEN PRINT TAB(10);"ITEM ";I
430 FOR K=1 TO 10
435 RIK(I,K)=0:NK(I,K)=0
439 FOR L=1 TO 32:NT=FPT(L)*LXK(L,K)*AK(K)/PL(L):RT=NT*U(L,I)
440 RIK(I,K)=RIK(I,K)+RT:NK(I,K)=NK(I,K)+NT
441 IF TRALL<>0 GOTO 448
442 PRINT "442";TAB(5);I;TAB(8);K;TAB(12);L;TAB(17);LXK(L,K);
443 PRINT TAB(33);AK(K);TAB(50);PL(L)
444 PRINT "444";TAB(5);U(L,I);TAB(8);NT;TAB(24);RT;TAB(39);RT
446 PRINT "446";TAB(5);RIK(I,K);TAB(25);NK(I,K)
448 NEXT L
449 IF (TRALL=0)OR(TRRN=0) THEN
    PRINT "449 ";TAB(8);"X(";K;")= ";X(K);TAB(25);
    PRINT "R= ";RIK(I,K);TAB(43);"N= ";NK(I,K)
    END IF
450 NEXT K
460 NEXT I
470 RETURN
REM ***SUBROUTINE TO COMPUTE P(X)***
600 DEV=-(CPT(I)+A(I)*X(K))
610 EP=EXP(DEV)
620 P=1/(1+EP):Q=1-P:RETURN
999 REM M-STEP--ITEMBIO ROUTINE FOR ESTIMATION OF ITEM PARAMETERS
1000 FOR NIT=1 TO 6
1040 SNW=0:SNWV=0:SNWX=0:SNWXV=0:SNWX2=0:SNWV2=0
1050 FOR K=1 TO 10:REM THETA LOOP
1060 IF NK(I,K)=0 GOTO 1180
1070 PI=RIK(I,K)/NK(I,K)
```

```
1135 REM CALCULATE NEWTON-RAPHSON TERMS
1140 DV=CPT(I)+A(I)*X(K):REM CPT=INTERCEPT:A=SLOPE
1142 PH=1/(1+EXP(-DV)):W=PH*(1-PH):IF W<.0000009 GOTO 1180
1145 V=(PI-PH)/W:IF TRALL<>0 GOTO 1160
1150 PRINT "X(";K;")= ";X(K);TAB(17);"PI= ";PI;
1151 PRINT TAB(30);"P HAT= ";PH
1155 PRINT TAB(17);"W= ";W;TAB(40);"V= ";V
1160 P1=NK(I,K)*W:P2=P1*V:P3=P2*V:P4=P1*X(K):P5=P4*X(K):P6=P4*V
1170 SNW=SNW+P1:SNWV=SNWV+P2:SNWX=SNWX+P4:SNWXV=SNWXV+P6
1171 SNWX2=SNWX2+P5:SNWV2=SNWV2+P3
1172 IF TRALL<>0 GOTO 1180
1173 PRINT TAB(5);"P1=";P1;TAB(25);"P2= ";P2;TAB(40);"P3= ";P3
1174 PRINT TAB(5);"P4=";P4;TAB(25);"P5="P5;TAB(40);"P6=";P6
1175 PRINT TAB(5);"SNW= ";SNW;TAB(25);"SNWV= ";SNWV;
1176 PRINT TAB(45);"SNWX= ";SNWX
1177 PRINT TAB(5);"SNWXV= ";SNWXV;TAB(25);"SNWX2= ";SNWX2;
1178 PRINT TAB(45);"SNWV2= ";SNWV2
1180 NEXT K
1190 IF SNW<=0 GOTO 1350
1200 DM=SNW*SNWX2-SNWX*SNWX:IF TRALL<>0 GOTO 1230
1210 PRINT "DENOMINATOR= ";DM
1230 IF DM<=.000099 GOTO 1350
1240 DCPT=(SNWV*SNWX2-SNWXV*SNWX)/DM
1250 DA=(SNW*SNWXV-SNWX*SNWV)/DM
1270 CPT(I)=CPT(I)+DCPT:A(I)=A(I)+DA
1275 IF (TRALL=0) OR (TRMLE=0) THEN
        PRINT "NIT= ";NIT;" ITEM= ";I
        PRINT TAB(9);"INTERCEPT= ";CPT(I);TAB(32);"CHANGE= ";DCPT
        PRINT TAB(9);"SLOPE= ";A(I);TAB(32);"CHANGE= ";DA
     ELSE
     END IF
1290 IF(ABS(CPT(I))>30)OR(ABS(A(I))>20) GOTO 1350
1300 IF(ABS(DCPT)<=.05)AND(ABS(DA)<=.05) GOTO 1340
1330 NEXT NIT
1340 RETURN
1350 PRINT "  OUT OF BOUNDS ERROR ITERATION ";NIT:GOTO 1340
REM**********READ CANNED DATA AND SET TRACES**********8
1500 FOR L=1 TO 32:FOR I=1 TO 5:READ U(L,I):NEXT I:NEXT L
1510 FOR K=1 TO 10:READ X(K):NEXT K
1520 FOR K=1 TO 10:READ AK(K):NEXT K
1530 FOR L=1 TO 32:READ FPT(L):NEXT L
1540 FOR I=1 TO 5:READ CPT(I),A(I):NEXT I
1550 REM SET UP TRACEING
1560 TRALL=-1:TRLP=-1:TRRN=-1:TRMLE=-1
1570 INPUT "TRACE ALL? Y/N ";YN$
1580 IF (YN$="Y") OR (YN$="y") THEN
     TRALL=0:TRLP=0:TRRN=0:TRMLE=0:GOTO 1650
     END IF
1590 INPUT "TRACE L,P? Y/N ";YN$
1600 IF (YN$="Y")OR(YN$="y") THEN TRLP=0
1610 INPUT "TRACE R AND N? Y/N ";YN$
1620 IF (YN$="Y") OR (YN$="y") THEN TRRN=0
1630 INPUT "TRACE MLE? Y/N ";YN$
```

```
1640 IF (YN$="Y") OR (YN$="y") THEN TRMLE=0
1650 INPUT "ENTER NUMBER OF CYCLES TO DO ";MNC
1670 CLS:PRINT " ALL   LP  TRRN  MLE    MAX CYCLES"
1680 PRINT TAB(2);TRALL;TAB(6);TRLP;TAB(11);TRRN;
1681 PRINT TAB(17);TRMLE;TAB(23);MNC
1690 RETURN
REM *************PRINT ITEM PRAMTER ESTIMATES*************
1700 PRINT "  ITEM";TAB(10);"INTERCEPT";TAB(28);"SLOPE ";
1701 PRINT TAB(43);"DIFFICULTY"
1710 FOR I=1 TO 5
1720 B(I)=-CPT(I)/A(I)
1730 PRINT TAB(5);I;TAB(10);CPT(I);TAB(25);A(I);TAB(40);B(I)
1740 NEXT I
1750 RETURN
REM **********CANNED DATA***************
9000 DATA 0,0,0,0,0
9010 DATA 0,0,0,0,1
9020 DATA 0,0,0,1,0
9030 DATA 0,0,0,1,1
9040 DATA 0,0,1,0,0
9050 DATA 0,0,1,0,1
9060 DATA 0,0,1,1,0
9070 DATA 0,0,1,1,1
9080 DATA 0,1,0,0,0
9090 DATA 0,1,0,0,1
9100 DATA 0,1,0,1,0
9110 DATA 0,1,0,1,1
9120 DATA 0,1,1,0,0
9130 DATA 0,1,1,0,1
9140 DATA 0,1,1,1,0
9150 DATA 0,1,1,1,1
9160 DATA 1,0,0,0,0
9170 DATA 1,0,0,0,1
9180 DATA 1,0,0,1,0
9190 DATA 1,0,0,1,1
9200 DATA 1,0,1,0,0
9210 DATA 1,0,1,0,1
9220 DATA 1,0,1,1,0
9230 DATA 1,0,1,1,1
9240 DATA 1,1,0,0,0
9250 DATA 1,1,0,0,1
9260 DATA 1,1,0,1,0
9270 DATA 1,1,0,1,1
9280 DATA 1,1,1,0,0
9290 DATA 1,1,1,0,1
9300 DATA 1,1,1,1,0
9310 DATA 1,1,1,1,1
9400 REM QUADRATURE POINTS AND WEIGHTS VIA BILOG
9410 DATA -4.0000,-3.1111,-2.2222,-1.3333,-.4444,.4444,1.3333
9420 DATA 2.2222,3.1111,4.0000
9430 DATA .000119,.002805,.03002,.1458,.3213,.3213,.1458,.03002
9450 DATA .002805,.000119
9500 DATA 3,6,2,11,1,1,3,4,1,8,0,16,0,3,2,15,10,29,14,81,3,28,15
```

```
9510 DATA 80,16,56,21,173,11,61,28,298
10000 DATA 0,1,0,1,0,1,0,1,0,1
```

E. Implementing The Bayesian Approach

E.1 Introduction

Under the Bayesian approach developed in Chapter 7, the item and examinee ability parameter estimates are obtained in separate procedures. As a result, this appendix will first present the implementation of item parameter estimation under the marginalized Bayesian approach. Then, the implementation of two different Bayesian techniques for estimating an examinee's ability will be presented. In all three cases, the implementation will be in the form of a BASIC computer program. The reader is forewarned that the numerical results obtained below will not agree perfectly with those yielded by the BILOG computer program for the LSAT-6 data. The appendix implements only the core procedures and lacks many features of the full BILOG program.

E.2 Marginal Bayesian Modal Item Parameter Estimation

To be consistent with the derivations presented in Chapter 7, the parameterizations $\alpha_i = \log a_i$ and b_i were employed in this section even though they differ from the slope and intercept form used in the marginal maximum likelihood estimation / expectation-maximization (MMLE/EM) program of Appendix D. Following the default specifications of the PC-BILOG program, the prior distribution for the α_i in the two-parameter logistic item characteristic curve model was assumed to be normal, and no prior was used for the b_i parameter. As was the case with MMLE/EM, the M-step employed the Newton-Raphson iterative approach for solving the nonlinear solution equations. Because the details of the solution for the two-parameter logistic item characteristic curve model were not presented in Chapter 7, the appropriate mathematics is given below.

The probability of a correct response under the two-parameter logistic item characteristic curve model is defined by

$$P_i(u_{ij} = 1|\theta_i) = \frac{\exp[e^{\alpha_i}(\theta_j - b_i)]}{1 + \exp[e^{\alpha_i}(\theta_j - b_i)]}. \tag{E.1}$$

The starting point for the estimation is the marginalized posterior probability distribution

$$g(\xi, \tau | U) \propto L(U|\xi, \tau)g(\xi|\eta). \tag{E.2}$$

The first, second, and cross derivatives of the logarithm of this posterior density are needed by the Newton-Raphson procedure. The first derivative of the marginalized likelihood component of Equation E.2 with respect to α_i is obtained by following Equations 7.10 and 7.11 with $c_i = 0$. The derivative is

$$\frac{\partial}{\partial \alpha_i} \{\log[L(U|\xi, \tau)]\} = e^\alpha \sum_{i=1}^{n} \int [u_{ij} - P_i(\theta_j)](\theta_j - b_i)[P(\theta_j|U_j, \xi, \tau)]d\theta. \tag{E.3}$$

The posterior term $[P(X_k|U_j, \xi, \tau)]$ corresponds to $\bar{n}_{ik}$ (Bock & Aitkin, 1981; Harwell, Baker, & Zwart, 1988). Substituting $\bar{n}_{ik}$ and putting Equation E.3 in numerical quadrature form yields

$$\frac{\partial}{\partial \alpha_i} \{\log[L(U|\xi, \tau)]\} = e^\alpha \sum_{k=1}^{q} [\bar{r}_{ik} - \bar{n}_{ik}P_i(X_k)](X_k - b_i). \tag{E.4}$$

The derivative of the prior term with respect to α_i is appended to the derivative of the likelihood component, yielding the Bayes modal expression

$$\frac{\partial}{\partial \alpha_i}[g(\xi, \tau | U)] = L_1 = e^\alpha \sum_{k=1}^{q} [\bar{r}_{ik} - \bar{n}_{ik}P_i(X_k)](X_k - b_i) - \frac{(\alpha_i - \mu_\alpha)}{\sigma_\alpha^2}. \tag{E.5}$$

The expression for the first derivative of the likelihood component with respect to b_i can be obtained by inspection of the three-parameter model result (Equation 7.17). The numerical quadrature form of this term is

$$\frac{\partial}{\partial b_i}[g(\xi, \tau | U)] = L_2 = \frac{\partial}{\partial b_i}\{\log[L(U|\xi, \tau)]\} = -e^\alpha \sum_{k=1}^{q} [\bar{r}_{ik} - \bar{n}_{ik}P_i(X_k)]. \tag{E.6}$$

No prior distribution was imposed on the difficulty parameter; hence, there is no appending term, and Equation E.6 also is the Bayes modal estimation expression.

The Newton-Raphson estimation procedure requires that the second and cross derivatives of the posterior distribution be taken with respect to the parameters α_i and b_i. Taking the derivative of Equation E.3 with respect to α_i yields

$$\frac{\partial^2}{\partial \alpha_i^2} \{\log[L(U|\xi, \tau)]\} = \sum_{i=1}^{n} \int \{e^{\alpha_i}[u_{ij} - P_i(\theta_j)]$$
$$- (e^{\alpha_i})^2 (\theta_j - b_i) P_i(\theta_j) Q_i(\theta_j)\} (\theta_j - b_i)[P(\theta_j|U_j, \xi, \tau)]d\theta. \tag{E.7}$$

In the usual two-parameter logistic item characteristic curve model, the second derivatives of the log-likelihood do not contain data terms. Because of the use of the logarithmic transformation of a_i, however, the terms involving the item score data (i.e., u_{ij}) do not disappear. Following standard practice (Garwood, 1941; Finney, 1952), the expectation of Equation E.7 is taken. This results in u_{ij} being replaced by $P_i(\theta_j)$, and then

$$E\left(\frac{\partial^2}{\partial \alpha_i^2}\{\log[L(U|\xi,\tau)]\}\right) =$$
$$-(e^{\alpha_i})^2 \sum_{i=1}^{n} \int (\theta_j - b_i)^2 P_i(\theta_j) Q_i(\theta_j)[P(\theta_j|U_j,\xi,\tau)]d\theta. \tag{E.8}$$

The second derivative of the prior component in Equation E.5 with respect to α_i yields $-1/\sigma_\alpha^2$, which is appended to the likelihood component. Substituting $\bar{n}_{ik}$ for $[P(\theta_j)|U_j,\xi,\tau)]$, the final expression in quadrature form is

$$L_{11} = -(e^{\alpha_i})^2 \sum_{k=1}^{q} \bar{n}_{ik}(X_k - b_i)^2 P_i(X_k) Q_i(X_k) - \frac{1}{\sigma_\alpha^2}. \tag{E.9}$$

The second derivative with respect to b_i, in quadrature form, is

$$E\left(\frac{\partial^2}{\partial b_i^2}\{\log[L(U|\xi,\tau)]\}\right) = L_{22} = -(e^{\alpha_i})^2 \sum_{k=1}^{q} \bar{n}_{ik} P_i(X_k) Q_i(X_k). \tag{E.10}$$

As no prior was imposed upon b_i, there is no appending term.

The cross-derivative term can be obtained by taking expectation of the derivative of Equation E.3 with respect to b_i;

$$E\left(\frac{\partial^2}{\partial \alpha_i b_i}\{\log[L(U|\xi,\tau)]\}\right) = e^{\alpha_j} \sum_{i=1}^{n} \int \{-[u_{ij} - P_i(\theta_j)]$$
$$-e^{\alpha_i}(\theta_j - b_i)P_i(\theta_j)Q_i(\theta_j)(\theta_j - b_i)\}[P(\theta_j|U_j,\xi,\tau)]d\theta. \tag{E.11}$$

The resulting numerical quadrature form, after taking the expectation, is

$$L_{12} = L_{21} = -(e^{\alpha_i})^2 \sum_{k=1}^{q} \bar{n}_{ik}(X_k - b_i)^2 P_i(X_k) Q_i(X_k). \tag{E.12}$$

Given L_1, L_2, L_{11}, L_{22}, and L_{12} the Newton-Raphson equation can be implemented.

The impact of the logarithmic transformation of the discrimination parameter can be seen in these derivatives. Each of the first derivatives has a leading e^{α_i} multiplier, and each of the second derivatives has an $(e^{\alpha_i})^2$ leading multiplier. If one examines the corresponding derivatives in a solution not involving the transformed discriminations (cf. Hambleton & Swaminathan, 1986, p. 132), only the terms involving the derivatives of b_i have leading multipliers. The first derivative has a multiplier of a_i as does the cross derivative. The second derivative with respect to b_i has a multiplier of a_i^2.

E.3 Implementation of Marginalized Bayesian Modal Item Parameter Estimation

Since the logic of the EM algorithm is unchanged by the use of a Bayesian estimation procedure, the program presented in Appendix D was modified to meet present needs. The E-step of the EM algorithm is unaffected by the use of prior distributions, and thus the associated subroutine (lines 300–470) of the previous computer program are unchanged. In the M-step, the subroutine that previously performed marginal maximum likelihood estimation of the item parameters has been modified to implement the marginalized Bayesian modal estimation approach. As was the case in Appendix D, the present computer program was tailored to the LSAT-6 data set. This data set involves five items and 1000 examinees. The expressions for $\bar{n}_{ik}$ and $\bar{r}_{ik}$ (given by Equations 7.14 and 7.15) involve a summation over the N examinees. In a given data set, many examinees can have the same vector of correct and incorrect responses to the I items. To reduce the computational demands, the N possible examinee item score vectors were grouped into 2^I possible item score patterns. These item score patterns are indexed by l, where $l = 1, 2, \ldots, 32$. The program initialization subroutine, beginning at line 1500, that is called at line 130 of the driver section establishes the environment for the estimation process. The 32 item-score patterns are prestored in the array U(I,J) via the DATA statements at lines 9000–9310 and statement 1500. The frequency of each pattern, FPT(J), appears in the DATA statements at lines 9500 and 9510. It was decided a priori to use $q = 10$ values of X_k along the ability scale. The 10 values of X_k and the corresponding weights $A(X_k)$ are retrieved from the DATA statements at lines 9410–9450. The initial values of $b_i = 0$ and $a_i = 1$ for each item were obtained from line 10000. The values of the hyperparameters of the normal prior distribution on α_i (i.e., $\mu_\alpha = 0$ and $\sigma_\alpha = 0.5$) are stored pairwise for each of the five items at line 10110 and set into the corresponding vectors at statement 1540. Essentially the same tracing options as those employed in Appendix D are available here, with the TRMLE being replace by TRBME to trace the Bayesian modal estimation process. The user is also able to select the BILOG FLOAT option and set the number of EM cycles to be performed.

The driver (kernel) section of the program (lines 100–290) implements the EM algorithm. The E-step is implemented in the subroutine in lines 300–470, and is the same as that of the computer program of Appendix D, which should be referred to for the internal details of this section.

The values of $\bar{r}_{ik}$ and $\bar{n}_{ik}$ for item 1 that were produced by the E-step subroutine in the first cycle of the EM algorithm are shown below:

```
CYCLE= 1  ITEM  1

  :

449    X( 1 )= -4        RIK= 6.591986E-04  NK= 4.662526E-03
449    X( 2 )= -3.111     RIK= 4.222471E-02  NK= .1457583
```

```
449    X( 3 )= -2.222    RIK=  1.36278     NK=  2.708543
449    X( 4 )= -1.333    RIK=  22.10459    NK=  31.07672
449    X( 5 )= -.4444    RIK=  155.1186    NK=  181.9732
449    X( 6 )=  .4444    RIK=  374.0979    NK=  402.3184
449    X( 7 )=  1.333    RIK=  286.3932    NK=  295.7766
449    X( 8 )=  2.222    RIK=  76.48853    NK=  77.55681
449    X( 9 )=  3.111    RIK=  8.033214    NK=  8.080218
449    X( 10 )=  4       RIK=  .3582704    NK=  .3591397
```

Lines 190–210 form a loop in which the item parameter estimates for each of the five items are computed using the $\bar{r}_{ik}$ and $\bar{n}_{ik}$ at each X_k as the data. Recall that the $\bar{n}_{ik}$ are the same for all five items. The actual parameter estimation is done in the M-step subroutine at lines 1000–1350, which is called at line 200 of the driver section. In the first cycle, the prestored values of a_i and b_i are used to initiate the iterative M-Step estimation process. However, the logarithm of a_i is taken at line 1030 so that the estimation is in terms of α_i. Because of the use of the α_i and b_i parameterization, there are some major differences between the M-step program in Appendix D and the present program. The changes appear in lines 1142–1170 and lines 1190–1196. In lines 1142 and 1145, three essential terms are calculated: W_{ij}, $[\bar{r}_{ik} - \bar{n}_{ik}P_i(\theta_j)]$, and $(X_k - b_i)$, denoted by the variable names W, RMN, and XMB, respectively. Four basic elements (P1–P4) are calculated from these terms and the L1, L2, L11, L22, and L12 terms of the Newton-Raphson equation are accumulated across quadrature nodes. These derivative terms can be displayed for each quadrature node at lines 1171–1174 via the TRALL variable. Only the values for the first and last quadrature node are shown below:

```
M-STEP  ITERATION= 1   ITEM= 1   Q-NODE= 1
    P=  1.798621E-02        W=  1.766271E-02
    R-N*P=  5.753374E-04    X-B= -4
    P1=  8.235282E-05       P2= -3.294113E-04
    P3=  1.317645E-03       P4= -2.30135E-03
    L1= -2.30135E-03        L2=  5.753374E-04
    L11=  1.317645E-03      L22=  8.235282E-05
    L12=L21= -3.294113E-04
```

⋮

```
M-STEP  ITERATION= 1   ITEM= 1   Q-NODE= 10
    P=  .9820138           W=  1.766273E-02
    R-N*P=  5.590297E-03   X-B=  4
    P1=  6.343388E-03      P2=  2.537355E-02
    P3=  .1014942          P4=  2.236119E-02
    L1=  81.78494          L2=  288.8938
    L11=  161.6884         L22=  200.4869
    L12=L21=  97.27988
```

Upon completion of the calculations for the last quadrature node, the appropriate signs are introduced, the leading multipliers are applied to the information matrix components, and the prior terms are appended. The results of these calculations and the determinant of the information matrix can be printed under the TRALL or TRBME options. The following was obtained for the first iteration of the first item:

```
AFTER MULTIPLYING BY EXP(ALPHA), ADDING BAYES TERMS
```

```
     L1=  81.78494          L2= -288.8938
     L11= 165.6884          L22=  200.4869
     L12=L21= -97.27988
DETERMINANT=  23754.96
```

Since the leading multiplier has a value of one in the first iteration and the prior term added to L2 is zero, only the magnitude of L11 was changed here by appending the prior terms.

Lines 1240 and 1250 compute the increments DB and DA, which reflect the change in the item parameter estimates from one iteration to the next within the M-step. The new values of α_i and b_i are computed at line 1270. The increments and new values at each iteration can be printed at line 1275 by either the TRALL or TRBME options. The following values were obtained for the first iteration of the first item:

```
NIT=  1        ITEM=  1
ALPHA(I)= -.4928127      DA= -.4928127
   B(I)= -1.680083       DB= -1.680083
```

Since both the initial values of α_i and b_i were zero, the new values of the indices are the same as the changes. Of course, this will differ as the iterative process proceeds. Line 1290 is a check to determine if α_i and b_i have assumed unreasonable values. If so, the "OUT OF BOUNDS ERROR" message at line 1350 is printed and the subroutine is exited. Line 1300 determines if both DB and DA are small enough to meet the convergence criterion of 0.05. If so, the M-step Bayesian modal estimation process is completed. It is important to recognize that, in the M-step, α_i and b_i are estimated. However, in both the E-step and the M-step, the calculation of $P_i(X_k)$ in the subroutine at line 600 is based upon a_i and b_i. Thus, before printing the results, the transformation of the discrimination parameter is reversed at line 1340, yielding a_i. The results of the M-step for the five items at the end of the first EM cycle were:

ITEM	DIFFICULTY	DISCRIMINATION
1	-2.289881	.9440868
2	-.4346133	.9428452
3	.3901892	.9713619
4	-.7777552	.9331672
5	-1.666262	.9172994

At the end of an EM cycle, the latest values of a_i and b_i are available as starting values in the next E-step and M-step. In a production program, an overall convergence criterion would be used to terminate the estimation process. In the present program, the maximum number of cycles performed is set by the user in response to a query in the initialization subroutine.

The program was run for 20 cycles, and item parameter estimates were obtained. In addition, the MMLE/EM program of Appendix D was also run for 20 cycles to obtain comparative results. Both sets of item parameter estimates are reported below.

Item	Bayes estimates		MMLE estimates	
	b_i	a_i	b_i	a_i
1	-3.283221	.8486590	-3.367784	0.8229333
2	-1.317096	.7584814	-1.364331	0.7264646
3	$-.2847672$	.8695322	$-.2821274$	0.8804284
4	-1.766085	.7357355	-1.858505	0.6918691
5	-2.858299	.7298232	-3.103260	0.6624422

The effect of the prior on a_i can be seen in these results. Compared to the MMLE results for this data set, the Bayesian estimates of a_i have generally been shrunk towards the prior mean of 1.13. In the present case, this means that the values of the discrimination estimates will be slightly higher than those of the MMLE/EM analysis. In addition, the range of the b_i estimates has been reduced, which is a reflection of the increased values of the a_i.

Under the FLOAT option of the PC-BILOG program (i.e., FOFLG), the hyperparameter μ_α is computed after the M-step as the average of the obtained item discrimination estimates. This hyperparameter value is computed by a subroutine at lines 1400–1440. The FLOAT option is selected in the initialization subroutine at line 1670 and is implemented at line 1192 of the M-step subroutine. The LSAT-6 data was reanalyzed using the FLOAT option and allowed to run for 20 cycles. The following item parameter estimates were obtained:

```
ITEM   DIFFICULTY      DISCRIMINATION
  1    -3.450861        .7987059
  2    -1.352925        .7339953
  3    -.289878         .8492491
  4    -1.821438        .7088962
  5    -2.988625        .6923823
```

The average values of the discrimination estimates are somewhat lower than those previously obtained. This reflects the fact that the average of the α_i ($-.282$) rather than a value of 0 was used as the mean of the prior distribution. The lower discrimination indices are also reflected in the somewhat wider range of values of the item difficulties.

It should be emphasized that the present program is a simple version of Mislevy's Bayesian procedure and does not involve many of the sophisticated features of PC-BILOG. Despite this, the present program produced marginalized Bayesian item parameter estimates that were quite similar to those produced by PC-BILOG for the LSAT-6 data.

E.4 Implementation of Bayesian Estimation of an Examinee's Ability

E.4.1 Bayesian Modal Estimation

In Chapter 7, the components of the Newton-Raphson equation solved for an examinee's ability estimate were presented for the three-parameter model.

In order to obtain the corresponding components under the two-parameter model, one simply sets the value of the c_i parameter to zero. Thus, the terms can be obtained by visual inspection from Equations 7.37 and 7.39, and they will not be presented here. Following the PC-BILOG program default specification, a normal prior distribution for the ability distribution was employed. The computer program implementing the Bayesian modal estimation of ability is a simplification and modification of the program in Appendix B for the maximum likelihood estimation of an examinee's ability. The capability to estimate ability under the one- or three-parameter item characteristic curve model was eliminated, and the Bayesian prior terms were appended to the terms in the numerator and denominator of the Newton-Raphson equation. The item parameters employed here are those yielded by the marginalized Bayesian estimation of Section E.3. The ability will be estimated for the twenty-second item response pattern (1,0,1,0,1) of the LSAT-6 data set.

At line 130 of the driver section, a subroutine, beginning at line 3000, is called that establishes the context of the estimation process. The number of items is set to five, and the bound on the size of an increment to an ability estimate is set to 0.5. The hyperparameters of the normal prior distribution were set to a mean (HPM) of zero and a variance (HPVAR) of one. The examinee's dichotomously scored responses to the five items are read into the vector UIJ(I) via statement 3010. The a_i, b_i parameterization was used for the item parameters stored in the DATA statements at lines 9510 and 9520. These are read into the vectors B(I) and A(I) via statements 3020 and 3030. Again, the TRALL and TRBME variables are used to control the amount of computational detail that the user desires to see. The user is also asked to specify the maximum number of iterations of the Newton-Raphson equation that are to be performed when the process does not converge. At line 150, the ability estimation subroutine beginning at line 2000 is called using an initial ability estimate of THETA= 0. The primary iterative loop of the Newton-Raphson procedure begins at line 2000 and ends at line 2460. The nested loop over the test items begins at line 2030 and ends at line 2190. At lines 2050 and 2060, the item parameters are used to compute the probability of correct response to the item PHAT at the ability level, THETA. The weighting coefficient W_{ij} and v_{ij} (i.e., W and VIJ are computed, and these are used to compute the components added to the numerator and denominator terms at lines 2110 and 2120. If TRALL is set to zero, the details of these computations are printed. To illustrate this, the detailed print for the first and last items are shown below:

```
ITERATION= 1
       ITEM= 1
     DIFFICULTY= -3.283     DISCRIMINATION= .849    THETA= 0
     UIJ= 1          DEV= 2.787267 P HAT= .9419839
     WIJ= 5.465025E-02 VIJ= 5.801612E-02
       NUMERATOR SUM= 4.925568E-02        DENOMINATOR SUM=
3.939195E-02

 ⋮

       ITEM= 5
     DIFFICULTY= -2.858     DISCRIMINATION= .73     THETA= 0
     UIJ= 1          DEV= 2.08634 P HAT= .8895684
```

```
WIJ=  9.823647E-02  VIJ=  .1104316
NUMERATOR SUM= -.6210175             DENOMINATOR SUM=  .4823222
```

Upon completion of the last item, the Bayesian prior terms are appended to the numerator and denominator terms. Since, in the first iteration, THETA=0, HPM=0, and HPVAR=1, the values of the appended terms are zero and minus one for the numerator and denominator, respectively. As the iterative process proceeds, these values will change. To be consistent with Equation 7.39, the denominator sum is made negative at line 2310, and the prior term is subtracted from it. If TRALL is set, the adjusted values will be printed as follows:

```
AFTER ADDING PRIOR TERMS
NUMERATOR SUM= -.6210175             DENOMINATOR SUM= -1.482322
```

After these adjustments, the increment DELTA to the initial ability estimate is computed at line 2380, and the following printed if either TRALL or TRBME is set to zero:

```
CHANGE IN THETA=  .418949
```

At line 2400, the absolute magnitude of the change is compared to the bound (BIGT). If greater than the bound, the change is limited to the appropriately signed value of BIGT at line 2420. Contrary to our usual practice, the negative signs were not factored out of the second derivative with respect to ability in Equation 7.39. Thus, the increment is subtracted from the present value of ability as per Equation 7.40. It should be noted that this differs from the practice in previous chapters and appendices. Thus, the new value of the ability estimate, THETA, is obtained at line 2430 by subtracting the change from the initial estimate. If either TRALL or TRBME is set to zero, the following is printed:

```
THETA=-.418949      CHANGE =  .418949
```

Next, the convergence of the overall Newton-Raphson procedure is determined. If the absolute value of DELTA is greater than 0.05, another iteration is initiated at lines 2010 that sets SUMNUM and SUMDEM to zero. The new value of the examinee's ability estimate will now be used in the calculations. To illustrate these, the detailed results for the first and fifth item in the second iteration are shown below:

```
ITERATION=  2
        ITEM=  1
    DIFFICULTY= -3.283    DISCRIMINATION=  .849    THETA=-.418949
    UIJ=  1         DEV=  2.431579 P HAT=  .9192039
    WIJ=  7.426811E-02 VIJ=  8.079612E-02
    NUMERATOR SUM=  6.859591E-02         DENOMINATOR SUM=
5.353253E-02

        :
        :

        ITEM=  5
    DIFFICULTY= -2.858    DISCRIMINATION=  .73    THETA=-.418949
    UIJ=  1         DEV=  1.780507 P HAT=  .8557595
    WIJ=  .1234352     VIJ=  .1442405
    NUMERATOR SUM= -.4058432             DENOMINATOR SUM=  .5430265
```

Upon completion of the fifth item, the prior terms are appended. Since the value of THETA is no longer zero, terms are appended to both the numerator and denominator sums, and the following were obtained:

```
AFTER ADDING PRIOR TERMS
NUMERATOR SUM= 1.310581E-02       DENOMINATOR SUM= -1.543026
```

Next, the increment DELTA to the ability estimate is computed at line 2380, and the following is printed if the traces are set to zero:

```
CHANGE IN THETA= -8.493574E-03
```

Since the absolute value of the change is less than 0.05, the Newton-Raphson procedure has converged. The estimation process is completed, and the normal return from the subroutine is taken at line 2480. If convergence had not been reached by the last iteration allowed, the "REACHED MAX ITERATIONS" message would have been printed before the subroutine return was taken. At line 160 in the driver section, the final value of the ability estimate is printed as:

```
ESTIMATED ABILITY= -.4104555
```

Two observations are in order: First, with the exception of the appended prior terms, the computer programs for Bayesian modal and maximum likelihood are the same. Second, the maximum likelihood ability estimate was -1.116383 while the Bayesian modal estimate from the same data was $-.4104555$. Thus, there was a considerable amount of shrinkage towards the prior mean of zero under the Bayesian approach, which is due primarily to the small number of items employed. With a large number of items, the effects of the appended prior terms would be much less.

E.4.2 Bayesian "Expected A Posteriori" Estimation

Inspection of Equation 7.43 for the expected a posteriori estimation of an examinee's ability reveals that it is similar to Equation 7.15 for the expected number of correct responses to an item. Both involve the likelihood of given pattern of item scores $L(X_k)$ and the quadrature weight $A(X_k)$. However, only the item score pattern of a given examinee is involved in Equation 7.43. Thus, the implementation of expected a posteriori (EAP) ability estimation has communalities with the E-step of both marginal maximum likelihood / expectation-maximization (MMLE/EM) and Bayesian modal estimation / expectation-maximization (MBE/EM) estimation of item parameters.

A short driver program, lines 100–170, is used to control the overall estimation process. The initialization subroutine, lines 1500–1620, is used to read the examinee's vector of scores on the five items of the LSAT-6 test into U(I). The quadrature nodes and weights are read into the vectors AK(K) and X(K). The item parameter estimates yielded by the Bayesian modal estimation procedure presented above are read into the vectors B(I) and A(I) and treated as the true values. The tracing inquiries are responded to and a summary of the choices is printed. The EAP subroutine, lines 300–530, is called at line 170 of the driver section. In order to make the computations a bit easier to understand, two major loops, lines 300–390 and lines 410–510, over the

10 quadrature nodes have been employed, even though one loop would have sufficed. In the first loop, the likelihood of the examinee's pattern of scores on the five items is calculated at each quadrature node X(K). To do this, an inner loop over the five items, lines 310–370, is used. At line 320 a subroutine is called to compute the probability of answering the item correctly, P, or incorrectly, Q. At line 320, the examinee's score U(I) on that item is used to set the value of PITEM. At line 340, the continuing product of the PITEM terms over the five items is calculated. If the TRALL variable is set, the value of PITEM and the running value of LXL(K) are printed. Only the values for the first and 10th quadrature nodes are shown below:

```
        K=  1  I=  1 U=  1  P=  .3523483       L(X)=  .3523483
        K=  1  I=  2 U=  0  P=  .8842916       L(X)=  .3115786
        K=  1  I=  3 U=  1  P=  3.797728E-02   L(X)=  1.183291E-02
        K=  1  I=  4 U=  0  P=  .8381089       L(X)=  9.917267E-03
        K=  1  I=  5 U=  1  P=  .3028717       L(X)=  3.00366E-03
  K=  1    L(X)=  3.00366E-03

  ⋮

        K= 10  I=  1 U=  1  P=  .9979406       L(X)=  .9979406
        K= 10  I=  2 U=  0  P=  1.745903E-02   L(X)=  1.742308E-02
        K= 10  I=  3 U=  1  P=  .9765224       L(X)=  1.701403E-02
        K= 10  I=  4 U=  0  P=  .0141502       L(X)=  2.407519E-04
        K= 10  I=  5 U=  1  P=  .9933492       L(X)=  2.391508E-04
  K= 10    L(X)=  2.391508E-04
```

It should be noted that the likelihood of an ability of -4.0000, X1, yielding this item score pattern is only .0030, while that of an ability of -0.4444, X5, is .0197. Upon completion of the last quadrature node, the terms in the numerator and denominator of Equation 7.43 can be calculated and summed over 10 quadrature nodes. This is accomplished in lines 410–510. At each quadrature node, the likelihood, LXK(K) is multiplied by the quadrature weight A(K), yielding the term LA. This term is then multiplied by the value of the quadrature node X(K), yielding LXA. At lines 440 and 450, the sums in the numerator and denominator of Equation 7.43 are accumulated. If the TRALL variable is set, the details of this process will be printed. Only the results for the first and last quadrature node are shown below:

```
  K=  1
        LXK(K)=  3.00366E-03    AK(K)=  .000119    X(K)=  -4
        LA=  3.574355E-07   XLA=  -1.429742E-06
  SUM NUMERATOR=  -1.429742E-06 SUM DENOMINATOR=  3.574355E-07

  ⋮

  K= 10
        LXK(K)=  2.391508E-04    AK(K)=  .000119    X(K)=   4
        LA=  2.845894E-08   XLA=  1.138358E-07
  SUM NUMERATOR=  -9.571826E-03 SUM DENOMINATOR=  2.497229E-02
```

Since the XLA term is the contribution of the kth quadrature node to the examinee's ability estimate, it is of interest to examine the full set of 10 values (n.b., rearranged from the output):

K=	1	X(K)=	-4	XLA=	-1.429742E-06
K=	2	X(K)=	-3.1111	XLA=	-9.715276E-05
K=	3	X(K)=	-2.2222	XLA=	-1.766667E-03
K=	4	X(K)=	-1.3333	XLA=	-7.457827E-03
K=	5	X(K)=	-.4444	XLA=	-4.859926E-03
K=	6	X(K)=	.4444	XLA=	2.811282E-03
K=	7	X(K)=	1.3333	XLA=	1.603798E-03
K=	8	X(K)=	2.2222	XLA=	1.87114E-04
K=	9	X(K)=	3.1111	XLA=	7.379402E-06
K=	10	X(K)=	4	XLA=	1.138358E-07

Note that the largest contributions to the ability estimate are coming from the central part of the distribution.

The examinee's ability estimate computed at line 520 is simply the ratio of the numerator sum to the denominator sum. The subroutine returns to the driver section via line 530 and at line 180 the ability estimate is printed as follows:

```
ESTIMATED THETA= -.3833693
```

Several observations are in order: First, the overall estimation process is noniterative. Second, although not capitalized upon here, the essential terms would be available from the last iteration of the Bayesian modal estimation calculations for items. Third, the numerator of Equation 7.43 is simply the sums of the XLA terms that weight the value of X_k by its likelihood and proportion of the prior ability distribution $A(X_k)$ at that node. In the BILOG program, the prior distribution is the final adjusted quadrature distribution rather than the unit normal used here. Fourth, the denominator of Equation 7.43 is the unconditional probability of the examinee's item score vector given the item parameters and the quadrature distribution. Thus, the ability estimate is normalized with respect to this probability.

E.5 BASIC Computer Programs

E.5.1 Program for marginalized Bayesian item parameter estimation

```
100 REM PROGRAM TO IMPLEMENT MARGINALIZED BAYESIAN ITEM
101 REM PARAMETER ESTIMATION
105 REM USES B=DIFFICULTY, A=DISCRIMINATION LSAT-6 DATA
110 DIM U(32,5),X(10),AK(10),FPT(32),LXK(32,10),PL(32)
120 DIM RIK(5,10),NK(5,10),PQ(10),A(5),B(5),PM(5),PV(5),ALPHA(5)
130 GOSUB 1500:REM READ CANNED DATA AND SET TRACES
140 FOR NC=1 TO MNC
150 PRINT " CYCLE=";NC
155 FOR L=1 TO 32:PL(L)=0:FOR K=1 TO 10:LXK(L,K)=1:NEXT K:NEXT L
160 REM E-STEP   GET EXPECTED R AND N
170 GOSUB 300
180 REM M-STEP   BAYES MODAL ESTIMATION OF ITEM PARAMETERS
190 FOR I=1 TO 5
```

```
200 GOSUB 1000
210 NEXT I
220 NEXT NC
230 REM DONE MAX CYCLES
240 PRINT "  ITEM";TAB(10);"DIFFICULTY";TAB(25);"DISCRIMINATION"
250 FOR I=1 TO 5
270 PRINT TAB(5);I;TAB(10);B(I);TAB(25);A(I)
280 NEXT I
290 STOP
REM **************END OF DRIVER SECTION*********
299 REM E-STEP SUBROUTINE
300 FOR L=1 TO 32
310 FOR K=1 TO 10
320 FOR I=1 TO 5
330 GOSUB 600
340 PQ(I)=P:IF U(L,I)=0 THEN PQ(I)=Q
342 IF TRALL<>0 GOTO 360
345 PRINT "345";TAB(5);L;TAB(10);K;TAB(15);I;TAB(20);PQ(I);
346 PRINT TAB(40);U(L,I)
360 LXK(L,K)=LXK(L,K)*PQ(I)
370 NEXT I
375 IF (TRALL=0)OR(TRLP=0) THEN
    PRINT "L= ";L;TAB(10);"K= ";K;TAB(20);"L(X)= ";LXK(L,K)
    END IF
380 NEXT K
390 PL(L)=0:FOR K=1 TO 10:PL(L)=PL(L)+LXK(L,K)*AK(K)
391 IF TRALL<>0 GOTO 395
392 PRINT "392";TAB(5);L;TAB(10);PL(L);TAB(35);LXK(L,K);
393 PRINT TAB(50);AK(K)
395 NEXT K
398 IF (TRALL=0) OR(TRLP=0) THEN PRINT TAB(10);"P(";L;")= ";PL(L)
400 NEXT L
410 REM R BAR AND N BAR LOOP
420 FOR I=1 TO 5
425 IF (TRALL=0)OR(TRRN=0) THEN PRINT TAB(10);"ITEM ";I
430 FOR K=1 TO 10
435 RIK(I,K)=0:NK(I,K)=0
440 FOR L=1 TO 32:NT=FPT(L)*LXK(L,K)*AK(K)/PL(L):RT=NT*U(L,I)
441 RIK(I,K)=RIK(I,K)+RT:NK(I,K)= NK(I,K)+NT
442 IF TRALL<>0 GOTO 448
443 PRINT "442";TAB(5);I;TAB(8);K;TAB(12);L;TAB(17);LXK(L,K);
444 PRINT TAB(33);AK(K); TAB(50);PL(L)
445 PRINT "444";TAB(5);U(L,I);TAB(8);NT;TAB(24);RT;TAB(39);RT
446 PRINT "446";TAB(5);RIK(I,K);TAB(25);NK(I,K)
448 NEXT L
449 IF (TRALL=0)OR(TRRN=0) THEN
    PRINT "449";TAB(8);"X(";K;")= ";X(K);
    PRINT TAB(25);"RIK= ";RIK(I,K);TAB(45);"NK= ";NK(I,K)
    END IF
450 NEXT K
460 NEXT I
470 RETURN
REM ******ROUTINE TO COMPUTE P(X)******
```

```
600 DEV=-A(I)*(X(K)-B(I))
610 P=1/(1+EXP(DEV)):Q=1-P
620 RETURN
999 REM M-STEP BAYESIAN MODAL ITEM PARAMETER ESTIMATION
1000 FOR NIT=1 TO 6
1030 ALPHA(I)=LOG(A(I))
1040 L1=0:L2=0:L11=0:L22=0:L12=0
1050 FOR K=1 TO 10:REM THETA LOOP
1060 IF NK(I,K)=0 GOTO 1180
1070 IF TRALL<>0 GOTO 1140
1075 PRINT "M-STEP  ITERATION= ";NIT;"  ITEM= ";I;"  Q-NODE= ";K
1135 REM CALC P,Q
1140 GOSUB 600:REM B= THRESHOLD A=DISCRIMINATION
1142 W=P*(1-P):IF W<.0000009 GOTO 1180
1145 RMN=RIK(I,K)-NK(I,K)*P:W=P*(1-P):XMB=X(K)-B(I)
1146 IF TRALL<>0 GOTO 1160
1150 PRINT TAB(5);"P= ";P;TAB(29);"W= ";W
1155 PRINT TAB(5);"R-N*P= ";RMN;TAB(29);"X-B= ";XMB
1160 P1=NK(I,K)*W:P2=P1*XMB:P3=P2*XMB:P4=RMN*XMB
1170 L1=L1+P4:L2=L2+RMN:L11=L11+P3:L22=L22+P1:L12=L12+P2
1171 IF TRALL<>0 GOTO 1180
1172 PRINT TAB(5);"P1= ";P1;TAB(29);"P2= ";P2
1173 PRINT TAB(5);"P3= ";P3;TAB(29);"P4= ";P4
1174 PRINT TAB(5);"L1= ";L1;TAB(29);"L2= ";L2
1175 PRINT TAB(5);"L11= ";L11;TAB(29);"L22= ";L22
1176 PRINT TAB(5);"L12=L21= ";L12
1180 NEXT K
1190 EA=EXP(ALPHA(I)):EA2=EA*EA
1192 IF FOFLG<>0 THEN LABAR=PM(I):GOTO 1195
1193 GOSUB 1400
1194 REM MULT BY EA AND ADD PRIOR TERMS
1195 L1=EA*L1-(ALPHA(I)-LABAR)/PV(I):L2=-EA*L2
1196 L11=EA2*L11+1/PV(I):L22=EA2*L22:L12=-EA2*L12
1197 REM -1 HAS BEEN INTRODUCED WHERE APPROPRIATE IN THE
1198 REM EXPRESSIONS  OF LINES 1195 AND 1196
1200 DM=L11*L22-L12*L12:IF TRALL<>0 GOTO 1230
1201 PRINT:PRINT TAB(5);"EA= ";EA;TAB(25);"EA2= ";EA2
1203 PRINT "AFTER MULTIPLYING BY EXP(ALPHA), ADDING BAYES TERMS"
1205 PRINT TAB(5);"L1= ";L1;TAB(29);"L2= ";L2
1207 PRINT TAB(5);"L11= ";L11;TAB(29)"L22= ";L22
1208 PRINT TAB(5);"L12=L21= ";L12
1210 PRINT "DETERMINANT= ";DM
1230 IF DM<=.000099 GOTO 1350
1240 DA=(L1*L22-L2*L12)/DM
1250 DB=(-L1*L12+L2*L11)/DM
1270 ALPHA(I)=ALPHA(I)+DA:B(I)=B(I)+DB
1275 IF (TRALL=0) OR (TRBME=0) THEN
     PRINT "NIT= ";NIT;TAB(15);"ITEM= ";I
     PRINT "ALPHA(I)= ";ALPHA(I);TAB(26)"DA= ";DA
     PRINT "  B(I)= ";B(I);TAB(26);"DB= ";DB
     ELSE
     END IF
1290 IF(ABS(ALPHA(I))>30)OR(ABS(B(I))>20) GOTO 1350
```

```
1300 IF(ABS(DA)<=.05)AND(ABS(DB)<=.05) GOTO 1340
1330 NEXT NIT
1340 A(I)=EXP(ALPHA(I)):RETURN
1350 PRINT "OUT OF BOUNDS ERROR":GOTO 1340
REM ********ROUTINE TO GET MEAN OF LOG(A)**********
1400 SUMLA=0
1420 FOR IJ=1 TO 5:SUMLA=SUMLA+ALPHA(IJ):NEXT IJ
1430 LABAR=SUMLA/5
1440 RETURN
REM**********READ CANNED DATA AND SET TRACES*******
1500 FOR L=1 TO 32:FOR I=1 TO 5:READ U(L,I):NEXT I:NEXT L
1510 FOR K=1 TO 10:READ X(K):NEXT K
1520 FOR K=1 TO 10:READ AK(K):NEXT K
1530 FOR L=1 TO 32:READ FPT(L):NEXT L
1540 FOR I=1 TO 5:READ B(I),A(I):NEXT I
1550 FOR I=1 TO 5:READ PM(I),PV(I):PV(I)=PV(I)*PV(I):NEXT I
1560 REM SET UP TRACEING
1570 TRALL=-1:TRLP=-1:TRRN=-1:TRBME=-1
1580 INPUT "TRACE ALL? Y/N ";YN$
1590 IF (YN$="Y")OR(YN$="y") THEN
     TRALL=0:TRLP=0:TRRN=0:TRBME=0:GOTO 1660
     END IF
1600 INPUT "TRACE L,P? Y/N ";YN$
1610 IF (YN$="Y")OR(YN$="y") THEN TRLP=0
1620 INPUT "TRACE R AND N? Y/N ";YN$
1630 IF (YN$="Y")OR(YN$="y") THEN TRRN=0
1640 INPUT "TRACE BME? Y/N ";YN$
1650 IF (YN$="Y")OR(YN$="y") THEN TRBME=0
1660 INPUT "USE FLOAT OPTION? Y/N ";YN$
1670 FOFLG=-1:IF (YN$="Y")OR(YN$="y") THEN FOFLG=0
1680 PRINT "FLOAT OPTION ";YN$
1690 INPUT "ENTER NUMBER OF CYCLES TO DO ";MNC
1700 PRINT " ALL  LP  RN  BME   MAX CYCLES"
1710 PRINT TAB(2);TRALL;TAB(6);TRLP;TAB(11);TRRN;TAB(15);TRBME;
1711 PRINT TAB(21);MNC
1720 CLS:RETURN
8999 REM ALL POSSIBLE ITEM SCORE PATTERNS
9000 DATA 0,0,0,0,0
9010 DATA 0,0,0,0,1
9020 DATA 0,0,0,1,0
9030 DATA 0,0,0,1,1
9040 DATA 0,0,1,0,0
9050 DATA 0,0,1,0,1
9060 DATA 0,0,1,1,0
9070 DATA 0,0,1,1,1
9080 DATA 0,1,0,0,0
9090 DATA 0,1,0,0,1
9100 DATA 0,1,0,1,0
9110 DATA 0,1,0,1,1
9120 DATA 0,1,1,0,0
9130 DATA 0,1,1,0,1
9140 DATA 0,1,1,1,0
9150 DATA 0,1,1,1,1
```

```
9160 DATA 1,0,0,0,0
9170 DATA 1,0,0,0,1
9180 DATA 1,0,0,1,0
9190 DATA 1,0,0,1,1
9200 DATA 1,0,1,0,0
9210 DATA 1,0,1,0,1
9220 DATA 1,0,1,1,0
9230 DATA 1,0,1,1,1
9240 DATA 1,1,0,0,0
9250 DATA 1,1,0,0,1
9260 DATA 1,1,0,1,0
9270 DATA 1,1,0,1,1
9280 DATA 1,1,1,0,0
9290 DATA 1,1,1,0,1
9300 DATA 1,1,1,1,0
9310 DATA 1,1,1,1,1
9400 REM QUADRATURE POINTS AND WEIGHTS VIA BILOG
9410 DATA -4.000,-3.111,-2.222,-1.333,-.4444,.4444,1.333,2.222
9420 DATA 3.111,4.000
9430 DATA .000119,.002805,.03002,.1458,.3213,.3213,.1458,.03002
9450 DATA .002805,.000119
9499 REM Item score PATTERN FREQUENCIES
9500 DATA 3,6,2,11,1,1,3,4,1,8,0,16,0,3,2,15,10,29,14,81,3,28,15
9510 DATA 80,16,56,21,173,11,61,28,298
9599 REM INITIAL VALUES FOR B(I) AND A(I)
10000 DATA 0,1,0,1,0,1,0,1,0,1
10100 REM PRIOR MEAN AND STD DEVIATION ON LOG(A)
10110 DATA 0,.5,0,.5,0,.5,0,.5,0,.5
```

E.5.2 Program for Bayesian Modal Estimation of an Examinee's Ability

```
100 REM PROGRAM FOR BAYESIAN MODAL ESTIMATION OF AN EXAMINEES
101 REM ABILITY UNDER THE TWO-PARAMETER LOGISTIC ITEM
110 REM CHARACTERISTIC CURVE MODEL LSAT-6 DATA
120 DIM A(10),B(10),JIRP(10)
130 GOSUB 3000:REM READ CANNED DATA AND SET TRACES
140 THETA=0:REM INITIAL ESTIMATE OF EXAMINEE'S ABILITY
150 GOSUB 2000:REM ESTIMATE ABILITY
160 PRINT "ESTIMATED ABILITY= ";THETA
170 STOP
REM ********END OF DRIVER SECTION
REM ********ABILITY ESTIMATION SUBROUTINE*******
2000 FOR NIT=1 TO MAXIT
2010 SUMNUM=0:SUMDEM=0:REM CLEAR SUMS
2020 PRINT:PRINT "ITERATION= ";NIT
2030 FOR I=1 TO NITEM:REM ITEM LOOP
2040 REM CACLUATE PHAT
2041 UIJ=JIRP(I)
2050 DEV=A(I)*(THETA-B(I))
2060 PHAT=1.0/(1.0+EXP(-DEV))
2090 WIJ=PHAT*(1.0-PHAT)
2100 VIJ=(UIJ-PHAT)
```

```
2110 SUMNUM=SUMNUM+A(I)*VIJ
2120 SUMDEM=SUMDEM+A(I)*A(I)*WIJ
2130 IF(TRALL=-1) THEN GOTO 2190
2140 PRINT TAB(10);"ITEM= ";I
2150 PRINT TAB(5);"DIFFICULTY= ";B(I);
2151 PRINT TAB(28);"DISCRIMINATION= ";A(I);
2152 PRINT TAB(52);"THETA=";THETA
2160 PRINT TAB(5);"UIJ= ";UIJ;TAB(20);"DEV= ";DEV;
2161 PRINT TAB(35);"P HAT= ";PHAT
2170 PRINT TAB(5);"WIJ= ";WIJ;TAB(25);"VIJ= ";VIJ
2180 PRINT TAB(5);"NUMERATOR SUM= ";SUMNUM;
2181 PRINT TAB(40);"DENOMINATOR SUM= ";SUMDEM:PRINT
2190 NEXT I
2300 SUMNUM=SUMNUM-(THETA-HPM)/HPVAR
2310 SUMDEM=-SUMDEM-(1.0/HPVAR)
2311 REM MINUS SIGNS TO MATCH EQUATON 7.39
2320 IF TRALL<>0 GOTO 2380
2330 PRINT TAB(5);"AFTER ADDING PRIOR TERMS"
2340 PRINT TAB(5);"NUMERATOR SUM= ";SUMNUM;
2341 PRINT TAB(40);"DENOMINATOR SUM= ";SUMDEM:PRINT
2380 DELTA=SUMNUM/SUMDEM
2390 IF (TRALL=0)OR(TRBME=0) THEN
     PRINT TAB(5);"CHANGE IN THETA= ";DELTA
     END IF
2400 IF(ABS(DELTA)<BIGT) GOTO 2430
2410 REM PROTECT AGAINST BIG CHANGE IN THETA
2420 IF(DELTA > 0.0) THEN
     DELTA=BIGT
     ELSE
     DELTA= -BIGT
     END IF
2430 THETA = THETA - DELTA
2431 REM CHANGE SUBTRACTED AS PER EQUATION 7.40
2440 IF (TRALL = 0) OR (TRBME = 0) THEN
     PRINT "THETA=";THETA;TAB(20);"CHANGE = ";DELTA
     END IF
2450 IF (ABS(DELTA) < .05) GOTO 2480
2460 NEXT NIT
2470 PRINT " REACHED MAX ITERATIONS"
2480 RETURN
REM *******READ CANNED DATA AND ESTABLISH TRACES*********
3000 NITEM = 5: BIGT = .5: HPM = 0: HPVAR = 1
3001 REM SET HYPERPARMETERS
3010 FOR I = 1 TO NITEM: READ JIRP(I): NEXT I
3020 FOR I = 1 TO NITEM: READ B(I): NEXT I
3030 FOR I = 1 TO NITEM: READ A(I): NEXT I
3050 TRALL = -1: TRBME = -1
3060 INPUT "TRACE ALL? Y/N "; YN$
3070 IF (YN$ = "Y") THEN TRALL = 0: TRBME = 0: GOTO 3100
3080 INPUT "TRACE BME? Y/N "; YN$
3090 IF (YN$ = "Y") THEN TRBME = 0
3100 INPUT "ENTER MAXIMUM NUMBER OF ITERATIONS TO DO "; MAXIT
3120 CLS : PRINT "ALL  BME  MAX ITERATIONS"
```

```
3130 PRINT TAB(2); TRALL; TAB(6); TRBME; TAB(14); MAXIT
3140 RETURN
9500 DATA 1,0,1,0,1
9520 DATA -3.283,-1.317,-.285,-1.766,-2.858
9530 DATA .849,.758,.870,.736,.730
```

E.5.3 Program for Bayesian Expected A Posteriori Estimation of an Examinee's Ability

```
100 REM PROGRAM TO IMPLMENT BAYESIAN EXPECTED A POSTERIORI
101 REM ABILITY ESTIMATION
105 REM USES B=DIFFICULTY A=DISCRIMNATION LSAT-6 DATA
110 DIM U(5),X(10),AK(10),LXK(10)
120 DIM RIK(5,10),NK(5,10),PITEM(10),A(5),B(5)
130 GOSUB 1500:REM READ CANNED DATA AND SET TRACES
155 FOR K=1 TO 10:LXK(K)=1:NEXT K
160 REM BAYESIAN EXPECTED A POSTERIORI ESTIMATION OF ABILITY
170 GOSUB 300
180 PRINT:PRINT "ESTIMATED THETA= ";THETA
190 STOP
REM **************END OF DRIVER SECTION*********
299 REM EAP ESTIMATION SUBROUTINE
300 FOR K=1 TO 10
310 FOR I=1 TO 5
320 GOSUB 600:REM EVALUATE TWO-PARAMETER ICC
330 PITEM=P:IF U(I)=0 THEN PITEM=Q
340 LXK(K)=LXK(K)*PITEM
350 IF TRALL<>0 GOTO 370
360 PRINT TAB(5);"K= ";K;TAB(12);"I= ";I;TAB(18);"U= ";U(I);
361 PRINT TAB(25);"P= ";PITEM;TAB(42);" L(X)= ";LXK(K)
370 NEXT I
380 IF (TRALL=0)OR(TRLS=0) THEN
    PRINT "K= ";K;TAB(10);"L(X)= ";LXK(K)
    END IF
390 NEXT K
400 SNUM=0:SDEN=0
410 FOR K=1 TO 10
420 LA=LXK(K)*AK(K)
430 XLA=X(K)*LA
440 SNUM=SNUM+XLA
450 SDEN=SDEN+LA
460 IF TRALL<>0 GOTO 500
470 PRINT "K= ";K
480 PRINT TAB(5);"LXK(K)= ";LXK(K);TAB(30);"AK(K)= ";AK(K);
481 PRINT TAB(50);"X(K)= ";X(K)
490 PRINT TAB(5);"LA= ";LA;TAB(25);"XLA= ";XLA
500 IF(TRALL=0) OR (TRLS=0) THEN
    PRINT "SUM NUMERATOR= ";SNUM;TAB(30);"SUM DENOMINATOR= ";SDEN
    END IF
510 NEXT K
520 THETA=SNUM/SDEN
530 RETURN
REM ******ROUTINE TO COMPUTE P(X)******
```

```
600 DEV=-A(I)*(X(K)-B(I))
610 P=1/(1+EXP(DEV)):Q=1-P
620 RETURN
REM**********READ CANNED DATA AND SET TRACES*******
1500 FOR I=1 TO 5:READ U(I):NEXT I
1510 FOR K=1 TO 10:READ X(K):NEXT K
1520 FOR K=1 TO 10:READ AK(K):NEXT K
1530 FOR I=1 TO 5:READ B(I),A(I):NEXT I
1540 REM SET UP TRACEING
1550 TRALL=-1:TRLS=-1
1560 INPUT "TRACE ALL? Y/N ";YN$
1570 IF (YN$="Y")OR(YN$="y") THEN TRALL=0:TRLS=0:GOTO 1600
1580 INPUT "TRACE L AND SUMS? Y/N ";YN$
1590 IF (YN$="Y")OR(YN$="y") THEN TRLS=0
1600 PRINT "TRACE: ALL    L AND SUMS"
1610 PRINT TAB(7);TRALL;TAB(15);TRLS
1620 CLS:RETURN
8999 REM EXAMINEE'S ITEM SCORE PATTERN
9000 DATA 1,0,1,0,1
9010 REM QUADRATURE POINTS AND WEIGHTS VIA BILOG
9020 DATA -4.0000,-3.1111,-2.2222,-1.3333,-0.4444,0.4444,1.3333
9030 DATA 2.2222,3.1111,4.0000
9040 DATA .000119,.002805,.03002,.1458,.3213,.3213,.1458,.03002
9050 DATA .002805,.000119
9060 REM VALUES FOR B(I) AND A(I)
9070 DATA -3.283,.849,-1.317,.758,-.285,.870,-1.766,.736
9071 DATA -2.858,.730
```

F. Implementation of Parameter Estimation Under the Graded Response Model

F.1 Introduction

The equations for the estimation of item and ability parameters under the graded response model were presented in Chapter 8. In the present appendix, these two building blocks for the graded response model will not be implemented within the joint maximum likelihood estimation (JMLE) paradigm. Separate implementation of the two building blocks reduces the complexity and the length of the required computer software. Because of the formulation of the parameter estimation process in terms of boundary curve parameters, the mathematics of the graded response case appears complex. However, the implementation is quite straightforward as the primary terms needed are simply the P and W yielded by a boundary curves at a given ability level. Section F.2 describes the implementation of estimation of item parameters under the graded response model that is based on Equation 8.7. The solution of this Newton-Raphson (Fisher scoring) equation yields the estimates of the item intercept parameters for the $m - 1$ boundary curves and the common slope. Using the relationships presented in Chapter 8, the estimates of the m item difficulty parameters can be obtained. Section F.3 describes the implementation of the estimation of a single examinee's ability using the responses to a five-item test scored under the graded response model.

F.2 Implementation of Item Parameter Estimation

The maximum likelihood estimation of the parameters of a single graded response item has been implemented in the form of a BASIC computer program. A contrived data set based on Table 8.1 was used. The item in this table had four response categories, and the location parameters of the three boundary curves were symmetrically placed about zero. Although this table was generated under a normal ogive model for the boundary curves, a logistic model will be assumed in the present example. Given a uniform distribution, from -3 to $+3$, over the ability scale, 7000 simulated examinees were grouped into seven equally spaced intervals. The number of examinees in an ability group selecting each of the four response categories was obtained by multiplying the probabilities in Table 8.1 by 1000.

In order to make the implementation of the item parameter estimation process more systematic, the computer program employs six subroutines to create the elements needed by terms in the Newton-Raphson procedure of Equation 8.7. As has been the practice in preceding appendices, these subroutines are under the control of a driver (kernel) section that is in lines 160–360 of the program. At line 160, the program initialization subroutine, lines 1000–1095, is called to establish the environment of the estimation process. The number of response categories MRC is set to four, and the number of boundary curves MBC is three. The maximum number of iterations, MAXIT, is set arbitrarily to four, the number of points on the ability scale, MAXGPS, is set to seven, and the initial value of the boundary curve slope A is set to 0.90. The seven midpoint values of the ability groups are read into the vector THETA(G) from the DATA statement at line 9010. The number of examinees in each group, F(G), and the number selecting each item response category, R(G,K), are read from the DATA statements in lines 9030–9090 via the FOR-NEXT loops at lines 1027 and 1030. The initial values of the intercept parameters for the three boundary curves are read via line 1040 from the DATA statement at line 9100. It should be noted that, throughout the computer program, the variable K indexes the response categories and the variable Y indexes the boundary curves. The user then responds to the tracing queries, and a summary of the selections is printed. Lines 170–330 of the driver section constitute the iterative Newton-Raphson (Fisher scoring) procedure, with the maximum possible number of cycles being defined by MAXIT. The subroutine, beginning at line 1100, is called at line 180 to set the vector of first derivatives and the matrix of second derivatives to zero before the estimation process starts. The FOR-NEXT loop defined by lines 200–270 contains the subroutines needed to compute the terms in the Newton-Raphson equation. The subroutine, beginning at line 1200, computes the cumulative probability PSTR(Y) and the weight W(Y) associated with each of the three boundary curves at the point THETA(G) on the ability scale. If TRALL is set, the current estimates of the curve parameters and the values of PSTAR and W are printed as follows:

```
MLE ITERATION  1
ABILITY LEVEL  1
BOUNDARY CURVE PSTAR'S
CPT( 1 )=  1.8  A=  1  THETA( 1 )= -3
   DEV= -1.2  PSTAR( 1 )=  .2314752   W( 1 )= .1778944
CPT( 2 )=  0  A=  1  THETA( 1 )= -3
   DEV= -3  PSTAR( 2 )=  4.742587E-02  W( 2 )= 4.517666E-02
CPT( 3 )= -1.8  A=  1  THETA( 1 )= -3
   DEV= -4.8  PSTAR( 3 )=  8.162569E-03  W( 3 )= 8.095942E-03
```

In the driver section, line 220 calls the subroutine, beginning at line 1300, that computes both the probability of selecting a response category using the PSTAR of the previous subroutine and the observed proportion POBS(K) of choosing each response category. The latter are computed using the vectors

R(G,K) and F(G). If TRALL is set, the following is printed for the current ability
level:

```
PK AND P OBSERVED
K=  1  PK=  .7685248   POBS=  .691
K=  2  PK=  .1840493   POBS=  .242
K=  3  PK=  .0392633   POBS=  .061
K=  4  PK=  8.162569E-03  POBS=   .006
```

After computing these terms for all response categories, a return is made
to the driver section of the program. The next subroutine, lines 1400–1470,
computes TERM1, TERM2, and the first derivative terms corresponding to the
intercept parameters. The first derivative term, SFD(Y), for each boundary
curve is the sum of the corresponding TERM2 elements. If TRALL is set, the
three results are printed as shown below:

```
INTERCEPT TERMS
Y=  1  TERM1 =  .4157396   TERM2=  73.95776
       FIRST DERIVATIVE =  73.95776
Y=  2  TERM1 =  .2387487   TERM2=  10.78587
       FIRST DERIVATIVE =  10.78587
Y=  3  TERM1 = -.8185508   TERM2= -6.62694
       FIRST DERIVATIVE = -6.62694
```

The subroutine beginning at line 1500 is called at line 240 of the driver. This
subroutine computes the terms involved in the first derivative with respect
to the common slope parameter. The FOR-NEXT loop in lines 1510–1560 is over
the number of response categories. At line 1520, TERM1 is computed as the
difference in the weights, W(K-1) and W(K). A TERM2 is computed at line 1530
and accumulated in SUMLS. If TRALL is set, the three terms for each category
are printed as follows:

```
COMMON SLOPE TERMS
K=  1  TERM1= -.1778944   TERM2= -.1599494
       SUMLS= -.1599494
K=  2  TERM1=  .1327178   TERM2=  .1745059
       SUMLS=  1.455657E-02
K=  3  TERM1=  3.708072E-02  TERM2=  .0576091
       SUMLS=  7.216568E-02
K=  4  TERM1=  8.095942E-03  TERM2=  5.951025E-03
       SUMLS=  .0781167
```

Upon completion of the loop, a TERM3 is computed at line 1570 that is the
product of F(G), THETA(G), and SUMLS. This term is accumulated in SFD(MRC).
If TRALL is set, the following is printed:

```
TERM3= -234.3501   FIRST DERIVATIVE TERM= -234.3501
```

The subroutine returns to the driver section via line 1595, and the first derivative terms at the current ability level are completed. The next two subroutines
compute the terms in the information matrix that are accumulated over the
ability levels. The subroutine beginning at line 1600 computes the strictly
intercept terms. The subroutine consists of two nested loops. The outer loop

is over the rows, and the inner loop is over columns of the matrix. It can
be seen in Equation 8.7 that only the diagonal and first off-diagonal terms
are needed. Lines 1650 and 1660 compute the on-diagonal terms. Lines 1680–
1700 compute the first off-diagonal elements. When the difference in row and
column indices are greater than one, no terms are computed. This bypass is
accomplished via lines 1630 and 1680. The result of each of the calculations is
in the variable TERM2, which is added to the appropriate cell in the information matrix. If TRALL is set, the intermediate terms and the matrix cell values
are printed as shown below:

```
INTERCEPT TERMS IN INFORMATION MATRIX
R=  1  C=  1 TERM1=  6.73452   TERM2=  213.1235
             MATRIX ELEMENT=  213.1235
R=  1  C=  2 TERM1=  8.036675  TERM2= -43.66587
             MATRIX ELEMENT= -43.66587
R=  1  C=  3 TERM1=  8.036675  TERM2= -43.66587
             MATRIX ELEMENT=  0
R=  2  C=  1 TERM1=  8.036675  TERM2= -43.66587
             MATRIX ELEMENT= -43.66587
R=  2  C=  2 TERM1=  30.9024   TERM2=  63.06965
             MATRIX ELEMENT=  63.06965
R=  2  C=  3 TERM1=  .3657476  TERM2= -9.315252
             MATRIX ELEMENT= -9.315252
R=  3  C=  1 TERM1=  .3657476  TERM2= -9.315252
             MATRIX ELEMENT=  0
R=  3  C=  2 TERM1=  .3657476  TERM2= -9.315252
             MATRIX ELEMENT= -9.315252
R=  3  C=  3 TERM1=  147.9795  TERM2=  9.699209
             MATRIX ELEMENT=  9.699209
```

The result here is that the 3×3 submatrix of the information matrix involving
only intercept derivatives has been computed at the current ability level.

The next subroutine, beginning at line 1800, computes the terms in the
fourth row that involve the slope. It should be noted that, although there are
only three boundary curves, there are four item parameters. Thus, the term
MRC can be used as the dimensionality of the information matrix. The cross-
derivative terms for the slope and intercept, as well as the second derivative
with respect to the slope, are computed at the current ability level. At line
1800, the row index is set to four, and a loop over the four columns of information matrix is initiated. If the column index does not equal four, the
values of TERM1 corresponding to the intercept and slope cross derivatives are
computed in lines 1820 and 1830. These terms
and the corresponding matrix element are printed as shown below:

```
SLOPE TERMS IN INFORMATION MATRIX
R=  4  C=  1
TERM1= -.9525741    TERM2= -508.3729   MATRIX ELEMENT= -508.3729
R=  4  C=  2
TERM1= -.2233127    TERM2= -30.26556   MATRIX ELEMENT= -30.26556
R=  4  C=  3
TERM1= -4.742588E-02 TERM2= -1.151871  MATRIX ELEMENT= -1.151871
```

If the column index equals the row index, the diagonal term is computed via lines 1860–1905. This involves a loop over the four response categories where TERM1 is computed from the category weights and accumulated in TERM2. If TRALL is set, the two terms computed for each response category are then printed as follows:

```
R=  4  C=  4
K=  1 TERM1=  4.117815E-02  TERM2=  4.117815E-02
K=  2 TERM1=  9.570263E-02  TERM2=  .1368808
K=  3 TERM1=  3.501946E-02  TERM2=  .1719002
K=  4 TERM1=  8.029858E-03  TERM2=  .1799301
```

The TERM2 is then multiplied by the number of examinees in the group and by the square of the ability level of the group and accumulated in MTRX(R,C). If TRALL is set, the following is printed:

```
TERM1=  8.029858E-03  TERM2=  1619.371  MATRIX ELEMENT=  1619.371
```

Since the information matrix is symmetric, lines 1950–1970 put the fourth row terms in the fourth column. If TRALL is set, the whole information matrix at the current ability level can be printed:

```
INFORMATION MATRIX
R=  1  C=  1  MATRIX ELEMENT=  213.1235
R=  1  C=  2  MATRIX ELEMENT= -43.66587
R=  1  C=  3  MATRIX ELEMENT=  0
R=  1  C=  4  MATRIX ELEMENT= -508.3729
R=  2  C=  1  MATRIX ELEMENT= -43.66587
R=  2  C=  2  MATRIX ELEMENT=  63.06965
R=  2  C=  3  MATRIX ELEMENT= -9.315252
R=  2  C=  4  MATRIX ELEMENT= -30.26556
R=  3  C=  1  MATRIX ELEMENT=  0
R=  3  C=  2  MATRIX ELEMENT= -9.315252
R=  3  C=  3  MATRIX ELEMENT=  9.699209
R=  3  C=  4  MATRIX ELEMENT= -1.151871
R=  4  C=  1  MATRIX ELEMENT= -508.3729
R=  4  C=  2  MATRIX ELEMENT= -30.26556
R=  4  C=  3  MATRIX ELEMENT= -1.151871
R=  4  C=  4  MATRIX ELEMENT=  1619.371
```

At this point, one pass has been completed over the six subroutines that compute the terms of the Newton-Raphson (Fisher scoring for parameters) equation. Since these six subroutines are called at each of the seven ability levels, the overall process is repeated six more times. Because of the length of the associated printing, the results for these remaining ability levels will not be shown here. Upon completion of the seventh ability level, the final information matrix for the first iteration was:

```
INFORMATION MATRIX
R=  1  C=  1  MATRIX ELEMENT=  1011.843
R=  1  C=  2  MATRIX ELEMENT= -389.6601
R=  1  C=  3  MATRIX ELEMENT=  0
R=  1  C=  4  MATRIX ELEMENT= -1079.99
```

```
R=  2  C=  1  MATRIX ELEMENT= -389.6601
R=  2  C=  2  MATRIX ELEMENT=  1317.279
R=  2  C=  3  MATRIX ELEMENT= -389.6601
R=  2  C=  4  MATRIX ELEMENT= -5.722046E-06
R=  3  C=  1  MATRIX ELEMENT=  0
R=  3  C=  2  MATRIX ELEMENT= -389.6601
R=  3  C=  3  MATRIX ELEMENT=  1011.843
R=  3  C=  4  MATRIX ELEMENT=  1079.99
R=  4  C=  1  MATRIX ELEMENT= -1079.99
R=  4  C=  2  MATRIX ELEMENT= -5.722046E-06
R=  4  C=  3  MATRIX ELEMENT=  1079.99
R=  4  C=  4  MATRIX ELEMENT=  5953.585
```

Notice the common values of cells involving the intercepts of the first and third boundary curves, which reflects the symmetrical placement of the underlying boundary curves.

Given the information matrix, its inverse is computed via the subroutine called at line 280 of the driver section. Since the inner details of the inversion process are not of interest, only the inverse matrix can be printed via the TRALL option and the following is obtained:

```
R=  1  COL=  1  INVER ELEMENT=  1.446373E-03
R=  1  COL=  2  INVER ELEMENT=  3.786012E-04
R=  1  COL=  3  INVER ELEMENT= -1.664794E-04
R=  1  COL=  4  INVER ELEMENT=  2.925739E-04
R=  2  COL=  1  INVER ELEMENT=  3.786012E-04
R=  2  COL=  2  INVER ELEMENT=  9.83126E-04
R=  2  COL=  3  INVER ELEMENT=  3.786011E-04
R=  2  COL=  4  INVER ELEMENT= -1.147661E-11
R=  3  COL=  1  INVER ELEMENT= -1.664794E-04
R=  3  COL=  2  INVER ELEMENT=  3.786011E-04
R=  3  COL=  3  INVER ELEMENT=  1.446373E-03
R=  3  COL=  4  INVER ELEMENT= -2.92574E-04
R=  4  COL=  1  INVER ELEMENT=  2.925739E-04
R=  4  COL=  2  INVER ELEMENT= -1.147661E-11
R=  4  COL=  3  INVER ELEMENT= -2.92574E-04
R=  4  COL=  4  INVER ELEMENT=  2.741128E-04
```

If the matrix can not be inverted, the variable IFLAG is set to zero, and the program terminates via the test at line 290. Next, the new values of the item parameter estimates are computed via the subroutine beginning at line 2300. This routine obtains the increments to the estimates by post-multiplying the inverse of the information matrix by the vector of first derivatives. This is accomplished at lines 2310–2355. The new values of the item parameter estimates are obtained via lines 2370–2390. If either TRALL or TRMLE are set, the increments and the new item parameter estimates are printed:

```
DELTA PARAMETER( 1 )= -.1013555
DELTA PARAMETER( 2 )= -2.012249E-08
DELTA PARAMETER( 3 )=  .1013555
DELTA PARAMETER( 4 )= -.1661691
INTERCEPT( 1 )=  1.698645
```

```
INTERCEPT( 2 )= -2.012249E-08
INTERCEPT( 3 )= -1.698644
COMMON SLOPE =  .8338309
```

At this point, one cycle of the iterative solution of the Newton-Raphson (Fisher scoring) method has been completed. The convergence of the solution is determined via the subroutine called at line 310 in the driver section. In this subroutine, the string variable CONVERG\$ is set to N. Then, the absolute value of each parameter estimate increment is compared to 0.05. If any value is greater than 0.05, the subroutine is exited. If all values are less than or equal to 0.05, the value of CONVERG\$ is set to Y and the subroutine is exited. At line 320 of the driver section, the value of CONVERG\$ is tested. If the value is N, another iteration of the Newton-Raphson equation solution is initiated via line 330 and line 180, where the appropriate sums, vectors, and matrices are set to zero. If the value is Y, the final results are printed via the subroutine beginning at line 3000. The final parameter estimates were obtained here in two iterations of the Newton-Raphson procedure. The results expressed in terms of the boundary curves are:

```
INTERCEPT( 1 ) =  1.720461
BOUNDARY CURVE B( 1 )= -2.021497
INTERCEPT( 2 ) = -1.353813E-09
BOUNDARY CURVE B( 2 )=  1.590695E-09
INTERCEPT( 3 ) = -1.720461
BOUNDARY CURVE B( 3 )=  2.021497
COMMON SLOPE=  .8510829
```

Using the relationships provided in Section 8.2, the difficulty parameter estimates for the four response categories are obtained via lines 3040–3080. The values are:

```
RESPONSE CATEGORY( 1 ) DIFFICULTY= -2.021497
RESPONSE CATEGORY( 2 ) DIFFICULTY= -1.010748
RESPONSE CATEGORY( 3 ) DIFFICULTY=  1.010748
RESPONSE CATEGORY( 4 ) DIFFICULTY=  2.021497
ITEM DISCRIMINATION=  .8510829
```

The data set used here was created for didactic purposes and involved large numbers of examinees within each of the seven ability groups. As a result, the Newton-Raphson equation converged in only two iterations while, with actual data, a greater number of iterations would be needed. In addition, the intercept parameters were symmetrically spaced about zero. This is reflected in the common values of cells in the information matrix involving the intercept terms of the first and third boundary curves.

F.3 Implementation of Ability Estimation

The terms involved in the Newton-Raphson (Fisher scoring for parameters) equation for estimating an examinees ability were presented in Section 8.3.

The iterative solution of this Equation 8.10 was implemented in a BASIC program for a test consisting of five four-response category items. Again, the iterative solution process was embedded in a driver section of the program. The subroutine called at line 130 establishes the environment of the estimation process. In line 4500, the number of response categories (MRC), the number of boundary curves (MBC), the number of items (NITEM), the maximum allowable number of iterations (MAXIT), and the initial estimate of the examinee's ability are established. The next two lines read in the boundary curve intercepts CPT(I,K) and the slope A(I) for the three boundary curves from the DATA statements in lines 9010–9050. The vector of the examinee's item responses, U(I), is read from the DATA statement at line 9060. The tracing options are set by the user, and a summary printed.

The implementation the Newton-Raphson equation is accomplished via a set of nested FOR-NEXT loops. The outer loop over the iterations of the solution equation is defined by lines 150–280 of the driver section. The next most inner loop, lines 170–240, is over the five items. The innermost loop, lines 180–220, is over the four item response categories. A test is performed at line 190 to determine if there is a match between the examinee's response to the item and the item response category being processed. If not, a branch to the end of the innermost loop is made. If the choice matches the item response category, the small probability flag PTSFLG$ is set to N, and the subroutine beginning at line 4700 is called. This subroutine computes the probability of a response being in the category, PK, and the weighting terms for the category, WK, and for the next lower category, WKM1. In order to do this, it must calculate the boundary probability up to the selected category, PSTRK, and that of the next lower curve, PSTRKM1. The probability of the elected category PK is the difference between these two probabilities. If the index K designates either the lowest or highest response category, it sets the appropriate lower or upper limit values of PK and the weighting terms. All of this is accomplished via lines 4710–4730. If TRALL is set, the details of the computations are printed on an item-by-item basis. The details for the first item are given below.

```
ITEM( 1 )    U(I)= 3
K-1= 2  DEV(K-1)= .34  PSTAR(K-1)= .5841905
         W(K-1)= .2429119
K= 3  DEV(K)= -1.38  PSTAR(K)= .201009  P(K)= .3831815
       W(K)= .1606044
```

At line 4780, the value of PK is tested to determine if it is so small as to cause a division by zero error. If so, PTSFLG$ is set to Y. The subroutine then returns to the driver section. If the value is Y, a branch is made to line 230 by passing the rest of the calculations for the item. If P is large enough, a subroutine is called that computes the elements of the first and second derivative terms of Equation 8.10 associated with an item response. The first derivative term for the item's response category is calculated at line 5000 and accumulated in the variable FDS. If TRALL is set, the value is printed as shown below:

```
FTERM=  .2148004  FIRST DERIVATIVE SUM=  .2148004
```

Computation of the contribution to the second derivative, involves the computation of three terms, TERM1, TERM2, and TERM3, which are functions of PK, PSTRK, PSTRKM1, WK, and WKM1. The TERM4 is the sum of these three terms, and it is multiplied by the square of the slope and the result accumulated in the variable SDS. If TRALL is set, the following is printed:

```
TERM1= -.2506346  TERM2= -.1067426  TERM3= -4.613923E-02
         TERM4= -.4035164  SDS= -.2915406
```

The computations within the item and response category loops are performed for each of the examinee's response choices. The results for the fifth item are shown below:

```
ITEM( 5 )    U(I)=  3
K-1=  2  DEV(K-1)=  1.26  PSTAR(K-1)=  .7790261
         W(K-1)=  .1721444
K=  3  DEV(K)= -1.14  PSTAR(K)=  .2423203  P(K)=  .5367057
         W(K)=  .1836012
FTERM= -2.134644E-02  FIRST DERIVATIVE SUM=  .1697909
TERM1= -.1762988  TERM2= -.1789912  TERM3= -4.556706E-04
         TERM4= -.3557456  SDS= -1.9407
```

After the fifth item response has been processed, the increment to the examinees ability estimate, DELTA, is simply the ratio of FDS to SDS. This and the new ability estimate is calculated at line 250 of the driver section. If either TRALL or TRMLE are set, then the values are printed as follows:

```
THETA=  .4874895  DELTA= -8.748952E-02
```

If the absolute value of DELTA is greater than 0.05, another iteration is performed via line 280, and the next iteration number is printed at line 160 as:

```
MLE ITERATION 2
```

This second iteration yielded DELTA= -.0047, which is less than the convergence criterion of 0.05, the iterative estimation process is terminated, and the following final estimated ability is printed:

```
THETA=  .5344664  DELTA= -4.697688E-02
THETA ESTIMATE=  .5344664
```

As was the case for dichotomous scoring, the implementation of ability estimation is straightforward and the computer program relatively short. In a production program, it and the program for item parameter estimation would be embedded with in the joint maximum likelihood estimation paradigm. In addition, software would be needed to read the examinee's item response vectors, group the examinees, obtain the values of F(G) and R(G,K), and compute initial estimates of the item and ability estimates. Also, a more extensive reporting of the results would be implemented.

F.4 BASIC Computer Programs

F.4.1 Item Parameter Estimation

```
100 REM MAXIMUM LIKELIHOOD ESTIMATION OF ITEM PARAMETERS FOR
110 REM A GRADED RESPONSE ITEM WITH 4 ORDERED RESPONSE CATEGORIES
120 DIM PSTR(4),W(4),PK(4),POBS(4),CPT(3),SFD(4),T(4),PDELTA(4)
130 DIM B(3),BK(4),THETA(7),F(7),R(7,4),MTRX(4,4),TK(4,4)
160 GOSUB 1000:REM READ CANNED DATA, ESTABLISH TRACE OF PROGRAM
170 FOR NIT=1 TO MAXIT
180 GOSUB 1100:REM CLEAR SUMS, VECTORS AND MATRIX
190 PRINT "MLE ITERATION ";NIT
200 FOR G= 1 TO MAXGPS:PRINT "ABILITY LEVEL ";G
210 GOSUB 1200:REM CALC P STAR
220 GOSUB 1300:REM CALC P OBSERVED
230 GOSUB 1400:REM CALC FIRST DERIVATIVE TERMS FOR INTERCEPTS
240 GOSUB 1500:REM CALC FIRST DERIVATIVE TERMS FOR SLOPE
250 GOSUB 1600:REM CALC INTERCEPT TERMS FOR INFORMATION MATRIX
260 GOSUB 1800:REM CALC SLOPE TERMS FOR INFORMATION MATRIX
270 NEXT G
280 GOSUB 2100:REM INVERT INFORMATION MATRIX
290 IF IFLAG=0 GOTO 370
300 GOSUB 2300:REM CALCU CHANGE VECTOR AND OBTAIN NEW ESTIMATES
310 GOSUB 2400:REM CALCULATE CONVERGENCE CRITERION
320 IF CONVERG$="Y" THEN GO TO 360
330 NEXT NIT
340 PRINT "REACHED MAXIMUM CYCLES"
360 GOSUB 3000:REM  PRINT ITEM PARAMETER ESTIMATES
370 END
REM ********READ CANNED DATA AND PRESTORE CONSTANTS ROUTINE******
1000 MRC=4:MBC=3:MAXIT=4:MAXGPS=7
1010 A=1:REM INITAL VALUE OF COMMON SLOPE
1020 FOR G=1 TO MAXGPS:READ THETA(G):NEXT G
1025 FOR G=1 TO MAXGPS:READ F(G):NEXT G
1027 FOR G=1 TO MAXGPS:FOR K=1 TO MRC
1030 READ R(G,K):NEXT K:NEXT G
1040 FOR Y=1 TO MBC:READ CPT(Y):NEXT Y
1050 INPUT "TRACE ALL COMPUTATIONS? Y/N ";YN$
1060 TRALL=-1:TRMLE=-1
1070 IF YN$="Y" THEN TRALL=0:TRMLE=0:GOTO 1091
1080 INPUT "TRACE MLE TERMS? Y/N ";YN$
1090 IF YN$="Y" THEN TRMLE=0
1091 PRINT "ALL  MLE  MAX CYCLES"
1092 PRINT TRMLE;TAB(6);TRMLE;TAB(13);MAXIT
1095 RETURN
REM *********CLEAR SUMS, VECTORS AND INFORMATION MATRIX*********
1100 FOR R=1 TO MRC:SFD(R)=0.0
1110 FOR C=1 TO MRC:MTRX(R,C)=0.0
1120 NEXT C:NEXT R
1130 RETURN
REM ********CALCULATE P STAR VALUES***********
1200 PSTR(0)=1.0:W(0)=0.0:PSTR(MRC)=0.0:W(MRC)=0.0
1205 IF TRALL=0 THEN PRINT "BOUNDARY CURVE PSTAR'S"
```

```
1210 FOR Y=1 TO MBC
1220 DEV=CPT(Y)+A*THETA(G)
1230 PSTR(Y)=1.0/(1+EXP(-DEV))
1240 W(Y)=PSTR(Y)*(1.0-PSTR(Y))
1250 IF TRALL=-1 GOTO 1280
1260 PRINT "CPT(";Y;")= ";CPT(Y);" A= ";A;
1261 PRINT " THETA(";G;")= ";THETA(G)
1270 PRINT TAB(5);"DEV= ";DEV;" PSTAR(";Y;")= ";PSTR(Y);
1271 PRINT " W(";Y;")=";W(Y)
1280 NEXT Y
1290 RETURN
REM ******CALCULATE OBSERVED PK AND PSTAR***********
1300 IF TRALL=0 THEN PRINT "PK AND P OBSERVED"
1305 FOR K=1 TO MRC
1310 IF(K=1) THEN PK(1)=1.0-PSTR(1): GOTO 1340
1320 IF(K=MRC) THEN PK(MRC)=PSTR(MBC):GOTO 1340
1330 PK(K)=PSTR(K-1)-PSTR(K)
1340 POBS(K)=R(G,K)/F(G)
1350 IF TRALL=0 THEN
     PRINT "K= ";K;" PK= ";PK(K);" POBS= ";POBS(K)
     END IF
1360 NEXT K
1370 RETURN
REM *******CALCULATE FIRST DERIVATVE TERMS FOR INTERCEPTS*******
1400 IF TRALL=0 THEN PRINT "INTERCEPT TERMS"
1405 FOR Y=1 TO MBC
1410 TERM1=-(POBS(Y)/PK(Y))+(POBS(Y+1)/PK(Y+1))
1420 TERM2=F(G)*W(Y)*TERM1
1430 SFD(Y)=SFD(Y)+TERM2
1440 IF TRALL= -1 THEN GOTO 1460
1450 PRINT "Y= ";Y;" TERM1 = ";TERM1;" TERM2= ";TERM2
1451 PRINT "        FIRST DERIVATIVE = ";SFD(Y)
1460 NEXT Y
1470 RETURN
REM ******CALCULATE FIRST DERIVATIVE TERMS FOR COMMON SLOPE******
1500 SUMLS=0.0:IF TRALL=0 THEN PRINT "COMMON SLOPE TERMS"
1510 FOR K=1 TO MRC
1520 TERM1= W(K-1)-W(K)
1530 TERM2= (POBS(K)/PK(K))*TERM1
1540 SUMLS=SUMLS+TERM2
1550 IF TRALL=0 THEN
     PRINT "K= ";K;" TERM1= ";TERM1;" TERM2= ";TERM2
     PRINT "        SUMLS= ";SUMLS
     END IF
1560 NEXT K
1570 TERM3=F(G)*THETA(G)*SUMLS
1580 SFD(MRC)= SFD(MRC)+TERM3
1590 IF TRALL=0 THEN
     PRINT "TERM3= ";TERM3;" FIRST DERIVATIVE TERM= ";SFD(MRC)
     END IF
1595 RETURN
REM ******CALCULATE INTERCEPT TERMS IN INFORMATION MATRIX********
1600 IF TRALL=0 THEN
```

```
       PRINT "INTERCEPT TERMS IN INFORMATION MATRIX"
       END IF
1610 FOR R=1 TO MBC
1620 FOR C=1 TO MBC
1630 IF R < > C THEN GO TO 1680
1640 REM DIAGONAL INTERCEPT TERM
1650 TERM1=1/PK(C)+1/PK(C+1)
1660 TERM2= F(G)*W(C)*W(C)*TERM1
1670 GOTO 1710
1680 IF ABS(R-C)>1 THEN GOTO 1720
1690 TERM1=F(G)*W(R)*W(C)
1700 IF R>C THEN
       TERM2=-(TERM1*(1/PK(R)))
     ELSE
       TERM2=-(TERM1*(1/PK(C)))
     END IF
1710 MTRX(R,C)=MTRX(R,C)+TERM2
1720 IF TRALL=0 THEN
     PRINT "R= ";R;" C= ";C;"TERM1= ";TERM1;" TERM2= ";TERM2
     PRINT "             MATRIX ELEMENT= ";MTRX(R,C)
     END IF
1730 NEXT C:NEXT R:RETURN
REM *******CALCULATE SLOPE TERMS FOR INFORMATION MATRIX*********
1800 IF TRALL=0 THEN PRINT "SLOPE TERMS IN INFORMATION MATRIX"
1805 R=MRC:FOR C=1 TO MRC
1810 IF TRALL=0 THEN PRINT "R= ";R;" C= ";C
1815 IF C=MRC THEN GOTO 1850
1820 TERM1=(W(C-1)-W(C))/PK(C)-(W(C)-W(C+1))/PK(C+1)
1830 TERM2= -(F(G)*W(C)*THETA(G)*TERM1)
1840 GOTO 1910
1850 TERM2=0.0
1860 FOR K=1 TO MRC
1870 TERM1=(W(K-1)-W(K))*(W(K-1)-W(K))*1.0/PK(K)
1880 TERM2=TERM2+TERM1
1890 IF TRALL=0 THEN
     PRINT "K= ";K;"TERM1= ";TERM1;" TERM2= ";TERM2
     END IF
1900 NEXT K
1905 TERM2=TERM2*F(G)*THETA(G)*THETA(G)
1910 MTRX(R,C)=MTRX(R,C)+TERM2
1920 IF TRALL=0 THEN
     PRINT "TERM1= ";TERM1;" TERM2= ";TERM2;
     PRINT " MATRIX ELEMENT= ";MTRX(R,C)
     END IF
1930 NEXT C
1940 REM PUT SLOPE ROW DERIVATIVES IN LAST COLUMN
1950 FOR C=1 TO MRC
1960 MTRX(C,R)=MTRX(R,C)
1970 NEXT C
1980 IF TRALL=0 THEN PRINT "INFORMATION MATRIX"
1981 FOR R=1 TO MRC:FOR C=1 TO MRC
1982 IF TRALL=0 THEN
     PRINT "R= ";R;" C= ";C;" MATRIX ELEMENT= ";MTRX(R,C)
```

```
      END IF
1984 NEXT C:NEXT R
1990 RETURN
REM*********MATRIX INVERSION ROUTINE*******************
2100 IFLAG=-1
2105 FOR KK= 1 TO MRC
2110 IF (MTRX(1,1)-.000001)<=0 GOTO 2220
2120 R=SQR(MTRX(1,1))
2130 FOR IK= 1 TO MBC:T(IK)=MTRX(IK+1,1)/R:NEXT IK
2140 T(MRC)=1.0/R
2150 FOR JK=1 TO MBC
2160 FOR IK=1 TO MBC:MTRX(IK,JK)=MTRX(IK+1,JK+1)-T(IK)*T(JK)
2170 NEXT IK:NEXT JK
2180 FOR IK=1 TO MRC:MTRX(IK,MRC)=-T(IK)*T(MRC):NEXT IK
2190 FOR JK=1 TO MBC:MTRX(MRC,JK)=MTRX(JK,MRC):NEXT JK
2195 NEXT KK
2200 FOR JK=1 TO MRC:FOR IK=1 TO MRC
2201 MTRX(IK,JK)=-MTRX(IK,JK)
2202 NEXT IK:NEXT JK
2210 GOTO 2240
2220 PRINT "THE MATRIX IS (VERY NEAR) SINGULAR, OR INDEFINITE"
2230 IFLAG=0:GOTO 2270
2240 IF TRALL=-1 GOTO 2270
2250 FOR JK=1 TO MRC:FOR IK=1 TO MRC
2255 PRINT "R= ";JK;" COL= ";IK;" INVER ELEMENT= ";MTRX(JK,IK)
2260 NEXT IK:NEXT JK
2270 RETURN
REM ****CALCULATE INCREMENTS TO ITEM PARMETER ESTIMATES*****
2300 FOR JK=1 TO MRC:PDELTA(JK)=0.0:NEXT JK
2310 FOR JK=1 TO MRC
2320 FOR IK=1 TO MRC
2330 PDELTA(JK)=PDELTA(JK)+MTRX(JK,IK)*SFD(IK)
2340 NEXT IK
2350 IF (TRALL=0) OR (TRMLE=0) THEN
      PRINT "DELTA PARAMETER(";JK;")= ";PDELTA(JK)
      END IF
2355 NEXT JK
2360 REM MAY NEED BIG DELTA PROTECTION HERE
REM *********OBTAIN NEW ITEM PARAMETER VALUES********
2370 FOR IK=1 TO MBC:CPT(IK)=CPT(IK)+PDELTA(IK)
2380 IF (TRALL=0) OR (TRMLE=0) THEN
      PRINT "INTERCEPT(";IK;")= ";CPT(IK)
      END IF
2385 NEXT IK
2390 A=A+PDELTA(MRC)
2392 IF (TRALL=0) OR (TRMLE=0) THEN PRINT "COMMON SLOPE = ";A
2395 RETURN
REM*********CHECK CONVERGENCE OF MLE ******************
2400 CONVERG$="N"
2410 FOR K=1 TO MRC
2420 IF ABS(PDELTA(K))>.05 GOTO 2450
2430 NEXT K:CONVERG$="Y"
2450 RETURN
```

```
REM*********PRINT FINAL SOLUTION **************
3000 FOR Y=1 TO MBC:
3010 B(Y)= -CPT(Y)/A
3020 PRINT "INTERCEPT(";Y;") = ";CPT(Y)
3021 PRINT "BOUNDARY CURVE B(";Y;")= ";B(Y)
3030 NEXT Y:PRINT "COMMON SLOPE= ";A
3040 FOR K=1 TO MRC:IF K=1 THEN BK(K)=B(1): GOTO 3070
3050 IF K=MRC THEN BK(K)=B(MBC):GOTO 3070
3060 BK(K)=(B(K)+B(K-1))/2
3070 PRINT "RESPONSE CATEGORY(";K;") DIFFICULTY= ";BK(K)
3080 NEXT K:PRINT "ITEM DISCRIMINATION= ";A:RETURN
9000 REM THETA(G)
9010 DATA -3,-2,-1,0.0,1,2,3
9020 DATA 1000,1000,1000,1000,1000,1000,1000
9030 DATA 691,242,61,6
9040 DATA 500,341,136,23
9050 DATA 309,382,242,67
9060 DATA 159,341,341,159
9070 DATA 67,242,382,309
9080 DATA 23,136,341,500
9090 DATA 6,61,242,691
9100 DATA 1.8,0.0,-1.8
```

F.4.2 Ability Estimation

```
100 REM PROGRAM TO ESTIMATE ABILTY UNDER GRADED RESPONSE MODEL
110 REM FIVE FOUR-RESPONSE TEST ITEMS
120 DIM A(5),CPT(5,3),U(5)
130 GOSUB 4500:REM READ CANNED DATA AND SET TRACES
140 SDS=0:FDS=0:PRINT TAB(5);"INITIAL THETA= ";THETA
150 FOR NIT= 1 TO MAXIT
160 PRINT:PRINT "MLE ITERATION ";NIT
170 FOR I= 1 TO NITEM:TERM4=0
180 FOR K=1 TO MRC
190 IF U(I) <> K GOTO 230
200 PTSFLG$="N":GOSUB 4700:REM CALC P'S AND W'S
210 IF PTSFLG$="Y" GOTO 230
220 GOSUB 5000:REM CALC FIRST AND SECOND DERIVATIVE TERMS
230 NEXT K
240 NEXT I
250 DELTA=FDS/SDS:THETA=THETA-DELTA
260 IF(TRALL=0) OR (TRMLE=0) THEN
    PRINT "THETA= ";THETA;" DELTA= ";DELTA
    END IF
270 IF ABS(DELTA) < .05 GOTO 300
280 NEXT NIT
290 PRINT "MAX CYCLES REACHED"
300 PRINT "THETA ESTIMATE= ";THETA
310 END
REM******* READ CANNED DATA AND SET TRACES***********
4500 MRC=4:MBC=3:NITEM=5:MAXIT=10:THETA=.4:REM INITIAL VALUE
4510 FOR I=1 TO NITEM:READ A(I)
4520 FOR K= 1 TO MBC:READ CPT(I,K)
```

```
4530 NEXT K:NEXT I
4540 FOR I= 1 TO NITEM:READ U(I):NEXT I
4550 TRALL=-1:INPUT "TRACE ALL? Y/N ";YN$
4560 IF YN$="N" GOTO 4580
4570 TRALL=0:TRMLE=0:GOTO 4595
4580 INPUT "TRACE MLE? Y/N ";YN$
4590 TRMLE=-1:IF YN$= "Y" THEN TRMLE=0
4595 PRINT "ALL    MLE    MAX CYCLES"
4596 PRINT TRALL;TAB(8);TRMLE;TAB(15);MAXIT
4599 RETURN
REM*******CALCULATE P'S  AND W'S*********************
4700 KM1=K-1
4710 IF KM1=0 THEN
       PSTRKM1=1.0:WKM1=0
     ELSE
       DEVKM1=CPT(I,KM1)+A(I)*THETA
       PSTRKM1=1.0/(1.0+EXP(-DEVKM1))
       WKM1=PSTRKM1*(1.0-PSTRKM1)
     END IF
4720 IF K=MRC THEN
       PSTRK=0:WK=0
     ELSE
       DEV=CPT(I,K)+A(I)*THETA
       PSTRK=1.0/(1.0+EXP(-DEV))
       WK=PSTRK*(1.0-PSTRK)
     END IF
4730 PK=PSTRKM1-PSTRK
4740 IF TRALL= -1 GOTO 4780
4750 PRINT "ITEM(";I;")";"   U(I)= ";U(I)
4760 PRINT "K-1= ";KM1;" DEV(K-1)= ";DEVKM1;
4761 PRINT " PSTAR(K-1)= ";PSTRKM1
4762 PRINT "         W(K-1)= ";WKM1
4770 PRINT "K= ";K;" DEV(K)= ";DEV;" PSTAR(K)= ";PSTRK;
4771 PRINT " P(K)= ";PK
4772 PRINT "         W(K)= ";WK
4780 IF (PK <.0001) OR (PK > .9999) THEN PTSFLG$="Y"
4790 RETURN
REM********FIRST DERIVATE************
5000 FTERM=(WKM1-WK)/PK
5010 FDS=FDS+FTERM
5020 IF TRALL =-1 GOTO 5040
5030 PRINT "FTERM= ";FTERM;" FIRST DERIVATIVE SUM= ";FDS
REM*******SECOND DERIVATIVE***********
5040 TERM1=-WK*(1.0-2*PSTRK)/PK
5050 TERM2= WKM1*(1.0-2*PSTRKM1)/PK
5060 TERM3= -(WKM1-WK)*(WKM1-WK)/(PK*PK)
5070 TERM4=TERM4+TERM1+TERM2+TERM3
5080 SDS=SDS+A(I)*A(I)*TERM4
5090 IF TRALL=-1 GOTO 5120
5100 PRINT "TERM1= ";TERM1;" TERM2= ";TERM2;" TERM3= ";TERM3
5110 PRINT TAB(10);"TERM4= ";TERM4;" SDS= ";SDS
5120 RETURN
REM*********CANNED DATA***********
```

```
9010 DATA .85,1.72,0.0,-1.720
9020 DATA 1.0,1.20,.2,-1.5
9030 DATA .90,2.1,1.1,-.70
9040 DATA 1.1,-.6,-.8,-1.2
9050 DATA 1.4,1.4,.7,-1.7
9060 DATA 3,2,3,4,3
```

G. Implementation of MLE Under Nominal Response Scoring

G.1 Introduction

As was the case with the graded response implementation of Appendix F, the estimation of the parameters of an item and an examinee's ability will be treated separately. Not embedding them within the joint maximum likelihood estimation paradigm will reduce the size of the program. The implementation of the maximum likelihood estimation of the item parameters will be provided in Section G.2, and the implementation of the maximum likelihood estimation of an examinee's ability in Section G.3.

G.2 Implementation of Item Parameter Estimation

In Chapter 9, the rather complex mathematics associated with the reparameterization of the item parameters under a nominal response scoring model was shown. When the Newton-Raphson equation is implemented for the estimation of the resulting item parameters, only the results of this mathematics are needed. As a further simplification, a contrived data set based on Table 9.1 will be employed. This table contains the probability of selecting each of three nominal response categories over the range of the ability scale from -4 to $+4$ in steps of 1.0. In the present case, the two extreme ability levels were eliminated and only the range -3 to $+3$ used. Assuming a uniform distribution over the ability scale, 7000 simulated examinees were grouped into the seven intervals. The probability of each response category, at a given ability, level was multiplied by 1000 to obtain the number of examinees selecting the category. The reparameterization process also reduces the number of item parameters from three to two intercepts and three to two slopes. This reduced dimensionality was used to advantage in the writing of the BASIC computer program.

As is our standard practice, a short driver program controls the parameter estimation process. The outer FOR-NEXT loop, lines 160–320, constitutes the iterative Newton-Raphson procedure. A nested FOR-NEXT loop, lines 190–250 controls the subroutines that compute the terms, at each ability level, needed by the components of the Newton-Raphson equation. The Newton-Raphson equation is solved via the subroutines called in lines 255–310.

The initialization subroutine is called a line 140 to establish the environment of the estimation process. At line 500, the number, MRC, of response categories is set to three; the number of transformed parameter, NTP, is set to two; the maximum allowable number of iterations of the solution, MAXIT, is set to five; and the number of ability groups, MAXGPS, is set to seven. The initial estimates of the item parameters, CPT(K) and A(K), for the three category multivariate logistic function (Equation 9.1) are read pairwise from the DATA statement at line 9000. At line 520, the seven midpoint values, THETA(G), of the ability groups are read from the DATA statement at line 9110. In the nested set of FOR-NEXT loops, at line 530 the number of examinees at each ability level is read in the outer loop. The number of examinees selecting each response category at the ability level is read by the inner loop. The tracing options are set and a summary of the selections are printed. Upon return to the driver section, the subroutine beginning at line 600 is called. This subroutine reparameterizes the initial intercept and slope parameters via simple contrasts between response category one and remaining response categories into the new intercepts, CP(I), and slopes, AP(I). If TRALL is set, the reduced set of parameter values are printed. The initial lines of the printout appear as follows:

```
REPARAMETERIZED INTERCEPTS AND SLOPES
CP( 1 )= -1.15   AP(I 1 )= .3
CP( 2 )= -1.25   AP(I 2 )=-1.2
```

The Newton-Rahpson iterative process is initiated via line 160 and the iteration identified by:

```
MLE ITERATION  1
```

A subroutine called at line 180 clears the vector of first derivative terms, FDS(R), and the vector of increments to the parameter estimates, PDELTA(R), and the information matrix. A loop over the seven ability levels is initiated at line 190. If TRALL is set, the following is printed:

```
ABILITY GROUP=  1
```

The subroutine called at line 210 evaluates the multivariate logistic function, Equation 9.1, using the initial values of the slope, A(K), and intercept, CPT(K), parameters for the three response categories. The denominator is the sum of the individual exponentials, EZ(K), for the three response categories. If TRALL is set, the following values are printed:

```
ABILITY GROUP 1
EVALUATE MULTIVARIATE LOGISTIC FUNCTION
   CPT= -.8   A= -.3   THETA= -3
   Z( 1 )=  .1   EZ=  1.105171  D= 1.105171
   CPT=  .35  A= -.6   THETA= -3
   Z( 2 )=  2.15  EZ=  8.584859  D= 9.69003
   CPT=  .45  A=  .9   THETA= -3
   Z( 3 )= -2.25  EZ=  .1053992  D= 9.795429
```

Then a loop over the three response categories computes the probability of the response category by dividing the EZ(K) terms by the denominator D. If TRALL is set, the values are printed:

```
P( 1 )=  .1128252
P( 2 )=  .8764148
P( 3 )=  1.076004E-02
```

The next subroutine called computes the terms in the first derivatives of the likelihood function with respect to the reparameterized slope and intercepts. Equation 9.18 is implemented for the two intercepts at lines 1210 and 1220. Equation 9.20 is implemented for the two slopes at lines 1230 and 1240. These terms are then added to the appropriate element in the first derivative vector, FDS(KK), of length four. If TRALL is set, this vector is printed:

```
COMPUTE FIRST DERIVATIVES
TERM( 1 )= -61.58522   SUM FIRST DERIV( 1 )=-61.58522
TERM( 2 )=  5.760041   SUM FIRST DERIV( 2 )= 5.760041
TERM( 3 )=  184.7557   SUM FIRST DERIV( 3 )= 184.7557
TERM( 4 )= -17.28012   SUM FIRST DERIV( 4 )=-17.28012
```

The first two rows correspond to the intercepts and the second two to the slopes. At line 240 of the driver section, a subroutine beginning at line 1400 is called that computes the matrix of weights, W(R,C), of dimensionality 2×2, that depend upon the category probabilities. This subroutine also computes the theta matrix, TM(R,C), of dimensionality 2×2 that involves the value of the current ability level. If TRALL is set, both matrices are printed as shown below:

```
WEIGHT AND THETA MATRICIES
R=  1  C=  1  W=  .1083119 TM=  1
R=  1  C=  2  W=  9.430259E-03 TM= -3
R=  2  C=  1  W=  9.430259E-03 TM= -3
R=  2  C=  2  W=  1.064426E-02 TM=  9
```

The next set of terms to be computed are those of the information matrix. This is accomplished via the Kronecker product of the theta and weight matrices implemented in the subroutine beginning at line 2000. Essentially, what happens in the nested FOR-NEXT loops at line 2000 is that the complete theta matrix is multiplied by a cell in the weight matrix to form a submatrix of dimensionality 2×2 in the information matrix of dimensionality 4×4. Before this is done, the cells of the TM matrix are multiplied by the number of examinees in the current ability group via line 2010. The IF statements in lines 2020–2050 determine which submatrices are being created, that is, the on- and off-diagonal 2×2 matrices. Each submatrix has its own pair of nested FOR-NEXT loops that creates the product of the TM cell and the W cell and sets the signs of the off-diagonals to minus and adds the product to the proper cell of the information matrix. When the lower right-hand submatrix is completed at line 2200, the TRALL variable controls the printing of the complete information matrix at the current ability level. The following is printed for the first ability level:

```
INFORMATION MATRIX
      MTRX( 1 , 1 )=  108.3119
      MTRX( 1 , 2 )= -9.430259
      MTRX( 1 , 3 )= -324.9357
      MTRX( 1 , 4 )=  28.29078
      MTRX( 2 , 1 )= -9.430259
      MTRX( 2 , 2 )=  10.64426
      MTRX( 2 , 3 )=  28.29078
      MTRX( 2 , 4 )= -31.93279
      MTRX( 3 , 1 )= -324.9357
      MTRX( 3 , 2 )=  28.29078
      MTRX( 3 , 3 )=  974.8073
      MTRX( 3 , 4 )= -84.87233
      MTRX( 4 , 1 )=  28.29078
      MTRX( 4 , 2 )= -31.93279
      MTRX( 4 , 3 )= -84.87233
      MTRX( 4 , 4 )=  95.79836
```

At this point, all the computations performed at the first ability level have been completed. This set of four subroutines will be called for each of the remaining six ability levels. Because of the length of the printing, only the final information matrix for ABILITY GROUP 7 is shown below:

```
INFORMATION MATRIX
      MTRX( 1 , 1 )=  903.8146
      MTRX( 1 , 2 )= -517.3434
      MTRX( 1 , 3 )= -605.2325
      MTRX( 1 , 4 )= -15.73852
      MTRX( 2 , 1 )= -517.3434
      MTRX( 2 , 2 )=  693.1353
      MTRX( 2 , 3 )= -15.73852
      MTRX( 2 , 4 )=  93.0495
      MTRX( 3 , 1 )= -605.2325
      MTRX( 3 , 2 )= -15.73852
      MTRX( 3 , 3 )=  2181.573
      MTRX( 3 , 4 )= -709.1055
      MTRX( 4 , 1 )= -15.73852
      MTRX( 4 , 2 )=  93.0495
      MTRX( 4 , 3 )= -709.1055
      MTRX( 4 , 4 )=  982.3619
```

Upon completion of the ability group loop, all the terms in the vector of first derivatives and in the information matrix have been computed. The subroutine called at line 255 computes the inverse of the information matrix. Again, we are not interested in the inner details of the inversion process. If TRALL is set, the inverse will be printed as shown below:

```
ROW=  1  COL=  1  INVERSE ELEMENT=  3.181909E-03
ROW=  1  COL=  2  INVERSE ELEMENT=  2.31563E-03
ROW=  1  COL=  3  INVERSE ELEMENT=  1.103695E-03
ROW=  1  COL=  4  INVERSE ELEMENT=  6.283291E-04
ROW=  2  COL=  1  INVERSE ELEMENT=  2.31563E-03
ROW=  2  COL=  2  INVERSE ELEMENT=  3.149887E-03
```

```
ROW=  2  COL=  3  INVERSE ELEMENT=  7.58099E-04
ROW=  2  COL=  4  INVERSE ELEMENT=  2.859652E-04
ROW=  3  COL=  1  INVERSE ELEMENT=  1.103695E-03
ROW=  3  COL=  2  INVERSE ELEMENT=  7.58099E-04
ROW=  3  COL=  3  INVERSE ELEMENT=  9.831281E-04
ROW=  3  COL=  4  INVERSE ELEMENT=  6.555338E-04
ROW=  4  COL=  1  INVERSE ELEMENT=  6.283291E-04
ROW=  4  COL=  2  INVERSE ELEMENT=  2.859652E-04
ROW=  4  COL=  3  INVERSE ELEMENT=  6.555338E-04
ROW=  4  COL=  4  INVERSE ELEMENT=  1.474123E-03
```

The subroutine called at line 260 postmultiplies the information matrix by the vector of first derivative terms to obtain the increments to the parameter estimates. This is accomplished in lines 2500–2520. If either TRALL or TRMLE is set, the increments are printed:

```
DELTA PARAMETER= .1133832
DELTA PARAMETER=-3.703428E-03
DELTA PARAMETER= .2354538
DELTA PARAMETER=-3.365469E-03
```

At lines 2550 and 2560 these increments are added to the intercepts and slopes, respectively. If either TRALL or TRMLE are set, the new values are printed:

```
CP(1)= -1.036617    AP(1)=  .5354539
CP(2)= -1.253703    AP(2)= -1.203366
```

It should be noted that these are the reparameterized intercepts and slopes. The next subroutine, beginning at line 2600 reverses the parameterization process and produces the intercept and slope parameters for all three nominal response categories. This is done in lines 2600–2610. If either TRALL or TRMLE are set, the results are printed:

```
INTERCEPT( 1 )= -.7634401  SLOPE( 1 )=-.2226372
INTERCEPT( 2 )=  .2731767  SLOPE( 2 )=-.7580911
INTERCEPT( 3 )=  .4902634  SLOPE( 3 )= .9807283
```

The convergence of the Newton-Raphson procedure is determined in the subroutine called at line 300 of the driver section. If the all the increments to the parameter estimates are less than or equal to 0.05, the CONVERG$ variable is set to Y. However, in the present iteration, only one increment is less than 0.05, and the value of CONVERG$ is left at N. This results in another iteration being performed.

Two additional iterations were required before convergence was attained. Because, the detailed printouts are rather lengthy, only the increments to the parameter estimates and the adjusted parameter estimates will be shown below for each iteration:

```
MLE ITERATION  2
 ...
DELTA PARAMETER= 3.412553E-02
DELTA PARAMETER= 5.257133E-03
DELTA PARAMETER= 5.976674E-02
```

```
DELTA PARAMETER=-2.597566E-03
CP(1)= -1.002491    AP(1)=  .5952206
CP(2)= -1.248446    AP(2)= -1.205963
INTERCEPT( 1 )= -.7503125  SLOPE( 1 )=-.2035809
INTERCEPT( 2 )=  .2521787  SLOPE( 2 )=-.7988014
INTERCEPT( 3 )=  .4981338  SLOPE( 3 )= 1.002382

MLE ITERATION 3
...
DELTA PARAMETER= 1.762912E-03
DELTA PARAMETER= 3.134565E-04
DELTA PARAMETER= 3.073689E-03
DELTA PARAMETER=-2.311668E-04
CP(1)= -1.000728    AP(1)=  .5982943
CP(2)= -1.248133    AP(2)= -1.206194
INTERCEPT( 1 )= -.7496204  SLOPE( 1 )=-.2026333
INTERCEPT( 2 )=  .2511079  SLOPE( 2 )=-.8009276
INTERCEPT( 3 )=  .4985125  SLOPE( 3 )= 1.003561
FINAL VALUES
INTERCEPT( 1 )= -.7496204  SLOPE( 1 )=-.2026333
INTERCEPT( 2 )=  .2511079  SLOPE( 2 )=-.8009276
INTERCEPT( 3 )=  .4985125  SLOPE( 3 )= 1.003561
```

These final values are within a few thousands of the values used to generate the probabilities in Table 9.1. Thus, the solution is a good one.

The interested reader can generalize the computer program to handle more item response categories and ability levels rather easily

G.3 Implementation of Ability Estimation

G.3.1 Introduction

The mathematics of the maximum likelihood estimation of ability under a nominal response model was presented in Section 9.3. Although an implementation of this procedure using matrix algebra would result in a compact computer program, this was not done. It was felt that the matrix algebra would hide much of the computational detail from the reader. The BASIC computer program employed here implements the algebra resulting from Equation 9.24 for the first derivative and that following Equation 9.29 for the second derivative of the log-likelihood with respect to an examinee's ability. In order to make the computations more regular and simpler to follow, a contrived data set was employed. The response category intercepts and slopes of the item in Table 9.1 were each permuted over five items, and their signs were also varied to create items with different configurations.

G.3.2 Implementation of Ability Estimation

The estimation of an examinee's ability is accomplished via the Newton-Raphson (Fisher scoring) procedure, which is represented in the driver section of the program. Lines 160–300 constitute the outer loop of iterative estimation procedure. The inner loop, lines 190–250, implements the subroutines that perform the calculations done for each item. The initialization subroutine, beginning at line 500, establishes the environment of the estimation process. The appropriate constants are set, the initial estimate of the examinee's ability is set to 1.0, the values of the item parameters are read from the data statements and the tracing options are set. A summary of the settings is printed. The subroutine at line 600 reparameterizes the intercept and slope parameters and is identical to that employed in the item parameter estimation program. If TRALL is set, the new intercepts and slopes are printed:

```
CP(1)= -.25      A(1)= -1.8
CP(1)=  1.0      A(1)= -.6
CP(2)= .25       A(2)= 1.8
CP(2)= 1.25      A(2)= 1.2
CP(3)= -1.25         A(3)=  1.2
CP(3)= -1.0          A(3)= 1.8
CP(4)= -1.25         A(4)= 1.2
CP(4)=  -1.0         A(4)= -.6
CP(5)= 1.0       A(5)= -1.8
CP(5)= -.25      A(5)= -.6
```

With these preliminaries out of the way, the iterative loop is initiated at line 160 and the item related computations performed. The subroutine called at line 210 computes the probability of selecting a response category by an examinee of ability THETA for all categories. The following is printed:

```
MLE ITERATION  1

ITEM=  1
EVALUATE MULTIVARIATE LOGISTIC FUNCTION
    CPT=  .25  A= -.8  THETA=  1
    Z( 1 )= -.55 EZ=   .5769498 D= .5769498
    CPT=  .5  A=  1  THETA=  1
    Z( 2 )=  1.5 EZ=  4.481689 D= 5.058639
    CPT= -.75  A= -.2  THETA=  1
    Z( 3 )= -.95 EZ=  .386741 D= 5.44538
            P( 1 )=  .1059522
            P( 2 )=  .823026
            P( 3 )=  7.102187E-02
```

Given these values and the examinee's item response vector IRP(I), the terms in the first derivative are calculated in the subroutine called at line 220. The FOR-NEXT loop over the three response categories sets the values of U(K) equal to one if the examinee choose the category and to zero otherwise. The contribution of the item to the first derivative is computed at line 1240 and accumulated in FDS at line 1250. If TRALL is set, the following is printed:

```
COMPUTE FIRST DERIVATIVES
TERM = -.9240599  SUM FIRST DERIV = -.9240599
```

The next subroutine called computes the weights used in the information term. If TRALL is set, they are printed:

```
WEIGHTS
   W11=  .0947263   W12=W21=  8.720139E-02  W22=  .1456542
   W13=  7.524921E-03   W23=  5.845284E-02   W33=  6.597777E-02
```

These weights are used in the subroutine beginning at line 2000 that computes the expected value of the second derivative of the log-likelihood with respect to ability. Lines 2010–2030 compute the three terms that, when added together, are the item's contribution to the second derivative term. Inspection of the terms following Equation 9.29 reveals that data values appear in these terms. As is the usual practice, the u_{ijk} were replaced by P_{ijk} as the leading multiplier of the term. Since there are now three components, they are added to SDS at line 2040. If TRALL is set, the following is printed:

```
SECOND DERIVATIVE
TERM1=  1.368711E-02
TERM2=  .2592593
TERM3= -3.564823E-02    SDS=  .2372982
```

This set of four subroutines is called for each of the five items and only the results for the fifth item are shown here:

```
ITEM=  5
EVALUATE MULTIVARIATE LOGISTIC FUNCTION
    CPT=  .25   A= -.8   THETA=  1
    Z( 1 )= -.55 EZ=  .5769498 D= .5769498
    CPT= -.75   A=  1   THETA=  1
    Z( 2 )=  .25 EZ=  1.284025 D= 1.860975
    CPT=  .5   A= -.2   THETA=  1
    Z( 3 )=  .3 EZ=  1.349859 D= 3.210834
        P( 1 )=  .1796884
        P( 2 )=  .399904
        P( 3 )=  .4204076
COMPUTE FIRST DERIVATIVES
TERM =  .8279282  SUM FIRST DERIV = -1.636441
WEIGHTS
   W11=  .1474005   W12=W21=  7.185813E-02  W22=  .2399808
   W13=  7.554238E-02   W23=  .1681227   W33=  .243665
SECOND DERIVATIVE
TERM1=  2.510895E-02
TERM2=  .4900413
TERM3= -4.124436E-02    SDS=  2.222292
```

Upon completion of the last item, the increment, TDELTA, to the initial ability estimate is the ratio of FDS to SDS. This is done at line 2500 of the subroutine called line 270 of the driver section. The increment is then added to the initial ability estimate to get the new value. If either TRALL or TRMLE are set, these two results are printed:

```
DELTA PARAMETER=-.7363752
ABILITY=  .2636248
```

The convergence of the iterative process is checked at line 290. If the increment is less than or equal to 0.05, convergence is achieved. However, the increment at the end of the first iteration was greater than 0.05, and another iteration was initiated via lines 300 and 170.

Since, the detailed printout is quite long, only the increments and the resulting ability estimates are reported here:

```
MLE ITERATION 2
...
DELTA PARAMETER= 7.975955E-02
ABILITY=  .3433844

MLE ITERATION  3
...
DELTA PARAMETER= 1.125995E-03
ABILITY=  .3445104
```

Since the increment from the third iteration is less than 0.05, the iterative procedure is stopped, and the obtained ability estimate is printed via line 320:

```
REACHED MAXIMUM ITERATIONS
FINAL ABILITY=  .3445104
```

Like the maximum likelihood estimation of many other item response theory parameters, the present procedure is somewhat unstable. This is in part due to the maximum likelihood estimation approach and partly due to the use of the multivariate logistic function to model the nominal response case. The problems due to the latter is a consequence of the fact that the probabilities for at least two response categories are asymptotic to one, see Figure 9.1 and Table 9.1. If, for example, an examinee chooses only response categories that have asymptotes at the positive end of the ability scale, such as (2,1,1,3,2) for the present example, the solution of the Newton-Raphson equation will diverge rather than converge. In a production program, techniques for handling such cases would be necessary. The only widely available computer program for the nominal response case is MULTILOG (Thissen, 1985), which employs a Bayesian approach rather than maximum likelihood estimation.

G.4 BASIC Computer Programs

G.4.1 Item Parameter Estimation

```
100 REM PROGRAM TO ESTIMATE ITEM PARAMETERS UNDER NOMINAL
110 REM RESPONSE MODEL FOR 3 CATEGORY ITEM, 7 ABILITY GROUPS
120 DIM P(3),CP(2),AP(2),CPT(3),A(3),Z(3),EZ(3),F(7),T(4)
125 DIM TERM(4),FDS(4),PDELTA(4),THETA(7)
130 DIM TM(2,2),W(2,2),MTRX(4,4),R(7,3)
140 GOSUB 500:REM READ CANNED DATA,SET TRACES
150 GOSUB 600:REM TRANSFORM ITEM PARAMETERS
160 FOR NIT=1 TO MAXIT
```

```
170 PRINT:PRINT "MLE ITERATION ";NIT
180 GOSUB 700:REM CLEAR SUMS
190 FOR G=1 TO MAXGPS
200 IF TRALL=0 THEN PRINT:PRINT TAB(1);"ABILITY GROUP= ";G
210 GOSUB 1000:REM CALCULATE P'S
220 GOSUB 1200:REM CALCULATE FIRST DERIVATIVES
230 GOSUB 1400:REM CALCULATE W'S
240 GOSUB 2000:REM CALCULATE SECOND DERIVATIVES
250 NEXT G
255 GOSUB 2300:REM INVERT MATRIX
256 IF IFLAG2=0 GOTO 350
260 GOSUB 2500:REM CALCULATE INCREMENTS AND NEW CP,AP
280 GOSUB 2600:REM CALCULATE NEW CPT AND A'S
290 CONVERGE$="N"
300 GOSUB 2800:REM CHECK CONVERGENCE
310 IF CONVERGE$ = "Y" GOTO 340
320 NEXT NIT
330 PRINT "REACHED MAXIMUM ITERATIONS"
340 PRINT "FINAL VALUES":GOSUB 2640
341 REM PRINT FINAL INTERCEPTS AND SLOPES
350 END
REM*******READ CANNED DATA AND SET TRACES*********
500 MRC=3:NTP=2:MAXIT=5:MAXGPS=7
510 FOR K=1 TO MRC:READ CPT(K):READ A(K):NEXT K
520 FOR G= 1 TO MAXGPS:READ THETA(G):NEXT G
530 FOR G= 1 TO MAXGPS:READ F(G):
531 FOR K=1 TO MRC:READ R(G,K):NEXT K:NEXT G
540 INPUT "TRACE ALL? Y/N ";YN$
550 TRALL=-1:TRMLE=-1
560 IF YN$ = "Y" THEN TRALL=0:TRMLE=0:GOTO 590
570 INPUT "TRACE MLE? Y/N ";YN$
580 IF YN$= "Y" THEN TRMLE=0
590 PRINT " ALL    MLE    MAX CYCLES"
593 PRINT TAB(3);TRALL;TAB(9);TRMLE;TAB(17);MAXIT
595 RETURN
REM********REPARAMETERIZE INTERCEPTS AND SLOPES ***********
600 IF TRALL=0 THEN
    PRINT "REPARAMETERIZED INTERCEPTS AND SLOPES"
    END IF
605 FOR I=1 TO 2
610 IF I=1 THEN
    CP(I)=CPT(1)-CPT(2):AP(I)=A(1)-A(2)
    ELSE
    CP(I)=CPT(1)-CPT(3):AP(I)=A(1)-A(3)
    END IF
620 IF TRALL=0 THEN
    PRINT "CP(";I;")= ";CP(I);" AP(I";I;")=";AP(I)
    END IF
630 NEXT I
640 RETURN
REM*******CLEAR SUMS EACH ITERATION***********
700 FOR R=1 TO 4:FDS(R)=0:PDELTA(R)=0.0
701 FOR C= 1 TO 4:MTRX(R,C)=0:NEXT C:NEXT R
```

```
799 RETURN
REM*******CALCULATE P'S**************
1000 IF TRALL=0 THEN
     PRINT "EVALUATE MULTIVARIATE LOGISTIC FUNCTION"
     END IF
1010 D=0:FOR K=1 TO MRC:IF TRALL=-1 GOTO 1020
1015 PRINT TAB(5);"CPT= ";CPT(K);" A= ";A(K);
1016 PRINT " THETA= ";THETA(G)
1020 Z(K)=CPT(K)+A(K)*THETA(G):EZ(K)=EXP(Z(K)):D=D+EZ(K)
1030 IF TRALL=0 THEN
     PRINT TAB(5);"Z(";K;")= ";Z(K);"EZ= ";EZ(K);"D=";D
     END IF
1040 NEXT K
1050 FOR K=1 TO MRC:P(K)=EZ(K)/D
1060 IF TRALL=0 THEN PRINT TAB(10);"P(";K;")= ";P(K)
1070 NEXT K
1080 RETURN
REM*****CALCULATE FIRST DERIVATES*************
1200 IF TRALL=0 THEN PRINT "COMPUTE FIRST DERIVATIVES"
1205 FOR KK=1 TO 4
1210 IF KK=1 THEN TERM(KK)=-R(G,2)+F(G)*P(2)
1220 IF KK=2 THEN TERM(KK)=-R(G,3)+F(G)*P(3)
1230 IF KK=3 THEN TERM(KK)=THETA(G)*(-R(G,2)+F(G)*P(2))
1240 IF KK=4 THEN TERM(KK)=THETA(G)*(-R(G,3)+F(G)*P(3))
1250 FDS(KK)=FDS(KK)+TERM(KK)
1270 IF TRALL=-1 GOTO 1290
1280 PRINT "TERM(";KK;")= ";TERM(KK);
1281 PRINT " SUM FIRST DERIV(";KK;")=";FDS(KK)
1290 NEXT KK:RETURN
REM********CALCULATE W'S*************
1400 W(1,1)=P(2)*(1.0-P(2))
1410 W(1,2)=P(2)*P(3)
1420 W(2,1)=P(3)*P(2)
1430 W(2,2)=P(3)*(1.0-P(3))
1440 TM(1,1)=1:TM(1,2)=THETA(G):TM(2,1)=THETA(G)
1445 TM(2,2)=THETA(G)*THETA(G)
1450 IF TRALL=-1 GOTO 1490
1455 PRINT "WEIGHT AND THETA MATRICIES"
1460 FOR R=1 TO 2:FOR C=1 TO 2
1470 PRINT TAB(5);"R= ";R;" C= ";C;" W= ";W(R,C);"TM= ";TM(R,C)
1480 NEXT C:NEXT R
1490 RETURN
REM ******KRONECKER PRODUCT ROUTINE FOR SECOND DERIVATIVES******
2000 FOR IB=1 TO 2:FOR JB= 1 TO 2
2010 MF=TM(IB,JB)*F(G)
2020 IF(IB=1 AND JB=1) GOTO 2100
2030 IF(IB=1 AND JB=2) GOTO 2120
2040 IF(IB=2 AND JB=1) GOTO 2140
2050 IF(IB=2 AND JB=2) GOTO 2160
2060 STOP
2100 FOR I=1 TO 2:FOR J=1 TO 2
2101 TEMP=MF*W(I,J):IF I < > J THEN TEMP=-TEMP
2103 MTRX(I,J)=MTRX(I,J)+TEMP
```

```
2105 NEXT J:NEXT I:GOTO 2200
2120 FOR I=1 TO 2:FOR J=1 TO 2
2121 TEMP=MF*W(I,J):IF I < > J THEN TEMP=-TEMP
2123 MTRX(I,J+JB)=MTRX(I,J+JB)+TEMP
2125 NEXT J:NEXT I:GOTO 2200
2140 FOR I=1 TO 2:FOR J=1 TO 2
2141 TEMP=MF*W(I,J):IF I < > J THEN TEMP=-TEMP
2143 MTRX(I+IB,J)=MTRX(I+IB,J)+TEMP
2145 NEXT J:NEXT I:GOTO 2200
2160 FOR I=1 TO 2:FOR J=1 TO 2
2161 TEMP=MF*W(I,J):IF I < > J THEN TEMP=-TEMP
2163 MTRX(I+IB,J+JB)=MTRX(I+IB,J+JB)+TEMP
2165 NEXT J:NEXT I:GOTO 2200
2200 NEXT JB:NEXT IB:IF TRALL=-1 GOTO 2250
2210 PRINT "INFORMATION MATRIX"
2220 FOR R=1 TO 4:FOR C=1 TO 4
2230 PRINT TAB(5);"MTRX(";R;",";C;")= ";MTRX(R,C)
2240 NEXT C:NEXT R
2250 RETURN
2300 REM********MATRIX INVERSION ROUTINE******************
2301 IFLAG2=-1
2303 MBC=3
2305 FOR KK= 1 TO 4
2310 IF (MTRX(1,1)-.000001)<=0 GOTO 2420
2320 R=SQR(MTRX(1,1))
2330 FOR IK= 1 TO MBC:T(IK)=MTRX(IK+1,1)/R:NEXT IK
2340 T(4)=1.0/R
2350 FOR JK=1 TO MBC
2360 FOR IK=1 TO MBC:MTRX(IK,JK)=MTRX(IK+1,JK+1)-T(IK)*T(JK)
2370 NEXT IK:NEXT JK
2380 FOR IK=1 TO 4:MTRX(IK,4)=-T(IK)*T(4):NEXT IK
2390 FOR JK=1 TO MBC:MTRX(4,JK)=MTRX(JK,4):NEXT JK
2395 NEXT KK
2400 FOR JK=1 TO 4:FOR IK=1 TO 4:MTRX(IK,JK)=-MTRX(IK,JK)
2405 NEXT IK:NEXT JK
2410 GOTO 2440
2420 PRINT "THE MATRIX IS (VERY NEAR ) SINGULAR OR INDEFINITE"
2430 IFLAG2=0:GOTO 2470
2440 IF TRALL < > 0 GOTO 2470
2450 FOR JK=1 TO 4:FOR IK=1 TO 4
2451 PRINT "ROW= ";JK;" COL= ";IK;
2352 PRINT " INVERSE ELEMENT= ";MTRX(JK,IK)
2460 NEXT IK:NEXT JK
2470 RETURN
REM************CALCULATE NEW CP,AP*******
2500 FOR R=1 TO 4:FOR C=1 TO 4
2510 PDELTA(R)=PDELTA(R)+MTRX(R,C)*FDS(C)
2520 NEXT C
2530 IF(TRALL=0) OR (TRMLE=0) THEN
     PRINT "DELTA PARAMETER=";PDELTA(R)
     END IF
2540 NEXT R
2550 CP(1)=CP(1)+PDELTA(1):CP(2)=CP(2)+PDELTA(2)
```

```
2560 AP(1)=AP(1)+PDELTA(3):AP(2)=AP(2)+PDELTA(4)
2570 IF (TRALL=0) OR (TRMLE=0) THEN GOSUB 2590
2580 RETURN
2590 PRINT "CP(1)= ";CP(1);"  AP(1)= ";AP(1)
2591 PRINT "CP(2)= ";CP(2);"  AP(2)= ";AP(2)
2592 RETURN
REM ************CALCULATE NEW INTERCEPTS AND SLOPES**********
2600 CPT(1)=(CP(1)+CP(2))/3:CPT(2)=(CP(2)-2*CP(1))/3
2605 CPT(3)=(CP(1)-2*CP(2))/3
2610 A(1)=(AP(1)+AP(2))/3:A(2)=(AP(2)-2*AP(1))/3
2611 A(3)=(AP(1)-2*AP(2))/3
2620 IF (TRALL=0) OR (TRMLE=0) GOTO 2640
2630 GOTO 2660
2640 FOR K=1 TO MRC
2650 PRINT "INTERCEPT(";K;")= ";CPT(K);
2651 PRINT " SLOPE(";K;")=";A(K):NEXT K
2660 RETURN
REM*********CHECK CONVERGENCE***************
2800 FOR R=1 TO 4
2810 IF ABS(PDELTA(R)) >.05 GOTO 2830
2820 NEXT R:CONVERGE$="Y"
2830 RETURN
REM*******************CANNED DATA***************
9000 DATA -.80,-.30,.35,-.60,.45,.90
9110 DATA -3.0,-2.0,-1.0,0.0,1.0,2.0,3.0
9120 DATA 1000,57,938,5
9130 DATA 1000,97,873,30
9140 DATA 1000,143,707,150
9150 DATA 1000,139,377,484
9160 DATA 1000,71,106,823
9170 DATA 1000,25,20,955
9180 DATA 1000,7,4,989
```

G.4.2 Ability Estimation

```
100 REM PROGRAM TO ESTIMATE EXAMINEE ABILITY PARAMETER
110 REM UNDER NOMINAL RESPONSE MODEL FOR FIVE 3-CATEGORY ITEMS
120 DIM P(3),CP(5,2),AP(5,2),CPT(5,3),A(5,3),Z(3),EZ(3)
130 DIM W,IRP(5),U(5):REM W NOT W(3,3)
140 GOSUB 500:REM READ CANNED DATA,SET TRACES
150 GOSUB 600:REM TRANSFORM ITEM PARAMETERS
160 FOR NIT=1 TO MAXIT
170 PRINT:PRINT "MLE ITERATION ";NIT
180 FDS=0:SDS=0:REM CLEAR SUMS
190 FOR I=1 TO NITEM
200 IF TRALL=0 THEN PRINT:PRINT TAB(1);"ITEM= ";I
210 GOSUB 1000:REM CALCULATE P'S
220 GOSUB 1200:REM CALCULATE FIRST DERIVATIVES
230 GOSUB 1400:REM CALCULATE W'S
240 GOSUB 2000:REM CALCULATE SECOND DERIVATIVES
250 NEXT I
270 GOSUB 2500:REM CALCULATE INCREMENTS AND NEW ABILITY
280 REM CHECK CONVERGENCE
```

```
290 IF ABS(TDELTA) <= .05  GOTO 310
300 NEXT NIT
310 PRINT "REACHED MAXIMUM ITERATIONS"
320 PRINT "FINAL ABILITY= ";THETA
330 END
REM*******READ CANNED DATA AND SET TRACES*********
500 MRC=3:NTP=2:NITEM=5:MAXIT=5:THETA=1:REM INITIAL THETA VALUE
510 FOR I=1 TO NITEM
520 FOR K=1 TO MRC:READ CPT(I,K):READ A(I,K)
530 NEXT K:NEXT I
540 FOR I=1 TO NITEM:READ IRP(I):NEXT I
550 INPUT "TRACE ALL? Y/N ";YN$
560 TRALL=-1:TRMLE=-1
570 IF YN$ = "Y" THEN TRALL=0:TRMLE=0:GOTO 593
580 INPUT "TRACE MLE? Y/N ";YN$
590 IF YN$= "Y" THEN TRMLE=0
593 PRINT " ALL   MLE   MAX CYCLES"
595 PRINT TAB(3);TRALL;TAB(9);TRMLE;TAB(17);MAXIT
597 RETURN
REM********REPARAMETERIZE INTERCEPTS AND SLOPES ***********
600 IF TRALL=0 THEN
    PRINT "REPARAMETERIZED INTERCEPTS AND SLOPES"
    END IF
610 FOR I=1 TO NITEM
620 FOR K=1 TO 2
630 IF K=1 THEN
    CP(I,K)=CPT(I,1)-CPT(I,2):AP(I,K)=A(I,1)-A(I,2)
    ELSE
    CP(I,K)=CPT(I,1)-CPT(I,3):AP(I,K)=A(I,1)-A(I,3)
    END IF
640 IF TRALL=0 THEN
    PRINT "CP(";I;")= ";CP(I,K);" A(";I;")=";AP(I,K)
    ENDIF
650 NEXT K:NEXT I
660 RETURN
REM*******CALCULATE P'S**************
1000 IF TRALL=0 THEN
     PRINT "EVALUATE MULTIVARIATE LOGISTIC FUNCTION"
     END IF
1010 D=0:FOR K=1 TO MRC:IF TRALL=-1 GOTO 1030
1020 PRINT TAB(5);"CPT= ";CPT(I,K);" A= ";A(I,K);
1021 PRINT " THETA= ";THETA
1030 Z(K)=CPT(I,K)+A(I,K)*THETA:EZ(K)=EXP(Z(K)):D=D+EZ(K)
1040 IF TRALL=0 THEN
     PRINT TAB(5);"Z(";K;")= ";Z(K);"EZ= ";EZ(K);"D=";D
     END IF
1050 NEXT K
1060 FOR K=1 TO MRC:P(K)=EZ(K)/D
1070 IF TRALL=0 THEN PRINT TAB(10);"P(";K;")= ";P(K)
1080 NEXT K
1090 RETURN
REM*****CALCULATE FIRST DERIVATES*************
1200 IF TRALL=0 THEN PRINT "COMPUTE FIRST DERIVATIVES"
```

```
1210 FOR K=1 TO MRC
1220 IF IRP(I)=K THEN
     U(K)=1
     ELSE
     U(K)=0
     END IF
1230 NEXT K
1240 TERM= -(AP(I,1)*(U(2)-P(2))+AP(I,2)*(U(3)-P(3)))
1250 FDS=FDS+TERM
1270 IF TRALL=-1 GOTO 1290
1280 PRINT "TERM = ";TERM;" SUM FIRST DERIV = ";FDS
1290 RETURN
REM********CALCULATE W'S*************
1400 W11=P(1)*(1.0-P(1))
1410 W12=P(1)*P(2)
1420 W21=W12
1430 W22=P(2)*(1.0-P(2))
1440 W13=P(1)*P(3)
1450 W23= P(2)*P(3)
1460 W33= P(3)*(1.0-P(3))
1470 IF TRALL=-1 GOTO 1495
1480 PRINT "WEIGHTS"
1481 PRINT TAB(5);"W11= ";W11;"  W12=W21= ";W12;" W22= ";W22
1490 PRINT TAB(5);"W13= ";W13;"  W23= ";W23;"  W33= ";W33
1495 RETURN
REM *****CALCULATE SECOND DERIVATIVES*********
2000 IF TRALL=0 THEN PRINT "SECOND DERIVATIVE"
2005 TEMP1=AP(I,1)*AP(I,1)*W12:TEMP2=AP(I,1)*AP(I,2)*W11
2006 TEMP3=AP(I,2)*AP(I,2)*W13:TEMP4=AP(I,1)*AP(I,1)*W22
2007 TEMP5=AP(I,1)*AP(I,2)*W21:TEMP6=AP(I,2)*AP(I,2)*W23
2008 TEMP7=AP(I,1)*AP(I,2)*W23:TEMP8=AP(I,1)*AP(I,1)*W23
2009 TEMP9=AP(I,1)*AP(I,2)*W13:TEMP10=AP(I,2)*AP(I,2)*W33
2010 TERM1=P(1)*(TEMP1+TEMP2+TEMP3)/MRC
2020 TERM2=P(2)*(-TEMP4-TEMP5+TEMP6)/MRC+TEMP4-TEMP7
2030 TERM3=P(3)*(TEMP8-TEMP9-TEMP10)/MRC+TEMP10-TEMP7
REM 2010 TERM1=P(1)*(AP(I,1)*AP(I,1)*W12+AP(I,1)*AP(I,2)*W11+
REM      AP(I,2)*AP(I,2)*W13)/MRC
REM 2020 TERM2=P(2)*(-AP(I,1)*AP(I,1)*W22-AP(I,1)*AP(I,2)*W21+
REM      AP(I,2)*AP(I,2)*W23)/MRC+AP(I,1)*AP(I,1)*W22-
REM      AP(I,1)*AP(I,2)*W23
REM 2030 TERM3=P(3)*(AP(I,1)*AP(I,1)*W23-AP(I,1)*AP(I,2)*W13-
REM      AP(I,2)*AP(I,2)*W33)/MRC+AP(I,2)*AP(I,2)*W33-
REM      AP(I,1)*AP(I,2)*W23
2040 SDS=SDS+TERM1+TERM2+TERM3
2050 IF TRALL < > 0 GOTO 2090
2060 PRINT "TERM1= ";TERM1
2070 PRINT "TERM2= ";TERM2
2080 PRINT "TERM3= ";TERM3;"   SDS= ";SDS
2090 RETURN
REM **********CALCULATE NEW ABILITY************
2500 TDELTA=FDS/SDS
2510 IF(TRALL=0) OR (TRMLE=0) THEN
     PRINT "DELTA PARAMETER=";TDELTA
```

```
      END IF
2520  THETA=THETA+TDELTA
2530  IF (TRALL=0) OR (TRMLE=0) THEN GOSUB 2550
2540  RETURN
2550  PRINT "ABILITY= ";THETA
2560  RETURN
REM*********CHECK CONVERGENCE***************
2800  IF ABS(TDELTA) >.05 GOTO 2820
2810  CONVERGE$="Y"
2820  RETURN
REM*******************CANNED DATA***************
9000  DATA   .25,-.8,.50,1.0,-.75,-.2
9001  DATA   .50,1.0,.25,-.8,-.75,-.2
9002  DATA  -.75,1.0,.50,-.2,.25,-.8
9003  DATA  -.75,.2,.50,-1.0,.25,.8
9004  DATA   .25,-.8,-.75,1.0,.50,-.2
9005  DATA  3,3,1,2,2
```

H. Implementation of MMLE/EM for the Rasch Model

H.1 Introduction

This appendix presents implementation of the marginal maximum likelihood estimation using the expectation and maximization algorithm for the Rasch model. All the derivations are based on Thissen (1982). Because the marginal maximum likelihood estimation procedures together with the expectation and maximization algorithm are presented several different places (e.g., Chapters 6, 7, 10, and 11), we will describe below the basic equations that are necessary for implementation.

The one-parameter logistic model of Thissen (1982) can be written as

$$P(u_i = 1|\theta) = \frac{1}{1 + \exp[-a(\theta - b_i)]}. \tag{H.1}$$

For the Rasch model, we may not use the response patterns. Instead, the score groups are used; $g = 0, 1, \ldots, n$, where n is the total number of items. Assuming that the marginalization of the nuisance ability parameters are performed using the Gaussian quadratures of the Bock-Aitkin (1981) algorithm, we may obtain the expected number of correct responses to item i at $\theta = X_l$ $(l = 1, \ldots, L)$ and the expected number of incorrect responses:

$$r_{li1}^* = \sum_{g=0}^{n} r_{gi+} L_g(X_l) \tag{H.2}$$

and

$$r_{li0}^* = \sum_{g=0}^{n} (N_g - r_{gi+}) L_g(X_l), \tag{H.3}$$

where r_{gi+} is the numbers of examinees in score group g who responded correctly to item i, N_g is the number of examinees in score group g, and $L_g(X_l)$ is the relative density for $\theta = X_l$ for score group g. Given provisional item parameter estimates,

$$L_g(X_l) = \frac{C_g(X_l)\phi(X_l)}{\sum_{l=1}^{L} C_g(X_l)\phi(X_l)}, \tag{H.4}$$

where $\phi(X_l)$ is the population distribution of $\phi(\theta)$ evaluated with X_l and

$$C_g(X_l) = \exp[-(n-g)aX_l] \prod_{i=1}^{n} P(u_i = 1 | \theta = X_l). \qquad (H.5)$$

With sets of expected numbers, the likelihood function can be written as

$$L(b_1, b_2, \ldots, b_n, a | \{r^*_{li1}\}, \{r^*_{li0}\}) \propto \prod_{i=1}^{n} P_{li}^{r^*_{li1}} (1 - P_{li})^{r^*_{li0}}, \qquad (H.6)$$

where

$$P_{li} = \frac{1}{1 + \exp[-a(X_l - b_i)]}. \qquad (H.7)$$

As a usual, the maximum likelihood estimates are the points of the parameters that yield the maximum of the likelihood function or the log of the likelihood function. The derivatives of the loglikelihood are

$$\frac{\partial \log L}{\partial b_i} = \sum_{l=1}^{L} a[r^*_{li0} P_{li} - r^*_{li1}(1 - P_{li})] \qquad (H.8)$$

and

$$\frac{\partial \log L}{\partial a} = \sum_{l=1}^{L} \sum_{i=1}^{n} (X_l - b_i)[r^*_{li1}(1 - P_{li}) - r^*_{li0} P_{li}]. \qquad (H.9)$$

We may use the Newton-Raphson iterations to solve the above system of equations. Two estimation methods are available and these are implemented in the BASIC programs. The first method requires the inversion of the matrix of the second derivatives. The second method, which is call the simpler relaxation solution, does not require the inversion.

H.2 Implementation

The first method that requires the inversion of the whole matrix of the second derivatives is presented in detail. The simpler relaxation solution is presented briefly in the end of this section.

The program begins with the question for the TRALL option. When set to zero, it causes nearly all the intermediate terms to be printed on the computer screen. When the variable TRALL is set to zero, only the current values of the item parameter estimates and the change in their values are printed after each iteration. Again, because the marginal maximum likelihood estimation procedures are presented several times, the TRALL option is set -1 (i.e., not invoked). The total number of the expectation and maximization cycles is to be specified next. One may set this number (i.e., MNC) to be 25.

The next sets of variables define the number of items NI, the number of score groups NG, the number of ability levels or quadrature points NL, the maximum number of the Newton-Raphson iterations NIT, the convergence criterion for the Newton-Raphson iteration CRITIT, and the convergence criterion for the expectation and maximization cycle CRITNC. In addition, the program define several dimensions of the variables. The ten ability levels are defined by X(L), ranged from -4.5 to 4.5 with an increment of 1.0.

The LSAT-6 data (Bock & Aitkin, 1981) were analyzed under the Rasch model via the BASIC programs. The response frequencies are represented as RP(G, I); each cell contains the number of examinees who got the correct response for item i who obtained the raw score g. The number of examinees who obtained respective total scores are represented as N(G). Data indicated that 3 examinees obtained a raw score of 0; 20 examinees obtained a raw score of 1; 85 examinees obtained a raw score of 2; 237 examinees obtained a raw score of 3; 357 examinees obtained a raw score of 4; and 298 examinees obtained a raw score of 5.

The initial values of item discrimination and item difficulty are specified next. The initial value for item discrimination is 1 and the initial values for item difficulty are based on the log odd ratios. The latent ability distribution is assumed to be a standard normal distribution. The $\phi(X_l)$ is defined by the variable PHIX(L).

Using the initial values of item parameters and the ordinates of the standard normal distribution, the expectation and maximization cycle begins. Note that all the variable names in the BASIC program correspond with term used in earlier equations. For example, P(L, I) is P_{li}, CX(G, L) is $C_g(X_l)$, SCXPHIX(G) is $\sum_{l=1}^{L} C_g(X_l)\phi(X_l)$, LX(G, L) is $L_g(X_l)$, R1(L, I) is r_{li1}^*, and R0(L, I) is r_{li0}^*.

The Newton-Raphson will then performed. The partial derivatives are represented as DB(I) is $\partial \log L/\partial b_i$ and DA(I) is $\partial \log L/\partial a$. The second derivatives are DDB(I) is $\partial^2 \log L/\partial b_i^2$, DDA(I) is $\partial^2 \log L/\partial a^2$, and DDC(I) is $\partial^2 \log L/\partial b_i \partial a$. The inverse of the matrix that contains all the parameters (i.e., $n+1=6$) is defined with DDI that is a 6×6 matrix. Note that with the TRALL = 0 option, all the intermediate item parameter estimates are obtained.

The last part of the Newton-Raphson procedure is to evaluate the increments or changes of the estimates. If the changes of the parameters are smaller than the criterion value (i.e., .0001), then the iteration will be terminated. The evaluation of the change in term of the expectation and maximization cycle will then be performed. If the changes of the estimates between two cycles are smaller than the criterion value (i.e., .001), then the cycle will be stopped.

The output from the BASIC program is as follows (note that portions, i.e., cycles 3–14, were removed):

```
TRACE ALL? Y/N ? N
ENTER NUMBER OF EM CYCLES ? 25

INITIAL B(I)

     I      B(I)
```

```
             1.00000   -2.49798
             2.00000   -0.89053
             3.00000   -0.21280
             4.00000   -1.16920
             5.00000   -1.90096

        CYCLE=  1

                   I       B(I)
             1.00000   -2.99579
             2.00000   -1.05491
             3.00000   -0.19687
             4.00000   -1.40196
             5.00000   -2.29211

                   A
             0.91189

                  NC     DIFFNC
             1.00000    0.49781

        CYCLE=  2

                   I       B(I)
             1.00000   -3.22059
             2.00000   -1.16343
             3.00000   -0.25640
             4.00000   -1.53067
             5.00000   -2.47378

                   A
             0.85486

                  NC     DIFFNC
             2.00000    0.22480

                   .
                   .
                   .

        CYCLE=  15

                   I       B(I)
             1.00000   -3.61347
             2.00000   -1.32179
             3.00000   -0.31748
             4.00000   -1.72926
             5.00000   -2.77881

                   A
             0.75557

        EM CYCLE CONVERGED AT NC=  15
```

```
        NC      DIFFNC
  25.00000    0.00090
```

As mentioned earlier, another BASIC program that implements the simpler relaxation solution is also available. The variables and procedures including the expectation and maximization cycles and the Newton-Raphson iterations are nearly identical to the previous program. The only exception is that the whole matrix is not inverted in this solution. The output from the program is as follows:

```
TRACE ALL? Y/N ? N
ENTER NUMBER OF EM CYCLES ? 25

INITIAL B(I)

           I        B(I)
     1.00000    -2.49798
     2.00000    -0.89053
     3.00000    -0.21280
     4.00000    -1.16920
     5.00000    -1.90096

CYCLE=  1

           I        B(I)
     1.00000    -2.89670
     2.00000    -1.01633
     3.00000    -0.16621
     4.00000    -1.35776
     5.00000    -2.22381

           A
     0.93749

          NC      DIFFNC
     1.00000    0.39872

CYCLE=  2

           I        B(I)
     1.00000    -3.01313
     2.00000    -1.09673
     3.00000    -0.25382
     4.00000    -1.43827
     5.00000    -2.31644

           A
     0.90389

          NC      DIFFNC
     2.00000    0.11643
```

```
      .
      .
      .

CYCLE=  25

        I       B(I)
  1.00000   -3.56481
  2.00000   -1.30517
  3.00000   -0.31314
  4.00000   -1.70626
  5.00000   -2.74178


        A
  0.76624

       NC    DIFFNC
 25.00000   0.00295
```

H.3 BASIC Computer Programs

H.3.1 Marginal Maximum Likelihood Estimation for the Rasch Model

```
'         PROGRAM TO IMPLEMENT MARGINAL MAXIMUM LIKELIHOOD RASCH
'         INVERSE WHOLE MATRIX

          TRALL = -1
          INPUT "TRACE ALL? Y/N "; YN$
          IF (YN$ = "Y") OR (YN$ = "y") THEN TRALL = 0
100       INPUT "ENTER NUMBER OF EM CYCLES "; MNC
          IF (MNC <= 0) THEN GOTO 100

          NI = 5
          NG = 6
'         G=0(1)5
          NL = 10
          NIT = 24
'         NIT = 4 * (NI + 1)
          CRITIT = .0001
          CRITNC = .001

          DIM RP(NG, NI), N(NG), X(NL), PHIX(NL), B(NI)
          DIM CX(NG, NL), PRODP(NL), SCXPHIX(NL), LX(NG, NL)
          DIM R1(NL, NI), RO(NL, NI), P(NL, NI), DB(NI), DDB(NI)
          DIM DDC(NI), DDI(NI + 1, NI + 1)
          DIM OLDB(NI), DELTAB(NI)

          FOR L = 1 TO NL
            READ X(L)
          NEXT L
```

```
    DATA -4.5, -3.5, -2.5, -1.5, -0.5, 0.5, 1.5, 2.5, 3.5, 4.5

    FOR G = 0 TO (NG - 1)
      FOR I = 1 TO NI
        READ RP(G, I)
      NEXT I
    NEXT G
    DATA 0.0, 0.0, 0.0, 0.0, 0.0
    DATA 10.0, 1.0, 1.0, 2.0, 6.0
    DATA 62.0, 24.0, 7.0, 28.0, 49.0
    DATA 212.0, 109.0, 63.0, 139.0, 188.0
    DATA 342.0, 277.0, 184.0, 296.0, 329.0
    DATA 298.0, 298.0, 298.0, 298.0, 298.0

    FOR G = 0 TO (NG - 1)
      READ N(G)
    NEXT G
    DATA 3.0, 20.0, 85.0, 237.0, 357.0, 298.0

    INITIAL VALUES
    A = 1!
    B(1) = 0!
    B(2) = 0!
    B(3) = 0!
    B(4) = 0!
    B(5) = 0!
    SUMN = 0!
    FOR G = 0 TO (NG - 1)
      SUMN = SUMN + N(G)
    NEXT G
    FOR I = 1 TO NI
      FOR G = 0 TO (NG - 1)
        B(I) = B(I) + RP(G, I)
      NEXT G
      B(I) = LOG((SUMN - B(I)) / B(I))
    NEXT I
    PRINT
    PRINT "INITIAL B(I)"
    PRINT
    PRINT "        I      B(I)"
    FOR I = 1 TO NI
      PRINT USING "####.#####"; I; B(I)
    NEXT I

    FOR L = 1 TO NL
      PI = 3.141592654#
      PHIX(L) = (1! / (SQR(2! * PI))) * _
        EXP(-(1! / 2!) * X(L) * X(L))
    NEXT L

    NC = 1
1000 PRINT
    PRINT "CYCLE= "; NC
```

```
      FOR I = 1 TO NI
        OLDB(I) = B(I)
      NEXT I
      OLDA = A
'     FOR CRITNC

      FOR L = 1 TO NL
        PRODP(L) = 1!
        FOR I = 1 TO NI
          P(L, I) = 1! / (1! + EXP(-A * (X(L) - B(I))))
          PRODP(L) = PRODP(L) * P(L, I)
        NEXT I
      NEXT L

      FOR G = 0 TO (NG - 1)
        FOR L = 1 TO NL
          CX(G, L) = PRODP(L) * EXP(-(NI - G) * A * X(L))
'         PRINT "G= "; G; "  L= "; L; "  CX(G,L)="; CX(G, L)
        NEXT L
      NEXT G

      FOR G = 0 TO (NG - 1)
      SCXPHIX(G) = 0!
        FOR L = 1 TO NL
          SCXPHIX(G) = SCXPHIX(G) + CX(G, L) * PHIX(L)
        NEXT L
      NEXT G

      FOR G = 0 TO (NG - 1)
        FOR L = 1 TO NL
          LX(G, L) = CX(G, L) * PHIX(L) / SCXPHIX(G)
        NEXT L
      NEXT G

      FOR L = 1 TO NL
        FOR I = 1 TO NI
          R1(L, I) = 0!
          R0(L, I) = 0!
          FOR G = 0 TO (NG - 1)
            R1(L, I) = R1(L, I) + RP(G, I) * LX(G, L)
            R0(L, I) = R0(L, I) + (N(G) - RP(G, I)) * LX(G, L)
          NEXT G
        NEXT I
      NEXT L

      FOR IT = 1 TO NIT

        IF TRALL = 0 THEN
          PRINT
          PRINT "ITERATION= "; IT
          PRINT
          PRINT "          I        B(I)       DB(I)       DDB(I)"
```

```
END IF
FOR I = 1 TO NI
  DB(I) = 0!
  DDB(I) = 0!
  FOR L = 1 TO NL
    P(L, I) = 1! / (1! + EXP(-A * (X(L) - B(I))))
    DB(I) = DB(I) + A * (R0(L, I) * P(L, I) - R1(L, I) * _
      (1! - P(L, I)))
    DDB(I) = DDB(I) - A * A * P(L, I) * (1! - P(L, I)) * _
      (R0(L, I) + R1(L, I))
  NEXT L
  IF TRALL = 0 THEN
    PRINT USING "######.#####"; I; B(I); DB(I); DDB(I)
  END IF
NEXT I

IF TRALL = 0 THEN
  PRINT
  PRINT "              A          DA          DDA"
END IF
DA = 0!
DDA = 0!
FOR L = 1 TO NL
  FOR I = 1 TO NI
    P(L, I) = 1! / (1! + EXP(-A * (X(L) - B(I))))
    DA = DA + (X(L) - B(I)) * (R1(L, I) * (1! - P(L, I)) _
      - R0(L, I) * P(L, I))
    DDA = DDA - (X(L) - B(I)) * (X(L) - B(I)) * P(L, I) _
      * (1! - P(L, I)) * (R0(L, I) + R1(L, I))
  NEXT I
NEXT L
IF TRALL = 0 THEN
  PRINT USING "######.#####"; A; DA; DDA
END IF

IF TRALL = 0 THEN
  PRINT
  PRINT "            I      DDC(I)"
END IF
FOR I = 1 TO NI
DDC(I) = 0!
  FOR L = 1 TO NL
    P(L, I) = 1! / (1! + EXP(-A * (X(L) - B(I))))
    DDC(I) = DDC(I) + (R0(L, I) * P(L, I) - R1(L, I) * _
      (1! - P(L, I))) + A * P(L, I) * (1! - P(L, I)) * _
      (X(L) - B(I)) * (R0(L, I) + R1(L, I))
  NEXT L
  IF TRALL = 0 THEN
    PRINT USING "######.#####"; I; DDC(I)
  END IF
NEXT I

IF TRALL = 0 THEN
```

```
    PRINT
    PRINT "INVERSE HESSIAN"
END IF
DDI(NI + 1, NI + 1) = 0!
FOR I = 1 TO NI
   DDI(NI + 1, NI + 1) = DDI(NI + 1, NI + 1) + DDC(I) * _
   (1! / DDB(I)) * DDC(I)
NEXT I
DDI(NI + 1, NI + 1) = 1! / (DDA - DDI(NI + 1, NI + 1))
FOR I = 1 TO NI
   DDI(I, NI + 1) = -(1! / DDB(I)) * DDC(I) * _
     DDI(NI + 1, NI + 1)
   DDI(NI + 1, I) = -DDI(NI + 1, NI + 1) * _
     DDC(I) / DDB(I)
NEXT I
FOR I = 1 TO NI
  FOR II = 1 TO NI
     DDI(I, II) = (1! / DDB(I)) * DDC(I) * _
       DDI(NI + 1, NI + 1) * (DDC(II) / DDB(II))
  NEXT II
NEXT I
FOR I = 1 TO NI
   DDI(I, I) = (1! / DDB(I)) + DDI(I, I)
NEXT I
IF TRALL = 0 THEN
  FOR I = 1 TO NI + 1
     PRINT USING "####.#####"; DDI(I, 1); DDI(I, 2); _
       DDI(I, 3); DDI(I, 4); DDI(I, 5); DDI(I, 6)
     ABOVE BASED ON NI
  NEXT I
END IF

FOR I = 1 TO NI
   DELTAB(I) = (DDI(I, 1) * DB(1) + DDI(I, 2) * DB(2) + _
     DDI(I, 3) * DB(3) + DDI(I, 4) * DB(4) + _
     DDI(I, 5) * DB(5) + DDI(I, 6) * DA)
   B(I) = B(I) - DELTAB(I)
NEXT I
DELTAA = (DDI(NI + 1, 1) * DB(1) + DDI(NI + 1, 2) * DB(2)_
   + DDI(NI + 1, 3) * DB(3) + DDI(NI + 1, 4) * DB(4) _
   + DDI(NI + 1, 5) * DB(5) + DDI(NI + 1, 6) * DA)
A = A - DELTAA

IF (TRALL = 0) THEN
   PRINT
   PRINT "            I       B(I)"
   FOR I = 1 TO NI
     PRINT USING "####.#####"; I; B(I)
   NEXT I
   PRINT
   PRINT "            A"
   PRINT USING "####.#####"; A
ELSE
```

```
      ENDIF

      DIFFIT = 0!
      FOR I = 1 TO NI
        IF (ABS(DELTAB(I)) > DIFFIT) THEN
          DIFFIT = ABS(DELTAB(I))
        END IF
      NEXT I
      IF (ABS(DELTAA) > DIFFIT) THEN
        DIFFIT = ABS(DELTAA)
      END IF
      IF (TRALL = 0) THEN
        PRINT
        PRINT "       IT    DIFFIT"
        PRINT USING "####.#####"; IT; DIFFIT
      END IF
      IF (DIFFIT < CRITIT) THEN GOTO 899

    NEXT IT

899 IF (TRALL = -1) THEN
      PRINT
      PRINT "       I       B(I)"
      FOR I = 1 TO NI
        PRINT USING "####.#####"; I; B(I)
      NEXT I
      PRINT
      PRINT "          A"
      PRINT USING "####.#####"; A
    END IF
    DIFFNC = ABS(A - OLDA)
    FOR I = 1 TO NI
      IF (ABS(B(I) - OLDB(I)) > DIFFNC) THEN
        DIFFNC = ABS(B(I) - OLDB(I))
      END IF
    NEXT I
    IF (DIFFNC < CRITNC) THEN
      PRINT
      PRINT "EM CYCLE CONVERGED AT NC= "; NC
      NC = MNC
    END IF
'   CRITNC MAY NOT BE EMPLOYED
    PRINT
    PRINT "       NC    DIFFNC"
    PRINT USING "####.#####"; NC; DIFFNC
    NC = NC + 1
    IF NC <= MNC GOTO 1000

    END
```

H.3.2 Marginal Maximum Likelihood Estimation for the Rasch Model: Simpler Relaxation Solution

```
'       PROGRAM TO IMPLEMENT MARGINAL MAXIMUM LIKELIHOOD RASCH
'       SIMPLER RELAXATION SOLUTION

        TRALL = -1
        INPUT "TRACE ALL? Y/N "; YN$
        IF (YN$ = "Y") OR (YN$ = "y") THEN TRALL = 0
100     INPUT "ENTER NUMBER OF EM CYCLES "; MNC
        IF (MNC <= 0) THEN GOTO 100

        NI = 5
        NG = 6
'       G=0(1)5
        NL = 10
        NIT = 24
'       N.B. NIT = 4 * (NI + 1)
        CRITIT = .0001
        CRITNC = .001

        DIM RP(NG, NI), N(NG), X(NL), PHIX(NL), B(NI)
        DIM CX(NG, NL), PRODP(NL), SCXPHIX(NL), LX(NG, NL)
        DIM R1(NL, NI), RO(NL, NI), P(NL, NI), DB(NI), DDB(NI)
        DIM OLDB(NI)

        FOR L = 1 TO NL
          READ X(L)
        NEXT L
        DATA -4.5, -3.5, -2.5, -1.5, -0.5, 0.5, 1.5, 2.5, 3.5, 4.5

        FOR G = 0 TO (NG - 1)
          FOR I = 1 TO NI
            READ RP(G, I)
          NEXT I
        NEXT G
        DATA 0.0, 0.0, 0.0, 0.0, 0.0
        DATA 10.0, 1.0, 1.0, 2.0, 6.0
        DATA 62.0, 24.0, 7.0, 28.0, 49.0
        DATA 212.0, 109.0, 63.0, 139.0, 188.0
        DATA 342.0, 277.0, 184.0, 296.0, 329.0
        DATA 298.0, 298.0, 298.0, 298.0, 298.0

        FOR G = 0 TO (NG - 1)
          READ N(G)
        NEXT G
        DATA 3.0, 20.0, 85.0, 237.0, 357.0, 298.0

'       INITIAL VALUES
        A = 1!
        B(1) = 0!
        B(2) = 0!
        B(3) = 0!
```

```
      B(4) = 0!
      B(5) = 0!
      SUMN = 0!
      FOR G = 0 TO (NG - 1)
        SUMN = SUMN + N(G)
      NEXT G
      FOR I = 1 TO NI
        FOR G = 0 TO (NG - 1)
          B(I) = B(I) + RP(G, I)
        NEXT G
        B(I) = LOG((SUMN - B(I)) / B(I))
      NEXT I
      PRINT
      PRINT "INITIAL B(I)"
      PRINT
      PRINT "         I        B(I)"
      FOR I = 1 TO NI
        PRINT USING "####.#####"; I; B(I)
      NEXT I

      FOR L = 1 TO NL
        PI = 3.141592654#
        PHIX(L) = (1! / (SQR(2! * PI))) * _
          EXP(-(1! / 2!) * X(L) * X(L))
      NEXT L

      NC = 1
1000  PRINT
      PRINT "CYCLE= "; NC

      FOR I = 1 TO NI
        OLDB(I) = B(I)
      NEXT I
      OLDA = A
,     FOR CRITNC

      FOR L = 1 TO NL
        PRODP(L) = 1!
        FOR I = 1 TO NI
          P(L, I) = 1! / (1! + EXP(-A * (X(L) - B(I))))
          PRODP(L) = PRODP(L) * P(L, I)
        NEXT I
      NEXT L

      FOR G = 0 TO (NG - 1)
        FOR L = 1 TO NL
          CX(G, L) = PRODP(L) * EXP(-(NI - G) * A * X(L))
,         PRINT "G= "; G; "  L= "; L; "  CX(G,L)="; CX(G, L)
        NEXT L
      NEXT G

      FOR G = 0 TO (NG - 1)
      SCXPHIX(G) = 0!
```

```
   FOR L = 1 TO NL
     SCXPHIX(G) = SCXPHIX(G) + CX(G, L) * PHIX(L)
   NEXT L
 NEXT G

 FOR G = 0 TO (NG - 1)
   FOR L = 1 TO NL
     LX(G, L) = CX(G, L) * PHIX(L) / SCXPHIX(G)
   NEXT L
 NEXT G

 FOR L = 1 TO NL
   FOR I = 1 TO NI
     R1(L, I) = 0!
     R0(L, I) = 0!
     FOR G = 0 TO (NG - 1)
       R1(L, I) = R1(L, I) + RP(G, I) * LX(G, L)
       R0(L, I) = R0(L, I) + (N(G) - RP(G, I)) * LX(G, L)
     NEXT G
   NEXT I
 NEXT L

 IF TRALL = 0 THEN
   PRINT
   PRINT _
"   ITERATION            I         B(I)        DB(I)        DDB(I)"
 END IF
 FOR I = 1 TO NI
   DB(I) = 0!
   DDB(I) = 0!
   FOR IT = 1 TO NIT
     FOR L = 1 TO NL
       P(L, I) = 1! / (1! + EXP(-A * (X(L) - B(I))))
       DB(I) = DB(I) + A * (R0(L, I) * P(L, I) - R1(L, I) * _
         (1! - P(L, I)))
       DDB(I) = DDB(I) - A * A * P(L, I) * (1! - P(L, I)) * _
         (R0(L, I) + R1(L, I))
     NEXT L
     IF TRALL = 0 THEN
       PRINT USING "######.#####"; IT; I; B(I); DB(I); DDB(I)
     END IF
     B(I) = B(I) - (DB(I) / DDB(I))
     IF (ABS(DB(I) / DDB(I)) < CRITIT) GOTO 700
   NEXT IT
700    NEXT I

 IF TRALL = 0 THEN
   PRINT
   PRINT "   ITERATION              A           DA          DDA"
 END IF
 FOR IT = 1 TO NIT
   DA = 0!
   DDA = 0!
```

```
      FOR L = 1 TO NL
        FOR I = 1 TO NI
          P(L, I) = 1! / (1! + EXP(-A * (X(L) - B(I))))
          DA = DA + (X(L) - B(I)) * (R1(L, I) * _
            (1! - P(L, I)) - R0(L, I) * P(L, I))
          DDA = DDA - (X(L) - B(I)) * (X(L) - B(I)) * _
            P(L, I) * (1! - P(L, I)) * (R0(L, I) + R1(L, I))
        NEXT I
      NEXT L
      IF TRALL = 0 THEN
        PRINT USING "######.#####"; IT; A; DA; DDA
      END IF
      A = A - (DA / DDA)
      IF (ABS(DA / DDA) < CRITIT) THEN
        GOTO 899
      END IF
    NEXT IT

899 PRINT
    PRINT "           I        B(I)"
    FOR I = 1 TO NI
      PRINT USING "####.#####"; I; B(I)
    NEXT I
    PRINT
    PRINT "             A"
    PRINT USING "####.#####"; A

    DIFFNC = ABS(A - OLDA)
    FOR I = 1 TO NI
      IF (ABS(B(I) - OLDB(I)) > DIFFNC) THEN
        DIFFNC = ABS(B(I) - OLDB(I))
      END IF
    NEXT I
    IF (DIFFNC < CRITNC) THEN
      PRINT
      PRINT "EM CYCLE CONVERGED AT NC= "; NC
      NC = MNC
    END IF
  ' CRITNC MAY NOT BE EMPLOYED
    PRINT
    PRINT "        NC    DIFFNC"
    PRINT USING "####.#####"; NC; DIFFNC
    NC = NC + 1
    IF NC <= MNC GOTO 1000

    END
```

I. Implementation of Multiple Groups Estimation

I.1 Introduction

The equations for the estimation of the latent ability distribution and both item and latent distribution parameters under the multiple group framework using the two-parameter logistic model were presented in Chapter 10. In the present appendix, three procedures are implemented; estimation of latent distribution parameters assuming the item parameters are known, estimation of both item and latent distribution parameters for the compact model, and estimation of both item and latent distribution parameters for the augmented model. The estimation for the compact and augmented models illustrate an application of the multiple group model to the detection of differential item function.

I.2 Estimation of Latent Distribution Parameters

The maximum likelihood estimation of the parameters of the latent distribution assuming the item parameters are known has been implemented in the form of a BASIC computer program. The program was written for the two-parameter logistic model, but the actual data analyzed were based on the one-parameter Rasch model. The LSAT-6 data converted to response patterns were used. There are five dichotomously scored items and 32 response patterns from the sample of 1,000 examinees. Because the BASIC program contains the data, these are not repeated here. The conditional maximum likelihood estimates of the item parameters are used. The difficulty estimates of the five items are -1.256, 0.475, 1.236, 0.168, and -0.623 (Andersen & Madsen, 1977). Using these values as known, the mean and variance of the population normal distribution of ability are computed from the patterned data. There are 20 fixed quadrature points ranging from -4.75 and 4.75 with the interval of 0.5.

In order to make the implementation of the estimation of the latent distribution parameters, the computer program is written in a way to have a sequential structure instead of employing a driver or kernel and subroutines. Except for the iterative routines, the program will be executed from the beginning to the end in a sequential manner.

The first line of the program contains constants that are to be used to define arrays of the variables. The total number of response pattern L is NL = 32; the number of items n is NI = 5; the total number of quadrature points Q is NQ = 20; and the maximum number of iterative cycles to obtain the population parameters is NC = 50. With these values, variable arrays are defined subsequently.

The estimates of the five item parameters are defined under the two-parameter logistic model by setting all item discrimination values to 1. The interval of the quadrature points is defined next as D = .5. The 32 response patters are read in as well as the number of cases for each response pattern R(L). With the initial values of the population mean and standard deviation, 0 and 1, respectively, the estimation cycle begins and continues until it reaches to the maximum number of the cycles, NC = 50.

Within each cycle, the fixed quadrature points X_q and the quadrature weights $A(X_q)$ are obtained. The quadrature weights are iteratively adjusted based on the population mean and standard deviation. Next, the terms that are needed to solve the likelihood equations are obtained. The major terms calculated in the cycles are LUX(L, Q), HU(L), PUX(L, Q), and SUMP(Q). These correspond with $L(U_j|X_q)$, $\sum_{q=1}^{Q} L(U_l|X_q)A(X_q)$, P_{lq}, and $\sum_{l=1}^{L} P_{lq}$, respectively.

In the end of each cycle, the mean and variance as well as standard deviation of the population distribution are reported. Deleting the middle portions, the first, the second, and the 50th iteration results are as follows:

MU	VAR	SIGMA
0.68319	0.82042	0.90577
1.00623	0.73066	0.85479
.		
.		
.		
1.47518	0.57155	0.75601

After finishing the last cycle specified by the maximum number of cycles, Sheppard's correction is applied to the final variance estimate (and hence to the final standard deviation also) to correct the histogram approximation of the continuous distribution. The quantity $d^2/12$ is subtracted from the variance for this purpose. The final estimates of the mean, variance, and standard deviation, and the variance and standard deviation from the Sheppard's correction are reported in the output:

MU	VAR	SIGMA	SHEPVAR	SHEPSIGMA
1.47518	0.57155	0.75601	0.55072	0.74210

The mean and Sheppard's variance from the BASIC program coincide with those reported in Andersen and Madsen (1977) and Mislevy (1984).

Next, the program will calculate the elements of the information matrix. The reported values are

I(MU) I(MU,VAR)   I(VAR)

```
515.32501 -89.55314 117.40747
```

and the information matrix is

$$I_{\mu,\sigma^2} = \begin{bmatrix} 515.32501 & -89.55314 \\ -89.55314 & 117.40747 \end{bmatrix}. \tag{I.1}$$

Based on the information matrix, the last part of the program will calculate and report the standard errors of the estimates and the covariance for the population parameters of the latent distribution of the LSAT-6 data. The results are:

```
SE(MU)    SE(VAR)    COVAR
0.04730   0.09909    0.00171
```

I.3 Estimation of Both Item and Latent Distribution Parameters

The data reported in Thissen, Steinberg, and Wainer (1993, p. 71, Table 4.1) are reanalyzed for the purpose of demonstrating the multiple group estimation in the differential item functioning application. The computer program MULTILOG (Thissen, 1991) also contains the analysis of the same data as an example. The detection of differential item functioning using the likelihood ratio is performed between 285 male students and 374 female students. There were four dichotomously scored items and the last item is the studied item. Performance of the female students (focal group) will be compared to that of the male students (reference group).

In the likelihood ratio test of differential item functioning, two models, a compact model and an augmented model, are compared using the -2 log likelihood values. Because two groups of examinees may have different latent distributions, both item parameters and the latent distribution parameters are estimated together. The estimates are used to calculate the log likelihood value that assesses the goodness of fit of the specified model and the data. Note that we will not present detailed explications of the detection of differential item functioning using the likelihood ratio test. Holland and Wainer (1993) contains many methods for differential item functioning.

Two BASIC programs are prepared; one for the compact model and the other for the augmented model. In the two BASIC programs, the first group is treated as the focal group and the second group is treated as the reference group. Hence, female is group 1 and male is group 2.

Let's look at the BASIC program that implemented the compact model first. In the compact model, all item parameters between two groups of examinees are assumed to be equal. The sets of the population distribution parameters may differ between two groups. Without loss of generality and for the sake of simplicity, we will assume that only the two population means are different. In addition, in order to establish the scale, it is assumed that $\mu_2 = 1$,

$\sigma_2 = 1$, and $\sigma_1 = \sigma_2$. We will analyze the data under the two-parameter logistic model. In the compact model, the parameters to be estimated are the set of item parameters ξ and the focal group population mean μ_1.

The first three lines are for the purpose of setting the trace option. The yes answer to the trace all question will create a bulky output that contains all results from the intermediate calculations. There are two groups of examinees (NG = 2), 16 response patterns for each group (NL = 16), four dichotomously scored items (NI = 4), and ten quadrature points (NQ = 10). Using these constants, variable arrays are specified.

The maximum number of the expectation and maximization cycles is set as 25 (MNC = 25), and the maximum Newton-Raphson iterations is set as 10 (MIT = 10). The initial item parameters are specified as $\zeta_i = 0$ and $\lambda_i = 1$, for $i = 1, \ldots, 4$. The program then reads in the response data U and the indicator variable Y.

The quadrature points are specified for the two groups. The initial parameters of the population distribution are specified, subsequently. Note that only the first group mean, μ_1, will be estimated and updated later. Next lines will be used to read in the number of cases of the response patterns for the two groups.

The line number 100 is the starting point of the expectation and maximization cycle. In the beginning part of the cycle, quadrature weights for the two groups are obtained based on the latent distribution parameters. When TRALL = 0, we can examine the quadrature points and corresponding weights as output.

The output from the program will be discussed assuming that the trace all option is not invoked because we had chances to inspect similar calculations in earlier chapters that used the expectation and maximization algorithm. In the expectation step of the program, the artificial data RB(G, I, Q) and NB(G, I, Q) are obtained. These quantities are $\bar{r}_{giq}$ and $\bar{N}_{gq}$ in Chapter 10.

After obtaining the artificial data, the Newton-Raphson procedures will be performed for each item. After obtaining the item paramter estimates for all four items, the focal group population mean, $\hat{\mu}_1$, will be obtained. At this point, one cycle of the estimation of both item and population parameters is finished. From the end of the first cycle, we obtain the following results:

```
TRACE ALL? Y/N ? N

CYCLE=  1

        I    ZETA(I) LAMBDA(I)      B(I)
  1.00000    1.44270   1.05081  -1.37294
  2.00000    0.47585   1.13293  -0.42001
  3.00000   -1.17952   1.12189   1.05137
  4.00000    0.00203   1.12861  -0.00180

    MUH(1)
   0.09728
```

```
      N(1)
374.00000
```

The number of examinees in the focal group, N(1), is printed out for only a verification purpose.

The cycles of the expectation and maximization steps are repeated until the maximum number of cycles are reached. In the final cycle, the estimates of the item and population parameters are as follows:

```
CYCLE=  25
```

I	ZETA(I)	LAMBDA(I)	B(I)
1.00000	1.65344	0.92716	-1.78333
2.00000	0.82343	1.28968	-0.63848
3.00000	-0.92799	1.24169	0.74736
4.00000	0.34661	1.43644	-0.24130

```
   MUH(1)
-0.21756

      N(1)
374.00000
```

In the last part of the BASIC program, the likelihood ratio goodness-of-fit statistic is calculated using

$$G^2 = 2 \sum_{g=1}^{G} \sum_{l=1}^{L} r_{gl} \log \left(\frac{r_{gl}}{N_g P_{gl}} \right). \tag{I.2}$$

All the terms necessary to obtain the likelihood ratio statistic can be found in the last portions of the output where r_{gl}, N_g, and P_{gl} are listed as R(G,L), SN(G), and PB(G,L), respectively. The G^2 for the compact model is 41.11972. The last portions of the output are as follows:

G	L	R(G,L)	SN(G)	PB(G,L)	SN*PB
1.00000	1.00000	29.00000	374.00000	0.08527	31.89230
1.00000	2.00000	7.00000	374.00000	0.02468	9.22903
1.00000	3.00000	50.00000	374.00000	0.14208	53.13703
1.00000	4.00000	30.00000	374.00000	0.07873	29.44338
1.00000	5.00000	15.00000	374.00000	0.04460	16.67935
1.00000	6.00000	4.00000	374.00000	0.03154	11.79546
1.00000	7.00000	6.00000	374.00000	0.00805	3.01199
1.00000	8.00000	0.00000	374.00000	0.00552	2.06270
1.00000	9.00000	67.00000	374.00000	0.13354	49.94247
1.00000	10.00000	63.00000	374.00000	0.17560	65.67517
1.00000	11.00000	12.00000	374.00000	0.02362	8.83283
1.00000	12.00000	10.00000	374.00000	0.03007	11.24547
1.00000	13.00000	2.00000	374.00000	0.00916	3.42492
1.00000	14.00000	6.00000	374.00000	0.01491	5.57511
1.00000	15.00000	22.00000	374.00000	0.04682	17.51201
1.00000	16.00000	51.00000	374.00000	0.14583	54.54078

G	L	R(G,L)	SN(G)	PB(G,L)	SN*PB
2.00000	1.00000	22.00000	285.00000	0.06435	18.33877
2.00000	2.00000	10.00000	285.00000	0.02174	6.19552
2.00000	3.00000	30.00000	285.00000	0.11875	33.84505
2.00000	4.00000	27.00000	285.00000	0.07606	21.67604
2.00000	5.00000	13.00000	285.00000	0.03871	11.03111
2.00000	6.00000	14.00000	285.00000	0.03159	9.00344
2.00000	7.00000	1.00000	285.00000	0.00696	1.98225
2.00000	8.00000	1.00000	285.00000	0.00550	1.56689
2.00000	9.00000	24.00000	285.00000	0.12715	36.23637
2.00000	10.00000	54.00000	285.00000	0.19335	55.10602
2.00000	11.00000	5.00000	285.00000	0.02238	6.37838
2.00000	12.00000	8.00000	285.00000	0.03294	9.38874
2.00000	13.00000	1.00000	285.00000	0.00900	2.56383
2.00000	14.00000	8.00000	285.00000	0.01697	4.83520
2.00000	15.00000	10.00000	285.00000	0.05053	14.40019
2.00000	16.00000	57.00000	285.00000	0.18404	52.45220

G	GOF(G)
1.00000	23.15011
2.00000	17.96962

LIKELIHOOD-RATIO GOODNESS-OF-FIT=   41.11972

The BASIC program for the augmented model is very similar to the one for the compact model. Earlier it was mentioned that the parameters to be estimated in the compact model are the set of item parameters ξ and the focal group population mean μ_1. In the augmented model, the parameters to be estimated are the common item parameters (ξ_1, ξ_2, ξ_3), the two sets of the studied item parameters for the reference and focal groups (ξ_{4R} and ξ_{4F}), and the focal group population mean (μ_1). The studied item is separately estimated for the reference group (as item 4) and for the focal group (as item 5).

The augmented model includes more parameters than the compact model. If the studied item is differentially functioning between the reference and focal groups, then the fit statistic G^2 of the augmented model should be significantly smaller than that of the compact model. The difference between the two G^2 values will be distributed as a chi-square distribution under the null hypothesis of no difference in the sets of item parameters with the degree of freedom based on the difference in the numbers of parameters in the two models.

The final cycle from the BASIC program for the augmented model yielded:

CYCLE= 25

I	ZETA(I)	LAMBDA(I)	B(I)
1.00000	1.57525	1.02347	-1.53913
2.00000	0.66321	1.31869	-0.50293
3.00000	-1.09056	1.26098	0.86485
4.00000	0.75745	1.57198	-0.48184
5.00000	-0.23592	1.25236	0.18838

```
        MUH(1)
       0.01029

         N(1)
     374.00000
```

The last portions of the output for the augmented model are as follows:

G	L	R(G,L)	SN(G)	PB(G,L)	SN*PB
1.00000	1.00000	29.00000	374.00000	0.08229	30.77570
1.00000	2.00000	7.00000	374.00000	0.01982	7.41260
1.00000	3.00000	50.00000	374.00000	0.14194	53.08501
1.00000	4.00000	30.00000	374.00000	0.06453	24.13511
1.00000	5.00000	15.00000	374.00000	0.04676	17.48682
1.00000	6.00000	4.00000	374.00000	0.02539	9.49499
1.00000	7.00000	6.00000	374.00000	0.00839	3.13649
1.00000	8.00000	0.00000	374.00000	0.00440	1.64511
1.00000	9.00000	67.00000	374.00000	0.15727	58.82008
1.00000	10.00000	63.00000	374.00000	0.15764	58.95790
1.00000	11.00000	12.00000	374.00000	0.02742	10.25563
1.00000	12.00000	10.00000	374.00000	0.02655	9.92926
1.00000	13.00000	2.00000	374.00000	0.01080	4.03957
1.00000	14.00000	6.00000	374.00000	0.01249	4.67229
1.00000	15.00000	22.00000	374.00000	0.06735	25.18995
1.00000	16.00000	51.00000	374.00000	0.14696	54.96350

G	L	R(G,L)	SN(G)	PB(G,L)	SN*PB
2.00000	1.00000	22.00000	285.00000	0.07193	20.49892
2.00000	2.00000	10.00000	285.00000	0.03156	8.99561
2.00000	3.00000	30.00000	285.00000	0.10896	31.05237
2.00000	4.00000	27.00000	285.00000	0.09887	28.17818
2.00000	5.00000	13.00000	285.00000	0.03431	9.77925
2.00000	6.00000	14.00000	285.00000	0.03816	10.87669
2.00000	7.00000	1.00000	285.00000	0.00621	1.77000
2.00000	8.00000	1.00000	285.00000	0.00664	1.89194
2.00000	9.00000	24.00000	285.00000	0.09601	27.36235
2.00000	10.00000	54.00000	285.00000	0.21824	62.19885
2.00000	11.00000	5.00000	285.00000	0.01693	4.82639
2.00000	12.00000	8.00000	285.00000	0.03694	10.52856
2.00000	13.00000	1.00000	285.00000	0.00628	1.78911
2.00000	14.00000	8.00000	285.00000	0.01693	4.82514
2.00000	15.00000	10.00000	285.00000	0.03103	8.84447
2.00000	16.00000	57.00000	285.00000	0.18099	51.58220

G	GOF(G)
1.00000	15.39481
2.00000	8.07150

```
LIKELIHOOD-RATIO GOODNESS-OF-FIT=    23.46631
```

The likelihood ratio statistic for item 4 is $18.65341 = 41.11972 - 23.46631$ with the two degrees of freedom.

I.4 BASIC Computer Programs

I.4.1 Estimation of Latent Distribution Parameters

```
        NL = 32: NI = 5: NQ = 20: NC = 50

        DIM ID$(NL)
        DIM U(NL, NI), X(NQ), AX(NQ), A(NI), B(NI), LUX(NL, NQ)
        DIF HU(NL), P(NI), PUX(NL, NQ), SUMP(NQ)
        DIM R(NL)

        A(1) = 1: A(2) = 1: A(3) = 1: A(4) = 1: A(5) = 1
        B(1) = -1.256: B(2) = .475: B(3) = 1.236: B(4) = .168: _
          B(5) = -.623
        SUMR = 1000

REM     D=INTERVAL FOR SHEPPARD'S CORRECTION
        D = .5

        FOR L = 1 TO NL
          READ ID$(L), R(L), U(L, 1), U(L, 2), U(L, 3), _
          U(L, 4), U(L, 5)
'         I = 1 TO NI
        NEXT L
        DATA 00000,  3,0,0,0,0,0
        DATA 00001,  6,0,0,0,0,1
        DATA 00010,  2,0,0,0,1,0
        DATA 00011, 11,0,0,0,1,1
        DATA 00100,  1,0,0,1,0,0
        DATA 00101,  1,0,0,1,0,1
        DATA 00110,  3,0,0,1,1,0
        DATA 00111,  4,0,0,1,1,1
        DATA 01000,  1,0,1,0,0,0
        DATA 01001,  8,0,1,0,0,1
        DATA 01010,  0,0,1,0,1,0
        DATA 01011, 16,0,1,0,1,1
        DATA 01100,  0,0,1,1,0,0
        DATA 01101,  3,0,1,1,0,1
        DATA 01110,  2,0,1,1,1,0
        DATA 01111, 15,0,1,1,1,1
        DATA 10000, 10,1,0,0,0,0
        DATA 10001, 29,1,0,0,0,1
        DATA 10010, 14,1,0,0,1,0
        DATA 10011, 81,1,0,0,1,1
        DATA 10100,  3,1,0,1,0,0
        DATA 10101, 28,1,0,1,0,1
        DATA 10110, 15,1,0,1,1,0
        DATA 10111, 80,1,0,1,1,1
        DATA 11000, 16,1,1,0,0,0
        DATA 11001, 56,1,1,0,0,1
        DATA 11010, 21,1,1,0,1,0
        DATA 11011,173,1,1,0,1,1
        DATA 11100, 11,1,1,1,0,0
```

```
DATA 11101, 61,1,1,1,0,1
DATA 11110, 28,1,1,1,1,0
DATA 11111,298,1,1,1,1,1

MUO = 0
SIGMAO = 1

PRINT
PRINT "        MU      VAR      SIGMA"

FOR C = 1 TO NC
CYCLE

  QNORMCON = 0
  FOR Q = 1 TO NQ
    X(Q) = -4.75 + (Q - 1) * D
    QNORMCON = QNORMCON + EXP(-(1 / 2) * ((X(Q) - MUO) _
      / SIGMAO) ^ 2)
  NEXT Q
  FOR Q = 1 TO NQ
    AX(Q) = EXP(-(1 / 2) * ((X(Q) - MUO) / SIGMAO) ^ 2) _
      / QNORMCON
  NEXT Q

  FOR L = 1 TO NL
    HU(L) = 0
    FOR Q = 1 TO NQ
      LUX(L, Q) = 1
    NEXT Q
  NEXT L

  FOR L = 1 TO NL
    FOR Q = 1 TO NQ
      FOR I = 1 TO NI
        P(I) = 1 / (1 + EXP(-(A(I) * (X(Q) - B(I)))))
        IF U(L, I) = 0 THEN P(I) = 1 - P(I)
        LUX(L, Q) = LUX(L, Q) * P(I)
      NEXT I
    NEXT Q
    HU(L) = 0
    FOR Q = 1 TO NQ
      HU(L) = HU(L) + LUX(L, Q) * AX(Q)
    NEXT Q
  NEXT L

  FOR L = 1 TO NL
    FOR Q = 1 TO NQ
      PUX(L, Q) = LUX(L, Q) * AX(Q) / HU(L)
    NEXT Q
  NEXT L

  MU = 0
  VAR = 0
```

```
            FOR Q = 1 TO NQ
              SUMP(Q) = 0
              FOR L = 1 TO NL
                SUMP(Q) = SUMP(Q) + PUX(L, Q) * R(L)
              NEXT L
              MU = MU + X(Q) * SUMP(Q)
            NEXT Q
            MU = MU / SUMR

            FOR Q = 1 TO NQ
              VAR = VAR + (X(Q) - MU) ^ 2 * SUMP(Q)
            NEXT Q
            VAR = VAR / SUMR

            SIGMA = SQR(VAR)

            PRINT USING "####.#####"; MU; VAR; SIGMA

            MUO = MU
            SIGMAO = SIGMA

        NEXT C

REM    SHEPPARD'S CORRECTION FOR FINAL ESTIMATE
       SHEVAR = VAR - D ^ 2 / 12
       SHESIGMA = SQR(VAR - D ^ 2 / 12)
       PRINT
       PRINT "        MU      VAR      SIGMA    SHEPVAR SHEPSIGMA"
       PRINT USING "####.#####"; MU; VAR; SIGMA; SHEVAR; SHESIGMA

REM    INFORMATION MATRIX
       SUMMU2 = 0
       SUMMUVAR = 0
       SUMVAR2 = 0

       FOR L = 1 TO NL
         TEMPMU2 = 0
         TEMPVAR2 = 0
         FOR Q = 1 TO NQ
           TEMPMU2 = TEMPMU2 + PUX(L, Q) * ((X(Q) - MU) / VAR)
           TEMPVAR2 = TEMPVAR2 + PUX(L, Q) * _
             (((X(Q) - MU) ^ 2 - VAR) / (2 * VAR ^ 2))
         NEXT Q
         SUMMU2 = SUMMU2 + R(L) * TEMPMU2 ^ 2
         SUMMUVAR = SUMMUVAR + R(L) * TEMPMU2 * TEMPVAR2
         SUMVAR2 = SUMVAR2 + R(L) * TEMPVAR2 ^ 2
       NEXT L

       PRINT
       PRINT "    I(MU) I(MU,VAR)    I(VAR)"
       PRINT USING "####.#####"; SUMMU2; SUMMUVAR; SUMVAR2
```

```
DET = SUMMU2 * SUMVAR2 - SUMMUVAR ^ 2
VARMU = SUMVAR2 / DET
VARVAR = SUMMU2 / DET
COVMUVAR = -SUMMUVAR / DET
SEMU = SQR(VARMU)
SEVAR = SQR(VARVAR)

PRINT
PRINT "   SE(MU)   SE(VAR)     COVAR"
PRINT USING "####.#####"; SEMU; SEVAR; COVMUVAR

END
```

I.4.2 Estimation of Both Item and Latent Distribution Parameters: The Compact Model

```
TRALL = -1
INPUT "TRACE ALL? Y/N "; YN$
IF (YN$ = "Y") OR (YN$ = "y") THEN TRALL = 0

NG = 2
NL = 16
NI = 4
NQ = 10

DIM U(NG, NL, NI), Y(NG, NL, NI), X(NG, NQ), AX(NG, NQ)
DIM SAX(NG), MU(NG), S2(NG), R(NG, NL)
DIM LX(NG, NL, NQ), PB(NG, NL), ZETA(NI), LAMBDA(NI)
DIM RB(NG, NI, NQ), NB(NG, NI, NQ)
DIM XB(NG, NL), SN(NG), MUH(NG)
DIM GOF(NG)

REM   Y=INDICATOR

MNC = 25
MIT = 10

ZETA(1) = 0: ZETA(2) = 0: ZETA(3) = 0: ZETA(4) = 0
LAMBDA(1) = 1: LAMBDA(2) = 1: LAMBDA(3) = 1: LAMBDA(4) = 1

FOR G = 1 TO NG
  FOR L = 1 TO NL
    FOR I = 1 TO NI
      READ U(G, L, I)
    NEXT I
  NEXT L
NEXT G
DATA 0,0,0,0
DATA 0,0,0,1
DATA 1,0,0,0
DATA 1,0,0,1
DATA 0,1,0,0
DATA 0,1,0,1
```

```
DATA 0,0,1,0
DATA 0,0,1,1
DATA 1,1,0,0
DATA 1,1,0,1
DATA 1,0,1,0
DATA 1,0,1,1
DATA 0,1,1,0
DATA 0,1,1,1
DATA 1,1,1,0
DATA 1,1,1,1
DATA 0,0,0,0
DATA 0,0,0,1
DATA 1,0,0,0
DATA 1,0,0,1
DATA 0,1,0,0
DATA 0,1,0,1
DATA 0,0,1,0
DATA 0,0,1,1
DATA 1,1,0,0
DATA 1,1,0,1
DATA 1,0,1,0
DATA 1,0,1,1
DATA 0,1,1,0
DATA 0,1,1,1
DATA 1,1,1,0
DATA 1,1,1,1

FOR G = 1 TO NG
  FOR L = 1 TO NL
    FOR I = 1 TO NI
      READ Y(G, L, I)
    NEXT I
  NEXT L
NEXT G
DATA 1,1,1,1
DATA 1,1,1,1
DATA 1,1,1,1
DATA 1,1,1,1
DATA 1,1,1,1
DATA 1,1,1,1
DATA 1,1,1,1
DATA 1,1,1,1
DATA 1,1,1,1
DATA 1,1,1,1
DATA 1,1,1,1
DATA 1,1,1,1
DATA 1,1,1,1
DATA 1,1,1,1
DATA 1,1,1,1
DATA 1,1,1,1
DATA 1,1,1,1
DATA 1,1,1,1
DATA 1,1,1,1
```

```
     DATA 1,1,1,1
     DATA 1,1,1,1
     DATA 1,1,1,1
     DATA 1,1,1,1
     DATA 1,1,1,1
     DATA 1,1,1,1
     DATA 1,1,1,1
     DATA 1,1,1,1
     DATA 1,1,1,1
     DATA 1,1,1,1
     DATA 1,1,1,1
     DATA 1,1,1,1
     DATA 1,1,1,1

     FOR G = 1 TO NG
       FOR Q = 1 TO NQ
         READ X(G, Q)
       NEXT Q
     NEXT G
     DATA -4.5,-3.5,-2.5,-1.5,-0.5,0.5,1.5,2.5,3.5,4.5
     DATA -4.5,-3.5,-2.5,-1.5,-0.5,0.5,1.5,2.5,3.5,4.5

     MU(1) = 0
     S2(1) = 1
     MU(2) = 0
     S2(2) = 1

     FOR G = 1 TO NG
       FOR L = 1 TO NL
         READ R(G, L)
       NEXT L
     NEXT G
     DATA 29,7,50,30,15,4,6,0,67,63,12,10,2,6,22,51
     DATA 22,10,30,27,13,14,1,1,24,54,5,8,1,8,10,57

     NC = 1

100  PRINT: PRINT "CYCLE= "; NC

     SAX(1) = 0
     SAX(2) = 0

     FOR G = 1 TO NG
       FOR Q = 1 TO NQ
         PI = 3.141592654#
         AX(G, Q) = (1 / (SQR(2 * PI))) * EXP(-(1 / 2) _
           *(X(G, Q) - MU(G))^2 / S2(G))
         SAX(G) = SAX(G) + AX(G, Q)
       NEXT Q
     NEXT G

     FOR G = 1 TO NG
       FOR Q = 1 TO NQ
```

```
          AX(G, Q) = AX(G, Q) / SAX(G)
      NEXT Q
    NEXT G

    IF TRALL = 0 THEN
      PRINT
      PRINT "          G         Q     X(G,Q)    AX(G,Q)"
      FOR G = 1 TO NG
        FOR Q = 1 TO NQ
          PRINT USING "####.#####"; G; Q; X(G, Q); AX(G, Q)
        NEXT Q
      NEXT G
    END IF

    FOR G = 1 TO NG
      FOR L = 1 TO NL
        PB(G, L) = 0
        FOR Q = 1 TO NQ
          LX(G, L, Q) = 1
        NEXT Q
      NEXT L
    NEXT G

REM   E STAGE

    IF TRALL = 0 THEN
      PRINT
      PRINT "          G         I         K       P(I)"
    END IF
    FOR G = 1 TO NG
      FOR L = 1 TO NL
        FOR I = 1 TO NI
          FOR Q = 1 TO NQ
            IF Y(G, L, I) = 0 GOTO 230
            PQ = 1 / (1 + EXP(-(ZETA(I) + LAMBDA(I) * _
              X(G, Q))))
            IF (L = 1) AND (TRALL = 0) THEN
              PRINT USING "####.#####"; G; I; Q; PQ
            END IF
            IF U(G, L, I) = 0 THEN PQ = 1 - PQ
            LX(G, L, Q) = LX(G, L, Q) * PQ
230         NEXT Q
        NEXT I
        PB(G, L) = 0
        FOR Q = 1 TO NQ
          PB(G, L) = PB(G, L) + LX(G, L, Q) * AX(G, Q)
        NEXT Q
      NEXT L
    NEXT G

    IF TRALL = 0 THEN
      PRINT
      PRINT "          G         L    PB(G,L)"
```

```
            FOR G = 1 TO NG
              FOR L = 1 TO NL
                PRINT USING "####.#####"; G; L; PB(G, L)
              NEXT L
            NEXT G
        END IF

        IF TRALL = 0 THEN
          PRINT
          PRINT "          G          L          Q LX(G,L,Q)"
          FOR G = 1 TO NG
            FOR L = 1 TO NL
              FOR Q = 1 TO NQ
                PRINT USING "####.#####"; G; L; Q; LX(G, L, Q)
              NEXT Q
            NEXT L
          NEXT G
        END IF

REM   N BAR AND R BAR

        FOR G = 1 TO NG
          FOR I = 1 TO NI
            FOR Q = 1 TO NQ
              RB(G, I, Q) = 0
              NB(G, I, Q) = 0
              FOR L = 1 TO NL
                IF Y(G, L, I) = 0 GOTO 320
                RB(G, I, Q) = RB(G, I, Q) + U(G, L, I) * _
                R(G, L) * LX(G, L, Q) * AX(G, Q) / PB(G, L)
                NB(G, I, Q) = NB(G, I, Q) + R(G, L) * _
                LX(G, L, Q) * AX(G, Q) / PB(G, L)
320             NEXT L
            NEXT Q
          NEXT I
        NEXT G

        IF TRALL = 0 THEN
          FOR G = 1 TO NG
            FOR I = 1 TO NI
              PRINT
              PRINT _
              "         G          I          Q RB(G,I,Q) NB(G,I,Q)"
              FOR Q = 1 TO NQ
                PRINT USING "####.#####"; G; I; Q; RB(G, I, Q); _
                NB(G, I, Q)
              NEXT Q
            NEXT I
          NEXT G
        END IF

REM    PROBIT STAGE M-STEP
```

```
FOR I = 1 TO NI

IF TRALL = 0 THEN
  PRINT
  PRINT "           I        IT   ZETA(I) LAMBDA(I)"
END IF

FOR IT = 1 TO MIT
  SSZ = 0
  SSL = 0
  SSZZ = 0
  SSZL = 0
  SSLL = 0
  FOR G = 1 TO NG
    FOR Q = 1 TO NQ
      IF NB(G, I, Q) <> 0 THEN
        PX = 1 / (1 + EXP(-(ZETA(I) + LAMBDA(I) * _
          X(G, Q))))
        W = PX * (1 - PX)
        SSZ = SSZ + RB(G, I, Q) - NB(G, I, Q) * PX
        SSL = SSL + (RB(G, I, Q) - NB(G, I, Q) * PX) * _
          X(G, Q)
        SSZZ = SSZZ - NB(G, I, Q) * W
        SSZL = SSZL - NB(G, I, Q) * W * X(G, Q)
        SSLL = SSLL - NB(G, I, Q) * W * X(G, Q) ^ 2
      END IF
    NEXT Q
  NEXT G
  DET = SSZZ * SSLL - SSZL ^ 2
  DZETA = (SSLL * SSZ - SSZL * SSL) / DET
  DLAMBDA = (-SSZL * SSZ + SSZZ * SSL) / DET
  ZETA(I) = ZETA(I) - DZETA
  LAMBDA(I) = LAMBDA(I) - DLAMBDA
  IF TRALL = 0 THEN
    PRINT USING "####.#####"; I; IT; ZETA(I); LAMBDA(I)
  END IF
  IF (ABS(DZETA) <= .01) AND (ABS(DLAMBDA) <= .01) GOTO 400
  NEXT IT
400   NEXT I

PRINT
PRINT "           I   ZETA(I) LAMBDA(I)       B(I)"
FOR I = 1 TO NI
  PRINT USING "####.#####"; I; ZETA(I); LAMBDA(I); _
    -ZETA(I) / LAMBDA(I)
NEXT I

FOR G = 1 TO NG
  SN(G) = 0
  FOR L = 1 TO NL
    SN(G) = SN(G) + R(G, L)
  NEXT L
NEXT G
```

```
REM   ESTIMATE MUH(1) ONLY

      MUH(1) = 0
      FOR L = 1 TO NL
        XB(1, L) = 0
        FOR Q = 1 TO NQ
          XB(1, L) = XB(1, L) + X(1, Q) * LX(1, L, Q) * _
          AX(1, Q)
        NEXT Q
        XB(1, L) = XB(1, L) / PB(1, L)
        MUH(1) = MUH(1) + R(1, L) * XB(1, L)
      NEXT L

      MUH(1) = (1 / SN(1)) * MUH(1)
      PRINT
      PRINT "    MUH(1)"
      PRINT USING "####.#####"; MUH(1)
      PRINT
      PRINT "       N(1)"
      PRINT USING "####.#####"; SN(1)
      MU(1) = MUH(1)

REM   ABOVE=DO NOT UPDATE

      NC = NC + 1
      IF NC <= MNC GOTO 100

REM   LIKELIHOOD-RATIO GOODNESS-OF-FIT STATISTIC

      FOR G = 1 TO NG
        FOR L = 1 TO NL
          FOR Q = 1 TO NQ
            LX(G, L, Q) = 1
          NEXT Q
        NEXT L
      NEXT G

      FOR G = 1 TO NG
        FOR L = 1 TO NL
          FOR I = 1 TO NI
            FOR Q = 1 TO NQ
              IF Y(G, L, I) = 0 GOTO 540
              PQ = 1 / (1 + EXP(-(ZETA(I) + LAMBDA(I) * _
              X(G, Q))))
              IF U(G, L, I) = 0 THEN PQ = 1 - PQ
              LX(G, L, Q) = LX(G, L, Q) * PQ
540         NEXT Q
          NEXT I
          PB(G, L) = 0
          FOR Q = 1 TO NQ
            PB(G, L) = PB(G, L) + LX(G, L, Q) * AX(G, Q)
          NEXT Q
```

```
      NEXT L
    NEXT G

    FOR G = 1 TO NG
      PRINT
      PRINT _
"       G       L    R(G,L)      SN(G)    PB(G,L)      SN*PB"
      FOR L = 1 TO NL
        PRINT USING "####.#####"; G; L; R(G, L); SN(G); _
          PB(G, L); SN(G) * PB(G, L)
      NEXT L
    NEXT G

    GOF(1) = 0
    GOF(2) = 0
    FOR G = 1 TO NG
      FOR L = 1 TO NL
        IF R(G, L) = 0 GOTO 560
        GOF(G) = GOF(G) + 2 * R(G, L) * LOG(R(G, L) / _
          (SN(G) * PB(G, L)))
560     NEXT L
    NEXT G

    PRINT
    PRINT "          G    GOF(G)"
    FOR G = 1 TO NG
      PRINT USING "####.#####"; G; GOF(G)
    NEXT G

    LRGOF = 0
    FOR G = 1 TO NG
      LRGOF = LRGOF + GOF(G)
    NEXT G

    PRINT
    PRINT "LIKELIHOOD-RATIO GOODNESS-OF-FIT=  "; _
      USING "####.#####"; LRGOF

    END
```

I.4.3 Estimation of Both Item and Latent Distribution Parameters: The Augmented Model

```
    TRALL = -1
    INPUT "TRACE ALL? Y/N "; YN$
    IF (YN$ = "Y") OR (YN$ = "y") THEN TRALL = 0

    NG = 2
    NL = 16
    NI = 5
    NQ = 10

    DIM U(NG, NL, NI), Y(NG, NL, NI), X(NG, NQ), AX(NG, NQ)
```

```
        DIM SAX(NG), MU(NG), S2(NG), R(NG, NL)
        DIM LX(NG, NL, NQ), PB(NG, NL), ZETA(NI), LAMBDA(NI)
        DIM RB(NG, NI, NQ), NB(NG, NI, NQ)
        DIM XB(NG, NL), SN(NG), MUH(NG)
        DIM GOF(NG)

REM     Y=INDICATOR

        MNC = 25
        MIT = 10

        ZETA(1) = 0: ZETA(2) = 0: ZETA(3) = 0: ZETA(4) = 0: _
          ZETA(5) = 0
        LAMBDA(1) = 1: LAMBDA(2) = 1: LAMBDA(3) = 1: _
          LAMBDA(4) = 1: LAMBDA(5) = 1

REM     ZETA(5) = 0 AND LAMBDA(5) = 1

        FOR G = 1 TO NG
          FOR L = 1 TO NL
            FOR I = 1 TO NI
              READ U(G, L, I)
            NEXT I
          NEXT L
        NEXT G
        DATA 0,0,0,0,0
        DATA 0,0,0,0,1
        DATA 1,0,0,0,0
        DATA 1,0,0,0,1
        DATA 0,1,0,0,0
        DATA 0,1,0,0,1
        DATA 0,0,1,0,0
        DATA 0,0,1,0,1
        DATA 1,1,0,0,0
        DATA 1,1,0,0,1
        DATA 1,0,1,0,0
        DATA 1,0,1,0,1
        DATA 0,1,1,0,0
        DATA 0,1,1,0,1
        DATA 1,1,1,0,0
        DATA 1,1,1,0,1
        DATA 0,0,0,0,0
        DATA 0,0,0,1,0
        DATA 1,0,0,0,0
        DATA 1,0,0,1,0
        DATA 0,1,0,0,0
        DATA 0,1,0,1,0
        DATA 0,0,1,0,0
        DATA 0,0,1,1,0
        DATA 1,1,0,0,0
        DATA 1,1,0,1,0
        DATA 1,0,1,0,0
        DATA 1,0,1,1,0
```

```
        DATA 0,1,1,0,0
        DATA 0,1,1,1,0
        DATA 1,1,1,0,0
        DATA 1,1,1,1,0

REM     FIVE ITEM DATA

        FOR G = 1 TO NG
          FOR L = 1 TO NL
            FOR I = 1 TO NI
              READ Y(G, L, I)
            NEXT I
          NEXT L
        NEXT G
        DATA 1,1,1,0,1
        DATA 1,1,1,0,1
        DATA 1,1,1,0,1
        DATA 1,1,1,0,1
        DATA 1,1,1,0,1
        DATA 1,1,1,0,1
        DATA 1,1,1,0,1
        DATA 1,1,1,0,1
        DATA 1,1,1,0,1
        DATA 1,1,1,0,1
        DATA 1,1,1,0,1
        DATA 1,1,1,0,1
        DATA 1,1,1,0,1
        DATA 1,1,1,0,1
        DATA 1,1,1,0,1
        DATA 1,1,1,1,0
        DATA 1,1,1,1,0
        DATA 1,1,1,1,0
        DATA 1,1,1,1,0
        DATA 1,1,1,1,0
        DATA 1,1,1,1,0
        DATA 1,1,1,1,0
        DATA 1,1,1,1,0
        DATA 1,1,1,1,0
        DATA 1,1,1,1,0
        DATA 1,1,1,1,0
        DATA 1,1,1,1,0
        DATA 1,1,1,1,0
        DATA 1,1,1,1,0
        DATA 1,1,1,1,0

REM     FIVE ITEM DATA

        FOR G = 1 TO NG
          FOR Q = 1 TO NQ
            READ X(G, Q)
          NEXT Q
```

```
    NEXT G
    DATA -4.5,-3.5,-2.5,-1.5,-0.5,0.5,1.5,2.5,3.5,4.5
    DATA -4.5,-3.5,-2.5,-1.5,-0.5,0.5,1.5,2.5,3.5,4.5

    MU(1) = 0
    S2(1) = 1
    MU(2) = 0
    S2(2) = 1

    FOR G = 1 TO NG
      FOR L = 1 TO NL
        READ R(G, L)
      NEXT L
    NEXT G
    DATA 29,7,50,30,15,4,6,0,67,63,12,10,2,6,22,51
    DATA 22,10,30,27,13,14,1,1,24,54,5,8,1,8,10,57

    NC = 1

100 PRINT: PRINT "CYCLE= "; NC

    SAX(1) = 0
    SAX(2) = 0

    FOR G = 1 TO NG
      FOR Q = 1 TO NQ
        PI = 3.141592654#
        AX(G,Q) = (1 / (SQR(2 * PI))) * EXP(-(1 / 2) * _
          (X(G, Q) - MU(G))^2 / S2(G))
        SAX(G) = SAX(G) + AX(G, Q)
      NEXT Q
    NEXT G

    FOR G = 1 TO NG
      FOR Q = 1 TO NQ
        AX(G, Q) = AX(G, Q) / SAX(G)
      NEXT Q
    NEXT G

    IF TRALL = 0 THEN
      PRINT
      PRINT "          G          Q     X(G,Q)    AX(G,Q)"
      FOR G = 1 TO NG
        FOR Q = 1 TO NQ
          PRINT USING "####.#####"; G; Q; X(G, Q); AX(G, Q)
        NEXT Q
      NEXT G
    END IF

    FOR G = 1 TO NG
      FOR L = 1 TO NL
        PB(G, L) = 0
        FOR Q = 1 TO NQ
```

```
        LX(G, L, Q) = 1
      NEXT Q
    NEXT L
  NEXT G

REM   E STAGE

    IF TRALL = 0 THEN
      PRINT
      PRINT "           G           I           K        P(I)"
    END IF
    FOR G = 1 TO NG
      FOR L = 1 TO NL
        FOR I = 1 TO NI
          FOR Q = 1 TO NQ
            IF Y(G, L, I) = 0 GOTO 230
            PQ = 1 / (1 + EXP(-(ZETA(I) + LAMBDA(I) * _
            X(G, Q))))
            IF (L = 1) AND (TRALL = 0) THEN
              PRINT USING "####.#####"; G; I; Q; PQ
            END IF
            IF U(G, L, I) = 0 THEN PQ = 1 - PQ
            LX(G, L, Q) = LX(G, L, Q) * PQ
230         NEXT Q
        NEXT I
        PB(G, L) = 0
        FOR Q = 1 TO NQ
          PB(G, L) = PB(G, L) + LX(G, L, Q) * AX(G, Q)
        NEXT Q
      NEXT L
    NEXT G

    IF TRALL = 0 THEN
      PRINT
      PRINT "           G           L     PB(G,L)"
      FOR G = 1 TO NG
        FOR L = 1 TO NL
          PRINT USING "####.#####"; G; L; PB(G, L)
        NEXT L
      NEXT G
    END IF

    IF TRALL = 0 THEN
      PRINT
      PRINT "           G           L           Q LX(G,L,Q)"
      FOR G = 1 TO NG
        FOR L = 1 TO NL
          FOR Q = 1 TO NQ
            PRINT USING "####.#####"; G; L; Q; LX(G, L, Q)
          NEXT Q
        NEXT L
      NEXT G
    END IF
```

```
REM   N BAR AND R BAR

      FOR G = 1 TO NG
        FOR I = 1 TO NI
          FOR Q = 1 TO NQ
            RB(G, I, Q) = 0
            NB(G, I, Q) = 0
            FOR L = 1 TO NL
              IF Y(G, L, I) = 0 GOTO 320
              RB(G, I, Q) = RB(G, I, Q) + U(G, L, I) * R(G, L) _
                * LX(G, L, Q) * AX(G, Q) / PB(G, L)
              NB(G, I, Q) = NB(G, I, Q) + R(G, L) * _
                LX(G, L, Q) * AX(G, Q) / PB(G, L)
320         NEXT L
          NEXT Q
        NEXT I
      NEXT G

      IF TRALL = 0 THEN
        FOR G = 1 TO NG
          FOR I = 1 TO NI
            PRINT
            PRINT _
            "          G          I          Q RB(G,I,Q) NB(G,I,Q)"
            FOR Q = 1 TO NQ
              PRINT USING "####.#####"; G; I; Q; _
                RB(G, I, Q); NB(G, I, Q)
            NEXT Q
          NEXT I
        NEXT G
      END IF

REM   PROBIT STAGE M-STEP

      FOR I = 1 TO NI

      IF TRALL = 0 THEN
        PRINT
        PRINT "          I          IT   ZETA(I) LAMBDA(I)"
      END IF

      FOR IT = 1 TO MIT
        SSZ = 0
        SSL = 0
        SSZZ = 0
        SSZL = 0
        SSLL = 0
        FOR G = 1 TO NG
          FOR Q = 1 TO NQ
            IF NB(G, I, Q) <> 0 THEN
              PX = 1 / (1 + EXP(-(ZETA(I) + LAMBDA(I) * _
                X(G, Q))))
```

```
              W = PX * (1 - PX)
              SSZ = SSZ + RB(G, I, Q) - NB(G, I, Q) * PX
              SSL = SSL + (RB(G, I, Q) - NB(G, I, Q) * PX) _
                * X(G, Q)
              SSZZ = SSZZ - NB(G, I, Q) * W
              SSZL = SSZL - NB(G, I, Q) * W * X(G, Q)
              SSLL = SSLL - NB(G, I, Q) * W * X(G, Q) ^ 2
        END IF
      NEXT Q
    NEXT G
    DET = SSZZ * SSLL - SSZL ^ 2
    DZETA = (SSLL * SSZ - SSZL * SSL) / DET
    DLAMBDA = (-SSZL * SSZ + SSZZ * SSL) / DET
    ZETA(I) = ZETA(I) - DZETA
    LAMBDA(I) = LAMBDA(I) - DLAMBDA
    IF TRALL = 0 THEN
      PRINT USING "####.#####"; I; IT; ZETA(I); LAMBDA(I)
    END IF
    IF (ABS(DZETA) <= .01) AND (ABS(DLAMBDA) <= .01) GOTO 400
      NEXT IT
400  NEXT I

    PRINT
    PRINT "        I   ZETA(I) LAMBDA(I)      B(I)"
    FOR I = 1 TO NI
      PRINT USING "####.#####"; I; ZETA(I); LAMBDA(I); _
        -ZETA(I) / LAMBDA(I)
    NEXT I

    FOR G = 1 TO NG
      SN(G) = 0
      FOR L = 1 TO NL
        SN(G) = SN(G) + R(G, L)
      NEXT L
    NEXT G

REM  ESTIMATE MUH(1) ONLY

    MUH(1) = 0
    FOR L = 1 TO NL
      XB(1, L) = 0
      FOR Q = 1 TO NQ
        XB(1, L) = XB(1, L) + X(1, Q) * LX(1, L, Q) * _
          AX(1, Q)
      NEXT Q
      XB(1, L) = XB(1, L) / PB(1, L)
      MUH(1) = MUH(1) + R(1, L) * XB(1, L)
    NEXT L

    MUH(1) = (1 / SN(1)) * MUH(1)
    PRINT
    PRINT "    MUH(1)"
    PRINT USING "####.#####"; MUH(1)
```

```
       PRINT
       PRINT "        N(1)"
       PRINT USING "####.#####"; SN(1)
       MU(1) = MUH(1)

REM    ABOVE=DO NOT UPDATE

       NC = NC + 1
       IF NC <= MNC GOTO 100

REM    LIKELIHOOD-RATIO GOODNESS-OF-FIT STATISTIC

       FOR G = 1 TO NG
         FOR L = 1 TO NL
           FOR Q = 1 TO NQ
             LX(G, L, Q) = 1
           NEXT Q
         NEXT L
       NEXT G

       FOR G = 1 TO NG
         FOR L = 1 TO NL
           FOR I = 1 TO NI
             FOR Q = 1 TO NQ
               IF Y(G, L, I) = 0 GOTO 540
               PQ = 1 / (1 + EXP(-(ZETA(I) + LAMBDA(I) * _
                 X(G, Q))))
               IF U(G, L, I) = 0 THEN PQ = 1 - PQ
               LX(G, L, Q) = LX(G, L, Q) * PQ
540            NEXT Q
           NEXT I
           PB(G, L) = 0
           FOR Q = 1 TO NQ
             PB(G, L) = PB(G, L) + LX(G, L, Q) * AX(G, Q)
           NEXT Q
         NEXT L
       NEXT G

       FOR G = 1 TO NG
         PRINT
         PRINT _
   "       G          L    R(G,L)    SN(G)    PB(G,L)      SN*PB"
         FOR L = 1 TO NL
           PRINT USING "####.#####"; G; L; R(G,L); SN(G); _
             PB(G,L); SN(G) * PB(G,L)
         NEXT L
       NEXT G

       GOF(1) = 0
       GOF(2) = 0
       FOR G = 1 TO NG
         FOR L = 1 TO NL
           IF R(G, L) = 0 GOTO 560
```

```
            GOF(G) = GOF(G) + 2 * R(G, L) * LOG(R(G, L) / _
              (SN(G) * PB(G, L)))
560     NEXT L
        NEXT G

        PRINT
        PRINT "          G    GOF(G)"
        FOR G = 1 TO NG
          PRINT USING "####.#####"; G; GOF(G)
        NEXT G

        LRGOF = 0
        FOR G = 1 TO NG
          LRGOF = LRGOF + GOF(G)
        NEXT G

        PRINT
        PRINT "LIKELIHOOD-RATIO GOODNESS-OF-FIT=  "; _
          USING "####.#####"; LRGOF

        END
```

J. Implementation of Estimation for Mixed Models

J.1 Introduction

In order to illustrate estimation of item parameters in the mixed models, a BASIC computer program was developed. The computer program, based on marginal maximum likelihood estimation of Bock and Aitkin (1981), has been written specifically to handle contrived data consist of three dichotomously-scored items and two polytomously-scored items. The two-parameter logistic model and the logistic form of the graded response model are used for parameter estimation.

It can be noted that a BASIC computer program can be developed for the graded response model under marginal maximum likelihood estimation because such a program can be seen as a simple variation of the program in the current appendix. (Appendix F contains implementation of the graded response model within the joint maximum likelihood estimation paradigm.)

J.2 Implementation

The equations for parameter estimation in the mixed model presented in Chapter 11 are implemented in the BASIC computer program. The first two lines of the program declare two subroutines, LUDCMP and LUBKSB (Sprott, 1991), for the matrix inversion. The two subroutines are used to have flexibility that readers can modify the BASIC computer program to handle items with different response categories. In the current program, the two-parameter logistic model has, of course, two response categories, and the graded response model has three response categories.

The variable TRALL, when set to zero, causes nearly all intermediate terms to be printed on the computer screen. When it is set to −1, only results from the maximization steps will be printed out. Note that the use of the TRALL option will produce a voluminous output.

The total number of items is five, NI = 5. There are three dichotomous items with two categories (NT = 2) and two polytomous items with three categories (NS = 3). The maximum number of response categories across all items is three, NK = 3. The total possible response patterns is $2^3 \times 3^2 = 72$, but only 58 patterns were observed, NL = 58. The total number of quadrature points is ten, NQ = 10.

Next, the program shows the dimension statements for the variables. We may explain some of these variables where theses are actually used. The maximum number of the expectation and maximization cycles is defined by MNC that shows a value of 25. The maximum possible number of the Newton-Raphson iterations for each item is set to 10, MIT = 10. The number of response categories for each item is defined by the NK array. Again, item 1 to 3 have two response categories and items 4 and 5 have three response categories.

The response data for the five items denoted as UU are read in using categories 1 and 2 for the three dichotomous items and using categories 1, 2, and 3 for the two polytomous items. Although it is certainly usual to use 0 as incorrect and 1 as correct for the dichotomous items, these are recoded to 1 as incorrect and 2 as correct to make the response categories more consistent with those of the polytomous scoring.

The two-dimensional Y array is used to take care of missing responses. But it is not actually used in the current analysis because there are no missing values. Assigning 1 indicates that the value of the UU is missing and assigning 0 indicates that the value is valid.

The number of cases for each response pattern is provided by R(L). The initial values of the parameters for the calculations of the terms in the expectation and maximization steps are listed next. It may be noted that the first three item are dichotomously scored and only one intercept parameter, ζ_{i1}, is required.

The U array is for converting the original response categories to dichotomous indicator variables. If $U_{li} = k$, then $u_{lik} = 1$ [i.e., U(L, I, K) = 1]; otherwise $u_{lik} = 0$ [i.e., U(L, I, K) = 0]. When the TRALL option is invoked, the program will print out all of the U values to the computer screen.

There are 10 quadrature points, X, ranged from -4.5 to 4.5 with an interval of 1.0. The corresponding 10 quadrature weights, AX, calculated assuming that the ability distribution is the standard normal distribution. When the TRALL option is used, the values of the quadrature points and the quadrature weights are printed.

Q	X(Q)	AX(Q)
1.00000	-4.50000	0.00002
2.00000	-3.50000	0.00087
3.00000	-2.50000	0.01753
4.00000	-1.50000	0.12952
5.00000	-0.50000	0.35207
6.00000	0.50000	0.35207
7.00000	1.50000	0.12952
8.00000	2.50000	0.01753
9.00000	3.50000	0.00087
10.00000	4.50000	0.00002

With NC = 1, the expectation and maximization steps begin. Using the TRALL option, values of many intermediate terms can be printed out. Especially, sets of terms P_l and $L_l(X_q)$ can be produced; the corresponding BASIC variable names are PB(L) and LX(L, Q).

In the end of the expectation step, the values of $\bar{r}_{ikq}$ and $\bar{N}_q$ are obtained; the variable names are RB(I, K, Q) and NB(I, K, Q). For example, these values for the two categories of the first item are shown below:

I	K	Q	RB(I,K,Q)	NB(I,K,Q)
1.00000	1.00000	1.00000	0.00029	0.00030
1.00000	1.00000	2.00000	0.01932	0.02114
1.00000	1.00000	3.00000	0.63241	0.78901
1.00000	1.00000	4.00000	9.39651	15.73378
1.00000	1.00000	5.00000	41.39193	130.44838
1.00000	1.00000	6.00000	40.24841	343.10156
1.00000	1.00000	7.00000	8.78273	231.07692
1.00000	1.00000	8.00000	0.51836	41.47307
1.00000	1.00000	9.00000	0.00998	2.31167
1.00000	1.00000	10.00000	0.00007	0.04419

I	K	Q	RB(I,K,Q)	NB(I,K,Q)
1.00000	2.00000	1.00000	0.00001	0.00030
1.00000	2.00000	2.00000	0.00181	0.02114
1.00000	2.00000	3.00000	0.15660	0.78901
1.00000	2.00000	4.00000	6.33727	15.73378
1.00000	2.00000	5.00000	89.05643	130.44838
1.00000	2.00000	6.00000	302.85318	343.10156
1.00000	2.00000	7.00000	222.29419	231.07692
1.00000	2.00000	8.00000	40.95471	41.47307
1.00000	2.00000	9.00000	2.30169	2.31167
1.00000	2.00000	10.00000	0.04412	0.04419

After obtaining the artificial data, $\bar{r}_{ikq}$ and $\bar{N}_q$ (or $\bar{N}_{ikq}$, but subscripts i and k can be dropped without loss of generality), the maximization step begins. Item parameters are estimated for one item at a time. The BASIC code is very similar to the portions of programs presented in earlier chapters. The main difference is the use of two subroutines, LUBKSB and LUBDCMP, to make the program more flexible to accommodate items with a different number of response categories.

The values from the Newton-Raphson iterations of the first item are reported below:

I	IT	ZETA(I,1)	LAMBDA(I)
1.00000	1.00000	0.20000	0.84185
1.00000	2.00000	0.40000	0.86053
1.00000	3.00000	0.60000	0.93549
1.00000	4.00000	0.80000	1.02669
1.00000	5.00000	1.00000	1.11122
1.00000	6.00000	1.20000	1.17468
1.00000	7.00000	1.38504	1.21138
1.00000	8.00000	1.39624	1.22310
1.00000	9.00000	1.39632	1.22325

After finishing all the iterations for all items, the item parameter estimates from the first cycle of the expectation and maximization steps are reported:

```
          I ZETA(I,1) LAMBDA(I)    B(I,1)
    1.00000    1.39632    1.22325  -1.14148

          I ZETA(I,1) LAMBDA(I)    B(I,1)
    2.00000    1.77634    1.07540  -1.65179

          I ZETA(I,1) LAMBDA(I)    B(I,1)
    3.00000    0.58474    1.03039  -0.56750

          I ZETA(I,1) ZETA(I,2) LAMBDA(I)    B(I,1)     B(I,2)
    4.00000    1.95903   -0.28708    1.04421  -1.87609    0.27493

          I ZETA(I,1) ZETA(I,2) LAMBDA(I)    B(I,1)     B(I,2)
    5.00000    1.17098   -1.20443    1.09183  -1.07250    1.10314
```

Note that $a_i = \lambda_i$ and $b_{ik} = -\zeta_{ik}/\lambda_i$.

Using the item paramter estimates from the maximization step as the new initial values, the second cycle of the expectation and maximization steps can be initiated. The similar results from the cycles are to be reported until the iteration reaches its maximum value, MNC = 25. The Newton-Raphson iteration results of the 25th cycle are:

```
CYCLE=  25

          I        IT ZETA(I,1) LAMBDA(I)
    1.00000    1.00000    2.45782    1.36048

          I        IT ZETA(I,1) LAMBDA(I)
    2.00000    1.00000    2.50898    0.90329

          I        IT ZETA(I,1) LAMBDA(I)
    3.00000    1.00000    1.28163    0.81360

          I        IT ZETA(I,1) ZETA(I,2) LAMBDA(I)
    4.00000    1.00000    2.80570    0.47615    1.06450

          I        IT ZETA(I,1) ZETA(I,2) LAMBDA(I)
    5.00000    1.00000    2.25490   -0.50945    1.46552
```

The final item parameter estimates obtained after 25 cycles are:

```
          I ZETA(I,1) LAMBDA(I)    B(I,1)
    1.00000    2.45782    1.36048  -1.80658

          I ZETA(I,1) LAMBDA(I)    B(I,1)
    2.00000    2.50898    0.90329  -2.77761

          I ZETA(I,1) LAMBDA(I)    B(I,1)
    3.00000    1.28163    0.81360  -1.57525

          I ZETA(I,1) ZETA(I,2) LAMBDA(I)    B(I,1)     B(I,2)
```

```
4.00000    2.80570    0.47615    1.06450   -2.63571   -0.44730

        I ZETA(I,1) ZETA(I,2) LAMBDA(I)     B(I,1)     B(I,2)
5.00000    2.25490   -0.50945    1.46552   -1.53863    0.34762
```

J.3 BASIC Computer Program

```basic
DECLARE SUB LUDCMP (A!(), N!, NP!, INDX!(), D!)
DECLARE SUB LUBKSB (A!(), N!, NP!, INDX!(), B!())

TRALL = -1
INPUT "TRACE ALL? Y/N "; YN$
IF (YN$ = "Y") OR (YN$ = "y") THEN TRALL = 0

        NL = 58
        NI = 5
        NQ = 10
        MK = 3
        NT = 2
        NS = 3

        DIM UU(NL, NI), U(NL, NI, MK)
        DIM Y(NL, NI), X(NQ), AX(NQ), R(NL), ZETA(NI, MK - 1)
        DIM LAMBDA(NI), LX(NL, NQ), PB(NL), PA(MK), P(MK)
        DIM RB(NI, MK, NQ), NB(NI, MK, NQ), DZ(MK), DDZZ(MK, MK)
        DIM WA(MK, NQ), PX(MK, NQ), NK(NI)
        DIM A1(NS, NS), Y1(NS, NS), INDX1(NS), DZETA(MK)
        DIM A2(NT, NT), Y2(NT, NS), INDX2(NT)

        MNC = 25
        MIT = 10
        NK(1) = 2: NK(2) = 2: NK(3) = 2: NK(4) = 3: NK(5) = 3

        FOR L = 1 TO NL
          FOR I = 1 TO NI
            READ UU(L, I)
          NEXT I
        NEXT L
        DATA 1,1,1,1,1
        DATA 1,1,1,1,3
        DATA 1,1,1,2,1
        DATA 1,1,1,2,2
        DATA 1,1,1,3,1
        DATA 1,1,1,3,2
        DATA 1,1,2,1,2
        DATA 1,1,2,2,1
        DATA 1,1,2,2,2
        DATA 1,1,2,2,3
        DATA 1,1,2,3,2
        DATA 1,1,2,3,3
        DATA 1,2,1,1,1
        DATA 1,2,1,1,2
```

```
DATA 1,2,1,2,1
DATA 1,2,1,2,2
DATA 1,2,1,3,1
DATA 1,2,1,3,2
DATA 1,2,1,3,3
DATA 1,2,2,1,1
DATA 1,2,2,1,2
DATA 1,2,2,2,1
DATA 1,2,2,2,2
DATA 1,2,2,2,3
DATA 1,2,2,3,1
DATA 1,2,2,3,2
DATA 1,2,2,3,3
DATA 2,1,1,1,1
DATA 2,1,1,2,1
DATA 2,1,1,2,2
DATA 2,1,1,2,3
DATA 2,1,1,3,1
DATA 2,1,1,3,3
DATA 2,1,2,1,1
DATA 2,1,2,1,2
DATA 2,1,2,1,3
DATA 2,1,2,2,1
DATA 2,1,2,2,2
DATA 2,1,2,2,3
DATA 2,1,2,3,2
DATA 2,1,2,3,3
DATA 2,2,1,1,1
DATA 2,2,1,1,2
DATA 2,2,1,2,1
DATA 2,2,1,2,2
DATA 2,2,1,2,3
DATA 2,2,1,3,1
DATA 2,2,1,3,2
DATA 2,2,1,3,3
DATA 2,2,2,1,1
DATA 2,2,2,1,2
DATA 2,2,2,1,3
DATA 2,2,2,2,1
DATA 2,2,2,2,2
DATA 2,2,2,2,3
DATA 2,2,2,3,1
DATA 2,2,2,3,2
DATA 2,2,2,3,3

'    Y(NL,NI)=MISSING INDICATOR
     FOR L = 1 TO NL
       FOR I = 1 TO NI
         READ Y(L, I)
       NEXT I
     NEXT L
     DATA 1,1,1,1,1
     DATA 1,1,1,1,1
```

```
DATA 1,1,1,1,1
DATA 1,1,1,1,1
DATA 1,1,1,1,1
DATA 1,1,1,1,1
DATA 1,1,1,1,1
DATA 1,1,1,1,1
DATA 1,1,1,1,1
DATA 1,1,1,1,1
DATA 1,1,1,1,1
DATA 1,1,1,1,1
DATA 1,1,1,1,1
DATA 1,1,1,1,1
DATA 1,1,1,1,1
DATA 1,1,1,1,1
DATA 1,1,1,1,1
DATA 1,1,1,1,1
DATA 1,1,1,1,1
DATA 1,1,1,1,1
DATA 1,1,1,1,1
DATA 1,1,1,1,1
DATA 1,1,1,1,1
DATA 1,1,1,1,1
DATA 1,1,1,1,1
DATA 1,1,1,1,1
DATA 1,1,1,1,1
DATA 1,1,1,1,1
DATA 1,1,1,1,1
DATA 1,1,1,1,1
DATA 1,1,1,1,1
DATA 1,1,1,1,1
DATA 1,1,1,1,1
DATA 1,1,1,1,1
DATA 1,1,1,1,1
DATA 1,1,1,1,1
DATA 1,1,1,1,1
DATA 1,1,1,1,1
DATA 1,1,1,1,1
DATA 1,1,1,1,1
DATA 1,1,1,1,1
DATA 1,1,1,1,1
DATA 1,1,1,1,1
DATA 1,1,1,1,1
DATA 1,1,1,1,1
DATA 1,1,1,1,1
DATA 1,1,1,1,1
DATA 1,1,1,1,1
DATA 1,1,1,1,1
DATA 1,1,1,1,1
DATA 1,1,1,1,1
DATA 1,1,1,1,1
DATA 1,1,1,1,1
DATA 1,1,1,1,1
DATA 1,1,1,1,1
DATA 1,1,1,1,1
DATA 1,1,1,1,1
DATA 1,1,1,1,1
DATA 1,1,1,1,1
```

```
DATA 1,1,1,1,1
DATA 1,1,1,1,1
DATA 1,1,1,1,1

FOR L = 1 TO NL
  READ R(L)
NEXT L
DATA 1,1,2,4,1,1,3,1,2,2
DATA 1,3,2,3,8,4,2,6,4,5
DATA 1,10,7,2,7,10,8,1,1,3
DATA 1,4,4,4,4,2,3,11,2,8
DATA 6,3,13,11,26,13,12,31,25,6
DATA 8,8,17,60,54,21,123,179

INITIAL VALUES
FOR I = 1 TO NI: READ ZETA(I, 1): NEXT I
DATA 0,0,0,.693,.693
FOR I = 1 TO NI: READ ZETA(I, 2): NEXT I
DATA 0,0,0,-.693,-.693
FOR I = 1 TO NI: READ LAMBDA(I): NEXT I
DATA 1,1,1,1,1

FOR L = 1 TO NL
  FOR I = 1 TO NI
    FOR K = 1 TO NK(I)
      U(L, I, K) = 0
    NEXT K
  NEXT I
NEXT L
IF TRALL = 0 THEN
  PRINT "          L          I          K  U(L,I,K)"
END IF
FOR L = 1 TO NL
  FOR I = 1 TO NI
    FOR K = 1 TO NK(I)
      IF UU(L, I) = K THEN U(L, I, K) = 1
      IF TRALL = 0 THEN
        PRINT USING "####.#####"; L; I; K; U(L, I, K)
      END IF
    NEXT K
  NEXT I
NEXT L

FOR Q = 1 TO NQ
  READ X(Q)
NEXT Q
DATA -4.5,-3.5,-2.5,-1.5,-.5,.5,1.5,2.5,3.5,4.5
MU = 0
S2 = 1
SAX = 0
FOR Q = 1 TO NQ
PI = 3.141592654#
AX(Q) = (1 / (SQR(2 * PI))) * EXP(-(1 / 2) * _
```

'

```
          (X(Q) - MU) ^ 2 / S2)
      SAX = SAX + AX(Q)
      NEXT Q
      FOR Q = 1 TO NQ
        AX(Q) = AX(Q) / SAX
      NEXT Q
      IF TRALL = 0 THEN
        PRINT
        PRINT "          Q      X(Q)      AX(Q)"
        FOR Q = 1 TO NQ
          PRINT USING "####.#####"; Q; X(Q); AX(Q)
        NEXT Q
      END IF

      NC = 1
100   PRINT
      PRINT "CYCLE= "; NC

      FOR L = 1 TO NL
        PB(L) = 0
        FOR Q = 1 TO NQ
          LX(L, Q) = 1
        NEXT Q
      NEXT L

'     E STAGE
      IF TRALL = 0 THEN
        PRINT
        PRINT "          I        Q      P(I)       PQ"
      END IF
      FOR L = 1 TO NL
        FOR I = 1 TO NI
          FOR Q = 1 TO NQ
            IF Y(L, I) = 0 GOTO 230
'     Y NOT EMPLOYED
            PA(0) = 1
            PA(NK(I)) = 0
            FOR K = 1 TO (NK(I) - 1)
              PA(K) = 1 / (1 + EXP(-(ZETA(I, K) + LAMBDA(I) _
              * X(Q))))
            NEXT K
            PQ = 0
            FOR K = 1 TO NK(I)
              P(K) = PA(K - 1) - PA(K)
              PQ = PQ + U(L, I, K) * P(K)
              IF TRALL = 0 THEN
                IF L = 1 THEN
                  PRINT USING "####.######"; I; Q; P(K); PQ
                END IF
              END IF
            NEXT K
            LX(L, Q) = LX(L, Q) * PQ
230       NEXT Q
```

```
        NEXT I
        PB(L) = 0
        FOR Q = 1 TO NQ
          PB(L) = PB(L) + LX(L, Q) * AX(Q)
        NEXT Q
      NEXT L
      IF TRALL = 0 THEN
        PRINT
        PRINT "        L      PB(L)"
        FOR L = 1 TO NL
          PRINT USING "####.#####"; L; PB(L)
        NEXT L
        PRINT "        L        Q    LX(L,Q)"
        FOR L = 1 TO NL
          FOR Q = 1 TO NQ
            PRINT USING "####.#####"; L; Q; LX(L, Q)
          NEXT Q
        NEXT L
      END IF

'     N BAR AND R BAR
      FOR I = 1 TO NI
        FOR K = 1 TO NK(I)
          FOR Q = 1 TO NQ
            RB(I, K, Q) = 0
            NB(I, K, Q) = 0
            FOR L = 1 TO NL
              IF Y(L, I) = 0 GOTO 320
              RB(I, K, Q) = RB(I, K, Q) + U(L, I, K) * R(L) _
              * LX(L, Q) * AX(Q) / PB(L)
              NB(I, K, Q) = NB(I, K, Q) + R(L) * LX(L, Q) * _
              AX(Q) / PB(L)
320           NEXT L
          NEXT Q
        NEXT K
      NEXT I
      IF TRALL = 0 THEN
        FOR I = 1 TO NI
          FOR K = 1 TO NK(I)
            PRINT
            PRINT _
            "        I        K        Q RB(I,K,Q) NB(I,K,Q)"
            FOR Q = 1 TO NQ
              PRINT USING "####.#####"; I; K; Q; RB(I, K, Q); _
                NB(I, K, Q)
            NEXT Q
          NEXT K
        NEXT I
      END IF

'     PROBIT STAGE M-STEP
      FOR I = 1 TO NI
        PRINT
```

```
        IF NK(I) = 2 THEN
          PRINT "           I           IT ZETA(I,1) LAMBDA(I)"
        END IF
        IF NK(I) = 3 THEN
          PRINT _
          "           I           IT ZETA(I,1) ZETA(I,2) LAMBDA(I)"
        END IF
'       THIS MAY BE MODIFIED TO INCLUDE MORE ZETA'S
        FOR IT = 1 TO MIT
          FOR K = 1 TO NK(I)
            DZ(K) = 0
            FOR KK = 1 TO NK(I)
              DDZZ(K, KK) = 0
            NEXT KK
          NEXT K
          FOR K = 1 TO NK(I)
            FOR Q = 1 TO NQ
              PA(0) = 1
              PA(NK(I)) = 0
              WA(0, Q) = 0
              WA(NK(I), Q) = 0
              FOR KK = 1 TO (NK(I) - 1)
                PA(KK) = 1 / (1 + EXP(-(ZETA(I, KK) + _
                  LAMBDA(I) * X(Q))))
              NEXT KK
              PX(K, Q) = PA(K - 1) - PA(K)
              WA(K, Q) = PA(K) * (1 - PA(K))
            NEXT Q
          NEXT K
          FOR K = 1 TO (NK(I) - 1)
            FOR Q = 1 TO NQ
              DZ(K) = DZ(K) + (-RB(I, K, Q) / PX(K, Q) + _
                RB(I, K + 1, Q) / PX(K + 1, Q)) * WA(K, Q)
            NEXT Q
          NEXT K
          FOR Q = 1 TO NQ
            FOR KK = 1 TO NK(I)
              DZ(NK(I)) = DZ(NK(I)) + (RB(I, KK, Q) / _
                PX(KK, Q)) * (WA(KK - 1, Q) - WA(KK, Q)) * X(Q)
            NEXT KK
          NEXT Q
          FOR K = 1 TO (NK(I) - 1)
            FOR Q = 1 TO NQ
              DDZZ(K, K) = DDZZ(K, K) - NB(I, K, Q) * _
                WA(K, Q) ^ 2 * (1 / PX(K, Q) + 1 / PX(K + 1, Q))
            NEXT Q
          NEXT K
          FOR Q = 1 TO NQ
            FOR KK = 1 TO NK(I)
              DDZZ(NK(I), NK(I)) = DDZZ(NK(I), NK(I)) - _
                NB(I, KK, Q) * X(Q) ^ 2 * (WA(KK - 1, Q) - _
                WA(KK, Q)) ^ 2 / PX(KK, Q)
            NEXT KK
```

```
NEXT Q
IF NK(I) >= 3 THEN
  FOR K = 2 TO (NK(I) - 1)
    FOR Q = 1 TO NQ
      DDZZ(K, K - 1) = DDZZ(K, K - 1) + NB(I, K, Q) _
        * WA(K, Q) * WA(K - 1, Q) / PX(K, Q)
    NEXT Q
  NEXT K
END IF
IF NK(I) >= 3 THEN
  FOR K = 1 TO (NK(I) - 2)
    FOR Q = 1 TO NQ
      DDZZ(K, K + 1) = DDZZ(K, K + 1) + NB(I, K, Q) _
        * WA(K, Q) * WA(K + 1, Q) / PX(K + 1, Q)
    NEXT Q
  NEXT K
END IF
FOR K = 1 TO (NK(I) - 1)
  FOR Q = 1 TO NQ
    DDZZ(K, NK(I)) = DDZZ(K, NK(I)) + NB(I, K, Q) _
      * WA(K, Q) * X(Q) * ((WA(K - 1, Q) - _
      WA(K, Q)) / PX(K, Q) - (WA(K, Q) - _
      WA(K + 1, Q)) / PX(K + 1, Q))
  NEXT Q
  DDZZ(NK(I), K) = DDZZ(K, NK(I))
NEXT K
IF NK(I) = 2 THEN
  FOR K = 1 TO NK(I)
    FOR KK = 1 TO NK(I)
      A2(K, KK) = DDZZ(K, KK)
    NEXT KK
  NEXT K
ELSE
  FOR K = 1 TO NK(I)
    FOR KK = 1 TO NK(I)
      A1(K, KK) = DDZZ(K, KK)
    NEXT KK
  NEXT K
END IF
NP = NK(I)
N = NP
IF NK(I) = 2 THEN
  CALL LUDCMP(A2(), N, NP, INDX2(), D)
  FOR II = 1 TO N
    FOR LL = 1 TO N
      XX2(LL) = 0
    NEXT LL
    XX2(II) = 1
    CALL LUBKSB(A2(), N, NP, INDX2(), XX2())
    FOR LL = 1 TO N
      Y2(LL, II) = XX2(LL)
    NEXT LL
  NEXT II
```

```
ELSE
  CALL LUDCMP(A1(), N, NP, INDX1(), D)
  FOR II = 1 TO N
    FOR LL = 1 TO N
      XX1(LL) = 0
    NEXT LL
    XX1(II) = 1
    CALL LUBKSB(A1(), N, NP, INDX1(), XX1())
    FOR LL = 1 TO N
      Y1(LL, II) = XX1(LL)
    NEXT LL
  NEXT II
END IF
FOR K = 1 TO NK(I)
  DZETA(K) = 0
NEXT K
IF NK(I) = 2 THEN
  FOR K = 1 TO NK(I)
    FOR KK = 1 TO NK(I)
      DZETA(K) = DZETA(K) + Y2(K, KK) * DZ(KK)
    NEXT KK
  NEXT K
ELSE
  FOR K = 1 TO NK(I)
    FOR KK = 1 TO NK(I)
      DZETA(K) = DZETA(K) + Y1(K, KK) * DZ(KK)
    NEXT KK
  NEXT K
END IF
BIG DZETA(K) PROTECTION
FOR K = 1 TO NK(I)
  FOR KK = 1 TO NK(I)
    IF DZETA(K) > .2 THEN DZETA(K) = .2
    IF DZETA(K) < -.2 THEN DZETA(K) = -.2
  NEXT KK
NEXT K
FOR K = 1 TO (NK(I) - 1)
  ZETA(I, K) = ZETA(I, K) - DZETA(K)
NEXT K
LAMBDA(I) = LAMBDA(I) - DZETA(NK(I))
IF NK(I) = 2 THEN
  PRINT USING "####.#####"; I; IT; ZETA(I, 1); _
    LAMBDA(I)
END IF
IF NK(I) = 3 THEN
  PRINT USING "####.#####"; I; IT; ZETA(I, 1); _
    ZETA(I, 2); LAMBDA(I)
END IF
DIFFZL = 0
FOR K = 1 TO NK(I)
  IF ABS(DZETA(K)) > DIFFZL THEN _
    DIFFZL = ABS(DZETA(K))
NEXT K
```

```
              IF DIFFZL <= .01 GOTO 400
          NEXT IT
400    NEXT I

       FOR I = 1 TO NI
         IF NK(I) = 2 THEN
           PRINT
           PRINT "        I ZETA(I,1) LAMBDA(I)    B(I,1)"
           B(I, 1) = -ZETA(I, 1) / LAMBDA(I)
           PRINT USING "####.#####"; I; ZETA(I, 1); LAMBDA(I); _
              B(I, 1)
         END IF
         IF NK(I) = 3 THEN
           PRINT
           PRINT _
"          I ZETA(I,1) ZETA(I,2) LAMBDA(I)    B(I,1)    B(I,2)"
           B(I, 1) = -ZETA(I, 1) / LAMBDA(I)
           B(I, 2) = -ZETA(I, 2) / LAMBDA(I)
           PRINT USING "####.#####"; I; ZETA(I, 1); _
              ZETA(I, 2); LAMBDA(I); B(I, 1); B(I, 2)
         END IF
       NEXT I
'      THIS MAY BE MODIFIED

       NC = NC + 1
       IF NC <= MNC GOTO 100

       END

SUB LUBKSB (A(), N, NP, INDX(), B())
II = 0
FOR I = 1 TO N
  LL = INDX(I)
  SUM = B(LL)
  B(LL) = B(I)
  IF II <> 0 THEN
    FOR J = II TO I - 1
      SUM = SUM - A(I, J) * B(J)
    NEXT J
  ELSEIF SUM <> 0! THEN
    II = I
  END IF
  B(I) = SUM
NEXT I
FOR I = N TO 1 STEP -1
  SUM = B(I)
  FOR J = I + 1 TO N
    SUM = SUM - A(I, J) * B(J)
  NEXT J
  B(I) = SUM / A(I, I)
NEXT I
END SUB
```

```
SUB LUDCMP (A(), N, NP, INDX(), D)
TINY = 1E-20
DIM VV(N)
D = 1!
FOR I = 1 TO N
  AAMAX = 0!
  FOR J = 1 TO N
    IF ABS(A(I, J)) > AAMAX THEN AAMAX = ABS(A(I, J))
  NEXT J
  IF AAMAX = 0! THEN PRINT "Singular matrix.": EXIT SUB
  VV(I) = 1! / AAMAX
NEXT I
FOR J = 1 TO N
  FOR I = 1 TO J - 1
    SUM = A(I, J)
    FOR K = 1 TO I - 1
      SUM = SUM - A(I, K) * A(K, J)
    NEXT K
    A(I, J) = SUM
  NEXT I
  AAMAX = 0!
  FOR I = J TO N
    SUM = A(I, J)
    FOR K = 1 TO J - 1
      SUM = SUM - A(I, K) * A(K, J)
    NEXT K
    A(I, J) = SUM
    DUM = VV(I) * ABS(SUM)
    IF DUM >= AAMAX THEN
      IMAX = I
      AAMAX = DUM
    END IF
  NEXT I
  IF J <> IMAX THEN
    FOR K = 1 TO N
      DUM = A(IMAX, K)
      A(IMAX, K) = A(J, K)
      A(J, K) = DUM
    NEXT K
    D = -D
    VV(IMAX) = VV(J)
  END IF
  INDX(J) = IMAX
  IF A(J, J) = 0! THEN A(J, J) = TINY
  IF J <> N THEN
    DUM = 1! / A(J, J)
    FOR I = J + 1 TO N
      A(I, J) = A(I, J) * DUM
    NEXT I
  END IF
NEXT J
ERASE VV
END SUB
```

K. Implementation of Gibbs Sampler

K.1 Introduction

A BASIC computer program was developed to illustrate estimation of item and ability parameters using Gibbs sampler. The computer program is based on the Baker's (1988) FORTRAN translation of the Albert's (1992) MATLAB program under the two-parameter normal ogive model.

It should be noted that in Gibbs sampler ability parameters can be estimated either jointly with item parameters or sequentially after obtaining the item parameters. If ability parameters are estimated after obtaining item parameters, other ability estimation procedures are needed (e.g., see Appendix E). Here, it is assumed that ability parameters are estimated together with item parameters.

K.2 Implementation

The first four lines of the program declare four functions, RANDN! (IDUM&), RAND! (IDUM&), PHIINV (P!), and PHI (X!). The function RANDN! (IDUM&) is used to generate a normally distributed deviate with zero mean and unit variance, where IDUM& is used for the purpose of initialization. The function RAND! (IDUM&) returns a uniform random deviate between 0 and 1. PHIINV (P!) returns the inverse normal deviate for the area P!. PHI (X!) is the normal distribution function for X!.

The program shows the dimension statements for the variables. Note that the brief specification information will be saved in a log file GIBBS.LOG. The input file is IRPB.DAT that contains the memory test data (Thissen, 1982). The data consist of the 40 examinees's responses to the 10 items. The memory test data are saved in IRPB.DAT as:

```
0,0,0,0,0,0,0,0,0,0
0,0,0,0,0,0,0,0,0,0
0,0,0,0,0,0,0,0,0,0
0,0,0,0,0,0,0,0,0,0
0,0,0,0,0,0,0,0,0,0
0,0,0,0,0,0,0,0,0,1
0,0,0,0,0,0,0,0,1,1
```

```
0,0,0,0,0,0,0,0,1,1
0,0,0,0,0,0,0,0,1,1
0,0,0,0,0,0,0,1,0,1
0,0,0,0,0,0,0,1,0,1
0,0,0,0,0,1,0,0,0,1
0,0,0,0,1,0,0,0,0,1
0,0,0,0,1,0,0,0,1,0
0,0,1,0,0,0,0,0,0,1
0,0,0,0,0,0,0,1,1,1
0,0,0,0,0,0,0,1,1,1
0,0,0,0,0,0,1,0,1,1
0,0,1,0,0,0,0,1,0,1
0,0,1,0,0,0,1,0,0,1
0,1,0,0,0,1,0,1,0,0
1,0,0,0,0,0,0,0,1,1
1,0,0,0,0,0,1,0,0,1
1,0,0,1,0,0,0,0,1,0
0,0,0,0,0,0,1,1,1,1
0,0,0,0,0,1,0,1,1,1
0,0,0,0,0,1,0,1,1,1
0,0,0,0,1,0,1,0,1,1
0,0,0,1,0,0,1,0,1,1
0,0,0,1,0,0,1,1,0,1
0,1,0,0,0,0,0,1,1,1
0,1,0,0,0,1,0,0,1,1
0,1,0,0,1,0,0,1,1,0
0,1,0,0,0,0,1,1,1,1
1,0,0,0,0,1,1,1,0,1
1,0,0,1,1,0,1,1,0,0
1,1,0,0,1,0,0,1,0,1
0,1,0,0,0,1,1,1,1,1
1,1,0,0,1,1,0,1,0,1
0,1,1,1,1,0,0,1,1,1
```

The computer program will ask several questions in order to read in the data. Appropriate answers to the questions are required to successfully run the program. For example, the first question is:

ENTER TITLE FOR RUN=

You can type in up to 64 characters as an informative title. Type the number of examinees (i.e., 40) for the following question:

ENTER NUMBER OF EXAMINEES=

Type the number of items (i.e., 10) for the following question:

ENTER NUMBER OF ITEMS=

After reading in the test data, the computer will print out:

ENTER SEED FOR RN GENERATORS USE 6 DIGITS=

You can type in a six digit random number (e.g., 192239). The next step is setting the hyperparameters of theta prior. The prior mean FMU = 0! and the prior variance VAR = 1!. The initialization of item and ability parameters are the next step. All item discrimination parameters A(I) are set to 1 and the γ_i parameters G(I) are initialized based on the classical item difficulty statistics. Ability parameters are initialized as 0.

The next three specifications control the sampling of parameters. First, the question,

ENTER ITERATION NUMBER TO START RECORDING=

indicates the number of iterations for burn-in. Albert (1992) used 10 (i.e., ITO = 10). The next question

ENTER NUMBER OF ITERATES BETWEEN RECORDINGS=

is for getting roughly independent samples. Albert (1992) used 5 (i.e., IWO = 5). The total number of samples are determined by the next question for which Albert (1992) used 200 (i.e., NUM = 200):

ENTER NUMBER OF RECORDED ITERATES=

The program will create four output files: ITEM.OUT contains generated item parameters; THET.OUT contains generated ability parameters; AVECT.COL contains item discrimination parameters; and GVECT.COL contains the negative item intercept parameters (i.e., γ_i). It can be noted that the BASIC program implemented the Bayesian version of Gibbs sampler (Johnson & Albert, 1999), and the prior standard deviation of item discrimination is SDA = .5 and the prior standard deviation of the negative item intercept is SDG = 2.

Table K.1. Estimated Item Parameters and Standard Errors (s.e.) of the Memory Test Items from Gibbs Sampler

Item	$\hat{\lambda}_i$ (s.e.)	$\hat{\zeta}_i$ (s.e.)
1	.635 (.341)	−1.019 (.278)
2	1.213 (.399)	−1.083 (.351)
3	.571 (.345)	−1.344 (.320)
4	.750 (.393)	−1.351 (.378)
5	.744 (.397)	−.989 (.317)
6	.777 (.359)	−.983 (.309)
7	.658 (.319)	−.685 (.258)
8	1.133 (.391)	−.056 (.307)
9	.611 (.336)	.025 (.242)
10	.757 (.338)	.830 (.275)

Using the above specifications, a sample run was performed. The GIBBS.LOG file contained the following:

```
ALBERT SETTING
NUMBER OF EXAMINEES=        40
NUMBER OF ITEMS=            10
```

```
RN SEED USED WAS=              123456
ITERATION NUMBER TO START RECORDING=        10
NUMBER OF ITERATES BETWEEN RECORDINGS (T)=              5
NUMBER OF RECORDED ITERATES (NUM)=        200
```

Table K.2. Ability Estimates and Standard Errors (s.e.) from Gibbs Sampler

Examinee	$\hat{\theta}$	(s.e.)	Examinee	$\hat{\theta}$	(s.e.)
1	−1.417	(.635)	21	.288	(.519)
2	−1.267	(.641)	22	−.156	(.533)
3	−1.328	(.696)	23	−.132	(.646)
4	−1.439	(.710)	24	−.132	(.617)
5	−1.271	(.630)	25	.304	(.555)
6	−.779	(.615)	26	.385	(.527)
7	−.456	(.593)	27	.374	(.568)
8	−.470	(.665)	28	.222	(.534)
9	−.551	(.604)	29	.173	(.508)
10	−.225	(.598)	30	.350	(.511)
11	−.221	(.558)	31	.554	(.505)
12	−.276	(.607)	32	.397	(.538)
13	−.357	(.597)	33	.503	(.511)
14	−.383	(.622)	34	.713	(.521)
15	−.448	(.589)	35	.612	(.615)
16	.045	(.575)	36	.512	(.569)
17	.033	(.543)	37	.761	(.505)
18	−.136	(.609)	38	1.024	(.523)
19	.053	(.593)	39	1.089	(.528)
20	−.106	(.595)	40	1.318	(.530)

The output files are relatively large. Using the values generated, two tables were obtained. Table K.1 contains item parameter estimates. Note that $a_i = \lambda_i$ and $\zeta_i = -\gamma_i$ in Table K.1. Ability estimates that were obtained jointly with the item parameter estimates are reported in Table K.2.

K.3 BASIC Computer Program

```
DECLARE FUNCTION RANDN! (IDUM&)
DECLARE FUNCTION RAND! (IDUM&)
DECLARE FUNCTION PHIINV (P!)
DECLARE FUNCTION PHI (X!)
DIM TITLE AS STRING * 64
DIM A(50), G(50), TH(100), Y(50, 100), Z(50, 100)
DIM XPX(2, 2), XPXINV(2, 2), AMAT(2, 2), AMATT(2, 2)
DIM PP(2, 2)
OPEN "GIBBS.LOG" FOR OUTPUT AS #9
INPUT "ENTER TITLE FOR RUN= "; TITLE
PRINT #9, TITLE
INPUT "ENTER NUMBER OF EXAMINEES= "; NEXAM
```

```
NEXAM$ = "NUMBER OF EXAMINEES= "
PRINT #9, NEXAM$, NEXAM
INPUT "ENTER NUMBER OF ITEMS= "; NITEM
NITEM$ = "NUMBER OF ITEMS= "
PRINT #9, NITEM$, NITEM
OPEN "IRPB.DAT" FOR INPUT AS #10
FOR J = 1 TO NEXAM
  FOR I = 1 TO NITEM
    INPUT #10, Y(I, J)
  NEXT I
NEXT J
CLOSE #10
INPUT "ENTER SEED FOR RN GENERATORS USE 6 DIGITS= "; IDUM&
IDUM$ = "RN SEED USED WAS= "
PRINT #9, IDUM$, IDUM
ISEED& = -IDUM&
TRN = RAND(ISEED&)
FMU = 0!
VAR = 1!
FOR I = 1 TO NITEM
  PMA = 1!
  A(I) = PMA
NEXT I
FOR I = 1 TO NITEM
  SUM1 = 0!
  FOR J = 1 TO NEXAM
    SUM1 = SUM1 + Y(I, J)
  NEXT J
  PHAT = (SUM1 + .5) / (CSNG(NEXAM) + 1!)
  G(I) = -PHIINV(PHAT) * SQR(1! + PMA)
NEXT I
FOR J = 1 TO NEXAM
  TH(J) = 0!
NEXT J
INPUT "ENTER ITERATION NUMBER TO START RECORDING= "; ITO
ITO$ = "ITERATION NUMBER TO START RECORDING= "
PRINT #9, ITO$, ITO
INPUT "ENTER NUMBER OF ITERATES BETWEEN RECORDINGS= "; IWO
IWO$ = "NUMBER OF ITERATES BETWEEN RECORDINGS (T)= "
PRINT #9, IWO$, IWO
INPUT "ENTER NUMBER OF RECORDED ITERATES= "; NUM
NUM$ = "NUMBER OF RECORDED ITERATES (NUM)= "
PRINT #9, NUM$, NUM
CLOSE #9
ITOTAL = ITO + (NUM - 1) * IWO
OPEN "ITEM.OUT" FOR OUTPUT AS #11
OPEN "THET.OUT" FOR OUTPUT AS #12
OPEN "AVECT.COL" FOR OUTPUT AS #13
OPEN "GVECT.COL" FOR OUTPUT AS #14
IDIFF = IWO - 1
FOR KK = 1 TO ITOTAL
  PRINT "ITERATION= "; KK
  FOR I = 1 TO NITEM
```

```
   FOR J = 1 TO NEXAM
     FLP = TH(J) * A(I) - G(I)
     BB = PHI(-FLP)
     U = RAND(IDUM&)
     TT = (BB * (1! - Y(I, J)) + (1! - BB) * Y(I, J)) * U + _
       BB * Y(I, J)
     Z(I, J) = PHIINV(TT) + FLP
   NEXT J
 NEXT I
 SUMA2 = 0!
 FOR I = 1 TO NITEM
   SUMA2 = SUMA2 + A(I) * A(I)
 NEXT I
 V = 1! / SUMA2
 PVAR = 1! / (1! / V + 1! / VAR)
 FOR J = 1 TO NEXAM
   FMN = 0!
   FOR I = 1 TO NITEM
     FMN = FMN + (A(I) * (Z(I, J) + G(I)))
   NEXT I
   PMEAN = (FMN + FMU / VAR) * PVAR
   TH(J) = RANDN(IDUM&) * SQR(PVAR) + PMEAN
 NEXT J
 XPX(1, 1) = 0!
 XPX(1, 2) = 0!
 XPX(2, 1) = 0!
 XPX(2, 2) = CSNG(NEXAM)
 FOR J = 1 TO NEXAM
   XPX(1, 1) = XPX(1, 1) + TH(J) * TH(J)
   XPX(1, 2) = XPX(1, 2) + TH(J)
 NEXT J
 XPX(1, 2) = -XPX(1, 2)
 XPX(2, 1) = XPX(1, 2)
 SDA = .5
 SDG = 2!
 PP(1, 1) = 1! / (SDA ^ 2)
 PP(1, 2) = 0!
 PP(2, 1) = 0!
 PP(2, 2) = 1! / (SDG ^ 2)
 XPX(1, 1) = XPX(1, 1) + PP(1, 1)
 XPX(1, 2) = XPX(1, 2) + PP(1, 2)
 XPX(2, 1) = XPX(2, 1) + PP(2, 1)
 XPX(2, 2) = XPX(2, 2) + PP(2, 2)
 DET = (XPX(1, 1) * XPX(2, 2) - XPX(1, 2) * XPX(2, 1))
 XPXINV(1, 1) = XPX(2, 2) / DET
 XPXINV(1, 2) = -XPX(1, 2) / DET
 XPXINV(2, 1) = -XPX(2, 1) / DET
 XPXINV(2, 2) = XPX(1, 1) / DET
 AMAT(1, 1) = SQR(XPXINV(1, 1))
 AMAT(1, 2) = XPXINV(1, 2) / AMAT(1, 1)
 AMAT(2, 1) = 0!
 AMAT(2, 2) = SQR(XPXINV(2, 2) - _
   (XPXINV(1, 2) * XPXINV(1, 2) / XPXINV(1, 1)))
```

```
FOR I = 1 TO NITEM
  XPZ1 = 0!
  XPZ2 = 0!
  FOR J = 1 TO NEXAM
    XPZ1 = XPZ1 + TH(J) * Z(I, J)
    XPZ2 = XPZ2 + Z(I, J)
  NEXT J
  XPZ2 = -XPZ2
  PMA = 1!
  PMB = 0!
  BZ1 = XPXINV(1, 1) * (XPZ1 + PP(1, 1) * PMA) + _
    XPXINV(1, 2) * (XPZ2 + PP(2, 2) * PMB)
  BZ2 = XPXINV(2, 1) * (XPZ1 + PP(1, 1) * PMA) + _
    XPXINV(2, 2) * (XPZ2 + PP(2, 2) * PMB)
  AMATT(1, 1) = AMAT(1, 1)
  AMATT(1, 2) = AMAT(2, 1)
  AMATT(2, 1) = AMAT(1, 2)
  AMATT(2, 2) = AMAT(2, 2)
  DO
    RN1 = RANDN(IDUM&)
    RN2 = RANDN(IDUM&)
    A(I) = AMATT(1, 1) * RN1 + AMATT(1, 2) * RN2 + BZ1
    G(I) = AMATT(2, 1) * RN1 + AMATT(2, 2) * RN2 + BZ2
  LOOP UNTIL A(I) > 0.
NEXT I
IF KK >= ITO THEN
  IDIFF = IDIFF + 1
  IF IDIFF = IWO THEN
    IDIFF = 0
    FOR I = 1 TO NITEM
      PRINT #11, USING "####.#####"; KK; I; A(I); G(I)
    NEXT I
    FOR J = 1 TO NEXAM
      PRINT #12, USING "####.#####"; KK; J; TH(J)
    NEXT J
    FOR I = 1 TO NITEM
      PRINT #13, USING "####.#####"; A(I)
    NEXT I
    FOR I = 1 TO NITEM
      PRINT #14, USING "####.#####"; G(I)
    NEXT I
  END IF
END IF
NEXT KK
CLOSE #12
CLOSE #11
END

FUNCTION PHI (X)
' (ABRAMOWITZ & STEGUN, 1972, P. 932)
IF X < -4.87726 THEN
  PHI = .00001
ELSEIF X > 4.87726 THEN
```

```
   PHI = .99999
ELSE
  Y = 0!
  IF X < 0! THEN
    Y = -1!
    X = -X
  END IF
  D1 = .049867347#
  D2 = .0211410061#
  D3 = .0032776263#
  D4 = .0000380036#
  D5 = .0000488906#
  D6 = .000005383#
  PHI = 1! - (1! / 2!) / (1! + D1 * X + D2 * X ^ 2 + _
    D3 * X ^ 3 + D4 * X ^ 4 + D5 * X ^ 5 + D6 * X ^ 6) ^ 16
  IF Y = -1! THEN
    PHI = (1! / 2!) / (1! + D1 * X + D2 * X ^ 2 + _
      D3 * X ^ 3 + D4 * X ^ 4 + D5 * X ^ 5 + D6 * X ^ 6) ^ 16
    X = -X
  END IF
END IF
END FUNCTION

FUNCTION PHIINV (P)
' (ABRAMOWITZ & STEGUN, 1972, P. 933)
IF P < .00001 THEN
  PHIINV = -4.87762
ELSEIF P > .99999 THEN
  PHIINV = 4.87762
ELSE
  Y = 0!
  IF P > .5 THEN
    Y = -1!
    P = 1 - P
  END IF
  T = SQR(LOG(1 / P ^ 2))
  C0 = 2.515517
  C1 = .802853
  C2 = .010328
  D1 = 1.432788
  D2 = .189269
  D3 = .001308
  PHIINV = -T + (C0 + C1 * T + C2 * T ^ 2) / _
    (1! + D1 * T + D2 * T ^ 2 + D3 * T ^ 3)
  IF Y = -1! THEN
    PHIINV = T - (C0 + C1 * T + C2 * T ^ 2) / _
      (1! + D1 * T + D2 * T ^ 2 + D3 * T ^ 3)
    P = 1 - P
  END IF
END IF
END FUNCTION

FUNCTION RAND (IDUM&) STATIC
```

```
' (CF. SPROTT, 1991, PP. 139-140; PRESS ET AL., 1992, P. 271)
IA& = 16807
IM& = 2147483647
AM = 4.65661287D-10
IQ& = 127773
IR& = 2836
NTAB& = 32
NDIV& = 67108865
EPS = .00000012#
RNMX = .99999988#
DIM IV&(32)
STATIC IV&, IY&
IF IDUM& <= 0 OR IY& = 0 THEN
  IDUM& = 1
  IF -IDUM& > 1 THEN IDUM& = -IDUM&
  FOR J = NTAB& + 8 TO 1 STEP -1
    K& = IDUM& / IQ&
    IDUM& = IA& * (IDUM& - K& * IQ&) - IR& * K&
    IF IDUM& < 0 THEN IDUM& = IDUM& + IM&
    IF J <= NTAB& THEN IV&(J) = IDUM&
  NEXT J
  IY& = IV&(1)
END IF
K& = IDUM& / IQ&
IDUM& = IA& * (IDUM& - K& * IQ&) - IR& * K&
IF IDUM& < 0 THEN IDUM& = IDUM& + IM&
J = 1 + IY& / NDIV&
IF J < 1 THEN J = 1
IF J > 32 THEN J = 32
IY& = IV&(J)
IV&(J) = IDUM&
RAND = AM * IY&
IF AM * IY& > RNMX THEN RAND = RNMX
END FUNCTION

FUNCTION RANDN (IDUM&)
' (CF. SPROTT, 1991, P. 145)
STATIC ISET, GSET
IF ISET = 0 THEN
  DO
    V1 = 2! * RAND(IDUM&) - 1!
    V2 = 2! * RAND(IDUM&) - 1!
    RSQ = V1 ^ 2 + V2 ^ 2
  LOOP UNTIL RSQ < 1! AND RSQ <> 0!
  FAC = SQR(-2! * LOG(RSQ) / RSQ)
  GSET = V1 * FAC
  RANDN = V2 * FAC
  ISET = 1
ELSE
  RANDN = GSET
  ISET = 0
END IF
END FUNCTION
```

References

1. Abbott, W. S. (1925). A method of computing the effectiveness of an insecticide. *Journal of Economic Entomology, 18*, 265–267.
2. Aitchison, S., & Brown, J. (1957). *The lognormal distribution.* Cambridge, England: Cambridge University Press.
3. Aitchison, J., & Silvey, S. D. (1957). The generalization of probit analysis to the case of multiple responses. *Biometrika, 44*, 131–140.
4. Albert, J. H. (1992). Bayesian estimation of normal ogive item response curves using Gibbs sampling. *Journal of Educational Statistics, 17*, 251–269.
5. Andersen, E. B. (1970). Asymptotic properties of conditional maximum-likelihood estimators. *Journal of the Royal Statistical Society, Series B, 32*, 283–301.
6. Andersen, E. B. (1972). The numerical solution of a set of conditional estimation equations. *Journal of the Royal Statistical Society, Series B, 34*, 42–54.
7. Andersen, E. B. (1973a). Conditional inference for multiple-choice questionnaires. *British Journal of Mathematical and Statistical Psychology, 26*, 31–44.
8. Andersen, E. B. (1973b). A goodness of fit test for the Rasch model. *Psychometrika, 38*, 123–140.
9. Andersen, E. B. (1977). Sufficient statistics and latent trait models. *Psychometrika, 42*, 69–81.
10. Andersen, E. B., & Madsen, M. (1977). Estimating the parameters of the latent population distribution. *Psychometrika, 42*, 357–374.
11. Andersen, E. B. (1997). The rating scale model. In W. J. van der Linden & R. K. Hambleton (Eds.), *Handbook of modern item response theory* (pp. 67–84). New York: Springer.
12. Anderson, T. W. (1984). *An introduction to multivariate statistical analysis* (2nd ed.). New York: Wiley.
13. Andrich, D. (1978). A rating formulation for ordered response categories. *Psychometrika, 43*, 561–573.
14. Andrich, D. (1978b). Relationships between the Thurstone and Rasch approaches to item scaling. *Applied Psychological Measurement, 2*, 451–462.
15. Andrich, D. (1979). A model for contingency tables having an ordered response classification. *Biometrics, 35*, 403–415.
16. Anscombe, F. J. (1956). On estimating binomial response relations. *Biometrika, 43*, 461–464.
17. Baker, F. B. (1959). Univac scientific computer program for test scoring and item analysis, (CPA1). *Behavioral Science, 4*, 254–255.
18. Baker, F. B. (1961). Empirical comparison of item parameters based on the logistic and normal functions. *Psychometrika, 26*, 239–246.
19. Baker, F. B. (1962). *Empirical determination of sampling distributions of item discrimination indices and a reliability coefficient* (Final Report, U.S.O.E., Contract OE-2-10-071).

20. Baker, F. B. (1967). The effect of criterion score grouping upon item parameter estimation. *British Journal of Mathematical and Statistical Psychology, 20,* 227–238.

21. Baker, F. B. (1977). Advances in item analysis. *Review of Educational Research, 47,* 151–178.

22. Baker, F. B. (1981). Log-linear, logit-linear models: A didactic. *Journal of Educational Statistics, 6,* 75–102.

23. Baker, F. B. (1985). *The basics of item response theory.* Portsmouth, NH: Heinemann.

24. Baker, F. B. (1986a). *MICRO-IRT: An item response theory test analysis computer program for the Apple II+ and IIe.* Unpublished manuscript, University of Wisconsin, Madison, WI.

25. Baker, F. B. (1986b). *GENIRV: A computer program for generating item responses.* Unpublished manuscript, University of Wisconsin.

26. Baker, F. B. (1987a). Methodology review: Item parameter estimation under the one-, two-, and three-parameter logistic models. *Applied psychological Measurement, 11,* 111–141.

27. Baker, F. B. (1987b). Item parameter estimation via minimum logit chi-square. *British Journal of Mathematical and Statistical Psychology, 40,* 50–60.

28. Baker, F. B. (1988). The item log-likelihood surface for two- and three-parameter item characteristic curve models. *Applied Psychological Measurement, 12,* 387–395.

29. Baker, F. B. (1990). Some observations on the metric of PC-BILOG results. *Applied Psychological Measurement, 14,* 139–150.

30. Baker, F. B. (1992). *Item response theory: Parameter estimation techniques.* New York: Marcel Dekker.

31. Baker, F. B. (1998). An investigation of the item parameter recovery characteristics of a Gibbs sampling approach. *Applied Psychological Measurement, 22,* 153–169.

32. Baker, F. B., & Martin, T. J. (1969). FORTAP: A FORTRAN test analysis package. *Educational and Psychological Measurement, 29,* 159–164.

33. Baker, F. B., & Subkoviak, M. J. (1981). Analysis of test results via log-linear models. *Applied Psychological Measurement, 5,* 503–515.

34. Baker, F. B., Cohen, A. S., & Barmish, B. R. (1988). Item characteristics of tests constructed via linear programming. *Applied Psychological Measurement, 12,* 189–199.

35. Baker, F. B., Al-Karni, A., & Al-Dosary, I. M. (1991). EQUATE: A computer program for the test characteristic curve method of IRT equating. *Applied Psychological Measurement, 50,* 529–549.

36. Béguin, A. A., & Glas, C. A. W. (2001). MCMC estimation and some model-fit analysis of multidimensional IRT models. *Psychometrika, 66,* 541–562.

37. Bennett, R. E., Rock, D. A., & Wang, M. (1991). Equivalence of free-response and multiple-choice items. *Journal of Educational Measurement, 28,* 77–92.

38. Berkson, J. (1944). Application of the logistic function to bio-assay. *Journal of the American Statistical Association, 39,* 357–365.

39. Berkson, J. (1949). Minimum χ^2 and maximum likelihood solution in terms of a linear transform, with particular reference to Bio-Assay. *Journal of the American Statistical Association, 44,* 273–278.

40. Berkson, J. (1951). Why I prefer logits to probits. *Biometrics, 7,* 327–329.

41. Berkson, J. (1954). Comments on R. A. Fisher: The analysis of variance with various binomial transformations. *Biometrics, 10,* 130–151.

42. Berkson, J. (1955a). Maximum likelihood and minimum χ^2 estimates of the logistic function. *Journal of the American Statistical Association, 50,* 130–162.

43. Berkson, J. (1955b). Estimate of the integrated normal curve by minimum normit chi square with particular reference to bio-assay. *Journal of the American Statistical Association,*

44. Bernardo, J. M., & Smith, A. F. M. (1994). *Bayesian theory.* Chichester, England: Wiley.

45. Billeaud, K., Swygert, K., Nelson, L., & Thissen, D. (1998). Some ideas about item response theory applied to combinations of multiple-choice and open-ended items: Scale scores for patterns of summed scores. In M. L. Bourque (Ed.), *Proceedings of Achievement Levels Workshop* (pp. 65–76). Washington, DC: National Assessment Governing Board.

46. Binet, A., & Simon, T. H. (1916). *The development of intelligence in young children.* Vineland, NJ: The Training School.

47. Birnbaum, A. (1957). *Efficient design and use of tests of a mental ability for various decision-making problems* (Series Report 58-16, Project No. 7755-23). USAF School of Aviation Medicine, Randolph Air Force Base, Texas.

48. Birnbaum, A. (1968). Some latent trait models and their use in inferring an examinee's ability. In F. M. Lord & M. R. Novick (Eds.), *Statistical theories of mental test scores* (pp. 397–479). Reading, MA: Addison-Wesley.

49. Bishop, Y. M. M., Fienberg, S. E., & Holland, P. W. (1975). *Discrete multivariate analysis: Theory and Practice.* Cambridge, MA: MIT Press.

50. Bliss, C. I. (1935). The comparison of the dosage-mortality curve. *Annals of Applied Biology, 22,* 134–167.

51. Bock, R. D. (1970). Estimating multinomial response relations. In R. C. Bose, I. M. Chakeavarti, P. C. Makalanobis, C. R. Rao, & K. J. C. Smith (Eds.), *Essays in probability and statistics* (pp. 111–132). Chapel Hill: The University of North Carolina Press.

52. Bock, R. D. (1972). Estimating item parameters and latent ability when responses are scored in two or more nominal categories. *Psychometrika, 37,* 29–51.

53. Bock, R. D. (1975). *Multivariate methods in behavioral research.* New York: McGraw–Hill.

54. Bock, R. D. (1983). The discrete Bayesian. In H. Wainer & S. Messick (Eds.), *Principals of modern psychological measurement* (pp. 103–115). Hillsdale, NJ: Erlbaum.

55. Bock, R. D. (1985). *Multivariate statistical methods in behavioral research.* Chicago: Scientific Software.

56. Bock, R. D. (1989). Measurement of human variation: A two-stage model. In R. D. Bock (Ed.), *Multilevel analysis of educational data* (pp. 319–342). San Diego, CA: Academic Press.

57. Bock, R. D. (1997). The nominal categories model. In W. J. van der Linden & R. K. Hambleton (Eds.), *Handbook of modern item response theory* (pp. 33–49). New York: Springer.

58. Bock, R. D., & Aitkin, M. (1981). Marginal maximum likelihood estimation of item parameters: Application of an EM algorithm. *Psychometrika, 46,* 443–459; *47,* 369 (Errata).

59. Bock, R. D., & Lieberman, M. (1970). Fitting a response model for n dichotomously scored items. *Psychometrika, 35,* 179–197.

60. Bock, R. D., & Mislevy, R. J. (1981). An item response curve model for matrix-sampling data: The California grade-three assessment. *New Directions for Testing and Measurement, 10,* 65–90.

61. Bock, R. D., & Yates, G. (1973). *MULTIQUAL: log-linear analysis of nominal or ordinal qualitative data by the method of maximum likelihood.* Ann Arbor, MI: National Educational Resources Inc.

62. Bock, R. D., & Zimowski, M. F. (1990). *Duplex design: Giving students a stake in educational development* (CSE Tech. Rep. No. 306). Los Angeles: UCLA Center for Research on Evaluation, Standards, and Student Testing.

63. Bock, R. D., & Zimowski, M. F. (1997). Multiple group IRT. In W. J. van der Linden & R. K. Hambleton (Eds.), *Handbook of modern item response theory* (pp. 433–448). New York: Springer.

64. Bock, R. D., Mislevy, R. J., & Thissen, D. (1991). *Item response theory.* Unpublished manuscript.

65. Bock, R. D., Muraki, E., & Pfiffenberger, W. (1988). Item pool maintenance in the presence of item parameter drift. *Journal of Educational Measurement, 25,* 275–285.

66. Bradlow, E. T., Wainer, H., & Wang, X. (1999). A Bayesian random effects model for testlets. *Psychometrika, 64,* 153–168.

67. Bridgeman, B., & Rock, D. A. (1993). Relationships among multiple-choice and open-ended analytical questions. *Journal of Educational Measurement, 30,* 313–329.

68. Brodgen, H. E. (1977). The Rasch model, the law of comparative judgment and additive conjoint measurement. *Psychometrika, 42,* 631–634.

69. Casella, G., & George, E. I. (1992). Explaining the Gibbs sampler. *The American Statistician, 46,* 167–174.

70. Cohen, L. (1979). Approximate expressions for parameter estimates in the Rasch model. *British Journal of Mathematical and Statistical Psychology, 32,* 113–120.

71. Conover, W. J. (1999). *Practical nonparametric statistics* (3rd ed.). New York: Wiley.

72. Cook, L. L., & Eignor, D. R. (1991). IRT equating methods. *Educational Measurement: Issues and Practice, 10*(3), 37–45.

73. Cornfield, J. (1969). The Bayesian outlook and its application (with discussion). *Biometrics, 25,* 617–657.

74. Cowles, M. K., & Carlin, B. P. (1996). Markov chain Monte Carlo convergence diagnostics: A comparative review. *Journal of the American Statistical Association, 91,* 883–904.

75. Cramer, E. M. (1962). A comparison of three methods of fitting the normal ogive. *Psychometrika, 27,* 183–192.

76. Cramér, H. (1946). *Mathematical methods of statistics.* Princeton, NJ: Princeton University Press.

77. David, F. N., Kendall, M. G., & Barton, D. E. (1966). *Symmetric Function and Allied Tables.* London: Cambridge University Press.

78. Day, N. E. (1969). Estimating the components of a mixture of normal distributions. *Biometrika, 56,* 463–474.

79. De Ayala, R. J. (1993). An introduction to polytomous item response theory models. *Measurement and Evaluation in Counseling and Development, 25,* 172–189.

80. De Ayala, R. J., & Sava-Bolesta, M. (1999). Item parameter recovery for the nominal response model. *Applied Psychological Measurement, 23,* 3–19.

81. de Finetti, B. (1974). Bayesianism: Its unifying role for both the foundations and applications of statistics. *International Statistical Review, 42,* 117–130.

82. de Gruijter, D. N. M. (1985). A note on the asymptotic variance–covariance matrix of item parameter estimates in the Rasch model. *Psychometrika, 50,* 247–249.

83. de Gruijter, D. N. M., & van der Kamp, L. J. T. (Eds.). (1976). *Advances in psychological and educational measurement.* London: Wiley.

84. de Leeuw, J., & Verhelst, N. (1986). Maximum likelihood estimation in generalized Rasch models. *Journal of Educational Statistics, 11*, 183–196.

85. Dempster, A. P., Laird, N. M., & Rubin, D. B. (1977). Maximum likelihood from incomplete data via the EM algorithm (with discussion). *Journal of the Royal Statistical Society, Series B, 39*, 1–38.

86. Dempster, A. P., Rubin, D. B., & Tsutakawa, R. K. (1981). Estimation in covariance components models. *Journal of the American Statistical Association, 76*, 341–353.

87. Dinero, T. E., & Haertel, E. (1977). Applicability of the Rasch Model with varying item discrimination parameters. *Applied Psychological Measurement, 1*, 581–592.

88. Dodd, B. G., Koch, W. R., & De Ayala, D. R. (1989). Operational characteristics of adaptive testing procedures using the graded response model. *Applied Psychological Measurement, 13*, 129–143.

89. Edwards, W., Lindman, H., & Savage, L. J. (1963). Bayesian statistical inference for psychological research. *Psychological Review, 70*, 193–242.

90. Ercikan, K., Schwarz, R. D., Julian, M. W., Burket, G. R., Weber, M. M., & Link, V. (1998). Calibration and scoring of tests with multiple-choice and constructed-response item types. *Journal of Educational Measurement, 35*, 137–154.

91. Everitt, B. S. (1977). *The analysis of contingency tables.* London: Chapman and Hall.

92. Fechner, G. T. (1860). *Elemente der Psychophysik.* Leipzig: Breitkopf und Härtel.

93. Ferguson, G. A. (1942). Item selection by the constant process. *Psychometrika, 7*, 19–29.

94. Fienberg, S. E. (1977). *The analysis of cross-classified categorical data.* Cambridge, MA: MIT Press.

95. Finn, J. D. (1974). *A general model for multivariate analysis.* New York: Holt.

96. Finney, D. J. (1944). The application of probit analysis to the results on mental tests. *Psychometrika, 19*, 31–39.

97. Finney, D. J. (1952). *Probit analysis* (2nd ed). Cambridge, Great Britain: Cambridge University Press.

98. Finney, D. J. (1971). *Probit analysis* (3rd ed). Cambridge, Great Britain: Cambridge University Press.

99. Fischer, G. H. (1974). *Einführung in die Theorie psychologischer Tests.* Bern: Huber.

100. Fischer, G. H., & Formann, A. K. (1972). *An algorithm and a FORTRAN program for estimating the item parameters of the linear logistic test model* (Research Bulletin No. 11). Vienna: Psychologisches Institute der Universitat Wien.

101. Fox, J.-P., & Glas, C. A. W. (2001). Bayesian estimation of a multilevel IRT model using Gibbs sampling. *Psychometrika, 66*, 271–288.

102. Garwood, F. (1941). The application of maximum likelihood to dosage-mortality curves. *Biometrika, 32*, 46–58.

103. Gelfand, A. E., & Smith, A. F. M. (1990). Sampling-based approaches to calculating marginal densities. *Journal of the American Statistical Association, 85*, 398–409.

104. Gelman, A. (1996). Inference and monitoring convergence. In W. R. Gilks, S. Richardson, & D. J. Spiegelhalter (Eds.), *Markov chain Monte Carlo in practice* (pp. 131–143). London: Chapman & Hall.

105. Gelman, A., Carlin, J. B., Stern, H. S., & Rubin, D. B. (1995). *Bayesian data analysis.* London: Chapman & Hall.

106. Gelman, A., & Rubin, D. B. (1992). Inference from iterative simulation using multiple sequences (with discussion). *Statistical Science, 7*, 457–511.
107. Geman, S., & Geman, D. (1984). Stochastic relaxation, Gibbs distributions, and the Bayesian restoration of images. *IEEE Transactions on Pattern Analysis and Machine Intelligence, 6*, 721–741.
108. Gilks, W. R. (1996). Full conditional distribution. In W. R. Gilks, S. Richardson, & D. J. Spiegelhalter (Eds.), *Markov chain Monte Carlo in practice* (pp. 75–88). London: Chapman & Hall.
109. Gilks, W. R., & Wild, P. (1992). Adaptive rejection sampling for Gibbs sampling. *Applied Statistics, 41*, 337–348.
110. Ghosh, M., Ghosh, A., Chen, M.-H., & Agresti, A. (1999). *Bayesian estimation for item response models* (Tech. Rep.). Gainsville: University of Florida, Department of Statistics.
111. Glas, C. A. W. (1988). The derivation of some tests for the Rasch model from the multinomial distribution. *Psychometrika, 53*, 525–546.
112. Goodman, L. A. (1968). The analysis of cross-classified data: Independence, quasi-independence, and interactions in contingency tables with or without missing entries. *Journal of the American Statistical Association, 63*, 1091–1131.
113. Grima, A. M., & Weichun, W. M. (2002, April). *Test scoring: Multiple-choice and constructed response items.* Paper presented at the annual meeting of the American Educational Research Association, New Orleans, LA.
114. Grizzle, J. E., Starmer, C. F., & Koch, G. G. (1969). Analysis of categorized data by linear models. *Biometrics, 25*, 489–504.
115. Gulliksen, H. (1950). *Theory of mental tests.* New York: Wiley.
116. Gumbel, E. J. (1961). Bivariate logistic distributions. *Journal of the American Statistical Association, 56*, 335–349.
117. Gurland, J., Lee, I., & Dahm, P. A. (1960). Polychotomous quantal response in biological assay. *Biometrics, 16*, 382–398.
118. Gustafsson, J.-E. (1977). *The Rasch model for dichotomous items: Theory, applications and a computer program.* Göteborg, Sweden: University of Göteborg, The Institute of Education.
119. Gustafsson, J.-E. (1980a). A solution of the conditional estimation problem for long tests in the Rasch model for dichotomous items. *Educatioanl and Psychological Measurement, 40*, 327–385.
120. Gustafsson, J.-E. (1980b). Testing and obtaining fit of data to the Rasch model. *British Journal of Mathematical and Statistical Psychology, 33*, 205–233.
121. Haberman, S. (1977). Maximum likelihood estimates in exponential response models. *Annals of Statistics, 5*, 815–841.
122. Haberman, S. J. (1978). *Analysis of qualitative data* (Vol. 1). New York: Academic Press.
123. Haley, D. C. (1952). *Estimation of the dosage mortality relationship when the dose is subject to error* (Tech. Rep. No. 15). Stanford, CA: Stanford University, Applied Mathematics and Statistics Laboratory.
124. Halmos, P. R., & Savage, L. J. (1949). Application of the Radon-Nikodym theorem to the theory of sufficient statistics. *Annals of Mathematical Statistics, 20*, 225.
125. Hambleton, R. K. (Ed.). (1983). *Applications of item response theory.* Vancouver, British Columbia, Canada: Educational Research Institute of British Columbia.
126. Hambleton, R. K., & Cook. L. L. (1977). Latent trait models and their use in the analysis of educational test data. *Journal of Educational Measurement, 14*, 75–96.

127. Hambleton, R. K., & Swaminathan, H. (1985). *Item response theory: Principles and applications.* Boston: Kluwer-Nijhoff.

128. Hambleton, R. K., & Traub, R. E. (1973). Analysis of empirical data using two logistic latent trait models. *British Journal of Mathematical and Statistical Psychology, 26,* 195–211.

129. Hambleton, R. K., & Rovinelli, R. (1973). A Fortran IV program for generating examinee response data from logistic test models. *Behavioral Science, 17,* 73–74.

130. Harwell, M. R., & Baker, F. B. (1991). The use of prior distributions in marginalized Bayesian item parameter estimation: A didactic. *Applied Psychological Measurement, 15,* 375–389.

131. Harwell, M. R. and Janosky, J. E. (1991). An Empirical study of the effects of small datasets and varying prior variances on item parameter estimation in BILOG. *Applied Psychological Measurement, 15,* 279–291.

132. Harwell, M. R., Baker, F. B., & Zwarts, M. (1988). Item parameter estimation via marginal maximum likelihood and an EM algorithm: A didactic. *Journal of Educational Statistics, 13,* 243–271.

133. Henrysson, S. (1962). The relation between factor loadings and biserial correlations in item analysis. *Psychometrika, 27,* 419–424.

134. Hildebrand, F. B. (1956). *Introduction to numerical analysis.* New York: McGraw–Hill.

135. Holland, P. W. (1990). The Dutch identity: a new tool for the study of item response models. *Psychometrika, 55,* 5–18.

136. Householder, A. S. (1953). *Principles of numerical analysis.* New York: McGraw–Hill.

137. Hulin, C. L., Drasgow, F., & Parsons, C. K. (1983). *Item response theory: Application to psychological measurement.* Homewood, IL: Dow Jones-Irwin.

138. Hulin, C. L., Lissak, R. I., & Drasgow, F. (1982). Recovery of two- and three-parameter logistic item characteristic curves: A Monte Carlo study. *Applied Psychological Measurement, 6,* 249–260.

139. Jansen, P. G. W. (1984). Relationships between the Thurstone, Coombs and Rasch approaches to item scaling. *Applied Psychological Measurement, 8,* 373–383.

140. Jansen, P. G. W., van den Wollenberg, A. L., & Wierda, F. W. (1988). Correcting unconditional parameter estimates in the Rasch model for inconsistency. *Applied Psychological Measurement, 12,* 297–306.

141. Jensema, C. J. (1976). A simple technique for estimating latent trait mental test parameters. *Educatioanl and Psychological Measurement, 36,* 705–715.

142. Johnson, N. L. (1949). Systems of frequency curves generated by methods of translation. *Biometrika, 36,* 149–176.

143. Johnson, V. E. (1996). On Bayesian analysis of multirater ordinal data: An application to automated essay grading. *Journal of the American Statistical Association, 91,* 42–51.

144. Johnson, V. E. (1997). Alternatives to GPA-based evaluation of student performance (with discussion). *Statistical Science, 12,* 247–277.

145. Johnson, V. E., & Albert, J. H. (1999). *Ordinal data modeling.* New York: Springer.

146. Kale, B. K. (1962). On the solution of likelihood equations by iteration processes: The multiparameter case. *Biometrika, 49,* 479–486.

147. Kelderman, H. (1984). Log linear Rasch model tests. *Psychometrika, 49,* 223–245.

148. Kendall, M. G., & Stuart, A. (1967). *The advanced theory of statistics* (Vol. 2). New York: Hafner.

149. Kendall, M. G., & Stuart, A. (1979). *The advanced theory of statistics* (4th ed., Vol. 2). New York: Oxford University Press.

150. Kendall, M. G., Stuart, A., & Ord, J. K. (1987). *Kendall's advanced theory of statistics* (5th ed., Vol. 1). New York: Oxford University Press.

151. Kiefer, J., & Wolfowitz, J. (1956). Consistency of the maximum likelihood estimator in the presence of infinitely many incidental parameters. *Annals of Mathematical Statistics, 27,* 887–906.

152. Kim, S.-H. (1998, June). *An evaluation of a Markov chain Monte Carlo method for the Rasch model.* Paper presented at the joint annual meeting of the Psychometric Society and the Classification Society of North America, Urbana, IL.

153. Kim, S.-H. (2001). An evaluation of a Markov chain Monte Carlo method for the Rasch model. *Applied Psychological Measurement, 25,* 163–176.

154. Kim, S.-H. (2002, June). *A continuation ratio model for ordered category items.* Paper presented at the annual meeting of the Psychometric Society, Chapel Hill, NC.

155. Kim, S.-H., Baker, F. B., & Subkoviak, M. J. (1989). The $1/kn$ rules in the minimum logit chi-square estimation procedure when small samples are used. *British Journal of Mathematical and Statistical Psychology, 40,* 50–60.

156. Kim, S.-H., & Cohen, A. S. (1999, April). *Accuracy of parameter estimation in Gibbs sampling under the two-parameter logistic model.* Paper presented at the annual meeting of the American Educational Research Association, Montreal, Canada.

157. Kim, S.-H., & Cohen, A. S. (2000, April). *An investigation of ability estimation in Gibbs sampling.* Paper presented at the annual meeting of the American Educational Research Association, New Orleans, LA.

158. Kim, S.-H., Cohen, A. S., Baker, F. B., Subkoviak, M. J., & Leonard, T. (1994). An investigation of hierarchical Bayes procedures in item response theory. *Psychometrika, 59,* 405–421.

159. Koch, W. R. (1983). Likert scaling using the graded response model. *Applied Psychological Measurement, 7,* 15–32.

160. Kolakowski, D., & Bock, R. D. (1973a). NORMOG–Maximum likelihood item analysis and test scoring: Normal ogive model. Ann Arbor: National Educational Resources, Inc.

161. Kolakowski, D., & Bock, R. D. (1973b). LOGOG: Maximum likelihood item analysis and test scoring: Logistic model for multiple item responses. Ann Arbor: National Educational Resources, Inc.

162. Kolen, M. (1981). Comparison of traditional and item response theory methods for equating tests. *Journal of Educational Measurement, 18,* 1–11.

163. Kolen, M. J., & Brennan, R. L. (1995). *Test equating: Methods and practices.* New York: Springer.

164. Lauritzen, S. L., Dawid, A. P., Larsen, B. N., & Leimer, H.-G. (1990). Independence properties of directed Markov fields. *Networks, 20,* 491–505.

165. Lawley, D. N. (1943). On problems connected with item selection and test construction. *Proceedings of the Royal Society of Edinburgh, 61A,* 273–287.

166. Lawley, D. N. (1944). The factorial analysis of multiple item tests. *Proceedings of the Royal Society of Edinburgh, 62A,* 74–82.

167. Lazarsfeld, P. F. (1954). A conceptual introduction to latent structure analysis. In P. F. Lazarsfeld (Ed.), *Mathematical thinking in the social sciences* (pp. 349–387). Glencoe, IL: The Free Press.

168. Lees, D. M., Wingersky, M. S., & Lord, F. M. (1972). *A computer program for estimating item characteristic curve parameters using Birnbaum's three parameter model* (ONR Tech. Rep., Contract No. N00014-69-C-0017). Princeton, NJ: Educational Testing Service.

169. Linacre, J. M. (2003). WINSTEPS Rasch measurement computer program [Computer software]. Chicago: Winsteps.com.

170. Lilliefors, H. W. (1967). On the Kolmogorov-Smirnov test for normality with mean and variance unknown. *Journal of the American Statistical Association, 62*, 399–402.

171. Lilliefors, H. W. (1969). On the Kolmogorov-Smirnov test for the exponential distribution with mean unknown. *Journal of the American Statistical Association, 64*, 387–389.

172. Lindley, D. V. (1970a). *Introduction to probability and statistics: From a Bayesian viewpoint, Part 1: Probability.* Cambridge, England: Cambridge University Press.

173. Lindley, D. V. (1970b). *Introduction to probability and statistics: From a Bayesian viewpoint, Part 2: Inference.* Cambridge, England: Cambridge University Press.

174. Lindley, D. V. (1971). *Bayesian statistics: A review.* Philadelphia: Society for Industrial and Applied Mathematics.

175. Lindley, D. V., & Smith, A. F. M. (1972). Bayesian estimates for the linear model. *Journal of the Royal Statistical Society, Series B, 34*, 1–41.

176. Linsay, B., Clogg, C. C. & Grego, J. (1991). Semiparametric estimation in the Rasch model and related exponential response models, including a simple latent class model for item analysis. *Journal of the American Statistical Association, 86*, 96–107.

177. Little, R. J. A., & Rubin, D. B. (1983). On jointly estimating parameters and missing data by maximizing the complete-data likelihood. *The American Statistician, 37*, 218–220.

178. Lord, F. M. (1952). A theory of test scores. *Psychometric Monograph*, No. 7.

179. Lord, F. M. (1953). An application of confidence intervals and of maximum likelihood to the estimation of an examinee's ability. *Psychometrika, 18*, 57–75.

180. Lord, F. M. (1968). An analysis of the Verbal Scholastic Aptitude Test using Birnbaum's three-parameter logistic model. *Educatioanl and Psychological Measurement, 28*, 989–1020.

181. Lord, F. M. (1970). Item characteristic curves as estimated without knowledge of their mathematical form—A confrontation of Birnbaum's logistic model. *Psychometrika, 35*, 43–50.

182. Lord, F. M. (1971). A theoretical study of two-stage testing. *Psychometrika, 36*, 227–242.

183. Lord, F. M. (1974a). Estimation of latent ability and item parameters when there are omitted responses. *Psychometrika, 39*, 247–264.

184. Lord, F. M. (1974b). Individualized testing and item characteristic curve theory. In D. H. Krantz, R. C. Atkinson, R. D. Luce, & P. Suppes (Eds.), *Contemporary developments in mathematical psychology* (Vol. 2, pp. 106–126). San Francisco: Freeman.

185. Lord, F. M. (1975a). *Evaluation with artificial data of a procedure for estimating ability and item characteristic curve parameters* (Research Bulletin RB-75-33). Princeton, NJ: Educational Testing Service.

186. Lord, F. M. (1975b). Relative efficiency of number-right and formula scores. *British Journal of Mathematical and Statistical Psychology, 28*, 46–50.

187. Lord, F. M. (1977). Practical applications of item characteristic curve theory. *Journal of Educational Measurement, 14*, 117–138.

188. Lord, F. M. (1980). *Applications of item response theory to practical testing problems*. Hillsdale, NJ: Erlbaum.

189. Lord, F. M. (1983). Statistical bias in maximum likelihood estimators of item parameters. *Psychometrika, 48*, 425–435.

190. Lord, F. M. (1986). Maximum likelihood and Bayesian parameter estimation in item response theory. *Journal of Educational Measurement, 23*, 157–162.

191. Lord, F. M., & Novick, M. R. (1968). *Statistical theories of mental test scores.* Reading, MA: Addison-Wesley.

192. Lord, F. M., & Wingersky, M. S. (1983). *Comparison of IRT observed-score and true-score "equatings"* (Research Bulletin 83-26). Princeton, NJ: Educational Testing Service.

193. Lord, F. M., & Wingersky, M. S. (1985). Sampling variances and covariances of parameter estimates in item response theory. In D. J. Weiss (Ed.), *Proceedings of the 1982 Item Response Theory and Computerized Adaptive Testing Conference* (pp. 69–88). Minneapolis, MN: University of Minnesota, Department of Psychology, Computerized Adaptive Testing Laboratory.

194. Loyd, B. H., & Hoover, H. D. (1980). Vertical equating using the Rasch model. *Journal of Educational Measurement, 17*, 179–193.

195. Luce, R. D., & Tukey, J. W. (1964). Simultaneous conjoint measurement: A new type of fundamental measurement. *Journal of Mathematical Psychology, 1*, 1–27.

196. Masters, G. N. (1982). A Rasch model for partial credit scoring. *Psychometrika, 47*, 149–174.

197. Masters, G. N., & Wright, B. D. (1984). The essential process in a family of measurement models. *Psychometrika, 49*, 529–544.

198. Masters, G. N., Wright, B. D. (1997). The partial credit model. In W. J. van der Linden & R. K. Hambleton (Eds.), *Handbook of modern item response theory* (pp. 101–121). New York: Springer.

199. Maxwell, A. E. (1959). Maximum likelihood estimates of item parameters using the logistic function. *Psychometrika, 24*, 221–227.

200. McKinley, R. L., & Reckase, M. D. (1980). *A comparison of the ANCILLES and LOGIST parameter estimation procedures for the three-parameter logistic model using goodness of fit as a criterion* (Research Rep. 80-2). Columbia: University of Missouri, Educational Psychology Department, Tailored Testing Research Laboratory.

201. Mellenbergh, G. J. , & Vijn, P. (1981). The Rasch model as a loglinear model. *Applied Psychological Measurement, 5*, 369–376.

202. Metropolis, N., Rosenbluth, A. W., Rosenbluth, M. N., Teller, A. H., & Teller, E. (1953). Equation of state calculations by fast computing machines. *The Journal of Chemical Physics, 21*, 1087–1092.

203. Metropolis, N., & Ulam, S. (1949). The Monte Carlo method. *Journal of the American Statistical Association, 44*, 335–341.

204. Mislevy, R. J. (1983). Item response models for grouped data. *Journal of Educational Statistics, 8*, 271–288.

205. Mislevy, R. J. (1984). Estimating latent distributions. *Psychometrika, 49*, 359–381.

206. Mislevy, R. J. (1986a). Recent developments in the factor analysis of categorical variables. *Journal of Educational Statistics, 11*, 3–31.

207. Mislevy, R. J. (1986b). Bayes modal estimation in item response models. *Psychometrika, 51*, 177–195.

208. Mislevy, R. J. (1987). Exploiting auxiliary information about examinees in the estimation of item parameters. *Applied Psychological Measurement, 11*, 81–91.

209. Mislevy, R. J. (1988). Exploiting auxiliary information about items in the estimation of Rasch item difficulty parameters. *Applied Psychological Measurement, 12*, 281–296.

210. Mislevy, R. J., & Bock, R. D. (1982a). BILOG: Maximum likelihood item analysis and test scoring with logistic models for binary items. Chicago: International Educational Services.

211. Mislevy, R. J., & Bock, R. D. (1982b). Adaptive EAP estimation of ability in a microcumputer environment. *Applied Psychological Measurement, 6*, 431–444.

212. Mislevy, R. J., & Bock, R. D. (1984). BILOG I: Maximum likelihood item analysis and test scoring: Logistic model. Mooresville, ID: Scientific Software.

213. Mislevy, R. J., & Bock, R. D. (1985). Implementation of the EM algorithm in the estimation of item parameters: The BILOG computer program. In D. J. Weiss (Ed.), *Proceedings of the 1982 Item Response Theory and Computerized Adaptive Testing Conference* (pp. 189–202). Minneapolis, MN: University of Minnesota, Department of Psychology, Computerized Adaptive Testing Conference.

214. Mislevy, R. J., & Bock, R. D. (1986). PC-BILOG: Item analysis and test scoring with binary logistic models [Computer software and manual]. Mooresville, IN: Scientific Software.

215. Mislevy, R. J., & Bock, R. D. (1989). BILOG 3: Item analysis and test scoring with binary logistic models. Mooresville, IN: Scientific Software.

216. Mislevy, R. J., & Stocking, M. L. (1989). A Consumer's Guide to LOGIST and BILOG. *Applied Psychological Measurement, 13*, 57–75.

217. Molenaar, I. W. (1983). Some improved diagnostics for failure of the Rasch model. *Psychometrika, 48*, 49–72.

218. Mood, A. F. (1950). *Introduction to the theory of statistics.* New York: McGraw–Hill.

219. Muraki, E. (1990). Fitting a polytomous item response model to Likert-type data. *Applied Psychological Measurement, 14*, 59–71.

220. Muraki, E. (1992). A generalized partial credit model: Application of an EM algorithm. *Applied Psychological Measurement, 16*, 159–176.

221. Muraki, E. (1993). Information functions of the generalized partial credit model. *Applied Psychological Measurement, 17*, 351–363.

222. Muraki, E. (1997). A generalized partial credit model. In W. J. van der Linden & R. K. Hambleton (Eds.), *Handbook of modern item response theory* (pp. 153–164). New York: Springer.

223. Muraki, E. (1999). Stepwise analysis of differential item functioning based on multiple-group partial credit model. *Journal of Educational Measurement, 36*, 217–232.

224. Muraki, E., & Bock, R. D. (2003). PARSCALE [Computer software]. Lincolnwood, IL: Scientific Software International.

225. Müller, G. E. (1904). *Die Gesichtspunkte und die Tatsachen der psychophysischen Methodik.* Wiesbaden: Bergmann.

226. Neyman, J., & Scott, E. L. (1948). Consistent estimates based on partially consistent observations. *Econometrica, 16*, 1–32.

227. Novick, M. R., & Jackson, P. H. (1974). *Statistical methods for educational and psychological research.* New York: McGraw–Hill.

228. Novick, M. R., Jackson, P. H., Thayer, D. T., & Cole, N. S. (1972). Estimating multiple regressions in m groups: A cross-validational study. *British Journal of Mathematical and Statistical Psychology, 5*, 33–50.

229. O'Hagan, A. (1976). On posterior joint and marginal modes. Biometrika, 63, 329–333.

230. Orchard, T., & Woodbury, M. A. (1972). A missing information principle: Theory and applications. *Proceedings of the Sixth Berkeley Symposium on Mathematical Statistics and Probability* (Vol. 1, 697-71). Berkeley: University of California Press.

231. Park, C., & Muraki, E. (2003). Bias of ability estimates using Warm's weighted likelihood estimator (WLE) in the generalized partial credit model (GPCM). In H. Yanai, A. Okada, K. Shigemasu, Y. Kano, & J. J. Meulman (Eds.), *New developments in psychometrics* (pp. 199-206). Tokyo: Springer.

232. Patz, R. J., & Junker, B. W. (1999a). A straightforward approach to Markov chain Monte Carlo methods for item response models. *Journal of Educational and Behavioral Statistics, 24*, 146-178.

233. Patz, R. J., & Junker, B. W. (1999b). Applications and extensions of MCMC in IRT: Multiple item types, missing data, and rated responses. *Journal of Educational and Behavioral Statistics, 24*, 342-366.

234. Pearl, R. (1922). *The biology of death*. Philadelphia: J. B. Lippincott.

235. Perline, R., Wright, B. D., & Wainer, H. (1977). *The Rasch model as additive conjoint measurement* (Research Memorandum No. 24). Chicago, IL: University of Chicago, Department of Education, Statistical Laboratory.

236. Raftery, A. E. (1996). Hypothesis testing and model selection. In W. R. Gilks, S. Richardson, & D. J. Spiegelhalter (Eds.), *Markov chain Monte Carlo in practice* (pp. 163-187). London: Chapman & Hall.

237. Ramsey, J. O. (1975). Solving implicit equations in psychometric data analysis. *Psychometrika, 40*, 337-360.

238. Rasch, G. (1960). *Probabilistic models for some intelligence and attainment tests*. Copenhagen: Danish Institute for Educational Research.

239. Rasch, G. (1961). On general laws and the meaning of measurement in psychology). In J. Neyman (Ed.), *Proceedings of the Fourth Berkeley Symposium on Mathematical Statistics and Probability* (Vol. 4, pp. 321-333). Berkeley: University of California Press.

240. Rasch, G. (1966a). An item analysis which takes individual differences into account. *British Journal of Mathematical and Statistical Psychology, 19*, 49-57.

241. Rasch, G. (1966b). An individualistic approach to item analysis. In P. F. Lazarsfeld , & N. W. Henry (Eds.), *Readings in mathematical social science* (pp. 89-107). Cambridge, MA: The MIT Press.

242. Rasch, G. (1980). *Probabilistic model for some intelligence and attainment tests* (With a foreword and afterword by B. D. Wright). Chicago: The University of Chicago Press.

243. Reckase, M. D. (1974). An interactive computer program for tailored testing based on the one-parameter logistic model. *Behavior Research Methods and Instrumentation, 6*, 208-212.

244. Reise, S. P., & Yu, J. (1990). Parameter recovery in the graded response model using MULTILOG. *Journal of Educational Measurement, 27*, 133-144.

245. Richardson, M. W. (1936). The relationship between difficulty and the differential validity of a test. *Psychometrika, 1*(2), 33-49.

246. Ripley, B. D. (1987). *Stochastic simulation*. New York: Wiley.

247. Robert, C. P., & Casella, G. (1999). *Monte Carlo statistical methods*. New York: Springer.

248. Rosa, K., Swygert, K. A., Nelson, L., & Thissen, D. (2001). Item response theory applied to combination of multiple-choice and constructed-response items— Scale scores for patterns of summed scores. In D. Thissen & H. Wainer (Eds.), *Test scoring* (pp. 253-292). Mahwah, NJ: Erlbaum.

249. Ross, J. , & Lumsden, J. (1968). Attribute and reliability. *British Journal of Mathematical and Statistical Psychology, 21*, 251-263.

250. Rubin, D. B. (1980). Using empirical Bayes techniques in the law school validity studies. *Journal of the American Statistical Association, 75*, 801–827.
251. Rudner, L. M. (2001). Informed test component weighting. *Educational Measurement: Issues and Practice, 20*(1), 16–19.
252. Samejima, F. (1969). Estimation of latent ability using a response pattern of graded scores. *Psychometrika Monograph Supplement*, No. 17.
253. Samejima, F. (1972). A general model for free-response data. *Psychometrika Monograph Supplement*, No. 18.
254. Samejima, F. (1973). A comment on Birnbaum's three-parameter logistic model in the latent trait theory. *Psychometrika, 38*, 221–233.
255. Samejima, F. (1976). The graded response model of the latent trait theory and tailored testing. In C. L. Clark (Ed.), *Proceedings of the First Conference on Computerized Adaptive Testing* (Professional Series 75-6, pp. 5–17). Washington DC: U.S. Civil Service Commission, Personnel Research and Development Center.
256. Samejima, F. (1977). Weakly parallel tests in latent trait theory with some criticisms of classical test theory. *Psychometrika, 42*, 193–198.
257. Samejima, F. (1979). *A new family of models for the multiple choice item* (Research Rep. No. 79-4). Knoxville: University of Tennessee, Department of Psychology.
258. Sanathanan, L. (1974). Some properties of the logistic model for dichotomous response. *Journal of the American Statistical Association, 69*, 744–749.
259. Sanathanan, L. & Blumenthal, S. (1978). The logistic model and estimation of latent structure. *Journal of the American Statistical Association, 73*, 794–799.
260. Scheiblechner, H. H. (1971). *A simple algorithm for C.M.L.-parameter estimation in Rasch's probabilistic measurement model with two or more categories of answers* (Research Bulletin, No. 5). Vienna: Psychologisches Institut der Universitat Wien.
261. Scheiblechner, H. (1977). Psychological models based upon conditional inference. In H. Spada, & W. F. Kempf (Eds.), *Structural models of thinking and learning* (pp. 185–202). Bern: Huber.
262. Seong, T.-J. (1990). Sensitivity of marginal maximum likelihood estimation of item and ability parameters to the characteristics of the prior ability distributions. *Applied Psychological Measurement, 14*, 299–311.
263. Smith, A. F. M., & Roberts, G. O. (1993). Bayesian computation via the Gibbs sampler and related Markov chain Monte Carlo methods. *Journal of the Royal Statistical Society, Series B, 55*, 3–24.
264. Spiegelhalter, D. J., Best, N. G., Gilks, W. R., & Inskip, H. (1996). Hepatitis B: a case study in MCMC methods. In W. R. Gilks, S. Richardson, & D. J. Spiegelhalter (Eds.), *Markov chain Monte Carlo in practice* (pp. 21–43). London: Chapman & Hall.
265. Spiegelhalter, D. J., Thomas, A., Best, N. G., & Gilks, W. R. (1996). *BUGS 0.5 examples* (Vol. 1, Version i). Cambridge, UK: University of Cambridge, Institute of Public Health, Medical Research Council Biostatistics Unit.
266. Spiegelhalter, D. J., Thomas, A., Best, N. G., & Gilks, W. R. (1997). BUGS: Bayesian inference using Gibbs sampling (Version 0.6) [Computer software]. Cambridge, UK: University of Cambridge, Institute of Public Health, Medical Research Council Biostatistics Unit.
267. Stocking, M. L., & Lord, F. M. (1983). Developing a common metric in item response theory. *Applied Psychological Measurement, 7*, 201–210.
268. Stroud, A. H., & Secrest, D. (1966). *Gaussian quadrature formulas.* Englewood Cliffs, NJ: Prentice-Hall.

269. Swaminathan, H., & Gifford, J. A. (1982). Bayesian estimation in the Rasch model. *Journal of Educational Statistics, 7,* 175–191.

270. Swaminathan, H., & Gifford, J. A. (1983). Estimation of parameters in the three-parameter latent trait model. In D. J. Weiss (Ed.), *New horizons in testing* (pp. 13–30). New York: Academic Press.

271. Swaminathan, H., & Gifford, J. A. (1985). Bayesian estimation in the two-parameter logistic model. *Psychometrika, 50,* 349–364.

272. Swaminathan, H., & Gifford, J. A. (1986). Bayesian estimation in the three-parameter logistic model. *Psychometrika, 51,* 589–601.

273. Swaminathan, H., Hambleton, R. K., Sireci, S. G., Xing, D., & Rizavi, S. M. (2003). Small sample estimation in dichotomous item response models: Effect of priors based on judgmental information on the accuracy of item parameter estimates. *Applied Psychological Measurement, 27,* 27–51.

274. Tanner, M. A. (1991). *Tools for statistical inference: Observed data and data augmentation methods.* New York: Springer-Verlag.

275. Tanner, M. A. (1996). *Tools for statistical inference: Methods for the exploration of posterior distributions and likelihood functions* (3rd ed.). New York: Springer.

276. Terman, L. M. (1916). *The measurement of intelligence.* Boston: Houghton Mifflin.

277. Terman, L., & Merrill, M. A. (1937). *Measuring intelligence.* Boston: Houghton Mifflin.

278. The MathWorks, Inc. (1996). MATLAB: The language of technical computing [Computer software]. Natick, MA: Author.

279. Theunissen, T. J. J. M. (1985). Binary programming and test design. *Psychometrika, 50,* 411–420.

280. Thissen, D. (1982). Marginal maximum likelihood estimation for the one-parameter logistic model. *Psychometrika, 47,* 175–186.

281. Thissen, D. (1986). *MULTILOG user's guide* (Version 5). Mooresville IN: Scientific Software.

282. Thissen, D. (1990, June). *Parameter estimation for item response models with (some) nonnormal population distributions.* Paper presented at the annual meeting of the Psychometric Society, Princeton, NJ. [see Appendix C of Thissen (1991)]

283. Thissen, D. (1991). *MULTILOG user's guide* (Version 6). Chicago: Scientific Software.

284. Thissen, D , & Steinberg, L. (1984). A response model for multiple choice items. *Psychometrika, 49,* 501–519.

285. Thissen, D., & Steinberg, L. (1986). A taxonomy of item response models. *Psychometrika, 51,* 567–577.

286. Thissen, D., & Steinberg, L. (1988). Data analysis using item response theory. *Psychological Bulletin, 104,* 385–395.

287. Thissen, D., & Wainer, H. (1982). Some standard errors in item response theory. *Psychometrika, 47,* 397–412.

288. Thissen, D., & Wainer, H. (Eds.). (2001). *Test scoring.* Mahwah, NJ: Erlbaum.

289. Thissen, D., Chen, W.-H., & Bock, R. D. (2003). MULTILOG [Computer software]. Lincolnwood, IL: Scientific Software International.

290. Thissen, D., Nelson, L., & Swygert, K. A. (2001). Item response theory applied to combination of multiple-choice and constructed-response items—Approximation methods for scale scores. In D. Thissen & H. Wainer (Eds.), *Test scoring* (pp. 293–341). Mahwah, NJ: Erlbaum.

291. Thissen, D., Steinberg, L, & Gerrard, M. (1986). Beyond group mean differences: The concept of item bias. *Psychological Bulletin, 99,* 118–128.

292. Thissen, D., Steinberg, L, & Wainer, H. (1988). Use of item response theory in the study of group differences in trace lines. In H. Wainer & H. I. Braun (Eds.), *Test validity* (pp. 147–169). Hillsdale, NJ: Erlbaum.

293. Thissen, D., Steinberg, L, & Wainer, H. (1993). Detection of differential item functioning using the parameters of item response models. In P. W. Holland & H. Wainer (Eds.), *Differential item functioning* (pp. 67–113). Hillsdale, NJ: Erlbaum.

294. Thissen, D., Wainer, H., & Wang, X. (1994). Are tests comprising both multiple-choice and free-response items necessarily less unidimensional than multiple-choice tests? an analysis of two tests. *Journal of Educational Measurement, 31*, 113–123.

295. Thomson, G. H. (1919). A direct deduction of the constant process used in the method of right and wrong cases. *Psychological Review, 26*, 454–464.

296. Thurstone, L. L. (1927). A law of comparative judgment. *Psychological Review, 34*, 273–286.

297. Tsutakawa, R. K. (1984). Estimation of two-parameter logistic item response curves. *Journal of Educational Statistics, 9*, 263–276.

298. Tsutakawa, R. K., & Lin, H. Y. (1986). Bayesian estimation of item response curves. *Psychometrika, 51*, 251–267.

299. Tucker, L. R. (1946). Maximum validity of a test with equivalent items. *Psychometrika, 11*, 1–13.

300. Tutz, G. (1997). Sequential models for ordered responses. In W. J. van der Linden & R. K. Hambleton (Eds.), *Handbook of modern item response theory* (pp. 139–152). New York: Springer.

301. Urban, F. M. (1910). Die psychophysischen Massmethoden als Grundlagen empirischer Messungen. *Archiv für die gesamte Psychologie, 15*, 261–355; *16*, 168–227.

302. Urry, V. W. (1974). Approximations to item parameters of mental test models and their uses. *Educatioanl and Psychological Measurement, 34*, 253–269.

303. Vale, C. D., & Weiss, D. J. (1977). *A comparison of information functions of multiple-choice and free-response vocabulary items* (Research Rep. 77-2). Minneapolis: University of Minnesota, Department of Psychology, Psychometric Methods Program.

304. van de Vijver F. J. R. (1986). The robustness of Rasch estimates. *Applied Psychological Measurement, 10*, 45–57.

305. van der Linden, W. J., & Hambleton, R. K. (Eds.). (1997). *Handbook of modern item response theory*. New York: Springer.

306. van den Wollenberg, A. L., Wierda, F. W., & Jansen, P. G. W. (1988). Consistency of Rasch model parameter estimation: A simulation study. *Applied Psycholgical Measurement, 12*, 307–313.

307. Verhelst, N. D., Glas, C. A. W., & de Vries, H. H. (1997). A steps model to analyze partial credit. In W. J. van der Linden & R. K. Hambleton (Eds.), *Handbook of modern item response theory* (pp. 123–138). New York: Springer.

308. Verhelst, N. D., Glas, C. A. W., & van der Sluis, A. (1984). Estimation problems in the Rasch-model: The basic symmetric functions. *Computational Statistics Quarterly, 1*, 245–262.

309. Verhelst, N. D., Glas, C. A. W., & Verstralen, H. H. F. M. (1995). One-parameter logistic model: OPLM [Computer program]. Arnhem, The Netherlands: CITO.

310. Verhulst, P.-F. (1844). Recherches mathématiques sur la loi d'accroissement de la population [Mathematical investigations on the law of growth of the population]. *Mémoires de l'Académie Royale de Sciences, des Lettres et des Beaux-arts de Belgique, Brussels; t. 18.*

311. Wainer, H. (1990). *Computerized adaptive testing: A primer.* Hillsdale, NJ: Erlbaum.

312. Wainer, H., & Thissen, D. (1993). Combining multiple-choice and constructed-response test scores: Toward a Marxist theory of test construction. *Applied Measurement in Education, 6,* 103–118.

313. Wainer, H., Bradlow, E. T., & Du, Z. (2000). Testlet response theory: An analog for the 3PL model useful in testlet-based adaptive testing. In W. J. van der Linden & C. A. W. Glas (Eds.), *Computerized adaptive testing: Theory and practice* (pp. 245–269). Dordrecht, The Netherlands: Kluwer Academic Publishers.

314. Wainer, H., Morgan, A., & Gustafsson, J.-E. (1980). A review of estimation procedures for the Rasch model with an eye toward longish tests. *Journal of Educational Statistics, 5,* 35–64.

315. Walker, H. M., & Lev, J. (1953). *Statistical inference.* New York: Holt.

316. Waller, M. I. (1974). *Removing the effects of random guessing from latent trait ability estimates* (Research Bulletin RB-74-32). Princeton, NJ: Educational Testing Service.

317. Wang, T., & Chen, W.-H. (1999, April). *Estimating latent distributions: A comparison of the MCMC and EM algorithm.* Paper presented at the annual meeting of the American Educational Research Association, Montreal, Canada.

318. Whitely, S. E. (1977). Models, meanings and misunderstandings: Some issues in applying Rasch's theory. *Journal of Educational Measurement, 14,* 227–235.

319. Whitely, S. E. (1980). Multicomponent latent trait model for ability tests. *Psychometrika, 45,* 479–494.

320. Whitely, S. E., & Dawis, R. V. (1974). The nature of objectivity with the Rasch model. *Journal of Educational Measurement, 11,* 163–178.

321. Wingersky, M. S. (1983). LOGIST: A program for computing maximum likelihood procedures for logistic test models. In R. K. Hambleton (Ed.), *Applications of item response theory* (pp. 45–56). Vancouver, British Columbia, Canada: Educational Research Institute of British Columbia.

322. Wingersky, M. S., & Lord, F. M. (1973). *A computer program for estimating examinee ability and item characteristic curve parameters when there are omitted responses* (Research Memorandum RM-73-2). Princeton, NJ: Educational Testing Service.

323. Wingersky, M. S., & Lord, F. M. (1984). An investigation of methods for reducing sampling error in certain IRT procedures. *Applied Psychological Measurement, 8,* 347–364.

324. Wingersky, M. S., Barton, M. A., & Lord, F. M. (1982). *LOGIST user's guide.* Princeton, NJ: Educational Testing Service.

325. Wingersky, M. S., Patrick, R., & Lord, F. M. (1999). *LOGIST user's guide* (Version 7.1). Princeton, NJ: Educational Testing Service.

326. Wingersky, M. S., Lees, D. M., Lennon, V., & Lord, F. M. (1969). A computer program for estimating true-score distributions and graduating observed-score distributions. *Educatioanl and Psychological Measurement, 29,* 689–692.

327. Wollack, J. A., Bolt, D. M., Cohen, A. S., & Lee, Y.-S. (2002). Recovery of item parameters in the nominal response model: A comparison of marginal maximum likelihood estimation and Markov chain Monte Carlo estimation. *Applied Psychological Measurement, 26,* 339–352.

328. Wood, R. L., Wingersky, M. S., & Lord, F. M. (1976). *LOGIST: A computer program for estimating examinee ability and item characteristic curve parameters* (RM-76-6). Princeton, NJ: Educational Testing Service.

329. Wright, B. D. (1968). Sample-free test calibration and person measurement. *Proceedings of the 1967 Invitational Conference on Testing Problems* (pp. 85–101). Princeton, NJ: Educational Testing Service.
330. Wright, B. D. (1977a). Solving measurement problems with the Rasch model. *Journal of Educational Measurement, 14,* 97–116.
331. Wright, B. D. (1977b). Misunderstanding the Rasch model. *Journal of Educational Measurement, 14,* 219–225.
332. Wright, B. D. (1988). The efficacy of unconditional maximum likelihood bias correction: Comment on Jansen, van den Wollenberg, and Wierda. *Applied Psychological Measurement, 12,* 315–318.
333. Wright, B. D., & Douglas, G. A. (1975). *Best test design and self-tailored testing* (Research Memorandum No. 19). Chicago: The University of Chicago, Department of Education, Statistical Laboratory.
334. Wright, B. D., & Douglas, G. A. (1977a). Best procedures for sample-free item analysis. *Applied Psychological Measurement, 1,* 281–295.
335. Wright, B. D., & Douglas, G. A. (1977b). Conditional versus unconditional procedures for sample-free item analysis. *Educational and Psychological Measurement, 37,* 573–586.
336. Wright, B. D., & Linacre, J. M. (1985). *Microscale manual.* Westport, CT: Mediax Interactive Technologies.
337. Wright, B. D., & Mead, R. J. (1975). *CALFIT: Sample-free calibration with a Rasch measurement model* (Research Memorandum No. 18). Chicago: The University of Chicago, Department of Education, Statistical Laboratory.
338. Wright, B. D., & Mead, R. J. (1978). *BICAL: Calibrating items and scales with the Rasch model* (Research Memorandum No. 23A). Chicago: The University of Chicago, Department of Education, Statistical Laboratory.
339. Wright, B. D., Mead, R. J., & Bell, S. R. (1980). *BICAL: Calibrating items with the Rasch model* (Research Memorandum No. 23C). Chicago: The University of Chicago, Department of Education, Statistical Laboratory.
340. Wright, B. D., Mead, R. J., & Draba, R. E. (1976). *Detecting and correcting test item bias with alogistic response model* (Research Memorandum No. 22). Chicago: University of Chicago, Department of Education, Statistical Laboratory.
341. Wright, B. D., & Panchapakesan, N. (1969). A procedure for sample-free item analysis. *Educatioanl and Psychological Measurement, 29,* 23–48.
342. Wright, B. D., & Stone, M. H. (1979). *Best test design.* Chicago: MESA Press.
343. Yen, M. W. (1981). Using simulation results to choose a latent trait model. *Applied Psychological Measurement, 5,* 245–262.
344. Yen, M. W. (1983). Use of the three-parameter logistic model in the development of a standardized achievement test. In R. K. Hambleton (Ed.), *Applications of item response theory* (pp. 123–141). Vancouver, British Columbia, Canada: Educational Research Institute of British Columbia.
345. Yen, W. M. (1987). A comparison of the efficiency and accuracy of BILOG and LOGIST. *Psychometrika, 52,* 275–291.
346. Yen, W. M. (1993). Scaling performance assessment: Strategies for managing local item dependence. *Journal of Educational Measurement, 30,* 187–214.
347. Zimowski, M. F., Muraki, E., Mislevy, R. J., & Bock, R. D. (1996). BILOG-MG: Multiple-group IRT analysis and test maintenance for binary items [Computer software]. Chicago: Scientific Software International.
348. Zimowski, M. F., Muraki, E., Mislevy, R. J., & Bock, R. D. (2003). BILOG-MG [Computer software]. Lincolnwood, IL: Scientific Software International.
349. Zwarts, M. (1986). Computer software for the Bock-Aitkin MMLE/EM item parameter estimation procedure (personal communication).

Index